Lecture Notes in Physics

The Lecture Notes in Physics

The series Lecture Notes in Physics (LNP), founded in 1969, reports new developments in physics research and teaching – quickly and informally, but with a high quality and the explicit aim to summarize and communicate current knowledge in an accessible way. Books published in this series are conceived as bridging material between advanced graduate textbooks and the forefront of research to serve the following purposes:

• to be a compact and modern up-to-date source of reference on a well-defined topic;

• to serve as an accessible introduction to the field to postgraduate students and nonspecialist researchers from related areas;

• to be a source of advanced teaching material for specialized seminars, courses and schools.

Both monographs and multi-author volumes will be considered for publication. Edited volumes should, however, consist of a very limited number of contributions only. Proceedings will not be considered for LNP.

Volumes published in LNP are disseminated both in print and in electronic formats, the electronic archive is available at springerlink.com. The series content is indexed, abstracted and referenced by many abstracting and information services, bibliographic networks, subscription agencies, library networks, and consortia.

Proposals should be sent to a member of the Editorial Board, or directly to the managing editor at Springer:

Dr. Christian Caron
Springer Heidelberg
Physics Editorial Department I
Tiergartenstrasse 17
69121 Heidelberg/Germany
christian.caron@springer-sbm.com

Joachim Asch   Alain Joye   (Eds.)

Mathematical Physics of Quantum Mechanics

Selected and Refereed Lectures from QMath9

Springer

Editors

Joachim Asch
Université du Sud Toulon Var
Centre de physique théorique
Département de Mathématiques
BP 20132
F-83957 La Garde Cedex
France
E-mail: asch@univ-tln.fr

Alain Joye
Institut Fourier
Université Grenoble 1
BP 74
38402 Saint-Martin-d'Hères Cedex
France
E-mail: alain.joye@ujf-grenoble.fr

J. Asch and A. Joye, *Mathematical Physics of Quantum Mechanics*,
Lect. Notes Phys. 690 (Springer, Berlin Heidelberg 2006), DOI 10.1007/b11573432

ISSN 0075-8450
ISBN-10 3-642-42157-1 Springer Berlin Heidelberg New York
ISBN-13 978-3-642-42157-0 Springer Berlin Heidelberg New York

Springer is a part of Springer Science+Business Media
springer.com
© Springer-Verlag Berlin Heidelberg 2006
Softcover re-print of the Hardcover 1st edition 2006

Typesetting: by the authors and TechBooks using a Springer LaTeX macro package

Printed on acid-free paper   SPIN: 11573432   54/TechBooks   5 4 3 2 1 0

Preface

The topics presented in this book were discussed at the conference "QMath9" held in Giens, France, September 12th-16th 2004. QMath is a series of meetings whose aim is to present the state of the art in the Mathematical Physics of Quantum Systems, both from the point of view of physical models and of the mathematical techniques developed for their study. The series was initiated in the early seventies as an attempt to enhance collaboration between mathematical physicists from eastern and western European countries. In the nineties it took a worldwide dimension. At the same time, due to engineering achievements, for example in the mesoscopic realm, there was a renewed interest in basic questions of quantum dynamics.

The program of QMath9, which was attended by 170 scientists from 23 countries, consisted of 123 talks grouped by the topics: *Nanophysics, Quantum dynamics, Quantum field theory, Quantum kinetics, Random Schrödinger operators, Semiclassical analysis, Spectral theory.* QMath9 was also the frame for the 2004 meeting of the European Research Group on "Mathematics and Quantum Physics" directed by Monique Combescure. For a detailed account of the program, see http://www.cpt.univ.mrs.fr/˜qmath9.

Expanded versions of several selected introductory talks presented at the conference are included in this volume. Their aim is to provide the reader with an easier access to the sometimes technical state of the art in a topic. Other contributions are devoted to a pedagogical exposition of quite recent results at the frontiers of research, parts of which were presented in "QMath9". In addition, the reader will find in this book new results triggered by discussions which took place at the meeting.

Hence, while based on the conference "QMath9", this book is intended to be a starting point for the reader who wishes to learn about the current research in quantum mathematical physics, with a general perspective. Effort has been made by the authors, editors and referees in order to provide contributions of the highest scientific standards to meet this goal.

We are grateful to Yosi Avron, Volker Bach, Stephan De Bièvre, Laszlo Erdös, Pavel Exner, Svetlana Jitomirskaya, Frédéric Klopp who mediated the scientific sessions of "QMath9".

We should like to thank all persons and institutions who helped to organize the conference locally: Sylvie Aguillon, Jean-Marie Barbaroux,

Nils Berglund, Jean-Michel Combes, Elisabeth Elophe, Jean-Michel Ghez, Corinne Roux, Corinne Vera, Université du Sud Toulon–Var and Centre de Physique Théorique Marseille.

We gratefully acknowledge financial support from: European Science Foundation (SPECT), International Association of Mathematical Physics, Ministère de l'Education Nationale et de la Recherche, Centre National de la Recherche Scientifique, Région Provence-Alpes-Côte d'Azur, Conseil Général du Var, Centre de Physique Théorique, Université du Sud Toulon–Var, Institut Fourier, Université Joseph Fourier.

Toulon   *Joachim Asch*
Grenoble   *Alain Joye*
January 2006

Contents

Part II Quantum Field Theory and Statistical Mechanics

Part V Semiclassical Analysis and Quantum Chaos

List of Contributors

W.K. Abou-Salem
Institute for Theoretical
Physics, ETH-Hönggerberg
CH-8093 Zürich, Switzerland
walid@phys.ethz.ch

M. Aizenman
Departments of Mathematics
and Physics
Jadwin Hall Princeton University
P. O. Box 708, Princeton
New Jersey, 08544
USA

A. Avila
CNRS UMR 7599
Laboratoire de Probabilités
et Modèles aléatoires
Université Pierre
et Marie Curie–Boite
courrier 188, 75252–Paris Cedex 05
France
artur@ccr.jussieu.fr

V. Betz
Institute for Biomathematics
and Biometry GSF
Forschungszentrum Postfach
1129 D-85758
Oberschleißheim
Germany
volker.betz@gsf.de

S. De Bièvre
Université des Sciences et
Technologies de Lille, UFR
de Mathématiques et Laboratoire
Painlevé, 59655 Villeneuve d'Ascq
Cedex, France
Stephan.De-Bievre@math.univ-
lille1.fr

O. Bokanowski
Laboratoire Jacques Louis Lions
(Paris VI)
UFR Mathématique B. P. 7012
& Université Paris VII
Paris, Paris Cedex 05
France
boka@math.jusssieu.fr

M. Büttiker
Département de Physique Théorique
Université de Genève
CH-1211 Genève 4, Switzerland
Markus.Buttiker@physics.unige.ch

J.-M. Combes
Département de Mathématiques
Université de Toulon-Var
BP 132, 83957 La Garde Cédex
France
combes@cpt.univ-mrs.fr

H.D. Cornean
Department of Mathematical
Sciences
Aalborg University
Fredrik Bajers Vej 7G
9220 Aalborg, Denmark
cornean@math.aau.dk

J. Dereziński
Department of Mathematics
and Methods in Physics
Warsaw University Hoża 74
00-682 Warsaw
Poland
Jan.Derezinski@fuw.edu.pl

A. Elgart
Department of Mathematics
Stanford University
Stanford, CA 94305-2125
USA
elgart@math.stanford.edu

L. Erdős
Institute of Mathematics
University of Munich
Theresienstr.
39 D-80333 Munich
Germany
lerdos@mathematik.uni-muenchen.
de

P. Exner
Department of Theoretical Physics
Nuclear Physics Institute
Academy of Sciences
25068 Řež near Prague
Czechia
and
Doppler Institute
Czech Technical University
Břehová 7, 11519
Prague, Czechia
exner@ujf.cas.cz

A. Fedotov
Department of Mathematical
Physics
St Petersburg State University 1
Ulianovskaja 198904
St Petersburg-Petrodvorets, Russia
mailto:fedotov@svs.ru
fedotov@svs.ru

J. Fröhlich
Institute for Theoretical
Physics
ETH-Hönggerberg
CH-8093 Zürich
Switzerland
juerg@itp.phys.ethz.ch

F. Germinet
Département de Mathématiques
Université de Cergy-Pontoise
Site de Saint-Martin
2 avenue Adolphe Chauvin
95302 Cergy-Pontoise Cédex
France
germinet@math.u-cergy.fr

S.D. Głazek
Institute of Theoretical Physics
Warsaw University
00-681 Hoza 69
Poland
stglazek@fuw.edu.pl

B. Helffer
Département de Mathématiques
Bât. 425, Université Paris
Sud UMR CNRS 8628, F91405
Orsay Cedex
France
Bernard.Helffer@math.u-psud.fr

P.D. Hislop
Department of Mathematics
University of Kentucky
Lexington KY 40506-0027
USA
hislop@ms.uky.edu

A. Jensen
Department of Mathematical
Sciences
Aalborg University
Fredrik Bajers Vej 7G
9220 Aalborg, Denmark
matarne@math.aau.dk

S. Jitomirskaya
University of California
Irvine Department of Mathematics
243 Multipurpose Science
& Technology Building
Irvine, CA 92697-3875
USA
szhitomi@uci.edu

F. Klopp
LAGA, Institut Galilée
U.M.R 7539 C.N.R.S
Université Paris-Nord
Avenue J.-B. Clément
F-93430 Villetaneuse, France
klopp@math.univ-paris13.fr

D. Lenz
Fakultät für Mathematik
Technische Universität
09107 Chemnitz, Germany
d.lenz@@mathematik.tu-chemnitz.de

E.H. Lieb
Departments of Mathematics
and Physics
Jadwin Hall Princeton University
P. O. Box 708, Princeton
New Jersey, 08544
USA

J.L. López
Departamento de Matemática
Aplicada Facultad de Ciencias
Universidad de Granada
Campus Fuentenueva
18071 Granada
Spain
jllopez@ugr.es

M. Măntoiu
"Simion Stoilow" Institute of
Mathematics, Romanian Academy
21, Calea Grivitei Street
010702-Bucharest, Sector 1
Romania
Marius.Mantoiu@imar.ro

K.A. Meissner
Institute of Theoretical Physics
Warsaw University
Hoża 69 00-681 Warsaw
Poland
Krzysztof.Meissner@fuw.edu

M. Merkli
Department of Mathematics and
Statistics, McGill University
805 Sherbrooke W., Montreal
QC Canada, H3A 2K6
and
Centre des Recherches
Mathématiques, Université de
Montréal, Succursale
centre-ville Montréal
QC Canada, H3C 3J7
merkli@math.mcgill.ca

V. Moldoveanu
National Institute of Materials
Physics, P.O. Box MG-7
Magurele, Romania
valim@infim.ro

J.S. Møller
Aarhus University Department of
Mathematical Sciences
8000 Aarhus C, Denmark
jacob@imf.au.dk

M. Moskalets
Department of Metal
and Semiconductor Physics
National Technical University
"Kharkiv Polytechnic Institute"
61002 Kharkiv, Ukraine
moskalets@kpi.kharkov.ua

B. Nachtergaele
Department of Mathematics
University of California
Davis One Shields Avenue
Davis, CA 95616-8366, USA
bxn@math.ucdavis.edu

S. Nonnenmacher
Service de Physique Théorique
CEA/DSM/PhT, Unité de recherche
associée au CNRS, CEA/Saclay
91191 Gif-sur-Yvette
France
snonnenmacher@cea.fr

J. Puig
Departament de Matemàtica
Aplicada I
Universitat Politècnica
de Catalunya. Av. Diagonal
647, 08028 Barcelona, Spain

J.V. Pulé
Department of Mathematical
Physics
University College
Dublin, Belfield, Dublin 4
Ireland
Joe.Pule@ucd.ie

R. Purice
"Simion Stoilow" Institute of
Mathematics, Romanian Academy
21, Calea Grivitei Street
010702-Bucharest, Sector 1
Romania
Radu.Purice@imar.ro

G. Raikov
Facultad de Ciencias
Universidad de Chile
Las Palmeras 3425
Santiago, Chile
graykov@uchile.cl

V. Rivasseau
Laboratoire de Physique Théorique
CNRS, UMR 8627
Université de Paris-Sud
91405 Orsay
France
rivass@th.u-psud.fr

M. Salmhofer
Theoretical Physics
University of
Leipzig Augustusplatz 10
D-04109 Leipzig
Germany
and
Max–Planck Institute for
Mathematics, Inselstr. 22
D-04103 Leipzig
Germany
Manfred.Salmhofer@itp.uni-
leipzig.de

Ó. Sánchez
Departamento de Matemática
Aplicada, Facultad de Ciencias
Universidad de Granada
Campus Fuentenueva
18071 Granada
Spain
ossanche@ugr.es

B. Schlein
Department of Mathematics
Stanford University
Stanford, CA 94305, USA
schlein@math.stanford.edu

H. Schulz-Baldes
Mathematisches Institut
Friedrich-Alexander-Universität
Erlangen-Nürnberg Bismarckstr. 1
D-91054, Erlangen
schuba@mi.uni-erlangen.de

R. Seiringer
Departments of Mathematics
and Physics, Jadwin Hall
Princeton University
P. O. Box 708, Princeton
New Jersey, 08544
USA

J. Soler
Departamento de Matemática
Aplicada, Facultad de Ciencias
Universidad de Granada
Campus Fuentenueva
18071 Granada
Spain
jsoler@ugr.es

J.P. Solovej
Department of Mathematics
University of Copenhagen
Universitetsparken 5
DK-2100 Copenhagen, Denmark
and
Institut für Theoretische Physik
Universität Wien
Boltzmanngasse 5
A-1090 Vienna, Austria
solovej@math.ku.dk

A. Soshnikov
University of California at Davis
Department of Mathematics
Davis, CA 95616, USA
soshniko@math.ucdavis.edu

H. Spohn
Zentrum Mathematik and
Physik Department
TU München D-85747 Garching
Boltzmannstr 3, Germany
spohn@ma.tum.de

S. Starr
Department of Mathematics
University of California
Los Angeles, Box 951555
Los Angeles, CA 90095-1555
USA
sstarr@math.ucla.edu

P. Stollmann
Fakultät für Mathematik
Technische Universität
09107 Chemnitz, Germany
p.stollmann@mathematik.tu-
chemnitz.de

S. Teufel
Mathematisches Institut
Universität Tübingen
Auf der Morgenstelle 10
72076 Tübingen
Germany
stefan.teufel@uni-tuebingen.de

A.F. Verbeure
Instituut voor Theoretische
Fysika, Katholieke Universiteit
Leuven Celestijnenlaan
200D, 3001 Leuven, Belgium
andre.verbeure@fys.kuleuven.
ac.be

H.-T. Yau
Department of Mathematics
Stanford University, CA-94305, USA
yau@math.stanford.edu

J. Yngvason
Institut für Theoretische Physik
Universität Wien
Boltzmanngasse 5
A-1090 Vienna, Austria
and
Erwin Schrödinger Institute for
Mathematical Physics
Boltzmanngasse 9
A-1090 Vienna, Austria

V.A. Zagrebnov
Université de la
Méditerranée and Centre de
Physique Théorique, Luminy-Case
907, 13288 Marseille, Cedex 09
France
zagrebnov@cpt.univ-mrs.fr

S. Zelditch
Department of Mathematics
Johns Hopkins University
Baltimore MD 21218
USA
zelditch@math.jhu.edu

Introduction

QMath9 gave a particular importance to summarize the state of the art of the field in a perspective to transmit knowledge to younger scientists. The main contributors to the field were gathered in order to communicate results, open questions and motivate new research by the confrontation of different view points. The edition of this book follows this spirit; the main effort of the authors and editors is to help finding an access to the variety of themes and the sometimes very sophisticated literature in Mathematical Physics.

The contributions of this book are organized in five topical groups: *Quantum Dynamics and Spectral Theory, Quantum Field Theory and Statistical Mechanics, Quantum Kinetics and Bose-Einstein Condensation, Disordered Systems and Random Operators, Semiclassical Analysis and Quantum Chaos.* This splitting is admittedly somewhat arbitrary, since there are overlaps between the topics and the frontiers between the chosen groups may be quite fuzzy. Moreover, there are close connections between the tools and techniques used in the analysis of quite different physical phenomena. An introduction to each theme is given. This plan is intended as a readers guide rather than as an attempt to put contributions into well defined categories.

Part I

Quantum Dynamics and Spectral Theory

Different aspects of the solution of a long-standing major problem in mathematical physics are reported in the contributions of Avila, Jitomirskaya and Puig. It had been conjectured for about thirty years by physicists and mathematicians that the problem of electrons confined to a plane under the influence of a periodic potential and a perpendicular magnetic field exhibits fractal spectral properties. Experimental evidence of Hofstadters butterfly-like energy spectrum was found about five years ago. Here the mathematical physicists Avila, Jitomirskaya and Puig report on the proof that the spectrum of a related operator is a Cantor set. Their proofs rely much on recent techniques in classical dynamical systems. We mention that the mathematical model still has fascinating unsolved aspects which are important to physics and especially the quantum hall effect for example the question whether the spectral gaps are open.

Building Micron-size robots which move much faster than bacteria is one of the visions of small scale physics. Y. Avron gave an introduction on recent results on the problem of designing an optimal micro-swimmer. These have been obtained using methods from geometry and linear response theory.

V. Betz and S. Teufel report on their progress in the old Landau–Zener problem. For a time dependent two state problem which is asymptotically constant, a detailed approximate solution which takes into account the adiabatic transitions is obtained for all times. It describes both the exponential smallness of the transition probability and the time scale over which it takes place.

The theory of transport in mesoscopic systems is addressed in two contributions. Büttiker and Moskalets treat quantum pumping. If the system is driven by several internal parameters oscillating slowly, a direct current may result. It can be calculated to leading order in terms of stationary scattering matrices. To take account the energy exchange with the environment the full time dependent scattering matrix is developed to next order. A mathematical proof of the formula relating conductance and transmittance has been given by H.D. Cornean, A. Jensen, V. Moldoveanu in the case of an adiabatically switched on external potential. The formula is applied numerically to a model.

Geometry meets physics again in the contribution of P. Exner who presents a conjecture about an interesting isoperimetric problem arising from the spectral analysis of a quantum model with point interactions.

S. Glazek discusses examples of renormalization group analysis applied to Schrödinger operators and in particular the occurrence of a limit cycle as a critical attractor instead of a fixed point.

Solving the Ten Martini Problem

Artur Avila[1] and Svetlana Jitomirskaya[2]

[1] CNRS UMR 7599, Laboratoire de Probabilités et Modèles aléatoires
Université Pierre et Marie Curie–Boite courrier 188
75252–Paris Cedex 05, France
artur@ccr.jussieu.fr

[2] University of California, Irvine, California
The research of S.J. was partially supported by NSF under DMS-0300974
szhitomi@uci.edu

Abstract. We discuss the recent proof of Cantor spectrum for the almost Mathieu operator for all conjectured values of the parameters.

1 Introduction

The almost Mathieu operator (a.k.a. the Harper operator or the Hofstadter model) is a Schrödinger operator on $\ell^2(\mathbb{Z})$,

$$(H_{\lambda,\alpha,\theta}u)_n = u_{n+1} + u_{n-1} + 2\lambda \cos 2\pi(\theta + n\alpha)u_n \,,$$

where $\lambda, \alpha, \theta \in \mathbb{R}$ are parameters (the *coupling*, the *frequency* and the *phase*). This model first appeared in the work of Peierls [21]. It arises in physics literature as related, in two different ways, to a two-dimensional electron subject to a perpendicular magnetic field [15, 23]. It plays a central role in the Thouless et al theory of the integer quantum Hall effect [27]. The value of λ of most interest from the physics point of view is $\lambda = 1$. It is called the critical value as it separates two different behaviors as far as the nature of the spectrum is concerned.

If $\alpha = \frac{p}{q}$ is rational, it is well known that the spectrum consists of the union of q intervals possibly touching at endpoints. In the case of irrational α the spectrum (which then does not depend on θ) has been conjectured for a long time to be a Cantor set for all $\lambda \neq 0$ [7]. To prove this conjecture has been dubbed the *Ten Martini problem* by Barry Simon, after an offer of Kac in 1981, see Problem 4 in [25].

In 1984 Bellissard and Simon [8] proved the conjecture for generic pairs of (λ, α). In 1987 Sinai [26] proved Cantor spectrum for a.e. α in the perturbative regime: for $\lambda = \lambda(\alpha)$ sufficiently large or small. In 1989 Helffer-Sjöstrand proved Cantor spectrum for the critical value $\lambda = 1$ and an explicitly defined generic set of α [16]. Most developments in the 90s were related to the following observation. For $\alpha = \frac{p}{q}$ the spectrum of $H_{\lambda,\alpha,\theta}$ can have at most $q - 1$ gaps. It turns out that all these gaps are open, except for the middle one for even q [11, 20]. Choi, Elliott, and Yui obtained in fact an exponential

A. Avila and S. Jitomirskaya: *Solving the Ten Martini Problem*, Lect. Notes Phys. **690**, 5–16
(2006)

lower bound on the size of the individual gaps from which they deduced Cantor spectrum for Liouville (exponentially well approximated by the rationals) α [11]. In 1994 Last, using certain estimates of Avron, van Mouche and Simon [6], proved zero measure Cantor spectrum for a.e. α (for an explicit set that intersects with but does not contain the set in [16]) and $\lambda = 1$ [18]. Just extending this result to the case of all (rather than a.e.) α was considered a big challenge (see Problem 5 in [25]).

A major breakthrough came recently with an influx of ideas coming from dynamical systems. Puig, using Aubry duality [1] and localization for $\theta = 0$ and $\lambda > 1$ [13], proved Cantor spectrum for Diophantine α and any noncritical λ [22]. At about the same time, Avila and Krikorian proved zero measure Cantor spectrum for $\lambda = 1$ and α satisfying a certain Diophantine condition, therefore extending the result of Last to all irrational α [3]. The solution of the Ten Martini problem as originally stated was finally given in [2]:

Main Theorem [2]. *The spectrum of the almost Mathieu operator is a Cantor set for all irrational α and for all $\lambda \neq 0$.*

Here we present the broad lines of the argument of [2]. For a much more detailed account of the history as well as of the physics background and related developments see a recent review [19].

While the ten martini problem was solved, a stronger version of it, dubbed by B. Simon the *Dry Ten Martini problem* is still open. The problem is to prove that all the gaps prescribed by the gap labelling theorem are open. This fact would be quite meaningful for the QHE related applications [4]. Dry ten martini was only established for Liouville α [2, 11] and for Diophantine α in the perturbative regime [22], using a theorem of Eliasson [12].

1.1 Rough Strategy

The history of the Ten Martini problem we described shows the existence of a number of different approaches, applicable on different parameter ranges.

Denote by $\Sigma_{\lambda,\alpha}$ the union over $\theta \in \mathbb{R}$ of the spectrum of $H_{\lambda,\alpha,\theta}$ (recall that the spectrum is actually θ-independent if $\alpha \in \mathbb{R} \setminus \mathbb{Q}$). Due to the obvious symmetry $\Sigma_{\lambda,\alpha} = -\Sigma_{-\lambda,\alpha}$, we may assume that $\lambda > 0$. Aubry duality gives a much more interesting symmetry, which implies that $\Sigma_{\lambda,\alpha} = \lambda \Sigma_{\lambda^{-1},\alpha}$. The critical coupling $\lambda = 1$ separates two very distinct regimes. The transition at $\lambda = 1$ can be clearly seen by consideration of the Lyapunov exponent $L(E) = L_{\lambda,\alpha}(E)$, for which we have the following statement.

Theorem 1. [9] *Let $\lambda > 0$, $\alpha \in \mathbb{R} \setminus \mathbb{Q}$. For every $E \in \Sigma_{\lambda,\alpha}$, $L_{\lambda,\alpha}(E) = \max\{\ln \lambda, 0\}$.*

With respect to the frequency α, one can broadly distinguish two approaches, applicable depending on whether α is well approximated by rationals or not (the Liouville and the Diophantine cases):

1. In the Liouvillian region, one can try to proceed by rational approximation, exploiting the fact that a significant part of the behavior at rational
   frequencies is accessible by calculation (this is a very special property of
   the cosine potential).
2. In the Diophantine region, one can attemp to solve two small divisor problems that have been linked with Cantor spectrum.
   (a) Localization (for large coupling), whose relevance to Cantor spectrum
       was shown in [22].
   (b) Floquet reducibility (for small coupling), which is connected to Cantor
       spectrum in [12,22].

Although Aubry duality relates both problems for $\lambda \neq 1$, it is important
to notice that the small divisor analysis is much more developed in the localization problem, where powerful non-perturbative methods are currently
available.

To decide whether α should be considered Liouville or Diophantine for
the Ten Martini problem, we introduce a parameter $\beta = \beta(\alpha) \in [0, \infty]$:

$$\beta = \limsup_{n \to \infty} \frac{\ln q_{n+1}}{q_n} , \tag{1}$$

where $\frac{p_n}{q_n}$ are the rational approximations of α (obtained by the continued
fraction algorithm). As β grows, the Diophantine approach becomes less and
less efficient, until it ceases to work, while the opposite happens for the
Liouville approach.

As discussed before, those lines of attack lead to the solution of the Ten
Martini problem in a very large region of the parameters, which is both
generic and of full Lebesgue measure. However there is no reason to expect
that one could cover the whole parameter range by this Liouville/Diophantine
dichotomy. Actually our analysis seems to indicate the existence of a critical
range, $\beta \leq |\ln \lambda| \leq 2\beta$, where one is close enough to the rationals to make the
small divisor problems intractable (so that, in particular, localization does
not hold in the full range of phases for which it holds for larger λ), but not
close enough so that one can borrow their gaps.

In order to go around the (seemingly) very real issues present in the
critical range, we will use a somewhat convoluted argument which proceeds by
contradiction. The contradiction argument allows us to exploit the following
new idea: roughly, absence of Cantor spectrum is shown to imply much better,
irrealistically good estimates. Still, those "fictitious" estimates are barely
enough to cover the critical range of parameters, and we are forced to push
the more direct approaches close to their technical limits.

We will need to apply this trick both in the Liouvillian side and in the
Diophantine side. In the Liouvillian side, it implies improved continuity estimates for the dependence of the spectrum on the frequency. In the Diophantine side, it immediately solves the "non-commutative" part of Floquet

reducibility: what remains to do is to solve the cohomological equation. Unfortunately, this can not be done directly. Instead, what we pick up from the ("soft") analysis of the cohomological equation is used to complement the ("hard") analysis of localization.

In the following sections we will succesively describe the analytic extension trick, the Liouville estimates, the two aspects of the Diophantine side (reducibility and localization), and we will conclude with some aspects of the proof of localization.

2 Analytic Extension

In Kotani theory, the complex analytic properties of Weyl's m-functions are used to describe the absolutely continuous component of the spectrum of an ergodic Schrödinger operator. However, it can also be interpreted as a theory about certain dynamical systems, *cocycles*.

We restrict to the case of the almost Mathieu operator. A formal solution of $H_{\lambda,\alpha,\theta} u = Eu$, $u \in \mathbb{C}^{\mathbb{Z}}$, satisfies the equation

$$\begin{pmatrix} E - 2\lambda \cos 2\pi(\theta + n\alpha) & -1 \\ 1 & 0 \end{pmatrix} \cdot \begin{pmatrix} u_n \\ u_{n-1} \end{pmatrix} = \begin{pmatrix} u_{n+1} \\ u_n \end{pmatrix} . \tag{2}$$

Defining $S_{\lambda,E}(x) = \begin{pmatrix} E - 2\lambda \cos 2\pi x & -1 \\ 1 & 0 \end{pmatrix}$, the importance of the products $S_{\lambda,E}(\theta + (n-1)\alpha) \cdots S_{\lambda,E}(\theta)$ becomes clear. Since $S_{\lambda,E}$ are matrices in $\mathrm{SL}(2,\mathbb{C})$, which has a natural action on $\overline{\mathbb{C}}$, $\begin{pmatrix} a & b \\ c & d \end{pmatrix} \cdot z = \frac{az+b}{cz+d}$, this leads to the consideration of the dynamical system

$$(\alpha, S_{\lambda,E}) : \mathbb{R}/\mathbb{Z} \times \overline{\mathbb{C}} \to \mathbb{R}/\mathbb{Z} \times \overline{\mathbb{C}} \tag{3}$$
$$(x, w) \mapsto (x + \alpha, S_{\lambda,E}(x) \cdot w) ,$$

which is the projective presentation of the almost Mathieu cocycle.

An invariant section for the cocycle $(\alpha, S_{\lambda,E})$ is a function $m : \mathbb{R}/\mathbb{Z} \to \overline{\mathbb{C}}$ such that $S_{\lambda,E}(x) \cdot m(x) = m(x + \alpha)$. The existence of a (sufficiently regular) invariant section is of course a nice feature, as it in a sense means that the cocycle does not see the whole complexity of the group $\mathrm{SL}(2,\mathbb{C})$: the cocycle is conjugate to a cocycle in a simpler group (of triangular matrices). The existence of two distinct invariant sections means that the simpler group is isomorphic to an even simpler, abelian group (of diagonal matrices).

It turns out that the cocycle is well behaved when E belongs to the resolvent set $\mathbb{C} \setminus \Sigma_{\lambda,\alpha}$: it is hyperbolic, which in particular means the existence of two continuous invariant sections. Moreover, the dependence of the invariant sections on E is analytic. Kotani showed that the existence of an open interval J in the spectrum where the Lyapunov exponent is zero allows one to use the Schwarz reflection principle with respect to E, and to conclude that the invariant sections can be analytically continued through J. Thus for $E \in J$, there are still two continuous invariant sections.

A crucial new idea is that those invariant sections are actually analytic also in the other variable.

Theorem 2. *Let* $0 < \lambda \le 1$, $\alpha \in \mathbb{R} \setminus \mathbb{Q}$. *Let* $J \in \Sigma_{\lambda,\alpha}$ *be an open interval. For* $E \in J$, *there exists an analytic map* $B_E : \mathbb{R}/\mathbb{Z} \to \mathrm{SL}(2, \mathbb{R})$ *such that*

$$B_E(x + \alpha) \cdot S_{\lambda,E}(x) \cdot B_E(x)^{-1} \in \mathrm{SO}(2, \mathbb{R}), \tag{4}$$

that is, $(\alpha, S_{\lambda,E})$ *is analytically conjugate to a cocycle of rotations. Moreover,* $(x, E) \mapsto B_E(x)$ *is analytic for* $(x, E) \in \mathbb{R}/\mathbb{Z} \times J$.

The proof uses the analyticity of the almost Mathieu cocycle $(\alpha, S_{\lambda,E})$ coupled with an analytic extension (Hartogs) argument.

3 The Liouvillian Side

The rational approximation argument centers around two estimates, on the size of gaps for rational frequencies, and on the modulus of continuity (in the Hausdorff topology) of the spectrum as a function of the frequency.

3.1 Gaps for Rational Approximants

The best effective estimate for the size of gaps had been given in [11], which established that all gaps of $\Sigma_{\lambda, \frac{p}{q}}$ (except the central collapsed gap for q even) have size at least $C(\lambda)^{-q}$, where $C(\lambda)$ is some explicit constant (for instance, $C(1) = 8$). Such effective constants are not good enough for our argument (for instance, it is important to have $C(\lambda)$ close to 1 when λ is close to 1). On the other hand, we only need asymptotic estimates, addressing rationals $\frac{p}{q}$ approximating some given irrational frequency for which we want to prove Cantor spectrum.

Theorem 3. *Let* $\alpha \in \mathbb{R} \setminus \mathbb{Q}$, $\lambda > 0$. *For every* $\epsilon > 0$, *if* $\frac{p}{q}$ *is close enough to* α *then all open gaps of* $\Sigma_{\lambda, \frac{p}{q}}$ *have size at least* $e^{-(|\ln \lambda| + \epsilon)q/2}$.

It was pointed out to us by Bernard Helffer (during the Qmath9 conference) that this asymptotic estimate does not hold under the sole assumption of $q \to \infty$, as is demonstrated by the analysis of Helffer and Sjostrand, so it is important to only consider approximations of a given irrational frequency.

The proof starts as in [11], which gives a global inequality relating all bands in the spectrum. We then use the integrated density of states to get a better (asymptotic) estimate on the position of bands in the spectrum. Using the Thouless formula, we get an asymptotic estimate for the size of gaps near a given frequency α and near a given energy $E \in \Sigma_{\lambda,\alpha}$ in terms of the Lyapunov exponent $L_{\lambda,\alpha}(E)$. Theorem 1 then leads to the precise estimate above.

3.2 Continuity of the Spectrum

The best general result on continuity of the spectrum was obtained in [6], 1/2-Hölder continuity. Coupled with the gap estimate for rational approximants, we get the following contribution to the Dry Ten Martini problem.

Theorem 4. *If $e^{-\beta} < \lambda < e^{\beta}$ then all gaps of $\Sigma_{\lambda, \frac{p}{q}}$ are open.*

Unfortunately this cannot be complemented by any Diophantine method that in one way or another requires localization, as it would miss the parameters such that $|\ln \lambda| = \beta > 0$. Indeed, there are certain reasons to believe that, for any θ operator $H_{\lambda,\alpha,\theta}$ has no exponentially decaying eigenfunctions for $\lambda \le e^{\beta}$.

Better estimates on continuity of the spectrum were obtained by [14] in the Diophantine range, but these estimates get worse in the critical range and can not be used. What we do instead is a "fictitious" improvement based on Theorem 2.

Theorem 5. *Let $\alpha \in \mathbb{R} \setminus \mathbb{Q}$, $\lambda > 0$. If $J \subset \text{int } \Sigma_{\lambda,\alpha}$ is a closed interval then there exists $C > 0$ such that for every $E \in J$, and for every $\alpha' \in \mathbb{R}$, there exists $E' \in \Sigma_{\lambda,\alpha'}$ with $|E - E'| < C|\alpha - \alpha'|$.*

This estimate, Lipschitz continuity, is obtained in the range $0 < \lambda \le 1$ using Theorem 2 and a direct dynamical estimate on perturbations of cocycles of rotations.

This result can be applied in an argument by contradiction:

Theorem 6. *If $e^{-2\beta} < \lambda < e^{2\beta}$ then $\Sigma_{\lambda,\alpha}$ is a Cantor set.*

4 The Diophantine Side

The Diophantine side is ruled by small divisor considerations. Two traditional small divisor problems are associated to quasiperiodic Schrödinger operators: localization for large coupling and Floquet reducibility for small coupling. Those two problems are largely related by Aubry duality.

While originally both problems were attacked by perturbative methods (very large coupling for localization and very small coupling for reducibility, depending on specific Diophantine conditions), powerful non-perturbative estimates are now available for the localization problem. For this reason, all the effective "hard analysis" we will do will be concentrated in the localization problem. However, those estimates by themselves are insufficient. We will need an additional soft analysis argument (again analytic extension), carried out for the reducibility problem under the assumption of non-Cantor spectrum, to improve (irrealistically) the localization results.

4.1 Reducibility

We say that $(\alpha, S_{\lambda,E})$ is reducible if it is analytically conjugate to a constant cocycle, that is, there exists an analytic map $B : \mathbb{R}/\mathbb{Z} \to \mathrm{SL}(2, \mathbb{R})$ such that $B(x + \alpha) \cdot S_{\lambda,E}(x) \cdot B(x)^{-1}$ is a constant A_*.

An important idea is that $(\alpha, S_{\lambda,E})$ is much more likely to be reducible if one assumes that $E \in \mathrm{int}\, \Sigma_{\lambda,\alpha}$, $0 < \lambda \le 1$. Indeed most of reducibility is taken care by Theorem 2, which simplifies the problem to proving reducibility for an analytic cocycle of rotations. This is a much easier task, which reduces to consideration of the classical cohomological equation

$$\phi(x) = \psi(x + \alpha) - \psi(x) , \tag{5}$$

which can be analysed via Fourier series: one has an explicit formula for the Fourier coefficients $\hat{\psi}(k) = \frac{1}{e^{2\pi i k\alpha} - 1}\hat{\phi}(k)$. The small divisors arise when $\|q\alpha\|_{\mathbb{R}/\mathbb{Z}}$ is small, where $\|\cdot\|_{\mathbb{R}/\mathbb{Z}}$ denotes the distance to the nearest integer.

This easily takes care of the case $\beta = 0$, but for $\beta > 0$ the information given by Theorem 2 is not quantitative enough to conclude. The analysis of the cohomological equation gives still the following interesting qualitative information.

Theorem 7. *Let $\alpha \in \mathbb{R} \setminus \mathbb{Q}$ and let $0 < \lambda \le 1$. Assume that $\beta < \infty$. Let $\Lambda_{\lambda,\alpha}$ be the set of $E \in \Sigma_{\lambda,\alpha}$ such that $(\alpha, S_{\lambda,E})$ is reducible. If $\Lambda_{\lambda,\alpha} \cap \mathrm{int}\, \Sigma_{\lambda,\alpha}$ has positive Lebesgue measure then $\Lambda_{\lambda,\alpha}$ has non-empty interior.*

The proof of this theorem uses again ideas from analytic extension.

Let $N = N_{\lambda,\alpha} : \mathbb{R} \to [0, 1]$ be the integrated density of states. One of the key ideas of [22] is that if $(\alpha, S_{\lambda,E})$ is reducible for some $E \in \Sigma_{\lambda,\alpha}$ such that $N(E) \in \alpha\mathbb{Z} + \mathbb{Z}$ then E is the endpoint of an open gap. The argument is particular to the cosine potential, and involves Aubry duality. It, in fact, extends to the case of any analytic function such that the dual model (which in general will be long-range) has simple spectrum.

Since an open subset of $\Sigma_{\lambda,\alpha}$ must intersect $\{E \in \Sigma_{\lambda,\alpha}, N(E) \in \alpha\mathbb{Z} + \mathbb{Z}\}$, we immediately obtain Cantor spectrum in the entire range of $\beta = 0$ just from the reducibility considerations alone. Note that $\beta = 0$ is strictly stronger than the Diophantine condition, and we did not use any localization result. As noted above, this $\beta = 0$ result extends to quasiperiodic potentials defined by analytic functions under the condition that the Lyapunov exponent is zero on the spectrum[1] and that the dual model has simple spectrum (it is actually enough to require that spectral multiplicities are nowhere dense).

For $0 < \beta < \infty$ it follows similarly that the hypothesis of the previous theorem must fail:

Corollary 1. *Let $\alpha \in \mathbb{R} \setminus \mathbb{Q}$ and let $0 < \lambda \le 1$. If $\beta < \infty$ then $\Lambda_{\lambda,\alpha} \cap \mathrm{int}\, \Sigma_{\lambda,\alpha}$ has zero Lebesgue measure.*

---

[1]This condition holds for all analytic functions for sufficiently small λ (in a nonperturbative way) so that the result of [10] applies, thus by [9] $L(E)$ is zero on the spectrum for all irrational α.

4.2 Localization and Reducibility

Aubry duality gives the following relation between reducibility and localization. If $E \in \Sigma_{\lambda,\alpha}$ is such that $N(E) \notin \alpha\mathbb{Z} + \mathbb{Z}$ then the following are equivalent:

1. $(\alpha, S_{\lambda,E})$ is reducible,
2. There exists $\theta \in \mathbb{R}$, such that $2\theta \in \pm N(E) + 2\alpha\mathbb{Z} + 2\mathbb{Z}$ and $\lambda^{-1}E$ is a localized eigenvalue (an eigenvalue for which the corresponding eigenfunction exponentially decays) of $H_{\lambda^{-1},\alpha,\theta}$.

Remark 1. When $N(E) \in \alpha\mathbb{Z} + \mathbb{Z}$, (1) still implies (2), but it is not clear that (2) implies (1) unless $\beta = 0$ (which covers the case treated in [22]). This is not however the main reason for us to avoid treating directly the case $N(E) \in \alpha\mathbb{Z} + \mathbb{Z}$.

Remark 2. The approach of [22] is to obtain a dense subset of $\{E \in \Sigma_{\lambda,\alpha}, N(E) \in \alpha\mathbb{Z} + \mathbb{Z}\}$ for which $(\alpha, S_{\lambda,E})$ is reducible, for $0 < \lambda < 1$ and α satisfying the Diophantine condition $\ln q_{n+1} = O(\ln q_n)$, as a consequence of localization for $H_{\lambda^{-1},\alpha,0}$ and Aubry duality. Such a localization result (for $\theta = 0$) is however not expected to hold in the critical range of α, see more discussion in the next section.

Thus proving localization of $H_{\lambda^{-1},\alpha,\theta}$ for a large set of θ allows one to conclude reducibility of $(\alpha, S_{\lambda,E})$ for a large set of E. Coupled with Corollary 1, we get the following criterium for Cantor spectrum.

Theorem 8. *Let $\alpha \in \mathbb{R} \setminus \mathbb{Q}$, $0 < \lambda \leq 1$. Assume that $\beta < \infty$. If $H_{\lambda^{-1},\alpha,\theta}$ displays localization for almost every $\theta \in \mathbb{R}$ then $\Sigma_{\lambda,\alpha}$ (and hence $\Sigma_{\lambda^{-1},\alpha}$) is a Cantor set.*

5 A Localization Result

In order to prove the Main Theorem, it remains to obtain a localization result that covers the pairs $\alpha \in \mathbb{R} \setminus \mathbb{Q}$ and $\lambda > 1$ which could not be treated by the Liouville method, namely the parameter region $\ln \lambda \geq 2\beta$.

In proving localization of $H_{\lambda,\alpha,\theta}$, two kinds of small divisors intervene,

1. The usual ones for the cohomological equation, arising from $q \in \mathbb{Z} \setminus \{0\}$ for which $\|q\alpha\|_{\mathbb{R}/\mathbb{Z}}$ is small,
2. Small denominators coming from $q \in \mathbb{Z}$ such that $\|2\theta + q\alpha\|_{\mathbb{R}/\mathbb{Z}}$ is small.

Notice that for any given α, a simple Borel-Cantelli argument allows one to obtain that for almost every θ the small denominators of the second kind satisfy polynomial lower bounds:

$$\|2\theta + q\alpha\|_{\mathbb{R}/\mathbb{Z}} > \frac{\kappa}{(1+q)^2} . \tag{6}$$

When $\theta = 0$, or more generally $2\theta \in \alpha\mathbb{Z} + \mathbb{Z}$, which is the case linked to Cantor spectrum in [22], the small divisors of the second type are exactly the same as the first type.[2] When $\beta > 0$, where the small denominators of the first type can be exponentially small, $\theta = 0$ is thus much worse behaved than almost every θ, leading to a smaller range where one should be able to prove localization. More precisely, one expects that localization holds for almost every θ if and only if $\ln \lambda > \beta$, and for $\theta = 0$ if and only if $\ln \lambda > 2\beta$. Even with all the other tricks, this would leave out the parameters such that $\ln \lambda = 2\beta > 0$.

In any case, the following localization result is good enough for our purposes.

Theorem 9. *Let $\alpha \in \mathbb{R} \setminus \mathbb{Q}$. Assume that $\ln \lambda > \frac{16}{9}\beta$. Then $H_{\lambda,\alpha,\theta}$ displays localization for almost every $\theta \in \mathbb{R}$.*

This is the most technical result of [2]. We use the general setup of [13], however our key technical procedure is quite different.

It is well known that to prove localization of $H_{\lambda,\alpha,\theta}$ it suffices to prove that all polynomially bounded solutions of $H_{\lambda,\alpha,\theta}\Psi = E\Psi$ decay exponentially.

We will use the notation $G_{[x_1,x_2]}(x,y)$ for matrix elements of the Green's function $(H - E)^{-1}$ of the operator $H_{\lambda,\alpha,\theta}$ restricted to the interval $[x_1, x_2]$ with zero boundary conditions at $x_1 - 1$ and $x_2 + 1$.

It can be checked easily that values of any formal solution Ψ of the equation $H\Psi = E\Psi$ at a point $x \in I = [x_1, x_2] \subset \mathbb{Z}$ can be reconstructed from the boundary values via

$$\Psi(x) = -G_I(x, x_1)\Psi(x_1 - 1) - G_I(x, x_2)\Psi(x_2 + 1) \ . \tag{7}$$

The strategy is to find, for every large integer x, a large interval $I = [x_1, x_2] \subset \mathbb{Z}$ containing x such that both $G(x, x_1)$ and $G(x, x_2)$ are exponentially small (in the length of I). Then, by using the "patching argument" of multiscale analysis, we can prove that $\Psi(x)$ is exponentially small in $|x|$. (The key property of Ψ, that it is a generalized eigenfunction, is used to control the boundary terms in the block-resolvent expansion.)

Fix $m > 0$. A point $y \in \mathbb{Z}$ will be called (m, k)-*regular* if there exists an interval $[x_1, x_2]$, $x_2 = x_1 + k - 1$, containing y, such that

$$|G_{[x_1,x_2]}(y, x_i)| < e^{-m|y-x_i|}, \text{ and } \operatorname{dist}(y, x_i) \geq \frac{1}{40}k; \ i = 1, 2 \ .$$

We now have to prove that every x sufficiently large is (m, k)-regular for appropriate m and k. The precise procedure to follow will depend strongly on the position of x with respect to the sequence of denominators q_n (we assume that $x > 0$ for convenience). Let $b_n = \max\{q_n^{8/9}, \frac{1}{20}q_{n-1}\}$. Let n be such that $b_n < x \leq b_{n+1}$. We distinguish between the two cases:

---

[2]Actually there is an additional very small denominator, 0 of the second type, which leads to special considerations, but is not in itself a show stopper.

1. **Resonant:** meaning $|x - \ell q_n| \leq b_n$ for some $\ell \geq 1$ and
2. **Non-resonant:** meaning $|x - \ell q_n| > b_n$ for all $\ell \geq 0$.

Theorem 9 is a consequence then of the following estimates:

Lemma 1. *Assume that θ satisfies (6). Suppose x is non-resonant. Let $s \in \mathbb{N} \cup \{0\}$ be the largest number such that $sq_{n-1} \leq \mathrm{dist}(x, \{\ell q_n\}_{\ell \geq 0})$. Then for any $\epsilon > 0$ for sufficiently large n,*

1. *If $s \geq 1$ and $\ln \lambda > \beta$, x is $(\ln \lambda - \frac{\ln q_n}{q_{n-1}} - \epsilon, 2sq_{n-1} - 1)$-regular.*
2. *If $s = 0$ then x is either $(\ln \lambda - \epsilon, 2[\frac{q_{n-1}}{2}] - 1)$ or $(\ln \lambda - \epsilon, 2[\frac{q_n}{2}] - 1)$ or $(\ln \lambda - \epsilon, 2q_{n-1} - 1)$-regular.*

Lemma 2. *Let in addition $\ln \lambda > \frac{16}{9}\beta$. Then for sufficiently large n, every resonant x is $(\frac{\ln \lambda}{50}, 2q_n - 1)$-regular.*

Each of those estimates is proved following a similar scheme, though the proof of Lemma 2 needs additional bootstrapping from the proof of Lemma 1. All small denominators considerations are entirely captured through the following concept:

We will say that the set $\{\theta_1, \ldots, \theta_{k+1}\}$ is ϵ-*uniform* if

$$\max_{z \in [-1,1]} \max_{j=1,\ldots,k+1} \prod_{\substack{\ell=1 \\ \ell \neq j}}^{k+1} \frac{|z - \cos 2\pi\theta_\ell|}{|\cos 2\pi\theta_j - \cos 2\pi\theta_\ell|} < e^{k\epsilon}. \tag{8}$$

The uniformity of some specific sequences can then be used to show that some $y \in \mathbb{Z}$ is regular following the scheme of [13]. In this approach, the goal is to find two non-intersecting intervals, I_1 around 0 and I_2 around y, of combined length $|I_1| + |I_2| = k + 1$, such that we can establish the uniformity of $\{\theta_i\}$ where $\theta_i = \theta + (x + \frac{k-1}{2})\alpha$, $i = 1, \ldots, k+1$, for x ranging through $I_1 \cup I_2$.

The actual proof of uniformity depends on the careful estimates of trigonometric products along arithmetic progressions $\theta + j\alpha$. Since $\int \ln|E - \cos 2\pi\theta|d\theta = -\ln 2$ for any $|E| \leq 1$ such estimates are equivalent to the analysis of large deviations in the appropriate ergodic theorem. A simple trigonometric expansion of (8) shows that uniformity involves equidistribution of the θ_i along with cumulative repulsion of $\pm\theta_i \pmod 1$'s, and thus involves both kinds of small divisors previously mentioned.

References

1. S. Aubry, The new concept of transition by breaking of analyticity. Solid State Sci. **8**, 264 (1978).
2. A. Avila and S. Jitomirskaya, The ten martini problem. Preprint (www.arXiv. org).

3. A. Avila and R. Krikorian, Reducibility or non-uniform hyperbolicity for quasi-periodic Schrödinger cocycles. Preprint (www.arXiv.org). To appear in Annals of Math.

4. J. E. Avron, D. Osadchy, and R. Seiler, A topological look at the quantum Hall effect, Physics today, 38–42, August 2003.

5. J. Avron and B. Simon, Almost periodic Schrödinger operators. II. The integrated density of states. Duke Math. J. **50**, 369–391 (1983).

6. J. Avron, P. van Mouche, and B. Simon, On the measure of the spectrum for the almost Mathieu operator. Commun. Math. Phys. **132**, 103–118 (1990).

7. M. Ya. Azbel, Energy spectrum of a conduction electron in a magnetic field. Sov. Phys. JETP **19**, 634–645 (1964).

8. J. Bellissard, B. Simon, Cantor spectrum for the almost Mathieu equation. J. Funct. Anal. **48**, 408–419 (1982).

9. J. Bourgain, S. Jitomirskaya, Continuity of the Lyapunov exponent for quasiperiodic operators with analytic potential, Dedicated to David Ruelle and Yasha Sinai on the occasion of their 65th birthdays. J. Statist. Phys. **108**, 1203–1218 (2002).

10. J. Bourgain, S. Jitomirskaya, Absolutely continuous spectrum for 1D quasiperiodic operators, Invent. math. **148**, 453–463 (2002).

11. M.D. Choi, G.A. Eliott, N. Yui, Gauss polynomials and the rotation algebra. Invent. Math. **99**, 225–246 (1990).

12. L. H. Eliasson, Floquet solutions for the one-dimensional quasi-periodic Schrödinger equation. Commun. Math. Phys. **146**, 447–482 (1992).

13. S. Jitomirskaya, Metal-Insulator transition for the almost Mathieu operator. Annals of Math. **150**, 1159–1175 (1999).

14. S. Ya. Jitomirskaya, I. V. Krasovsky, Continuity of the measure of the spectrum for discrete quasiperiodic operators. Math. Res. Lett. **9**, no. 4, 413–421 (2002).

15. P.G. Harper, Single band motion of conduction electrons in a uniform magnetic field, Proc. Phys. Soc. London **A 68**, 874–892 (1955).

16. B. Helffer and J. Sjöstrand, Semiclassical analysis for Harper's equation. III. Cantor structure of the spectrum. Mm. Soc. Math. France (N.S.) **39**, 1–124 (1989).

17. S. Kotani, Lyapunov indices determine absolutely continuous spectra of stationary random one-dimensional Schrödinger operators. Stochastic analysis (Katata/Kyoto, 1982), 225–247, North-Holland Math. Library, 32, North-Holland, Amsterdam, 1984.

18. Y. Last, Zero measure of the spectrum for the almost Mathieu operator, CMP **164**, 421–432 (1994).

19. Y. Last, Spectral theory of Sturm-Liouville operators on infinite intervals: a review of recent developments. Preprint 2004.

20. P.M.H. van Mouche, The coexistence problem for the discrete Mathieu operator, Comm. Math. Phys., **122**, 23–34 (1989).

21. R. Peierls, Zur Theorie des Diamagnetismus von Leitungselektronen. Z. Phys., **80**, 763–791 (1933).

22. J. Puig, Cantor spectrum for the almost Mathieu operator, Comm. Math. Phys, **244**, 297–309 (2004).

23. A. Rauh, Degeneracy of Landau levels in chrystals, Phys. Status Solidi **B 65**, K131-135 (1974).

24. B. Simon, Kotani theory for one-dimensional stochastic Jacobi matrices, Comm. Math. Phys. **89**, 227–234 (1983).

25. B. Simon, Schrödinger operators in the twenty-first century, Mathematical Physics 2000, Imperial College, London, 283–288.
26. Ya. Sinai, Anderson localization for one-dimensional difference Schrödinger operator with quasi-periodic potential. J. Stat. Phys. **46**, 861–909 (1987).
27. D.J. Thouless, M. Kohmoto, M.P. Nightingale and M. den Nijs, Quantised Hall conductance in a two dimensional periodic potential, Phys. Rev. Lett. **49**, 405–408 (1982).

Swimming Lessons for Microbots

Y. Avron

Dept of Physics, Technion, Haifa, Israel

The powerpoint file of this lecture is available at
http://physics.technion.ac.il/~avron/files/ppt/robots-chaos5.htm

Y. Avron: *Swimming Lessons for Microbots*, Lect. Notes Phys. **690**, 17–17 (2006)
www.springerlink.com

Landau-Zener Formulae
from Adiabatic Transition Histories

Volker Betz[1] and Stefan Teufel[2]

[1] Institute for Biomathematics and Biometry, GSF Forschungszentrum,
   Postfach 1129, D-85758 Oberschleißheim, Germany
   volker.betz@gsf.de
[2] Mathematisches Institut, Universität Tübingen, Auf der Morgenstelle 10,
   72076 Tübingen, Germany
   stefan.teufel@uni-tuebingen.de

Abstract. We use recent results on precise coupling terms in the optimal superadiabatic basis in order to determine exponentially small transition probabilities in the adiabatic limit of time-dependent two-level systems. As examples, we discuss the Landau-Zener and the Rosen-Zener models.

Key words: Superadiabatic basis, exponential asymptotics, Darboux principle.

1 Introduction

Transitions between separated energy levels of slowly time-dependent quantum systems are responsible for many important phenomena in physics, chemistry and even biology. In the mathematical model the slow variation of the Hamiltonian is expressed by the smallness of the *adiabatic parameter* ε in the Schrödinger equation

$$\big(i\partial_s - H(\varepsilon s)\big)\phi(t) = 0 \,, \tag{1}$$

where $H(t)$ is a family of self-adjoint operator on a suitable Hilbert space. In order to see in (1) nontrivial effects from the time-variation of the Hamiltonian, one has to follow the solutions up to times s of order ε^{-1}. Alternatively one can transform (1) to the *macroscopic time scale* $t = \varepsilon s$, resulting in the equation

$$\big(i\varepsilon\partial_t - H(t)\big)\phi(t) = 0 \,, \tag{2}$$

and study solutions of (2) for times t of order one. Often one is interested in the situation where the Hamiltonian is time-independent for large negative and positive times. Then one can consider the scattering limit and the aim is to compute the scattering amplitudes. In the simplest and at the same time paradigmatic example the Hamiltonian is just a 2×2 matrix

$$H(t) = \begin{pmatrix} Z(t) & X(t) \\ X(t) & -Z(t) \end{pmatrix} \,,$$

which can be chosen real symmetric and traceless without essential loss of generality [1]. With this choice for $H(t)$, the Schrödinger equation (2) is

V. Betz and S. Teufel: *Landau-Zener Formulae from Adiabatic Transition Histories*, Lect. Notes Phys. **690**, 19–32 (2006)
www.springerlink.com

just an ordinary differential equation for the $\mathbb{C}^2$-valued function $\phi(t)$. But even this simple system displays a very interesting behavior, of which we will give an informal description here in the introduction. The mathematical mechanism which generates this behavior will be explained in the main body of this paper.

We will assume that $H(t)$ has two distinct eigenvalues $\{E_+(t), E_-(t)\}$ for any t and approaches constant matrices as $t \to \pm\infty$. Then also the eigenvalues $\{E_+(t), E_-(t)\}$ and the orthonormal basis $\{v_+(t), v_-(t)\}$ of $\mathbb{R}^2$ consisting of the real eigenvectors of $H(t)$ have limits as $t \to \pm\infty$. By definition, the transition probability from the "upper" to the "lower" eigenstate is given by

$$P = \lim_{t\to\infty} |\phi_-(t)|^2 := \lim_{t\to\infty} |\langle v_-(t), \phi(t)\rangle_{\mathbb{C}^2}|^2 , \qquad (3)$$

where $\phi(t)$ is a solution of (2) with

$$\lim_{t\to-\infty} |\phi_-(t)|^2 = 1 - \lim_{t\to-\infty} |\phi_+(t)|^2 = 0 . \qquad (4)$$

Despite the presence of a natural small parameter, the adiabatic parameter $\varepsilon \ll 1$, it is far from obvious how to compute P even to leading order in ε. This is because the transition amplitudes connecting different energy levels are exponentially small with respect to ε, i.e. of order $\mathcal{O}(e^{-c/\varepsilon})$ for some $c > 0$, and thus have no expansion in powers of ε.

The result of a numerical computation of $\phi_-(t)$ for a typical Hamiltonian $H(t)$ is displayed in Fig. 1a. After rising to a value which is of order ε, $|\phi_-|$ falls off again and finally, in the regime where $H(t)$ is approximately constant, settles for a value of order $e^{-c/\varepsilon}$.

It is no surprise that $\sup_{t\in\mathbb{R}} |\phi_-(t)|$ is of order ε: this is just a consequence of the proof of the adiabatic theorem [8], and in fact we perform the relevant calculation in Sect. 2. There we see that the size of $\phi_-(t)$ is determined by

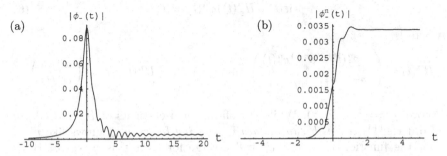

Fig. 1. This figure shows the lower components of a numerical solution of (5) for $\varepsilon = 1/6$. In (a), the lower component in the adiabatic basis rises to a value order ε before approaching its exponentially small asymptotic value. In (b), the lower component in the optimal superadiabatic basis rises monotonically to its final value. Note the different axes scalings, as the asymptotic values in both pictures agree

the size of the off-diagonal elements of the adiabatic Hamiltonian $H_{ad}(t)$. The latter is obtained by expressing (2) in the *adiabatic basis* $\{v_+(t), v_-(t)\}$. More precisely, let $U_0(t)$ be the orthogonal matrix that takes the adiabatic basis into the canonical basis. Then multiplication of (2) with $U_0(t)$ from the left leads to

$$\big(i\varepsilon\partial_t - H_{ad}(t)\big)\phi_{ad}(t) := U_0(t)\big(i\varepsilon\partial_t - H(t)\big)U_0^*(t)U_0(t)\phi(t) = 0 , \quad (5)$$

where $H_{ad}(t) = \mathrm{diag}(E_+(t), E_-(t)) - i\varepsilon U_0(t)\dot{U}_0^*(t)$. Clearly, the off-diagonal elements of the matrix H_{ad} are of order ε, and $\phi_-(t)$ is just the second component of $\phi_{ad}(t)$. However, the $\mathcal{O}(\varepsilon)$ smallness of the coupling in the adiabatic Hamiltonian does not explain the exponentially small scattering regime in Fig. 1a. In the adiabatic basis, there is no easy way to see why this effect should take place, although with some goodwill it may be guessed by a heuristic calculation to be presented in the next section.

A natural strategy to understand the exponentially small scattering amplitudes goes back to M. Berry [1]: the solution of (2) with initial condition (4) remains in the positive adiabatic subspace spanned by $v_+(t)$ only up to errors of order ε. Hence one should find a better subspace, the *optimal superadiabatic subspace*, in which the solution remains up to exponentially small errors for all times. Since we are ultimately interested in the transition probabilities, at the same time this subspace has to coincide with the adiabatic subspace as $t \to \pm\infty$. One way to determine the superadiabatic subspaces is to optimally truncate the asymptotic expansion of the true solution in powers of ε, as Berry [1] did. Alternatively one can look for a time-dependent basis of $\mathbb{C}^2$ such that the analogues transformation to (5) yields a Hamiltonian with exponentially small off-diagonal terms. To do so, one first constructs the n-th *superadiabatic basis* recursively from the adiabatic basis for any $n \in \mathbb{N}$. Let us write $U_\varepsilon^n(t)$ for the transformation taking the n-th superadiabatic basis into the canonical one. Then as in (5) the Schrödinger equation takes the form

$$\big(i\varepsilon\partial_t - H_\varepsilon^n(t)\big)\phi^n(t) = 0 , \quad (6)$$

where

$$H_\varepsilon^n(t) = \begin{pmatrix} \rho_\varepsilon^n(t) & \varepsilon^{n+1}c_\varepsilon^n(t) \\ \varepsilon^{n+1}\bar{c}_\varepsilon^n(t) & -\rho_\varepsilon^n(t) \end{pmatrix} \quad \text{and} \quad \phi^n(t) = U_\varepsilon^n(t)\phi(t) = \begin{pmatrix} \phi_+^n(t) \\ \phi_-^n(t) \end{pmatrix} .$$
$$(7)$$

Above, $\rho_\varepsilon^n = \frac{1}{2} + \mathcal{O}(\varepsilon^2)$. While the off-diagonal elements of H_ε^n indeed are of order ε^{n+1}, the *n-th superadiabatic coupling function* c_ε^n grows like $n!$ so that the function $n \mapsto \varepsilon^{n+1}c_\varepsilon^n$ will diverge for each ε as $n \to \infty$. However, for each $\varepsilon > 0$ there is an $n_\varepsilon \in \mathbb{N}$ such that $\varepsilon^{n+1}c_\varepsilon^n$ takes its minimal value for $n = n_\varepsilon$. This defines the *optimal superadiabatic basis*. In this basis the off-diagonal elements of $H_\varepsilon^n(t)$ are exponentially small for all t. As a consequence, also the lower component $\phi_-^n(t)$ of the solution with $\lim_{t\to-\infty}\phi_-^n(t) = \lim_{t\to-\infty}\phi_-(t) = 0$ is exponentially small, as illustrated in

Fig. 1b, and one can compute the scattering amplitude by first order perturbation theory.

Berry and Lim [1, 2] showed on a non-rigorous level that $\phi_-^n(t)$ is not only exponentially small in ε but has the universal form of an error function, a feature also illustrated in Fig. 1(b). A rigorous derivation of the optimal superadiabatic Hamiltonian and of the universal transition histories has been given recently in [3] and [4].

The aim of this note is to explain certain aspects of the results from [4] and to show how to obtain scattering amplitudes from them. In Sect. 2 we basically give a more detailed and also more technical introduction to the problem of exponentially small non-adiabatic transitions. Section 3 contains a concise summary of the results obtained in [4]. In order to apply these results to the scattering situation, we need some control on the time decay of the error estimates appearing in our main theorem. In Sect. 4 we use standard Cauchy estimates to obtain such bounds and give a general recipe for obtaining rigorous proofs of scattering amplitudes. We close with two examples, the Landau-Zener model and the Rosen-Zener model. While the Landau-Zener model displays, in a sense to be made precise, a generic transition point, the Rosen-Zener model is of a non-generic type, which is not covered by existing rigorous results.

2 Exponentially Small Transitions

From now on we study the Schrödinger equation (2) with the Hamiltonian

$$H_{\mathrm{ph}}(t) = \begin{pmatrix} Z(t) & X(t) \\ X(t) & -Z(t) \end{pmatrix} = \rho(t) \begin{pmatrix} \cos\theta_{\mathrm{ph}}(t) & \sin\theta_{\mathrm{ph}}(t) \\ \sin\theta_{\mathrm{ph}}(t) & -\cos\theta_{\mathrm{ph}}(t) \end{pmatrix}. \tag{8}$$

Thus $H_{\mathrm{ph}}(t)$ is a traceless real-symmetric 2×2-matrix, and the eigenvalues of $H_{\mathrm{ph}}(t)$ are $\pm\rho(t) = \pm\sqrt{X(t)^2 + Z(t)^2}$. We assume that the gap between them does not close, i.e. that $2\rho(t) \geq g > 0$ for all $t \in \mathbb{R}$. As to be detailed below, we assume that X and Z are real-valued on the real axis and analytic on a suitable domain containing the real axis. Moreover, in order to be able to consider the scattering limit it is assumed that $H_{\mathrm{ph}}(t)$ approaches limits $H_\pm$ sufficiently fast as $t \to \pm\infty$.

Before proceeding we simplify (8) by switching to the *natural time scale*

$$\tau(t) = 2 \int_0^t \mathrm{d}s\, \rho(s). \tag{9}$$

Since $\rho(t)$ is assumed to be strictly positive, the map $t \mapsto \tau$ is a bijection of $\mathbb{R}$. In the natural time scale the Schrödinger equation (2) becomes

$$\big(\mathrm{i}\varepsilon\partial_\tau - H_{\mathrm{n}}(\tau)\big)\phi(\tau) = 0 \tag{10}$$

with Hamiltonian

$$H_n(\tau) = \frac{1}{2}\begin{pmatrix} \cos\theta_n(\tau) & \sin\theta_n(\tau) \\ \sin\theta_n(\tau) & -\cos\theta_n(\tau) \end{pmatrix}, \tag{11}$$

where $\theta_n(\tau) = \theta_{ph}(t(\tau))$. As a consequence we now deal with a Hamiltonian with constant eigenvalues equal to $\pm\frac{1}{2}$, which is completely defined through the single real-analytic function θ_n.

The transformation (5) to the adiabatic basis, i.e. the orthogonal matrix that diagonalizes $H_n(\tau)$, is

$$U_0(\tau) = \begin{pmatrix} \cos(\theta_n(\tau)/2) & \sin(\theta_n(\tau)/2) \\ \sin(\theta_n(\tau)/2) & -\cos(\theta_n(\tau)/2) \end{pmatrix}. \tag{12}$$

Multiplying (10) from the left with $U_0(\tau)$ yields the Schrödinger equation in the *adiabatic representation*

$$\bigl(i\varepsilon\partial_\tau - H_\varepsilon^a(\tau)\bigr)\phi^a(\tau) = 0, \tag{13}$$

where

$$H_\varepsilon^a(\tau) = \begin{pmatrix} \dfrac{1}{2} & \dfrac{i\varepsilon}{2}\theta_n'(\tau) \\ -\dfrac{i\varepsilon}{2}\theta_n'(\tau) & -\dfrac{1}{2} \end{pmatrix} \quad \text{and} \quad \phi^a(\tau) = U_0(\tau)\phi(\tau) = \begin{pmatrix} \phi_+(\tau) \\ \phi_-(\tau) \end{pmatrix}. \tag{14}$$

θ_n' is called the *adiabatic coupling function*.

The exponentially small scattering amplitude in Fig. 1a) can be guessed by a heuristic calculation. We solve (13) for $\phi_-(\tau)$ using $\phi_+(\tau) = e^{-\frac{i\tau}{2\varepsilon}} + \mathcal{O}(\varepsilon)$, which holds according to the adiabatic theorem [8], and variation of constants, i.e.

$$\begin{aligned}
\phi_-(\tau) &= \frac{i}{\varepsilon}e^{\frac{i\tau}{2\varepsilon}}\int_{-\infty}^{\tau} d\sigma\, e^{-\frac{i\sigma}{2\varepsilon}}\left(-\frac{i\varepsilon}{2}\theta_n'(\sigma)\right)\phi_+(\sigma) \\
&= \frac{1}{2}e^{\frac{i\tau}{2\varepsilon}}\int_{-\infty}^{\tau} d\sigma\, \theta_n'(\sigma)\,e^{-\frac{i\sigma}{\varepsilon}} + \mathcal{O}(\varepsilon).
\end{aligned} \tag{15}$$

Integration by parts yields

$$\phi_-(\tau) = \frac{i\varepsilon}{2}\theta_n'(\tau) - \frac{i\varepsilon}{2}e^{\frac{i\tau}{2\varepsilon}}\int_{-\infty}^{\tau} d\sigma\, \theta_n''(\sigma)\,e^{-\frac{i\sigma}{\varepsilon}} + \mathcal{O}(\varepsilon). \tag{16}$$

The first term in this expression is of order ε and not smaller. This strongly suggests that the $\mathcal{O}(\varepsilon)$ error estimate in the adiabatic theorem is optimal, which we have seen to be indeed the case. However, no conclusion can be inferred from (16) for the scattering regime $\tau \to \infty$ since θ_n' vanishes there.

The key to the heuristic treatment of the scattering amplitude is to calculate the integral in (15) not by integration by parts but by contour integration in the complex plane. For the sake of a simple argument let us assume here that θ_n' is a meromorphic function. Let τ_c be the location of the pole in the lower complex half plane closest to the real line and γ its residue, then from (15) and contour integration around the poles in the lower half plane we read off

$$\lim_{\tau \to \infty} |\phi_-(\tau)|^2 = \pi^2 \gamma^2 \, e^{-\frac{2\mathrm{Im}\tau_c}{\varepsilon}} + \mathcal{O}(\varepsilon^2) \,. \tag{17}$$

Strictly speaking (17) tells us nothing new: while the explicit term is exponentially small in ε, the error term is of order ε^2 and thus the statement is not better than what we know from the adiabatic theorem already. Nevertheless it turns out that the exponential factor appearing here actually yields the correct asymptotic behavior of the transition probability. Our heuristic argument also correctly attributes the dominant part of the transition to the pole of θ_n' closest to the real axis. The prefactor in (17), however, is wrong, the correct answer being

$$\lim_{\tau \to \infty} |\phi_-(\tau)|^2 = 4\sin^2\left(\frac{\pi\gamma}{2}\right) e^{-\frac{2\mathrm{Im}\tau_c}{\varepsilon}} (1 + \mathcal{O}(\varepsilon^\alpha)) \,, \tag{18}$$

for some $\alpha > 0$. Expression (18) is a generalization of the Landau-Zener formula and was first rigorously derived in [7].

The problem to solve when trying to rigorously treat exponentially small transitions and to arrive at the correct result (18) is to control the solution of (2) up to errors that are not only exponentially small in ε, but smaller than the leading order transition probability. As a consequence a naive perturbation calculation in the adiabatic basis will not do the job.

The classical approach [7] to cope with this is to solve (2) not on the real axis but along a certain path in the complex plane, where the lower component of the solution is always exponentially small. The comparison with the solution on the real line is made only in the scattering limit at $\tau = \pm\infty$. The trick is to choose the path in such a way that it passes through the relevant singularity of θ_n' in the complex plane. In a neighborhood of the singularity one can solve (13) explicitly and thereby determine the leading order contribution to the transition probability. Moreover, away from the transition point the path must be chosen such that the lower component $\phi_-(\tau)$ remains smaller than the exponentially small leading order contribution from the transition point for all τ along this path. There are two drawbacks of this approach: the technical one is that there are examples (see the Rosen-Zener model below), where such paths do not exist. On the conceptual side, this approach yields only the scattering amplitudes, but gives no information whatsoever about the solution for finite times.

Our approach is motivated by the findings of Berry [1] and of Berry and Lim [2]. Instead of solving (13) along a path in the complex plane we solve the problem along the real axis but in a super-adiabatic basis instead of the adiabatic one, i.e. we solve (6) with the Hamiltonian (7) and the optimal $n(\varepsilon)$. While the off-diagonal elements of the Hamiltonian in the adiabatic basis are only of order ε, cf. (14), the off-diagonal elements of the Hamiltonian in the optimal superadiabatic basis are exponentially small, i.e. of order $e^{-c/\varepsilon}$.

In order to control the exponentially small transitions, we will give precise exponential bounds on the coupling $\varepsilon^{n_\varepsilon+1}c_\varepsilon^{n_\varepsilon}(\tau)$ away from the transition regions and explicitly determine the asymptotic form of $c_\varepsilon(\tau)$ within each transition region. Since the superadiabatic bases agree asymptotically for $t \to \pm\infty$ with the adiabatic basis, the scattering amplitudes agree in all these bases. In the optimal superadiabatic basis the correct transition probabilities (18) now follow from a first order perturbation calculation analogous to the one leading to (17) in the adiabatic basis. However, in addition to the scattering amplitudes we obtain approximate solutions for all times, i.e. "histories of adiabatic quantum transitions" [1]. As is illustrated in Fig. 1b, these are monotonous and asymptotically take the form of an error function.

3 The Hamiltonian in the Super-Adiabatic Representation

In [4] we formulate our results for the system (10) and (11). However, we have to keep in mind that (10) and (11) arise from the physical problem (2) and (8) through the transformation to the natural time scale (9). Therefore, to be physically relevant, the assumptions must be satisfied by all θ_n arising from generic Hamiltonians of the form (8). As observed in [2], see also [4], for such θ_n the adiabatic coupling function θ'_n is real analytic and at its complex singularities z_0 closest to the real axis it has the form

$$\theta'_n(z - z_0) = \frac{-i\gamma}{z - z_0} + \sum_{j=1}^{N}(z - z_0)^{-\alpha_j}h_j(z - z_0) , \qquad (19)$$

where $|\mathrm{Im}z_0| > 0$, $\gamma \in \mathbb{R}$, $\alpha_j < 1$ and h_j is analytic in a neighborhood of 0 for $j = 1, \ldots, N$.

The following norms on the real line capture exactly the behavior (19) of the complex singularities of θ'_n. They are at the heart of the analysis in [4].

Definition 1. Let $\tau_c > 0$, $\alpha > 0$ and $I \subset \mathbb{R}$ be an interval. For $f \in C^\infty(I)$ we define

$$\|f\|_{(I,\alpha,\tau_c)} := \sup_{t\in I} \sup_{k\geq 0} |\partial^k f(t)| \frac{\tau_c^{\alpha+k}}{\Gamma(\alpha+k)} \leq \infty \qquad (20)$$

and

$$F_{\alpha,\tau_c}(I) = \left\{ f \in C^\infty(I) : \|f\|_{(I,\alpha,\tau_c)} < \infty \right\}.$$

The connection of these norms with (19) relies on the Darboux Theorem for power series and is described in [4]. Let us just note here that θ'_n as given in (19) is an element of $F_{1,\tau_c}(\{\tau_r\})$ for $\tau_c = \mathrm{Im}(z_0)$ and $\tau_r = \mathrm{Re}(z_0)$, while the second term of (19) is in $F_{\beta,\tau_c}(\{\tau_r\})$ with $\beta = \max_j \alpha_j$. In order to control the transitions histories, the real line will be segmented into intervals I, which are either considered to be a small neighborhood of a transition point or to contain no transition point. Assumption 1 below thus applies to intervals without transition point and Assumption 2 generically holds near a transition point. In the rest of this section we drop the subscript n for the natural time scale in order not to overburden our notation.

Assumption 1: *For a compact interval I and $\delta \geq 0$ let $\theta'(\tau) \in F_{1,\tau_c+\delta}(I)$.*

Assumption 2: *For γ, τ_r, $\tau_c \in \mathbb{R}$ let*

$$\theta_0'(t) = i\gamma \left(\frac{1}{\tau - \tau_r + i\tau_c} - \frac{1}{\tau - \tau_r - i\tau_c} \right)$$

be the sum of two complex conjugate first order poles located at $\tau_r \pm i\tau_c$ with residues $\mp i\gamma$. On a compact interval $I \subset [\tau_r - \tau_c, \tau_r + \tau_c]$ with $\tau_r \in I$ we assume that

$$\theta'(\tau) = \theta_0'(\tau) + \theta_r'(\tau) \quad \text{with} \quad \theta_r'(\tau) \in F_{\alpha,\tau_c}(I) \tag{21}$$

for some γ, τ_c, $\tau_r \in \mathbb{R}$, $0 < \alpha < 1$.

It turns out that under Assumption 2 the optimal superadiabatic basis is given as the $n_\varepsilon^{\text{th}}$ superadiabatic basis where $0 \leq \sigma_\varepsilon < 2$ is such that

$$n_\varepsilon = \frac{\tau_c}{\varepsilon} - 1 + \sigma_\varepsilon \quad \text{is an even integer.} \tag{22}$$

The two main points of the following theorem are: outside the transition regions, the off-diagonal elements of the Hamiltonian in the optimal superadiabatic basis are bounded by (24), while within each transition region they are asymptotically equal to $g(\varepsilon, \tau)$ as given in (ii).

Theorem 1. *(i) Let H satisfy Assumption 1. Then there exists $\varepsilon_0 > 0$ such that for all $\varepsilon \in (0, \varepsilon_0]$ and all $\tau \in I$ the elements of the superadiabatic Hamiltonian (7) and the unitary $U_\varepsilon^{n_\varepsilon}(\tau)$ with n_ε as in (22) satisfy*

$$\left| \rho_\varepsilon^{n_\varepsilon}(\tau) - \frac{1}{2} \right| \leq \varepsilon^2 \phi_1\left(\|\theta'\|_{(I,1,\tau_c+\delta)} \right) \tag{23}$$

$$\left| \varepsilon^{n_\varepsilon+1} c_\varepsilon^{n_\varepsilon}(\tau) \right| \leq \sqrt{\varepsilon}\, e^{-\frac{\tau_c}{\varepsilon}(1+\ln\frac{\tau_c+\delta}{\tau_c})} \phi_1\left(\|\theta'\|_{(I,1,\tau_c+\delta)} \right) \tag{24}$$

and

$$\|U_\varepsilon^{n_\varepsilon}(\tau) - U_0(\tau)\| \leq \varepsilon\phi_1\left(\|\theta'\|_{(I,1,\tau_c+\delta)} \right). \tag{25}$$

Here $\phi_1 : \mathbb{R}^+ \to \mathbb{R}^+$ is a locally bounded function with $\phi_1(x) = \mathcal{O}(x)$ as $x \to 0$ which is independent of I and δ.

(ii) Let H satisfy Assumption 2 and define

$$g(\varepsilon, \tau) = 2\mathrm{i} \sqrt{\frac{2\varepsilon}{\pi \tau_c}} \, \sin\left(\frac{\pi\gamma}{2}\right) \mathrm{e}^{-\frac{\tau_c}{\varepsilon}} \, \mathrm{e}^{-\frac{(\tau - \tau_r)^2}{2\varepsilon \tau_c}} \, \cos\left(\frac{\tau - \tau_r}{\varepsilon} - \frac{(\tau - \tau_r)^3}{3\varepsilon \tau_c^2} + \frac{\sigma_\varepsilon \tau}{\tau_c}\right).$$

There exists $\varepsilon_0 > 0$ and a constant $C < \infty$ such that for all $\varepsilon \in (0, \varepsilon_0]$ and all $\tau \in I$

$$\left| \varepsilon^{n_\varepsilon + 1} c_\varepsilon^{n_\varepsilon}(\tau) - g(\varepsilon, \tau) \right| \leq C \varepsilon^{\frac{3}{2} - \alpha} \mathrm{e}^{-\frac{\tau_c}{\varepsilon}}. \tag{26}$$

Furthermore, the assertions of part (i) hold with $\delta = 0$.

Remark 1. In [4] we show in addition that the error bounds in Theorem 1 are locally uniform in the parameters α, γ and τ_c. This generality is not needed here and thus omitted from the statement.

In order to pass to the scattering limit it is now necessary to show that the errors in part (i) of Theorem 1, i.e. in the regions away from the transition points, are integrable.

4 The Scattering Regime

We will treat the scattering regime by using first order perturbation theory on the equation in the optimal superadiabatic basis. As in (15), variation of constants yields

$$\phi_-^{n_\varepsilon}(\tau) = \frac{\mathrm{i}}{\varepsilon} \mathrm{e}^{\frac{\mathrm{i}}{\varepsilon} \int_{-\infty}^{\tau} \mathrm{d}\sigma \, \rho(\sigma)} \int_{-\infty}^{\tau} \mathrm{d}\sigma \, \mathrm{e}^{-\frac{\mathrm{i}}{\varepsilon} \int_{-\infty}^{\sigma} \mathrm{d}\nu \, \rho(\nu)} \, c(n_\varepsilon, \sigma) \, \phi_+^{n_\varepsilon}(\sigma), \tag{27}$$

where we put $c(n_\varepsilon, \tau) = \varepsilon^{n_\varepsilon + 1} c_\varepsilon^{n_\varepsilon}(\tau)$. We now replace $\rho_\varepsilon^{n_\varepsilon}(\tau)$ and $c(n_\varepsilon, \tau)$ in (27) by the explicit asymptotic values given in Theorem 1, and use the adiabatic approximation $\phi_+^{n_\varepsilon}(\tau) = \mathrm{e}^{-\frac{\mathrm{i}\tau}{2\varepsilon}} + \mathcal{O}(\varepsilon)$. To this end we assume that θ_n' has k poles of the form (19) at distance τ_c from the real axis and none closer to the real axis. Let $g_j(\varepsilon, \tau)$ be the associated coupling functions of Theorem 1 for $j = 1, \ldots, k$ and

$$f_1(\tau) = \left| \rho_\varepsilon^{n_\varepsilon}(\tau) - \frac{1}{2} \right|, \quad f_2(\tau) = c(n_\varepsilon, \tau) - \sum_{j=1}^{k} g_j(\varepsilon, \tau), \quad f_3(\tau) = \phi_+^{n_\varepsilon}(\tau) - \mathrm{e}^{-\frac{\mathrm{i}\tau}{2\varepsilon}}.$$

Then

$$\mathrm{e}^{\frac{\mathrm{i}}{\varepsilon} \int_{-\infty}^{\tau} \mathrm{d}\sigma \, \rho(\sigma)} = \mathrm{e}^{\frac{\mathrm{i}\tau}{2\varepsilon}} \left(1 + \mathcal{O}\left(\varepsilon \underbrace{\int_{-\infty}^{\tau} \mathrm{d}\sigma \, f_1(\sigma)}_{=: F_1(\tau)} \right) \right)$$

and

$$\phi_-^{n_\varepsilon}(\tau) = \frac{i}{\varepsilon} e^{\frac{i\tau}{2\varepsilon}} \left(1 + \mathcal{O}(\varepsilon F_1(\tau))\right) \int_{-\infty}^{\tau} d\sigma \, e^{-\frac{i\sigma}{2\varepsilon}} \left(1 + \mathcal{O}(\varepsilon F_1(\sigma))\right)$$

$$\times \left(\sum_{j=1}^{k} g_j(\varepsilon, \sigma) - f_2(\sigma)\right) \phi_+^{n_\varepsilon}(\sigma)$$

$$= \frac{i}{\varepsilon} e^{\frac{i\tau}{2\varepsilon}} \int_{-\infty}^{\tau} d\sigma \, e^{-\frac{i\sigma}{\varepsilon}} \sum_{j=1}^{k} g_j(\varepsilon, \sigma)$$

$$+ \mathcal{O}\left((\|F_1\|_\infty + \varepsilon^{-1}\|f_3\|_\infty) \int_{-\infty}^{\tau} d\sigma |c(n_\varepsilon, \sigma)| + \varepsilon^{-1} \int_{-\infty}^{\tau} d\sigma |f_2(\sigma)|\right).$$

Assuming integrability of the error terms in (23) and (24), the following lemma can be established by straightforward computations.

Proposition 1. *Let $\theta_n'(\tau)$ be as above and let $\tau \mapsto \|\theta_n'\|_{(\{\tau\},1,\tau_c+\delta)}$ be integrable outside of some bounded interval and for some $\delta > 0$. Then*

$$\phi_-^{n_\varepsilon}(\tau) = \frac{i}{\varepsilon} e^{\frac{i\tau}{2\varepsilon}} \int_{-\infty}^{\tau} d\sigma \, e^{-\frac{i\sigma}{\varepsilon}} \sum_{j=1}^{k} g_j(\varepsilon, \sigma) + \mathcal{O}(\varepsilon^{\frac{1}{2}-\alpha} e^{-\frac{\tau_c}{\varepsilon}}).$$

Note that the leading term in Proposition 1 is of order $e^{-\frac{\tau_c}{\varepsilon}}$. Thus for $\alpha \geq \frac{1}{2}$ the estimate is too weak. However, a more careful analysis of the error near the transition points allows one to replace $\varepsilon^{\frac{1}{2}-\alpha}$ by $\varepsilon^{1-\alpha}$ in Proposition 1, see [5], and thus to obtain a nontrivial estimate for all $\alpha < 1$.

Since the functions $g_j(\varepsilon, \tau)$ are explicitly given in Theorem 1, the leading order expression for $\phi_-^{n_\varepsilon}(\tau)$ can be computed explicitly as well. A simple computation, c.f. [3], yields for $k = 1$ that

$$\phi_-^{n_\varepsilon}(\tau) = \frac{i}{\varepsilon} e^{\frac{i\tau}{2\varepsilon}} \int_{-\infty}^{\tau} d\sigma \, e^{-\frac{i\sigma}{\varepsilon}} g(\varepsilon, \sigma) + \mathcal{O}(\varepsilon^{\frac{1}{2}-\alpha} e^{-\frac{\tau_c}{\varepsilon}})$$

$$= \sin\left(\frac{\pi\gamma}{2}\right) e^{-\frac{\tau_c}{\varepsilon}} e^{\frac{i\tau}{2\varepsilon}} \left(\text{erf}\left(\frac{\tau}{\sqrt{2\varepsilon\tau_c}}\right) + 1\right) + \mathcal{O}(\varepsilon^{\frac{1}{2}-\alpha} e^{-\frac{\tau_c}{\varepsilon}}).$$

For more than one transition point the same computation reveals interference effects, c.f. [5]. In the limit $\tau \to \infty$ we recover the Landau-Zener formula for the transition probability:

$$|\phi_-^{n_\varepsilon}(\infty)|^2 = 4\sin^2\left(\frac{\pi\gamma}{2}\right) e^{-\frac{2\tau_c}{\varepsilon}} + \mathcal{O}(\varepsilon^{\frac{1}{2}-\alpha} e^{-\frac{2\tau_c}{\varepsilon}}). \tag{28}$$

Proposition 1 yields the transition histories as well as the transition probabilities in the scattering limit for a large class of Hamiltonians under the assumption that $\|\theta_n'\|_{(\{\tau\},1,\tau_c+\delta)}$ is integrable at infinity for some $\delta > 0$. At first sight it might seems hard to establish integrability of this norm, since it involves derivatives of θ_n' of all orders. However, the following proposition shows that $\|\theta_n'\|_{(\{\tau\},1,\tau_c+\delta)}$ can be bounded by the supremum of the function θ_n' in a ball around τ with radius slightly larger than $\tau_c + \delta$.

Proposition 2. *Let* $\alpha > 0$ *and* $r > 0$. *Assume for some* $\delta > 0$ *that* f *is analytic on*

$$B_{r+\delta} = \{z \in \mathbb{C} : |z| \le r + \delta\}.$$

Then

$$\|f\|_{(\{0\},\alpha,r)} \le \frac{r^\alpha}{e \ln((r+\delta)/r)} \sup_{z \in B_{r+\delta}} |f(z)|.$$

Proof. Put $M = \sup_{z \in B_{r+\delta}} |f(z)|$. By the Cauchy formula,

$$\partial_t^k f(0) = k! \oint_{|z|=r+\delta} dz \, \frac{f(z)}{z^{k+1}} \le 2\pi \, k! \, M (r+\delta)^{-k}.$$

Therefore

$$\partial_t^k f(0) \frac{r^{\alpha+k}}{\Gamma(\alpha+k)} \le M r^\alpha \frac{\Gamma(1+k)}{\Gamma(\alpha+k)} \left(\frac{r}{r+\delta}\right)^k. \tag{29}$$

The k-dependent part of the right hand side above is obviously maximal for $\alpha = 0$, and then is equal to $\phi(k) := k(r/(r+\delta))^k$. $\phi(k)$ is maximal at $k = 1/\ln((r+\delta)/r)$ with value $1/(e \ln((r+\delta)/r)$, and the claim follows by taking the supremum over k in (29). $\qquad \blacksquare$

Hence, integrability of $\|\theta_n'\|_{(\{\tau\},1,\tau_c+\delta)}$ follows if we can establish sufficient decay of $\sup_{|z-\tau|<\tau_c+2\delta} |f(z)|$ as $\tau \to \infty$. We will demonstrate how to do this for two simple examples. More elaborate examples including interference effects can be found in [5]. We will use the transformation formula

$$\theta_n'(\tau(t)) = \frac{\theta_{ph}'(t)}{2\rho(t)} = \frac{1}{2\rho(t)} \frac{d}{dt} \arctan\left(\frac{X}{Z}\right)(t) = \frac{X'Z - Z'X}{2\rho^3}(t). \tag{30}$$

Example 1 (Landa-Zener model). The paradigmatic example is the Landau-Zener Hamiltonian

$$H(t) = \begin{pmatrix} a & t \\ t & -a \end{pmatrix},$$

which is explicitly solvable [10] and for which the transition probabilities are well-known. Nevertheless it is instructive to exemplify our method on this simple model. We have $X(t) = t$ and $Z(t) = a > 0$. Thus $\rho^2(t) = a^2 + t^2$, and the transformation to the natural time scale reads

$$\tau(t) = 2 \int_0^t \sqrt{a^2 + s^2} \, ds. \tag{31}$$

From (30) one reads off that complex zeros of ρ give rise to complex singularities of θ_n'. In the Landau-Zener model, ρ has two zeros at $t_c = \pm ia$. Thus (31) yields $\tau_c = \frac{a^2 \pi}{2}$, and expansion of $\theta_n'(\tau)$ around τ_c shows $\gamma = \frac{1}{3}$ and $\alpha = \frac{1}{3}$, cf. [2,4]. We now apply Proposition 1 in order to pass to the scattering limit. According to Proposition 2 we need to control the decay of $|\theta_n'|$ in a finite strip around the real axis for large $|\tau|$. From (31) one reads off that

$$|\tau(t)| \le 2 \cdot 2|t|\sqrt{a^2 + |t|^2} \le 4(a^2 + |t|^2) \,,$$

and thus $|t^2| \ge |\tau|/4 - a^2$. From (30) and the estimates above we infer for $|\tau|$ sufficiently large that

$$|\theta'_n(\tau)| = \frac{a}{2|a^2 + t(\tau)^2|^{3/2}} \le \frac{a}{2(|t(\tau)|^2 - a^2)^{3/2}} \le \frac{a}{2(|\tau|/4 - 2a^2)^{3/2}} \,.$$

Consequently, Proposition 2 yields for every $r, \delta > 0$ and $\tau \in \mathbb{R}$ sufficiently large that

$$\|\theta'_n\|_{(\{\tau\},1,r)} \le \frac{r}{e \ln((r+\delta)/r)} \frac{a}{((|\tau| - r - \delta)/4 - 2a^2)^{3/2}} \,.$$

Thus $\tau \mapsto \|\theta'_n\|_{(\{\tau\},1,r)}$ is integrable at infinity for any $r > 0$, and in particular for $r = \tau_c + 2\delta$. According to (28), we have therefore shown the classical Landau-Zener formula

$$|\phi^{n_\varepsilon}_-(\infty)|^2 = e^{-\frac{a^2\pi}{\varepsilon}} + \mathcal{O}(\varepsilon^{\frac{1}{6}} e^{-\frac{a^2\pi}{\varepsilon}}) \,.$$

For the Landau-Zener model, the transition probabilities can also be proved by the method of [7]. There, the anti-Stokes lines, i.e. the level lines $\mathrm{Im}(\tau(t)) = \mathrm{Im}(\tau(t_c))$, play an essential role. In particular, the method requires that an anti-Stokes line emanating from the critical point t_c of τ stays in a strip of finite width around the real axis as $\mathrm{Re}(t) \to \pm\infty$. As shown in Fig. 2, this is the case in the present example.

The previous example also shows a useful general strategy: One can use (30) in order to find upper bounds on $\tau(t)$, which in turn yield lower bounds on the inverse function $t(\tau)$. These can then be used in (30) to estimate the decay at infinity of θ'_n in a strip around the real axis. It is clear that this strategy also works in cases where the Hamiltonian is not given in closed form. Of course, things are much easier when we know θ'_n explicitly. This is the case in the following example.

Example 2 (Rosen-Zener model). In this model $X(t) = \frac{1}{2(t^2+1)}$ and $Z(t) = \frac{t}{2(t^2+1)}$. Therefore $\tau(t) = \mathrm{arsinh}(t)$, $\tau_c = \mathrm{Im}(\mathrm{arsinh}(i)) = \pi$, and (30) yields $\theta'_n(\tau) = 1/\cosh(\tau)$ in the natural time scale. It is immediate that $|\theta'_n(\tau)| \le c\exp(-|\tau|)$ for large $|\tau|$ in each fixed strip around the real axis, and that $\gamma = 1$. Since θ'_n is meromorphic, $0 < \alpha$ can be chosen arbitrarily small. In summary, Propositions 1 and 2 yield

$$|\phi^{n_\varepsilon}_-(\infty)|^2 = 4e^{-\frac{\pi}{\varepsilon}} + \mathcal{O}(\varepsilon^{\frac{1}{2}-\alpha} e^{-\frac{\pi}{\varepsilon}}) \,.$$

Although the Rosen-Zener example is very easy in our picture, it is not clear how to prove it using the methods of [7]. The reason is that there are no anti-Stokes lines emanating from the singularity of θ'_n and staying in a bounded strip around the real axis as $\mathrm{Re}(t) \to \pm\infty$. In fact, the only relevant anti-Stokes line remains on the imaginary axis, cf. Fig. 2.

(a)

(b)

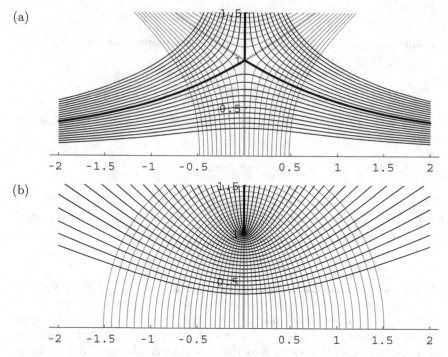

Fig. 2. This figure shows the Stokes and anti-Stokes lines for $\tau(t)$ in the Landau-Zener model (Fig. 2a) and the Rosen-Zener model (Fig. 2b). Level lines of $\mathrm{Re}\tau(t)$ are grey, while the lines of $\mathrm{Im}\tau(t)$ are black. The fat lines correspond to the Stokes and anti-Stokes lines emanating from the critical point t_c of $\tau(t)$ in the upper complex half-plane. In both examples, $t_c = i$. While in the Landau-Zener model, two anti-Stokes lines remain in a finite strip around the real axis, the anti-Stokes line of the Rosen-Zener model remains on the imaginary axis

References

1. M.V. Berry. *Histories of adiabatic quantum transitions*, Proc. R. Soc. Lond. A **429**, 61–72 (1990).
2. M.V. Berry and R. Lim. *Universal transition prefactors derived by superadiabatic renormalization*, J. Phys. A **26**, 4737–4747 (1993).
3. V. Betz and S. Teufel. *Precise coupling terms in adiabatic quantum evolution*, to appear in Annales Henri Poincaré (2005).
4. V. Betz and S. Teufel. *Precise coupling terms in adiabatic quantum evolution: the generic case*, to appear in Commun. Math. Phys. (2005). (2004).
5. V. Betz and S. Teufel. *Exponentially small transitions in adiabatic quantum evolution*, in preparation.
6. G. Hagedorn and A. Joye. *Time development of exponentially small non-adiabatic transitions*, Commun. Math. Phys. **250**, 393–413 (2004).
7. A. Joye. *Non-trivial prefactors in adiabatic transition probabilities induced by high order complex degeneracies*, J. Phys. A **26**, 6517–6540 (1993).

8. T. Kato. *On the adiabatic theorem of quantum mechanics*, Phys. Soc. Jap. **5**, 435–439 (1950).
9. S. Teufel. *Adiabatic perturbation theory in quantum dynamics*, Springer Lecture Notes in Mathematics 1821, 2003.
10. C. Zener. *Non-adiabatic crossing of energy levels*, Proc. Roy. Soc. London **137** 696–702, (1932).

Scattering Theory
of Dynamic Electrical Transport

M. Büttiker[1] and M. Moskalets[2]

[1] Département de Physique Théorique, Université de Genève, CH-1211 Genève 4,
Switzerland
Markus.Buttiker@physics.unige.ch
[2] Department of Metal and Semiconductor Physics, National Technical University
"Kharkiv Polytechnic Institute", 61002 Kharkiv, Ukraine
moskalets@kpi.kharkov.ua

Abstract. We have developed a scattering matrix approach to coherent transport
through an adiabatically driven conductor based on photon-assisted processes. To
describe the energy exchange with the pumping fields we expand the Floquet scattering matrix up to linear order in driving frequency.

Key words: Emissivity; Instantaneous currents; Internal response; Photon–assisted
transport; Quantum pump effect; Scattering matrix

1 From an Internal Response
to a Quantum Pump Effect

The possibility to vary several parameters at the same frequency but with
different phases [1] of a coherent (mesoscopic) system opens up new prospects
for the investigation of dynamical quantum transport. The adiabatic variation
of parameters is of particular interest since at small frequencies the conductor stays close to an equilibrium state: the opening of inelastic conduction
channels is avoided and quantum mechanical phase coherence is preserved to
the fullest extend possible. The relevant physics has a simple and transparent
explanation within the scattering matrix approach.

The variation of parameters leads to a dynamic scattering geometry. Quite
generally we consider changes in the scattering geometry as an *internal* response [2] in contrast to the external response generated by voltages applied
to the contacts of the conductor, see Fig. 1. In general a linear response consists both of a response to an external potential oscillations, and response to
internal potentials. The internal response can be expressed with the help of
the emissivity $\nu(\alpha, \boldsymbol{r})$. The emissivity $\nu(\alpha, \boldsymbol{r})$ is the portion of the density of
states at $\boldsymbol{r}$ of carriers that will exit the conductor through contact α. The
emissivity relates the amplitude $I_\alpha(\omega)$ of the current in lead α to the amplitude $U(\boldsymbol{r}, \omega)$ of a small and slowly oscillating internal potential. At zero
temperature we have [2]

$$I_\alpha(\omega) = \mathrm{i}e^2\omega \int \mathrm{d}^3 r \nu(\alpha, \boldsymbol{r}) U(\boldsymbol{r}, \omega) \,. \tag{1}$$

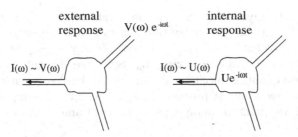

Fig. 1. External and internal response: An ac current with amplitude $I(\omega)$, say, in the left lead can arise as a response either to an oscillating potential $V(t) = V(\omega)e^{-i\omega t} + V(-\omega)e^{i\omega t}$ at one of the external reservoirs, or as a response to an oscillating potential profile $\delta U(r,t) = \delta u(r)\left(U(\omega)e^{-i\omega t} + U(-\omega)e^{i\omega t}\right)$ inside the mesoscopic sample

Here e is an electron charge and $i^2 = -1$. The integral in (1) runs over the region in which the potential deviates from its equilibrium value (typically the volume occupied by the scatterer).

The emissivity is expressed in terms of the scattering matrix S (of the stationary scatterer) and its functional derivative with respect to the internal potential variation $\delta U(r,t) = \delta u(r)(U(\omega)e^{-i\omega t} + U(-\omega)e^{i\omega t})$:

$$\nu(\alpha, r) = -\frac{1}{4\pi i} \sum_{\beta=1}^{N_r} \left[S_{\alpha\beta}^* \frac{\delta S_{\alpha\beta}}{\delta eu(r)} - \frac{\delta S_{\alpha\beta}^*}{\delta eu(r)} S_{\alpha\beta} \right] . \tag{2}$$

The scattering matrix is evaluated at the Fermi energy $E = \mu$. The summation runs over all the leads (for simplicity here assumed to be single channel) connecting the sample to external $N_r = 1, 2, 3...$ reservoirs.

Applying the inverse Fourier transformation to (1) gives the current $I_\alpha(t)$ flowing in response to a time-dependent internal potential $U(r,t)$. Equation (1) can easily be generalized to find the response to an arbitrary field or parametric variation of the scattering geometry [3]. To arrive at a general expression we remember that the scattering matrix depends on the internal potential U. That in turn makes S time-dependent, $S(t) \equiv S[U(t)]$. Thus alternatively we can express (1) in the form

$$I_\alpha(t) = \frac{ie}{2\pi} \sum_{\beta=1}^{N_r} S_{\alpha\beta}^* \frac{dS_{\alpha\beta}}{dt} . \tag{3}$$

Originally in [2] the potential $U(r,t)$ is the self-consistent Coulomb potential. However the form of (3) tells us that the current $I_\alpha(t)$ can arise in response to a slow variation of any quantity (parameter) which affects the scattering properties of a mesoscopic sample. For instance the current generated by a slowly varying vector potential [4] permits to derive the Landauer dc-conductance from (3). But (3) is not limited to the linear response regime.

The current $I_\alpha(t)$ given by (3) is a nonlinear functional of the scattering matrix. Therefore, the mesoscopic system can exhibit an internal rectification effect, i.e., an oscillating internal potential (or any appropriate oscillating parameter) can result in a dc current I_{dc}. Since the elements of the scattering matrix are quantum mechanical amplitudes the dc-current is the result of a *quantum* rectification process. This is a quantum pump effect [3–9]. An approach to quantum pumping based on (1) and (3) was put forth by Brouwer [3].

Since the time derivative enters (3), quantum rectification will work only under special conditions [3]. Let one or several parameters affecting the scattering properties of a mesoscopic sample change adiabatically and periodically in time. Then the scattering matrix changes periodically as well. Consider the point representing the scattering matrix in the space of all scattering matrices. During the completion of one time period this point will move along a closed line $\mathcal{L}$. Then the dc current $I_{dc,\alpha}$, which is the current $I_\alpha(t)$ averaged over the time period $T = 2\pi/\omega$, can be represented as a contour integral in the above mentioned abstract space [9]:

$$I_{dc,\alpha} = \frac{ie\omega}{4\pi^2} \oint_{\mathcal{L}} (dSS^\dagger)_{\alpha\alpha} .$$ (4)

The dc current $I_{dc,\alpha}$, is non-zero if and only if the line $\mathcal{L}$ encloses a non-vanishing area $\mathcal{F}$. The easy way to see this is to consider a two parameter space with parameters being S and $S^\dagger$. Since S and $S^\dagger$ depend on the same set of parameters, $S = S(\{X_j\})$, $S^\dagger = S^\dagger(\{X_j\})$, $j = 1, 2, \ldots, N_p$, (which includes, for instance, the shape of a sample, the internal potential, the magnetic field, the temperature, the pressure, the Fermi energy, etc.) then to get the cycle with $\mathcal{F} \neq 0$ it is necessary to have at least two parameters $X_1(t) = X_1 \cos(\omega t + \varphi_1)$ and $X_2(t) = X_2 \cos(\omega t + \varphi_2)$ varying with the same frequency ω and with a phase lag $\Delta\varphi \equiv \varphi_1 - \varphi_2 \neq 0$. In particular, if the oscillating amplitudes are small, i.e., if the scattering matrix changes only a little across $\mathcal{F}$, then the pumped current is proportional to the square of the cycle area [3]:

$$I_{dc,\alpha} = \frac{e\omega \sin(\Delta\varphi) X_1 X_2}{2\pi} \sum_{\beta=1}^{N_r} \Im \left(\frac{\partial S^*_{\alpha\beta}}{\partial X_1} \frac{\partial S_{\alpha\beta}}{\partial X_2} \right)_{X_1=0, X_2=0} .$$ (5)

The very simple and compact expression (4) allows to find the pumped current for a wide range of situations. Illustrative examples can be found in [10]. In the following we will now discuss the pumping process from the point of view of photon-assisted transport through a mesoscopic system. We will show how the interlay between photon–assisted transport and quantum mechanical interference results in a quantum pump effect [11, 12].

2 Quantum Coherent Pumping: A Simple Picture

By nature, the quantum pump effect is a rectification effect. A single parameter variation only leads to an ac-current. In a two parameter variation the modulation of the scatterer due to the second parameter rectifies the ac-currents generated by the first parameter. Rectification is only achieved if the driven system can scatter electrons in an asymmetric way. Here the asymmetry means that the probability $T_{\alpha\beta}$ for an electron to pass through the sample, say, from the lead β to the lead α and the probability $T_{\beta\alpha}$ to transit the scatterer in the reverse direction differ, $T_{\alpha\beta} \neq T_{\beta\alpha}$. Then the flux of electrons entering the scatterer through lead α and the flux of electrons scattered and leaving the system through the same lead α differ from each other, resulting in a net electron flow in lead α. The resulting current can be viewed as a result of an asymmetrical redistribution of incoming flows between the outgoing leads.

To clarify the physical mechanism which can lead to asymmetric scattering we now emphasize the essential difference between the driven scatterer and a stationary one. The key difference is the possibility of photon–assisted transport. In the stationary case if an electron with energy E enters the phase coherent system then it leaves the system with the same energy E. In contrast, in the driven case while traversing the system an electron can absorb (or emit) energy quanta $n\hbar\omega$ and thus it can leave the system with an energy $E_n = E + n\hbar\omega$.

It is important that the electron changes its energy interacting with a system which is modulated deterministically. As a consequence the inelastic processes is coherent. If there are several possibilities for transmission through the system absorbing or emitting the same energy (say, $n\hbar\omega$) the corresponding quantum–mechanical amplitudes will interfere. Such an interference of photon–assisted amplitudes can lead to directional asymmetry of electron propagation through a driven mesoscopic sample.

To illustrate this process we consider a simple but generic example. It is a system consisting of two regions with oscillating potentials $V_1(t) = 2V \cos(\omega t + \varphi_1)$ and $V_2(t) = 2V \cos(\omega t + \varphi_2)$ separated by the distance L. For the sake of simplicity we assume that both potentials oscillate with the same small amplitude $2V$. Consider an electron with energy E incident on the system. In leading order in the oscillating amplitudes only absorption/emission of a single energy quantum $\hbar\omega$ needs to be taken into account. So, there are only three scenarios to traverse the system. In the first case, an electron does not change its energy, the outgoing energy is $E^{(\text{out})} = E$. In the second case, it absorbs one energy quantum, $E^{(\text{out})} = E + \hbar\omega$. In the third case, it emits an energy $\hbar\omega$, $E^{(\text{out})} = E - \hbar\omega$. Since all these processes correspond to different final states (which differ in energy $E^{(\text{out})}$ from each other) then the full probability T to pass through the system is a sum of three contributions:

$$T = T^{(0)}(E; E) + T^{(+)}(E + \hbar\omega; E) + T^{(-)}(E - \hbar\omega; E) . \tag{6}$$

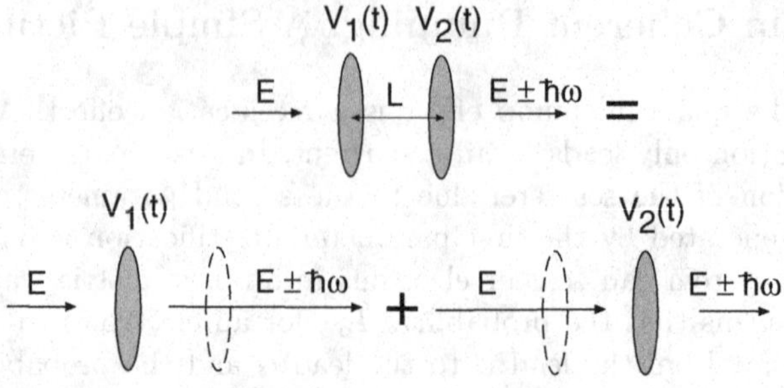

Fig. 2. For the process in which a carrier with energy E traverses two oscillating potentials $V_1(t)$ and $V_2(t)$, a distance L apart, and absorbs/emits an energy quantum $\hbar\omega$ there are two interfering alternatives. The modulation quantum can be either absorbed/emitted at the first barrier or at the second one

Here the first argument is an outgoing electron energy while the second argument is an incoming energy.

The probability $T^{(0)}$, like the tranmission probability of stationary scatterer, is insensitive to the propagation direction. In contrast $T^{(+)}$ and $T^{(-)}$ are directionally sensitive. Therefore we concentrate on the last two.

Let us consider $T^{(+)}$. There are two possibilities to pass through the system and to absorb an energy $\hbar\omega$, see, Fig. 2. The first possibility is to absorb an energy due to the oscillation of $V_1(t)$. The second possibility is to absorb an energy interacting with $V_2(t)$. In these two processes an electron has the same initial state and the same final state. Therefore, we can not distinguish between these two possibilities and according to quantum mechanics to calculate the corresponding probability we first have to add up the corresponding amplitudes and only then take the square. Let us denote the amplitude corresponding to the propagation through the system with absorbing $\hbar\omega$ at V_j as $\mathcal{A}^{(j,+)}$, $j = 1, 2$. Then the corresponding full probability is:

$$T^{(+)} = \left| \mathcal{A}^{(1,+)} + \mathcal{A}^{(2,+)} \right|^2 . \tag{7}$$

Each amplitude (either $\mathcal{A}^{(1,+)}$ or $\mathcal{A}^{(2,+)}$) is a product of two terms, the amplitude $\mathcal{A}^{(\mathrm{free})}(E) = e^{ikL}$ (here $k = \sqrt{2mE}/\hbar$ is an electron wave number) of free propagation in between the potentials and the amplitude $\mathcal{A}_j^{(+)}$ describing the absorption of an energy quantum $\hbar\omega$ at the potential V_j. The amplitude $\mathcal{A}_j^{(+)}$ is proportional to the corresponding Fourier coefficient of $V_j(t)$. The proportionality constant is denoted by α. Therefore we have $\mathcal{A}_j^{(+)} = \alpha V_j e^{-i\varphi_j}$.

The probability for an electron going from the left to the right is denoted by $T_{\rightarrow}^{(+)}$. The probability corresponding to the reverse direction – by $T_{\leftarrow}^{(+)}$. Our aim is to show that

$$T_{\rightarrow}^{(+)} \neq T_{\leftarrow}^{(+)} . \tag{8}$$

First we consider $T_{\rightarrow}^{(+)}$. Scattering from the left to the right an electron first meets the potential $V_1(t)$ and only then the potential $V_2(t)$. Therefore if an electron absorbs the energy $\hbar\omega$ at the first potential it traverses the remaining part of the system with enhanced energy $E_{+1} = E + \hbar\omega$. The corresponding amplitude is $\mathcal{A}_{\rightarrow}^{(1,+)} = \mathcal{A}_1^{(+)} \mathcal{A}^{(\text{free})}(E_{+1})$. While if an electron absorbed $\hbar\omega$ at the second potential then it goes through the system with the initial energy E. The quantum mechanical amplitude corresponding to such a process reads: $\mathcal{A}_{\rightarrow}^{(2,+)} = \mathcal{A}^{(\text{free})}(E)\mathcal{A}_2^{(+)}$. If the energy quantum is much smaller then the electron energy, $\hbar\omega \ll E$, then we can expand the phase factor corresponding to free propagation with enhanced energy E_{+1} to first order in the driving frequency: $k_{+1}L \approx \left(k + \frac{\omega}{v}\right)L$, here $v = \hbar k/m$ is an electron velocity. Thus we have:

$$\mathcal{A}_{\rightarrow}^{(1,+)} \approx \alpha V e^{-i\varphi_1} e^{i\left(k+\frac{\omega}{v}\right)L},$$

$$\mathcal{A}_{\rightarrow}^{(2,+)} = e^{ikL} \alpha V e^{-i\varphi_2}. \tag{9}$$

Substituting these amplitudes into (7) we obtain the probability to pass through the system from the left to the right with the absorption of an energy quantum $\hbar\omega$:

$$T_{\rightarrow}^{(+)} = 2\alpha^2 V^2 \left\{1 + \cos\left(\varphi_1 - \varphi_2 - \frac{\omega L}{v}\right)\right\}. \tag{10}$$

Now we consider the probability $T_{\leftarrow}^{(+)}$. Going from the right to the left an electron first meets the potential $V_2(t)$ and then the potential $V_1(t)$. Therefore, the corresponding amplitudes are:

$$\mathcal{A}_{\leftarrow}^{(1,+)} = e^{ikL} \alpha V e^{-i\varphi_1},$$

$$\mathcal{A}_{\leftarrow}^{(2,+)} \approx \alpha V e^{-i\varphi_2} e^{i\left(k+\frac{\omega}{v}\right)L} \tag{11}$$

Using (7) and (11) we find:

$$T_{\leftarrow}^{(+)} = 2\alpha^2 V^2 \left\{1 + \cos\left(\varphi_1 - \varphi_2 + \frac{\omega L}{v}\right)\right\}. \tag{12}$$

Comparing (10) and (12) we see that the transmission probability depends on the direction of electron propagation as we announced in (8).

Let us characterize the asymmetry in transmission probability by the difference $\Delta T^{(+)} = T_{\rightarrow}^{(+)} - T_{\leftarrow}^{(+)}$:

$$\Delta T^{(+)} = 4\alpha^2 V^2 \sin(\Delta\varphi) \sin\left(\frac{\omega L}{v}\right). \tag{13}$$

In our simple case the emission leads to the same asymmetry in the photon–assisted transmission probability: $\Delta T^{(-)} = \Delta T^{(+)}$. Therefore if there are the same electron flows I_0 coming from the left and from the right, then a net current $I = I_0 \left(\Delta T^{(-)} + \Delta T^{(+)} \right)$ is generated. We assume a positive current to be directed to the right.

We see that the induced current I depends separately on $\Delta\varphi = \varphi_1 - \varphi_2$ and on the factor $\omega L/v$. This is an additional dynamical phase due to absorption of an energy quantum $\hbar\omega$. Such a separation of phase factors can be interpreted in the following way. The presence of the phase lag $\Delta\varphi$ (by modulo 2π) between the oscillating potentials $V_1(t)$ and $V_2(t)$ breaks the time reversal invariance of a problem and hence potentially permits the existence of a steady particle flow. The second term emphasizes a spatial asymmetry of a model consisting of two inequivalent oscillating regions separated by a distance L, and tells us that the interference of photon–assisted amplitudes is the mechanism inducing electron flow. Neither a phase lag nor a photon-assisted process can separately lead to a dc current.

The simple model presented here reproduces several generic properties of a periodically driven quantum system. First, a driven spatially asymmetric system can pump a current between external electron reservoirs to which this system is coupled. Second, the pumped current is periodic in the phase lag between the driving parameters. Third, at small driving frequency, $\omega \to 0$, the current generated is linear in ω. In contrast the oscillatory dependence on ω of (13) is a special property of the simple resonant tunneling structure [13].

3 Beyond the Frozen Scatterer Approximation: Instantaneous Currents

In experiments the external electrical circuit to which the pump is connected is important [14, 15]. Consequently, it is of interest to understand the workings of a quantum pump connected to contacts which are not at equilibrium but support dc-voltages and ac-potentials. We now present a number of results of a rigorous calculation of dynamically generated currents within the scattering matrix approach for spinless non-interacting particles [16]. These results permit the investigation of pumping also in experimentally more realistic non-ideal situations.

We generalize the approach of [2] to the case of strong periodic driving when many photon processes are of importance. To take them into account we use the Floquet scattering matrix whose elements $S_{F,\alpha\beta}(E_n, E)$ are the quantum mechanical amplitudes (times $\sqrt{k_n/k}$, where $k_n = \sqrt{2mE_n}/\hbar$, $E_n = E + n\hbar\omega$) for scattering of an electron with energy E from lead β to lead α with absorption ($n > 0$) or emission ($n < 0$) of $|n|$ energy quanta $\hbar\omega$. Like in [2] we deal with low frequencies (adiabatic driving) and calculate the current linear in ω. In this limit we can expand the Floquet scattering matrix in powers of ω.

To zero-th order the Floquet sub-matrices $S_F(E_n, E)$ are merely the matrices of Fourier coefficients S_n of the stationary scattering matrix with time-dependent (pump)-parameters, $S(t) \equiv S(\{X_j(t)\})$. The matrix $S(t)$ is the *frozen* scattering matrix, in the sense that it describes the time moment t and hence stationary scattering. If all the reservoirs are kept at the same conditions (potential, temperature..) then the knowledge of only the frozen scattering matrix is sufficient to calculate the current flowing through the scatterer, see (3). Under more general conditions knowledge of the frozen scattering matrix is not sufficient. We stress that the frozen scattering matrix does not describe the scattering of electrons by a time-dependent scatterer: only the Floquet scattering matrix does. Since we are interesting in a current linear in ω then, in general, it is necessary to know the Floquet scattering matrix with the same accuracy. Thus we have to go beyond the frozen scattering matrix approximation and to take into account the corrections of order ω. As we illustrated in Sect. 2 such corrections are due to interference between photon–assisted amplitudes.

We use the following ansatz:

$$S_F(E_n, E) = S_n(E) + \frac{n\hbar\omega}{2}\frac{\partial S_n(E)}{\partial E} + \hbar\omega A_n(E) + O(\omega^2) . \tag{14}$$

The matrix $A(E, t)$ [with Fourier coefficients $A_n(E)$] introduced here is a key ingredient. As we will see this matrix reflects the asymmetry in scattering from one lead to the other and back. The unitarity condition for the Floquet scattering matrix leads to the following equation for the matrix $A(E, t)$:

$$\hbar\omega \left(S^\dagger(E, t)A(E, t) + A^\dagger(E, t)S(E, t) \right) = \frac{1}{2}\mathcal{P}\{S^\dagger; S\} , \tag{15}$$

where $\mathcal{P}$ is the Poisson bracket with respect to energy and time

$$\mathcal{P}\{S^\dagger; S\} = i\hbar \left(\frac{\partial S^\dagger}{\partial t}\frac{\partial S}{\partial E} - \frac{\partial S^\dagger}{\partial E}\frac{\partial S}{\partial t} \right) .$$

The matrix A can not be expressed in terms of the frozen scattering matrix $S(t)$ and it has to be calculated (like S itself) in each particular case. Nevertheless there are several advantages in using (14).

First, the matrix A has a much smaller number of elements than the Floquet scattering matrix. The matrix A depends on only one energy, E, and, therefore, it has $N_r \times N_r$ elements like the stationary scattering matrix S. In contrast, the Floquet scattering matrix S_F depends on two energies, E and E_n, and, therefore, it has $(2n_{max} + 1) \times N_r \times N_r$ relevant elements. Here n_{max} is the maximum number of energy quanta $\hbar\omega$ absorbed/emitted by an electron interacting with the scatterer which we have to take into account to correctly describe the scattering process. For small amplitude driving we have $n_{max} \approx 1$, whereas if the parameters vary with a large amplitude then $n_{max} \gg 1$.

Second, the Floquet scattering matrix has no definite symmetry with respect to a magnetic field H reversal. In contrast both the frozen scattering matrix S and A have. The analysis of the micro-reversibility of the equations of motion gives the following symmetry:

$$S(-H) = S^T(H) ,$$

$$A(-H) = -A^T(H) ,$$
(16)

where the upper index "T" denotes the transposition. In the absence of a magnetic field, $H = 0$, the matrix A is antisymmetric in lead indices, $A_{\alpha\beta} = -A_{\beta\alpha}$.

Next, using the adiabatic expansion, (14), we calculate the full time-dependent current $I_\alpha(t)$ flowing in lead α as follows [16]:

$$I_\alpha(t) = \int\limits_0^\infty dE \sum_\beta \left\{ \frac{e}{h} [f_{0,\beta} - f_{0,\alpha}] |S_{\alpha\beta}(E,t)|^2 \right.$$

$$\left. - e\frac{\partial}{\partial t} \left[f_{0,\beta} \frac{dN_{\alpha\beta}(E,t)}{dE} \right] + f_{0,\beta} \frac{dI_{\alpha\beta}(E,t)}{dE} \right\} .$$
(17)

Here $f_{0,\alpha}$ is the Fermi distribution function for electrons in reservoir α. We assume a current in a lead to be positive if it is directed from the scatterer to the corresponding reservoir. Equation (17) generalizes (3) to the case with external reservoirs being nonidentical (e.g., having different chemical potentials, temperatures, etc.). The three parts in the curly brackets on the RHS of (17) can be interpreted as follows: The first part defines the currents injected from the external reservoirs. It depends on the time-dependent conductance $G_{\alpha\beta}(t) = \frac{e^2}{h}|S_{\alpha\beta}(t)|^2$ of the frozen scatterer and hence it describes a classical rectification contribution to the dc current $I_{dc,\alpha}$. The second part defines the current generated by the oscillating charge $Q(t)$ of the scatterer:

$$Q(t) = e \sum_\alpha \sum_\beta \int\limits_0^\infty dE f_{0,\beta} \frac{dN_{\alpha\beta}(E,t)}{dE} .$$
(18)

Here $dN_{\alpha\beta}/dE$ is the global partial density of states for a frozen scatterer:

$$\frac{dN_{\alpha\beta}}{dE} = \frac{i}{4\pi} \left(\frac{\partial S_{\alpha\beta}^*}{\partial E} S_{\alpha\beta} - S_{\alpha\beta}^* \frac{\partial S_{\alpha\beta}}{\partial E} \right) .$$
(19)

Apparently this part gives no contribution to the dc current. The third part describes the currents generated by the oscillating scatterer. The ability to generate these ac currents differentiates a non-stationary dynamical scatterer from a merely frozen scatterer.

The *instantaneous spectral currents* $dI_{\alpha\beta}/dE$ pushed by the oscillating scatterer from lead β to lead α read:

$$\frac{dI_{\alpha\beta}}{dE} = \frac{e}{h} \left(2\hbar\omega Re[S_{\alpha\beta}^* A_{\alpha\beta}] + \frac{1}{2}\mathcal{P}\{S_{\alpha\beta}; S_{\alpha\beta}^*\} \right) .$$
(20)

The two terms in this equation have different symmetry properties with respect to the interchange of lead indices. That is most evident in the absence of a magnetic field, $H = 0$. In this case the first term on the RHS of (20) is antisymmetric in lead indices while the second term has the symmetry of the stationary scattering matrix S and is thus symmetric. Therefore, the matrix A is responsible for the directional asymmetry of dynamically generated currents:

$$\frac{dI_{\alpha\beta}}{dE} \neq \frac{dI_{\beta\alpha}}{dE} \,. \qquad (21)$$

We remark, that if one calculates the dc current generated in the particular case when all the reservoirs are at same potentials and temperatures ($f_{0,\alpha} = f_{0,\beta}$, $\forall \, \alpha, \beta$) then (17) (after averaging over a pump period) generates Brouwer's result, (4). In this case the matrix A plays no role. Different contributions of this matrix can be combined as in (15). In contrast, the matrix A is important in less symmetrical situations, when the electron flows arriving at the scatterer from different leads are different.

The spectral currents $dI_{\alpha\beta}/dE$ are subject to the following conservation law:

$$\sum_{\alpha=1}^{N_{\mathrm{r}}} \frac{dI_{\alpha\beta}(E,t)}{dE} = 0 \,. \qquad (22)$$

Such a property supports the point of view that these currents arise inside the scatterer. They are generated by the non-stationary scatterer without any external current source. The appearance of currents subject to the conservation law (22) can be easily illustrated within the quasi–particle picture [12]. Let all the reservoirs have the same chemical potential $\mu_\alpha = \mu$, $\alpha = 1, \ldots, N_{\mathrm{r}}$. We introduce quasi–particles: the quasi–electrons corresponding to filled states with energy $E > \mu$ and holes corresponding to empty states with energy $E < \mu$. Then at zero temperature there are no incoming quasi–particles. In other words, from each lead the vacuum of quasi–particles is falling upon the scatterer. Interacting with the oscillating scatterer, the system of (real) electrons can gain, say, n energy quanta $\hbar\omega$. In the quasi–particle picture this process corresponds to the creation of a quasi–electron–hole pair with energy $n\hbar\omega$. The pair dissolves and the quasi–particles are scattered separately to the same or different leads, see, Fig. 3. If the scattering matrix depends on energy then the quasi–electron and hole are scattered, on average, into different leads since they have different energies. Suppose the electron leaves the scattering region through lead α and the hole leaves through lead β. Since electrons and holes have opposite charge the current pulses created in the leads α and β have different sign. As a result a current pulse arises between the α and β reservoirs. In this picture it is evident that there is no incoming current and the sum of outgoing currents does satisfy the conservation law (22).

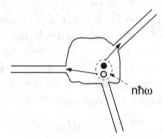

Fig. 3. Interacting with an oscillating potential the electron system gains modulation quanta of energy. Absorption of an energy $n\hbar\omega$ leads to the creation of a non-equilibrium quasi–electron–hole pair. The quasi–electron (*black circle*) and hole (*open circle*) can leave the scattering region through different leads. This leads to current pulses of different signs in the corresponding leads

Note, from (15) and (20) it is obvious that a conductor with strictly energy independent scattering matrix does not produce current and thus it does not show a quantum pump effect.

The current $I_\alpha(t)$, calculated with (17), satisfies the continuity equation:

$$\sum_\alpha I_\alpha(t) + \frac{\partial Q(t)}{\partial t} = 0 \,, \tag{23}$$

and thus conserves charge. To demonstrate this we use the unitarity of the frozen scattering matrix, $\sum_\alpha |S_{\alpha\beta}|^2 = \sum_\beta |S_{\alpha\beta}|^2 = 1$, and the definition of the charge $Q(t)$ of the scatterer, see (18).

It follows from (22) that the dynamically generated currents $dI_{\alpha\beta}/dE$ do not contribute to (23). Therefore they have nothing to do with charging/discharging of a scatterer. The existence of these currents is an intrinsic property of dynamical scatterer.

In conclusion, we clarified the role played by photon–assisted processes in adiabatic electron transport through a periodically driven mesoscopic system. The interference between the corresponding photon–assisted amplitudes makes the transmission probability dependent on the electron transmission direction in striking contrast with stationary scattering. To consider properly this effect we introduced an adiabatic (in powers of a driving frequency) expansion of the Floquet scattering matrix and demonstrated that already linear in ω terms exhibit the required asymmetry.

The ability to generate currents is only one interesting aspect of quantum pumps. Recent works point to the possibility of dynamical controlled generation of entangled electron-hole states [17, 18]. This brings into focus the dynamic quantum state of pumps. This and other properties will likely assure a continuing lively interest in quantum pumping.

References

1. M. Switkes, C.M. Marcus, K. Campman et al: Science **283**, 1905 (1999)
2. M. Büttiker, H. Thomas, A. Prêtre: Z. Phys. B **94**, 133 (1994)
3. P.W. Brouwer: Phys. Rev. B **58**, R10135 (1998)
4. D. Cohen: Phys. Rev. B 68, 201303 (2003)
5. D.J. Thouless: Phys. Rev. B **27**, 6083 (1983)
6. B. Spivak, F. Zhou, M.T. Beal Monod: Phys. Rev. B **51**, 13226 (1995)
7. I.L. Aleiner, A.V. Andreev: Phys. Rev. Lett. **81**, 1286 (1998)
8. F. Zhou, B. Spivak, B. Altshuler: Phys. Rev. Lett. **82**, 608 (1999)
9. J.E. Avron, A. Elgart, G.M. Graf et al: Phys. Rev. B **62**, R10618 (2000)
10. J.E. Avron, A. Elgart, G.M. Graf et al: J. Stat. Phys. **116**, 425 (2004)
11. B. Wang, J. Wang, H. Guo: Phys. Rev. B **65**, 073306 (2002)
12. M. Moskalets, M. Büttiker: Phys. Rev. B **66**, 035306 (2002)
13. M. Moskalets, M. Büttiker: Phys. Rev. B **66**, 205320 (2002)
14. M.L. Polianski, P.W. Brouwer: Phys. Rev. B **64**, 075304 (2001)
15. M.G. Vavilov, L. DiCarlo, C.M. Marcus: Phys. Rev. B 71, 241309 (2005)
16. M. Moskalets, M. Büttiker: Phys. Rev. B **69**, 205316 (2004)
17. P. Samuelsson, M. Büttiker: Phys. Rev. B **71**, 245317 (2005)
18. C.W.J. Beenakker, M. Titov, B. Trauzettel: Phys. Rev. Lett. **94**, 186804 (2005)

The Landauer-Büttiker Formula and Resonant Quantum Transport

Horia D. Cornean[1], Arne Jensen[2], and Valeriu Moldoveanu[3]

[1] Department of Mathematical Sciences, Aalborg University, Fredrik Bajers Vej 7G, 9220 Aalborg, Denmark
cornean@math.aau.dk
[2] Department of Mathematical Sciences, Aalborg University, Fredrik Bajers Vej 7G, 9220 Aalborg, Denmark
matarne@math.aau.dk
[3] National Institute of Materials Physics, P.O.Box MG-7, Magurele, Romania
xvalim@infim.ro

We give a short presentation of two recent results. The first one is a rigorous proof of the Landauer-Büttiker formula, and the second one concerns resonant quantum transport. The detailed results are in [2]. In the last section we present the results of some numerical computations on a model system.

Concerning the literature, then see the starting point of our work, [6]. In [4] a related, but different, problem is studied. See also [5] and the recent work [1].

1 The Landauer-Büttiker Formula

We start by introducing the notation and the assumptions. The model used here describes a finite sample coupled to a finite number of leads. The leads may be finite or semi-infinite. We use a discrete model, i.e. the tight-binding approximation. The sample is modeled by a finite set $\Gamma \subset \mathbf{Z}^2$. Each lead is modeled by $\mathcal{N} = \{0, 1, \ldots, N\} \subseteq \mathbf{N}$. The case $\mathcal{N} = \mathbf{N}$ ($N = +\infty$) is the semi-infinite lead. We assume that we have $M \geq 2$ leads. The one-particle Hilbert space is then

$$\mathcal{H} = \ell^2(\Gamma) \oplus \underbrace{\ell^2(\mathcal{N}) \oplus \cdots \oplus \ell^2(\mathcal{N})}_{M \text{ copies}} . \tag{1}$$

The Hamiltonian is denoted by H. It is the sum of the following components. For the sample we can take any selfadjoint operator H^S on $\ell^2(\Gamma)$. In each lead we take the discrete Laplacian with Dirichlet boundary conditions. The leads are numbered by $\alpha \in \{1, 2, \ldots, M\}$. Thus

$$H^L = \sum_{\alpha=1}^M H_\alpha^L, \quad H_\alpha^L = \sum_{n_\alpha \in \mathcal{N}} t_L(|n_\alpha\rangle\langle n_\alpha + 1| + |n_\alpha\rangle\langle n_\alpha - 1|) \tag{2}$$

H.D. Cornean et al.: *The Landauer-Büttiker Formula and Resonant Quantum Transport*, Lect. Notes Phys. **690**, 45–53 (2006)
www.springerlink.com

Functions in $\ell^2(\mathcal{N})$ are by convention extended to be zero at -1 and $N+1$. The parameter t_L is the hopping integral. The coupling between the leads and the sample is described by the tunneling Hamiltonian

$$H^T = H^{LS} + H^{SL}, \quad \text{where} \quad H^{LS} = \tau \sum_{\alpha=1}^{M} |0_\alpha\rangle\langle\mathcal{S}^\alpha|, \tag{3}$$

and H^{SL} is the adjoint of H^{LS}. Here $|0_\alpha\rangle$ denotes the first site on lead α, and $|\mathcal{S}^\alpha\rangle$ is the contact site on the sample. The parameter τ is the coupling constant. It is arbitrary in this section, but will be taken small in the next section. The total one-particle Hamiltonian is then

$$H = H^S + H^L + H^T \quad \text{on } \mathcal{H}. \tag{4}$$

First we consider electronic transport through the system. Initially the leads are finite, all of length N, with N arbitrary. We work exclusively in the grand canonical ensemble. Thus our system is in contact with a reservoir of energy and particles. We study the linear response of a system of non-interacting Fermions at temperature T and with chemical potential μ. The system is subjected adiabatically to a perturbation, defined as follows.

Let χ_η be a smooth switching function, i.e. $0 \le \chi_\eta(t) \le 2$, $\chi_\eta(t) = e^{\eta t}$ for $t \le 0$, while $\chi_\eta(t) = 1$ for $t > 1$. The time-dependent perturbation is then given by

$$V(N,t) = \chi_\eta(t) \sum_{\alpha=1}^{M} V_\alpha I_\alpha(N).$$

Here $I_\alpha(N) = \sum_{n_\alpha=0}^{N} |n_\alpha\rangle\langle n_\alpha|$ is the identity on the α-copy of $\ell^2(\mathcal{N})$. This perturbation models the adiabatic application of a constant voltage V_α on lead α, which will generate a charge transfer between the leads via the sample.

We are interested in deriving the current response of the system due to the perturbation. In the grand canonical ensemble we need to look at the second quantized operators. We omit the details and state the result. The current at time $t = 0$ in lead α is given by

$$\mathcal{I}_\alpha(0) = \sum_{\beta=1}^{M} g_{\alpha\beta}(T, \mu, \eta, N) V_\beta + \mathcal{O}(V^2). \tag{5}$$

The $g_{\alpha\beta}(T, \mu, \eta, N)$ are the conductance coefficients [3]. It is clear from the above formula that we work in the linear response regime. Below we are going to take the limit $N \to \infty$, followed by the limit $\eta \to 0$. The limits have to be taken in this order, since the error term is in fact $\mathcal{O}(V^2/\eta^2)$.

The next step is to look at the transmittance, which is obtained from scattering theory, applied to the pair of operators (K, H_0), where $H_0 = H^L$ ($N = +\infty$ case) and $K = H_0 + H^S + H^T$. Properly formulated this is done in the two space scattering framework, see [7]. Since the perturbation

$H^S + H^T$ is of finite rank, and since we have explicitly a diagonalization of the operator H_0, the stationary scattering theory gives an explicit formula for the scattering matrix, which is an $M \times M$ matrix, depending on the spectral parameter $\lambda = 2t_L \cos(k)$ of H_0. The T-operator is then given by an $M \times M$ matrix $t_{\alpha\beta}(\lambda)$, and the transmittance is given by

$$T_{\alpha\beta}(\lambda) = |t_{\alpha\beta}(\lambda)|^2 . \tag{6}$$

It follows from the explicit formulas that $T_{\alpha\beta}(\lambda)$ is real analytic on $(-2t_L, 2t_L)$, and zero outside this interval.

With these preparations we can state the main result.

Theorem 1. *Consider* $\alpha \neq \beta$, $T > 0$, $\mu \in (-2t_L, 2t_L)$, *and* $\eta > 0$. *Assume that the point spectrum of K (corresponding to the $N = +\infty$ case) is disjoint from $\{-2t_L, 2t_L\}$. Then taking first the limit $N \to \infty$, and then $\eta \to 0$, we have*

$$g_{\alpha,\beta}(T, \mu) = \lim_{\eta \to 0}\Big[\lim_{N \to \infty} g_{\alpha,\beta}(T, \mu, \eta, N)\Big]$$

$$= -\frac{1}{2\pi} \int_{-2t_L}^{2t_L} \frac{\partial f_{\text{F-D}}(\lambda)}{\partial \lambda} T_{\alpha\beta}(\lambda) d\lambda . \tag{7}$$

Here $f_{\text{F-D}}(\lambda) = 1/(e^{(\lambda-\mu)/T} + 1)$ is the Fermi-Dirac function. If we finally take the limit $T \to 0$, we obtain the Landauer formula

$$g_{\alpha,\beta}(0_+, \mu) = \frac{1}{2\pi} T_{\alpha\beta}(\mu) . \tag{8}$$

The proof of this main result is quite long and technical. One has to study the two sides of the equality above. The scattering part (the transmittance) is quite straightforward, using the Feshbach formula. The conductance part is a fairly long chain of arguments, as is the proof of the equality statement in the theorem. We refer to [2] for the details.

2 Resonant Transport in a Quantum Dot

In the previous section we have allowed the coupling constant τ (see (3)) to be arbitrarily large. The only assumption was that $\{-2t_L, 2t_L\}$ was not in the point spectrum of K. We now look at the small coupling case, $\tau \to 0$. In this case we will assume that the sample Hamiltonian H^S does not have eigenvalues $\{-2t_L, 2t_L\}$. It then follows from a perturbation argument, using the Feshbach formula, that the same is true for K, provided τ is sufficiently small.

Since H^S is an operator on the finite dimensional space $\ell^2(\Gamma)$, is has a purely discrete spectrum. We enumerate the eigenvalues in the interval $(-2t_L, 2t_L)$:

$$\sigma(H^S) \cap (-2t_L, 2t_L) = \{E_1, \ldots, E_J\} \,.$$

Let $\beta \neq \gamma$ be two different leads. The conductance between these two is now denoted by $\mathcal{T}_{\beta,\gamma}(\lambda, \tau)$, making the dependence on the coupling constant explicit, see (6).

Theorem 2. *Assume that the eigenvalues $\{E_1, \ldots, E_J\}$ are nondegenerate, and denote by $\phi_1, \ldots \phi_J$ the corresponding normalized eigenfunctions. We then have the following results:*

(i) *For every $\lambda \in (-2t_L, 2t_L) \setminus \{E_1, \ldots, E_J\}$ we have*

$$\lim_{\tau \to 0} \mathcal{T}_{\beta,\gamma}(\lambda, \tau) = 0 \,. \tag{9}$$

(ii) *Let $\lambda = E_j$. If either $\langle \mathcal{S}^\beta, \phi_j \rangle = 0$ or $\langle \mathcal{S}^\gamma, \phi_j \rangle = 0$, then*

$$\lim_{\tau \to 0} \mathcal{T}_{\beta,\gamma}(E_j, \tau) = 0 \,. \tag{10}$$

(iii) *Let $\lambda = E_j$. If both $\langle \mathcal{S}^\beta, \phi_j \rangle \neq 0$ and $\langle \mathcal{S}^\gamma, \phi_j \rangle \neq 0$, then there exist positive constants $C(E_j)$, such that*

$$\lim_{\tau \to 0} \mathcal{T}_{\beta,\gamma}(E_j, \tau) = C(E_j) \left| \frac{\langle \mathcal{S}^\beta, \phi_j \rangle \cdot \langle \mathcal{S}^\gamma, \phi_j \rangle}{\sum_{\alpha=1}^{M} |\langle \mathcal{S}^\alpha, \phi_j \rangle|^2} \right|^2 . \tag{11}$$

This result can be interpreted as follows. Case (i): If the energy of the incident electron is not close to the eigenvalues of H^S, it will not contribute to the current. Case (ii): If the incident energy is close to some eigenvalue of H^S, but the eigenfunction is not localized along both contact points $\mathcal{S}^\beta$ and $\mathcal{S}^\gamma$, again there is no current. Case (iii): In order to have a peak in the current it is necessary for H^S to have extended edge states, which couple to several leads.

3 A Numerical Example

We end this contribution with some numerical results on the transport through a noninteracting quantum dot described by a discrete lattice containing 20×20 sites and coupled to two leads connected to two opposite corners. The magnetic flux is fixed and measured in arbitrary units, while the lead-dot coupling was set to $\tau = 0.2$. The sample Hamiltonian H^S is given by the Dirichlet restriction to the above mentioned finite domain of

$$H^S(V_g) = \sum_{(m,n) \in \mathbf{Z}^2} \left((E_0 + V_g)|m,n\rangle\langle m,n| + t_1(e^{-i\frac{Bm}{2}}|m,n\rangle\langle m,n+1| + h.c.) \right.$$

$$\left. + t_2(e^{-i\frac{Bn}{2}}|m,n\rangle\langle m+1,n| + h.c.) \right) . \tag{12}$$

Here *h.c.* means hermitian conjugate, E_0 is the reference energy, B is a magnetic field, from which the magnetic phases appear (the symmetric gauge was used), while t_1 and t_2 are hopping integrals between nearest neighbor sites.

The constant denoted V_g adds to the on-site energies E_0, simulating the so-called "plunger gate voltage" in terms of which the conductance is measured in the physical literature. The variation of V_g has the role to "move" the dot levels across the *fixed* Fermi level of the system (recall that the latter is entirely controlled by the semi-infinite leads). Otherwise stated, the eigenvalues of $H^S(V_g)$ equal the ones of $H^S(V_g = 0)$ (we denote them by $\{E_i\}$), up to a global shift V_g. Using the Landauer-Büttiker formula (8), and the formulas (3.8) and (4.6) in [2], it turns out that the computation of the conductivity between the two leads (or equivalently, of T_{12}) reduces to the inversion of an effective Hamiltonian.

Moreover, when V_g is fixed such that there exists an eigenvalue E_i of $H^S(V_g = 0)$ obeying $E_i + V_g = E_F$, the transmittance behavior is described by (11). Thus one expects to see a series of peaks as V_g is varied. Here the Fermi level was fixed to $E_F = 0.0$ and the hopping constants in the lattice $t_1 = 1.01$ and $t_2 = 0.99$. We have taken the magnetic flux (which equals B for the unit lattice) to 0.15. Then the resonances appear, whenever $V_g = -E_i$ (since the spectrum of our discrete operator $H^S(0)$ is a subset of $[-4, 4]$, the suitable interval for varying V_g is the same).

Before discussing the resonant transport let us analyze the spectrum of our dot at $V_g = 0$, in order to emphasize the role of the magnetic field. We recall that we used Dirichlet boundary conditions (DBC) and the magnetic field appears in the Peierls phases of H^S (see (12)). In Fig. 1 we plot the first 200 eigenvalues (this suffices since the spectrum is symmetrically located with respect to 0, i.e both E_i and $-E_i$ belong to $\sigma(H_S(0))$). One notices two things.

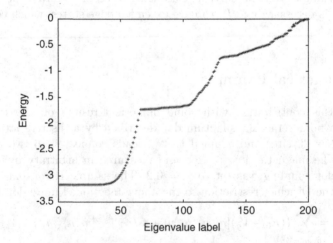

Fig. 1. The dot spectrum

First, there are two narrow energy intervals ($[-3.17, -3.16]$ and $[-1.75, 1.65]$) covered by many eigenvalues ($\sim$33 and 45 respectively). Secondly, the much larger ranges $[-3.16, -1.72]$ and $[-1.65, -0.8]$ contain only 25 and 30 eigenvalues. This particular structure of the spectrum is due to both the magnetic field and the DBC. The dense regions are reminescences of the Landau levels of the infinite system while the largely spaced eigenvalues appear *between* the Landau levels due to the DBC. As we shall see below their corresponding eigenfunctions are mostly located on the edge of the sample. As the energy approaches zero, the distinction between edge and bulk states is not anymore clear and one can have quite complex topologies for eigenfunctions. We point out that the "clarity", the length, and the number of edge states regions intercalated in the Landau gaps, increase as the sample gets bigger.

Now let us again comment on (11). Here E_i must be replaced by $E_i + V_g$, where E_i are eigenvalues of $H^S(0)$. Remember that we took $\mu = 0$. The number of leads is $M = 2$. By inspecting formula (4.6) in [2], one can show that the constant $C(E_i + V_g)$ will always equal 4 (we have $k_\mu = \pi/2$ and $t_L = 1$). Therefore, each time we fulfill the condition $V_g = -E_i$, we obtain a peak in the transmittance, which for small τ should be close to

$$4 \left| \frac{\langle \mathcal{S}^1, \phi_j \rangle \cdot \langle \mathcal{S}^2, \phi_j \rangle}{\sum_{\alpha=1}^2 |\langle \mathcal{S}^\alpha, \phi_j \rangle|^2} \right|^2 \leq 1 . \tag{13}$$

We have equality with 1, if and only if $|\langle \mathcal{S}^1, \phi_j \rangle| = |\langle \mathcal{S}^2, \phi_j \rangle|$, and this does not depend on the magnitude of these quantities. Therefore, even for weakly coupled, but completely symmetric eigenfunctions, we can expect to have a strong signal. In fact, in this case the relevant parameter is

$$\min \left\{ \frac{|\langle \mathcal{S}^1, \phi_j \rangle|}{|\langle \mathcal{S}^2, \phi_j \rangle|}, \frac{|\langle \mathcal{S}^2, \phi_j \rangle|}{|\langle \mathcal{S}^1, \phi_j \rangle|} \right\} . \tag{14}$$

Now let us investigate how the transmittance behaves, when V_g is varied. Figure 2a shows the peaks corresponding to the first six (negative) eigenvalues of $H^S(V_g = 0)$. Their amplitude is very small because the associated eigenvectors are (exponentially) small at the contact sites, and not completely symmetric (since $t_1 \neq t_2$). In fact, a few eigenvectors with more symmetry do generate some small peaks. The spatial localisation of the second and the sixth eigenvector is shown in Figs. 2b,c.

The peak aspect changes drastically at lower gate potentials as the Fermi level encounters levels whose eigenstates have a strong component on the contact subspace (see Figs. 3b and 3c for the spatial localisation of the 38th and the 49th eigenstate). The transmittance is close to unity in this regime, since the parameter in (14) is also nearly one. This is explained by the fact that t_1 and t_2 have very close values, and the relative perturbation induced by the lack of symmetry is much smaller than for the bulk states. One notices that the width of the peaks increases as V_g is decreased as well as their

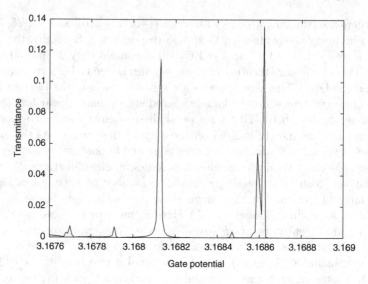

2nd state

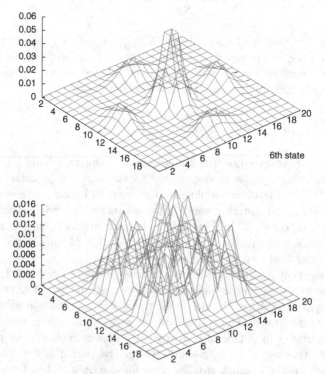

6th state

Fig. 2. *Top* to *bottom*: parts a, b, c

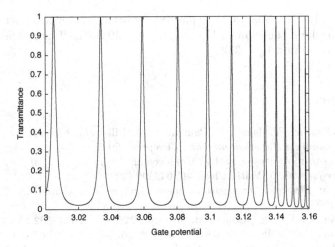

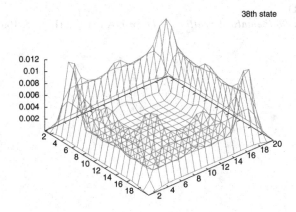

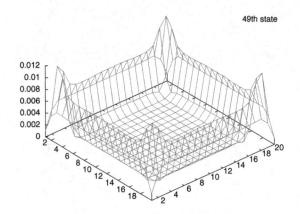

Fig. 3. *Top* to *bottom*: parts a, b, c

separation. In Figs. 3b,c we have plotted the 38th eigenfunction, which gives the first peak on the right of Fig. 3a, and the 49th eigenfunction associated to the peak around $V_g = 3.06$.

References

1. W. Aschbacher, V. Jaksic, Y. Pautrat, C.-A. Pillet: *Introduction to non-equilibrium quantum statistical mechanics* Preprint, 2005.
2. H.D. Cornean, A. Jensen, and V. Moldoveanu: *A rigorous proof of the Landauer-Büttiker formula.* J. Math. Phys. **46**, 042106 (2005).
3. S. Datta: *Electronic transport in mesoscopic systems* Cambridge University Press, 1995.
4. V. Jaksic, C.-A. Pillet, Comm. Math. Phys. **226**, no. 1, 131–162 (2002).
5. V. Jaksic, C.-A. Pillet, J. Statist. Phys. **108** no. 5-6, 787–829 (2002).
6. V. Moldoveanu, A. Aldea, A. Manolescu, M. Niţă, Phys. Rev. B **63**, 045301-045309 (2001).
7. D.R. Yafaev: *Mathematical Scattering Theory*, Amer. Math. Soc. 1992.

Point Interaction Polygons:
An Isoperimetric Problem

Pavel Exner

Department of Theoretical Physics, Nuclear Physics Institute, Academy of
Sciences, 25068 Řež near Prague, Czechia, and Doppler Institute, Czech
Technical University, Břehová 7, 11519 Prague, Czechia
exner@ujf.cas.cz

We will discuss a new type of an isoperimetric problem concerning a Hamiltonian with N point interactions in $\mathbb{R}^d$, $d = 2, 3$, all with the same coupling constant, placed at vertices of an equilateral polygon $\mathcal{P}_N$. We show that the ground state energy is locally maximized by a regular polygon and conjecture that the maximum is global; on the way we encounter an interesting geometric inequality. We will also mention some extensions of this problem.

1 Introduction

In my talk at the QMath9 conference I discussed three recent mathematical results inspired by investigations of quantum waveguides. Two of them are either already published or will be soon [2,11] while the third one [6] is still not at the time when I am writing this text; this is the reason why I have chosen to devote this proceedings contribution to it. In addition, the problem is amusing and has various extensions worth to explore.

Relations between geometrical configurations and extremal values of spectral quantities are a trademark topic of mathematical physics. Let me recall the Faber-Krahn inequality [12,14] which states that among all regions of a fixed volume the principal eigenvalue of the Dirichlet Laplacian is minimized by a ball. What is intriguing in this and similar results, however, is that they are sensitive to the topology. If you consider nonsimply connected regions of annular type, for instance, then a rotationally symmetric shape on the contrary typically *maximizes* the ground-state energy [8,13].

The question is what happens if the particle confinement is much weaker than a hard wall modelled by Dirichlet boundary condition, so that speaking about topology loses meaning. Since I do not adhere to the Bourbaki philosophy and believe that a good example is vital for any theory, I am going to discuss a simple model situation in which the confinement is extremely weak being realized by a loop-shaped "polymer-type" point interaction array defined as in [1]. It has been demonstrated recently that spectral properties of such arrays depend substantially on their geometry, in particular, that curvature leads to an effective attractive interaction [4,10]. In this context it is natural to ask whether the mentioned result about Dirichlet annuli [8] has

P. Exner: *Point Interaction Polygons: An Isoperimetric Problem*, Lect. Notes Phys. **690**,
55–64 (2006)
www.springerlink.com

an analogue in the situation when the point interactions are arranged along a closed curve of a fixed length. We will state below this *isoperimetric problem* properly and show that the shape close as possible to the circular one, namely a regular polygon, is a *local* maximizer for the principal eigenvalue.

I am unable for the time being to answer the question about the *global uniqueness* of this maximizer. What I can do, however, and what I will demonstrate is its reformulation in terms of a purely geometric problem which you would expect to be solved in Euclid's Στοιχεῖον, or at least not much later. Surprisingly, it seems to be open, and it has an appeal which problems expressible in terms of grammar-school mathematics usually have. At the same time, the problem has various extensions which are no less interesting.

2 The Local Result in Geometric Terms

Let $\mathcal{P}_N \subset \mathbb{R}^d$, $d = 2,3$, be a polygon which we will for the present purpose identify with an ordered set of its vertices, $\mathcal{P}_N = \{y_1, \ldots, y_N\}$; if the vertex indices exceed this range they are understood mod N. We suppose that $\mathcal{P}_N$ is *equilateral*, $|y_{i+1} - y_i| = \ell$ for a fixed $\ell > 0$ and any i. By $\tilde{\mathcal{P}}_N$ we denote a *regular* polygon of the edge length ℓ, which means planar (this is trivial, of course, if $d = 2$) with vertices lying on a circle of radius $\ell \left(2 \sin \frac{\pi}{N}\right)^{-1}$.

We are interested in the operator $-\Delta_{\alpha, \mathcal{P}_N}$ in $L^2(\mathbb{R}^d)$ with N point interactions in the sense of [1], all of the same coupling α, placed at the vertices of $\mathcal{P}_N$. We suppose that it has a non-empty discrete spectrum,

$$\epsilon_1 \equiv \epsilon_1(\alpha, \mathcal{P}_N) := \inf \sigma\left(-\Delta_{\alpha, \mathcal{P}_N}\right) < 0 \,,$$

which is satisfied for any $\alpha \in \mathbb{R}$ if $d = 2$, while in the case $d = 3$ it is true below a certain critical value of α – cf. [1, Sect. II.1]. The main result which I am going to report here can be then stated as follows.

Theorem 1. *Under the stated conditions, $\epsilon_1(\alpha, \mathcal{P}_N)$ is for fixed α and ℓ locally sharply maximized by a regular polygon, $\mathcal{P}_N = \tilde{\mathcal{P}}_N$.*

It goes without saying that speaking about uniqueness of the maximizer, we have always in mind the family of regular polynomials related mutually by Euclidean transformations of the underlying space $\mathbb{R}^d$.

To prove Theorem 1 we are going to show first that the task can be reformulated in purely geometric terms. Using $k = i\kappa$ with $\kappa > 0$, we find the eigenvalues $-\kappa^2$ solving the following spectral condition,

$$\det \Gamma_k = 0 \qquad \text{with} \qquad (\Gamma_k)_{ij} := (\alpha - \xi^k)\delta_{ij} - (1 - \delta_{ij})g_{ij}^k \,,$$

where $g_{ij}^k := G_k(y_i - y_j)$, or equivalently

$$g_{ij}^k = \begin{cases} \frac{1}{2\pi} K_0(\kappa|y_i - y_j|) & \ldots \quad d = 2 \\ \frac{e^{-\kappa|y_i - y_j|}}{4\pi|y_i - y_j|} & \ldots \quad d = 3 \end{cases} \tag{1}$$

and the regularized Green's function at the interaction site is

$$\xi^k = \begin{cases} -\frac{1}{2\pi}\left(\ln\frac{\kappa}{2} + \gamma_E\right) & \cdots \quad d = 2 \\ -\frac{\kappa}{4\pi} & \cdots \quad d = 3 \end{cases}$$

The matrix $\Gamma_{i\kappa}$ has N eigenvalues counting multiplicity which are increasing in $(0, \infty)$ as functions of κ – see [15] and recall that they are real-analytic and non-constant in view of their known asymptotic behavior [1]. The sought quantity, $\epsilon_1(\alpha, \mathcal{P}_N)$, corresponds to the point κ where the lowest of these eigenvalues vanishes. Consequently, we have to check that

$$\min \sigma(\Gamma_{i\tilde{\kappa}_1}) < \min \sigma(\tilde{\Gamma}_{i\tilde{\kappa}_1}) \tag{2}$$

holds locally for $\mathcal{P}_N \neq \tilde{\mathcal{P}}_N$, where $-\tilde{\kappa}_1^2 = \epsilon_1(\alpha, \tilde{\mathcal{P}}_N)$.

Next we notice that the lowest eigenvalue of $\tilde{\Gamma}_{i\tilde{\kappa}_1}$ corresponds to the eigenvector $\tilde{\phi}_1 = N^{-1/2}(1, \ldots, 1)$. Indeed, by [1] there is a bijective correspondence between an eigenfunction $c = (c_1, \ldots, c_N)$ of $\Gamma_{i\kappa}$ at the point, where the corresponding eigenvalue equals zero, and the corresponding eigenfunction of $-\Delta_{\alpha, \mathcal{P}_N}$ given by $c \leftrightarrow \sum_{j=1}^{N} c_j G_{i\kappa}(\cdot - y_j)$, up to a normalization. Again by [1], the principal eigenvalue of $-\Delta_{\alpha, \mathcal{P}_N}$ is simple, so it has to be associated with a one-dimensional representation of the corresponding discrete symmetry group of $\tilde{\mathcal{P}}_N$; it follows that $c_1 = \cdots = c_N$. Hence

$$\min \sigma(\tilde{\Gamma}_{i\tilde{\kappa}_1}) = (\tilde{\phi}_1, \tilde{\Gamma}_{i\tilde{\kappa}_1}\tilde{\phi}_1) = \alpha - \xi^{i\tilde{\kappa}_1} - \frac{2}{N}\sum_{i<j} \tilde{g}_{ij}^{i\tilde{\kappa}_1} . \tag{3}$$

On the other hand, for the l.h.s. of (2) we have a variational estimate

$$\min \sigma(\Gamma_{i\tilde{\kappa}_1}) \leq (\tilde{\phi}_1, \Gamma_{i\tilde{\kappa}_1}\tilde{\phi}_1) = \alpha - \xi^{i\tilde{\kappa}_1} - \frac{2}{N}\sum_{i<j} g_{ij}^{i\tilde{\kappa}_1} ,$$

and therefore it is certainly sufficient to check that the inequality

$$\sum_{i<j} G_{i\kappa}(y_i - y_j) > \sum_{i<j} G_{i\kappa}(\tilde{y}_i - \tilde{y}_j) \tag{4}$$

holds *for all* $\kappa > 0$ and $\mathcal{P}_N \neq \tilde{\mathcal{P}}_N$ in the vicinity of the regular polygon $\tilde{\mathcal{P}}_N$. For brevity we introduce the symbol ℓ_{ij} for the diagonal length $|y_i - y_j|$ and $\tilde{\ell}_{ij} := |\tilde{y}_i - \tilde{y}_j|$. We define the function $F : (\mathbb{R}_+)^{N(N-3)/2} \to \mathbb{R}$ by

$$F(\{\ell_{ij}\}) := \sum_{m=2}^{[N/2]} \sum_{|i-j|=m} \left[G_{i\kappa}(\ell_{ij}) - G_{i\kappa}(\tilde{\ell}_{ij})\right] ;$$

notice that $m = 1$ does not contribute because $\ell_{i,i+1} = \tilde{\ell}_{i,i+1} = \ell$ by assumption. Our aim is to show that $F(\{\ell_{ij}\}) > 0$ except if $\{\ell_{ij}\} = \{\tilde{\ell}_{ij}\}$. We use the fact that the function $G_{i\kappa}(\cdot)$ is *convex* for any fixed $\kappa > 0$ and $d = 2, 3$ as it can be seen from cf. (1); this yields the inequality

$$F(\{\ell_{ij}\}) \geq \sum_{m=2}^{[N/2]} \nu_m \left[G_{i\kappa}\left(\frac{1}{\nu_m} \sum_{|i-j|=m} \ell_{ij} \right) - G_{i\kappa}(\tilde{\ell}_{1,1+m}) \right],$$

where ν_n is the number of the appropriate diagonals,

$$\nu_m := \begin{cases} N & \dots & m = 1, \dots, [\frac{1}{2}(N-1)] \\ \frac{1}{2}N & \dots & m = \frac{1}{2}N \quad \text{for } N \text{ even} \end{cases}$$

At the same time, $G_{i\kappa}(\cdot)$ is monotonously decreasing in $(0, \infty)$, so the sought claim would follow if we demonstrate the inequality

$$\tilde{\ell}_{1,m+1} \geq \frac{1}{\nu_n} \sum_{|i-j|=m} \ell_{ij}$$

and show that it is sharp for at least one value of m if $\mathcal{P}_N \neq \tilde{\mathcal{P}}_N$.

In this way we have reduced the task to *verification of a geometric inequality*. Since it may be of independent interest we will state it more generally, without dimensional restrictions. Let $\mathcal{P}_N$ be an equilateral polygon in $\mathbb{R}^d$, $d \geq 2$. Given a fixed integer $m = 2, \dots, [\frac{1}{2}N]$ we denote by $\mathcal{D}_m$ the *sum of lengths of all m-diagonals*, i.e. the diagonals jumping over m vertices.

(P_m) The quantity $\mathcal{D}_m$ is, in the set of equilateral polygons $\mathcal{P}_N \subset \mathbb{R}^d$ with a fixed edge length $\ell > 0$, uniquely maximized by $\tilde{\mathcal{D}}_m$ referring to the (family of) regular polygon(s) $\tilde{\mathcal{P}}_N$.

3 Proof of Theorem 1

We have thus to demonstrate the following claim for $d = 2, 3$.

Theorem 2. *The property (P_m) holds locally for any $m = 2, \dots, [\frac{1}{2}N]$.*

Proof. Let us look for local maxima of the function

$$f_m : f_m(y_1, \dots, y_N) = \frac{1}{N} \sum_{i=1}^{N} |y_i - y_{i+m}|$$

under the constraints $g_i(y_1, \dots, y_n) = 0$, where

$$g_i(y_1, \dots, y_n) := \ell - |y_i - y_{i+1}|, \quad i = 1, \dots, N.$$

There are $(N-2)(d-1)-1$ independent variables here because $2d-1$ parameters are related to Euclidean transformations and can be fixed. We put

$$K_m(y_1, \ldots, y_N) := f_m(y_1, \ldots, y_N) + \sum_{r=1}^{N} \lambda_r g_r(y_1, \ldots, y_n) \qquad (5)$$

and compute the derivatives $\nabla_j K_m(y_1, \ldots, y_N)$ which are equal to

$$\frac{1}{N} \left\{ \frac{y_j - y_{j+m}}{|y_j - y_{j+m}|} + \frac{y_j - y_{j-m}}{|y_j - y_{j-m}|} \right\} - \lambda_j \frac{y_j - y_{j+1}}{\ell} - \lambda_{j-1} \frac{y_j - y_{j-1}}{\ell} \, .$$

We want to show that these expressions vanish for a regular polygon. Let us introduce a parametrization for any planar equilateral polygon. Without loss of generality we may suppose that it lies in the plane of the first two axes. The other coordinates are then zero and we neglect them writing

$$y_j = \ell \left(\sum_{n=0}^{j-1} \cos \left(\sum_{i=1}^{n} \beta_i - \varphi \right), \sum_{n=0}^{j-1} \sin \left(\sum_{i=1}^{n} \beta_i - \varphi \right) \right), \qquad (6)$$

where $\varphi \in \mathbb{R}$ is a free parameter and β_i is the "bending angle" at the ith vertex (modulo 2π); the family of these angles satisfies naturally the condition

$$\sum_{i=1}^{N} \beta_i = 2\pi w \qquad (7)$$

for some $w \in \mathbb{Z} \setminus \{0\}$. Choosing $\tilde{\varphi} = \frac{\pi}{N}$ and $\tilde{\beta}_i = \frac{2\pi i}{N}$, we get in particular

$$\tilde{y}_{\pm m} = \ell \left(\pm \sum_{n=0}^{m-1} \cos \frac{\pi}{N}(2n+1), \sum_{n=0}^{m-1} \sin \frac{\pi}{N}(2n+1) \right) \, .$$

Then we have

$$|\tilde{y}_j - \tilde{y}_{j \pm m}| = \ell \left[\left(\sum_{n=0}^{m-1} \cos \frac{\pi}{N}(2n+1) \right)^2 + \left(\sum_{n=0}^{m-1} \sin \frac{\pi}{N}(2n+1) \right)^2 \right] =: \ell \Upsilon_m \, ,$$

and consequently, $\nabla_j K_m(\tilde{y}_1, \ldots, \tilde{y}_N) = 0$ holds for $j = 1, \ldots, N$ if we choose all the Lagrange multipliers λ_r in (5) equal to the expression

$$\lambda = \frac{\sigma_m}{N \Upsilon_m} \qquad \text{with} \qquad \sigma_m := \frac{\sum_{n=0}^{m-1} \sin \frac{\pi}{N}(2n+1)}{\sin \frac{\pi}{N}} = \frac{\sin^2 \frac{\pi m}{N}}{\sin^2 \frac{\pi}{N}} \, . \qquad (8)$$

The second partial derivatives, $\nabla_{k,r} \nabla_{j,s} K_m(y_1, \ldots, y_N)$, are computed to be

$$\frac{1}{N}\left\{\frac{\delta_{kj} - \delta_{k,j+m}}{|y_j - y_{j+m}|}\delta_{rs} - \frac{(y_j - y_{j+m})_r(y_j - y_{j+m})_s(\delta_{kj} - \delta_{k,j+m})}{|y_j - y_{j+m}|^3}\right.$$

$$+ \frac{\delta_{kj} - \delta_{k,j-m}}{|y_j - y_{j-m}|}\delta_{rs} - \frac{(y_j - y_{j-m})_r(y_j - y_{j-m})_s(\delta_{kj} - \delta_{k,j-m})}{|y_j - y_{j-m}|^3}$$

$$\left. + \frac{\lambda}{\ell}\left(\delta_{k,j+m} + \delta_{k,j-m} - 2\delta_{kj}\right)\delta_{rs}\right\}.$$

This allows us to evaluate the Hessian at the stationary point corresponding to $\tilde{\mathcal{P}}_N$. After a long but straightforward calculation we arrive at the expression

$$\sum_{k,j,r,s} \nabla_{k,r}\nabla_{j,s}K_m(\tilde{y}_1,\ldots,\tilde{y}_N)\xi_{k,r}\xi_{j,s} \tag{9}$$

$$= \frac{1}{N\ell\Upsilon_m}\sum_{j=1}^{N}\left\{|\xi_j - \xi_{j+m}|^2 - \frac{(\xi_j - \xi_{j+m}, \tilde{y}_j - \tilde{y}_{j+m})^2}{|\tilde{y}_j - \tilde{y}_{j+m}|^2} - \sigma_m|\xi_j - \xi_{j+1}|^2\right\}.$$

We observe that the form depends on vector differences only, in other words, it is invariant with respect to translations and transforms naturally when the system is rotated. Furthermore, the sum of the first two terms in the bracket at the r.h.s. of (9) is non-negative by Schwarz inequality.

Since the second term in non-positive, it will be sufficient to establish negative definiteness of the "reduced" quadratic form

$$\xi \mapsto S_m[\xi] := \sum_j \left\{|\xi_j - \xi_{j+m}|^2 - \sigma_m|\xi_j - \xi_{j+1}|^2\right\} \tag{10}$$

on $\mathbb{R}^{Nd}$. Moreover, it is enough to consider here the case $d = 1$ only because S_m is a sum of its "component" forms. We observe that the matrices corresponding to the two parts of (10) can be simultaneously diagonalized; the corresponding eigenfunctions are $\{\binom{\sin}{\cos}(\mu_r j)\}_{j=1}^N$, where $\mu_r = \frac{2\pi r}{N}$ and $r = 0, 1, \ldots, m - 1$. Taking the corresponding eigenvalues we see that it is necessary to establish the inequalities

$$4\left(\sin^2\frac{\pi m r}{N} - \sigma_m \sin^2\frac{\pi r}{N}\right) < 0 \tag{11}$$

for $m = 2, \ldots, [\frac{1}{2}N]$ and $r = 2, \ldots, m-1$. We left out here the case $r = 1$ when the l.h.s. of (11) vanishes, however, the above explicit form of the eigenfunctions shows that the corresponding $\xi_j - \xi_{j+m}$ are in this case proportional to $\tilde{y}_j - \tilde{y}_{j+m}$ so the second term at the r.h.s. of (9) is negative unless $\xi = 0$.

Using the expression (8) for σ_m we can rewrite the condition (11) in terms of Chebyshev polynomials of the second kind as

$$U_{m-1}\left(\cos\frac{\pi}{N}\right) > \left|U_{m-1}\left(\cos\frac{\pi r}{N}\right)\right|. \tag{12}$$

One can check this inequality directly, because (12) is equivalent to

$$\sin\frac{\pi m}{N}\sin\frac{\pi r}{N} > \left|\sin\frac{\pi}{N}\sin\frac{\pi m r}{N}\right|, \qquad 2 \le r < m \le \left[\frac{N}{2}\right].$$

We have $\sin x \, \sin(\eta^2/x) \ge \sin\eta$ for a fixed $\eta \in (0,\frac{1}{2}\pi)$ and $2\eta^2/\pi \le x \le \frac{1}{2}\pi$, and moreover, this inequality is sharp if $x \ne \eta$, hence the desired assertion follows from the inequality $\sin^2 x - \sin\frac{\pi}{N}\sin\frac{Nx^2}{\pi} \ge 0$ valid for $x \in (0,\frac{1}{2}\pi)$. This concludes the proof of Theorem 2, and by that also of Theorem 1.

4 About the Global Maximizer

I have said that the question whether the maximizer represented by regular polygons is global remain open. Using the "geometrization" argument described above, we see that it is necessary to prove the following claim.

Conjecture 1. The property (P_m) holds globally for any $m = 2,\ldots,[\frac{1}{2}N]$.

Let us look at the problem in more detail in the particular case of *planar polygons*, $d = 2$. We employ a parametrization analogous to (6): for a fixed i we identify y_i with the origin and set for simplicity $\varphi = 0$, i.e.

$$y_{i+m} = \ell\left(1 + \sum_{n=1}^{m-1}\cos\sum_{j=1}^{n}\beta_{j+i}, \; \sum_{n=1}^{m-1}\sin\sum_{j=1}^{n}\beta_{j+i}\right);$$

in addition to the "winding number" condition (7) we require naturally also that $y_i = y_{i+N}$, or in other words

$$1 + \sum_{n=1}^{N-1}\cos\sum_{j=1}^{n}\beta_{j+i} = \sum_{n=1}^{N-1}\sin\sum_{j=1}^{n}\beta_{j+i} = 0 \tag{13}$$

for any $i = 1,\ldots,N$. The mean length of all m-diagonals is easily found,

$$M_m = \frac{\ell}{N}\sum_{i=1}^{N}\left[\left(1 + \sum_{n=1}^{m-1}\cos\sum_{j=1}^{n}\beta_{j+i}\right)^2 + \left(\sum_{n=1}^{m-1}\sin\sum_{j=1}^{n}\beta_{j+i}\right)^2\right]^{1/2},$$

or alternatively

$$M_m = \frac{\ell}{N}\sum_{i=1}^{N}\left[m + 2\sum_{n=1}^{m-1}\sum_{r=1}^{n}\cos\sum_{j=r}^{n}\beta_{j+i}\right]^{1/2}. \tag{14}$$

This result allows us to prove the claim in the simplest nontrivial case.

Proposition 1. *The property (P_2) holds globally if $d = 2$.*

Proof. By (14) the mean length of the 2-diagonals equals

$$M_2 = \frac{\sqrt{2}\ell}{N} \sum_{i=1}^{N} (1 + \cos\beta_i)^{1/2} = \frac{2\ell}{N} \sum_{i=1}^{N} \cos\frac{\beta_i}{2} \;;$$

notice that $\cos\frac{\beta_i}{2} > 0$ because $\beta_i \in (-\pi, \pi)$. Using now convexity of the function $u \mapsto -\cos\frac{u}{2}$ in $(-\pi, \pi)$ together with the condition (7) we find

$$-\sum_{i=1}^{N} \cos\frac{\beta_i}{2} \geq -N \cos\left(\sum_{i=1}^{N} \frac{\beta_i}{2}\right) = -N \cos\frac{\pi}{N} \;,$$

and therefore $M_2 \leq 2\ell \cos\frac{\pi}{N} = \tilde{M}_2$. Moreover, since the said function is strictly convex, the inequality is sharp unless all the β_i's are the same.

For $m \geq 3$ the situation is more complicated and one has to take into account also the condition (13); for the moment the problem remains open.

5 Some Extensions

The main task which this lecture raises is, of course, to prove Conjecture 1 and by that the global uniqueness of the maximizer. At the same time, the present problem offers various other extensions. One may ask, for instance, what will be the maximizer when we replace the equilaterality by a prescribed ordered N-tuple of polygon lengths $\{\ell_j\}$ and/or coupling constants $\{\alpha_j\}$. In both cases the task becomes more difficult because we loose the ground state symmetry which yielded the relation (3) and consequently the geometric reformulation based on the inequality (4). Using a perturbative approach, one may expect that for small symmetry violation the maximizer will not be far from $\tilde{\mathcal{P}}_N$, while in the general case we have no guiding principle.

One can also attempt to extend the result to point interaction family of point interactions in $\mathbb{R}^3$ placed on a closed surface. In this case, however, there is no unique counterpart to the equilaterality and one has to decide first what the "basic cell" of such a polyhedron surface should be.

Another extensions of our isoperimetric problem concern "continuous" versions of the present situation, i.e. Schrödinger operators with singular interactions supported by closed curves or surfaces – cf. [5, 9] and references therein – or with a "transverse" potential well extended along a closed curve. Let us restrict here to describing the situation in the simplest case when we have an operators in $L^2(\mathbb{R}^2)$ given formally by the expression

$$H_{\alpha,\Gamma} = -\Delta - \alpha\delta(x - \Gamma), \tag{15}$$

where $\alpha > 0$ and Γ is a C^2 smooth loop in the plane of a fixed length $L > 0$ without cusp-shaped self-intersections; for a proper definition of the operator (15) see [9]. In analogy with Theorem 1 we have the following claim.

Theorem 3. *Within the specified class of curves, $\epsilon_1(\alpha, \Gamma)$ is for any fixed $\alpha > 0$ and $L > 0$ locally sharply maximized by a circle.*

One can also conjecture that the circle is a sharp global maximizer, even under weaker regularity assumptions.

The proof is based on the generalized Birman-Schwinger principle [3] which makes it again possible to rewrite the problem in purely geometric terms – see [7] for details – namely as the task to check the inequality

$$\int_0^L |\Gamma(s+u) - \Gamma(s)|^p \, ds \; \leq \; \frac{L^{1+p}}{\pi^p} \sin^p \frac{\pi u}{L} \qquad (16)$$

for $p = 1$ and all $u \in (0, \frac{1}{2}L]$ and to show that is sharp unless Γ is a circle. Moreover, it follows from the convexity of the function $x \mapsto x^2$ that it is sufficient to prove the inequality (16) with $p = 2$ instead.

While we believe that these inequalities hold globally, we have at the moment a proof only for a particular case when u is small enough; the situation is similar to that in Proposition 1. To prove the local result, one has to express Γ through its (signed) curvature $\gamma := \dot{\Gamma}_2 \ddot{\Gamma}_1 - \dot{\Gamma}_1 \ddot{\Gamma}_2$; up to a Euclidean transformations we have

$$\Gamma(s) = \left(\int_0^s \cos \beta(s') \, ds', \int_0^s \sin \beta(s') \, ds' \right), \qquad (17)$$

where $\beta(s) := \int_0^s \gamma(s') \, ds'$ is the bending angle relative to the tangent at the chosen initial point, $s = 0$; to ensure that the curve is closed, the conditions $\int_0^L \cos \beta(s') \, ds' = \int_0^L \sin \beta(s') \, ds' = 0$ must be satisfied. Then the left-hand side of the inequality (16) with $p = 2$ becomes

$$\int_0^L \left[\left(\int_s^{s+u} \cos \beta(s') \, ds' \right)^2 + \left(\int_s^{s+u} \sin \beta(s') \, ds' \right)^2 \right] ds := c_\Gamma^2(u); \quad (18)$$

the relation (17) is analogous to (6), and similarly for the chord length expressions. A sequence of integral transformations described in detail in [7] brings the quantity in question to the form

$$c_\Gamma^2(u) = 2 \int_0^u dx \, (u - x) \int_0^L dz \cos \left(\int_{z-\frac{1}{2}x}^{z+\frac{1}{2}x} \gamma(s) \, ds \right). \qquad (19)$$

Gentle deformations of a circle can be characterized by the curvature $\gamma(s) = \frac{2\pi}{L} + g(s)$, where g is continuous and small in the sense that $\|g\|_\infty \ll L^{-1}$ and satisfies the condition $\int_0^L g(s) \, ds = 0$. Expanding in powers of g and using the last named condition, we find

$$c_\Gamma^2(u) = \frac{L^3}{\pi^2} \sin^2 \frac{\pi u}{L} - I_g(u) + \mathcal{O}(g^3) ,$$

where the error term is a shorthand for $\mathcal{O}(\|Lg\|_\infty^3)$. One has to show that $I_g(u) > 0$ unless $g = 0$ identically which is trivial for $u \leq \frac{1}{4}L$ while for $u \in (\frac{1}{4}L, \frac{1}{2}L]$ one can employ expansion of g into a Fourier series – cf. [7]. Naturally, a further extension of the present problem concerns a maximizer for the generalized Schrödinger operator in $\mathbb{R}^3$ with an attractive δ interaction supported by a closed surface of a fixed area A, and its generalization to closed hypersurfaces of codimension one in $\mathbb{R}^d$, $d > 3$. In the case of $d = 3$ we have again a heuristic argument relying on [5,8] which suggests that the problem is solved by the sphere provided the discrete spectrum is not empty, of course, which is a nontrivial assumption in this case. The Birman-Schwinger reduction of the problem can be performed again leading to to new modifications of the geometric inequalities discussed above.

Acknowledgments

The research has been partially supported by ASCR within the project IRP AV0Z10480505, its presentation by the QMath9 organizing committee.

References

1. S. Albeverio, F. Gesztesy, R. Høegh-Krohn, H. Holden: *Solvable Models in Quantum Mechanics*, 2nd edn (AMS Chelsea, Providence, R.I., 2005).
2. D. Borisov, P. Exner: J. Phys. A: Math. Gen. **37**, 3411 (2004).
3. J.F. Brasche, P. Exner, Yu.A. Kuperin, P. Šeba: J. Math. Anal. Appl. **184**, 112 (1994).
4. P. Exner: Lett. Math. Phys. **57**, 87 (2001).
5. P. Exner: Spectral properties of Schrödinger operators with a strongly attractive δ interaction supported by a surface. In *Contemporary Mathematics*, vol 339, ed by P. Kuchment (AMS, Providence, R.I., 2003) pp 25–36.
6. P. Exner: An isoperimetric problem for point interactions, J. Phys. A: Math. Gen. **38**, 4795 (2005).
7. P. Exner: An isoperimetric problem for leaky loops and related mean-chord inequalities, J. Math. Phys. **46**, 062105 (2005).
8. P. Exner, E.M. Harrell, M. Loss: Optimal eigenvalues for some Laplacians and Schrödinger operators depending on curvature. In *Operator Theory : Advances and Applications*, vol 108, ed by J. Dittrich et al. (Birkhäuser, Basel 1998) pp 47–53.
9. P. Exner, T. Ichinose: J. Phys. A: Math. Gen. **34**, 1439 (2001).
10. P. Exner, K. Němcová: J. Phys. A: Math. Gen. **36**, 10173 (2003).
11. P. Exner, O. Post: Convergence of spectra of graph-like thin manifolds, J. Geom. Phys. **54**, 77 (2005).
12. G. Faber: Sitzungber. der math.-phys. Klasse der Bayerische Akad. der Wiss. zu München, 169 (1923).
13. E.M. Harrell, P. Kröger, K. Kurata: SIAM J. Math. Anal. **33**, 240 (2001).
14. E. Krahn: Ann. Math. **94**, 97 (1925).
15. M.G. Krein, G.K. Langer: Sov. J. Funct. Anal. Appl. **5**, 59 (1971).

Limit Cycles in Quantum Mechanics

Stanisław D. Głazek

Institute of Theoretical Physics, Warsaw University, Poland
stglazek@fuw.edu.pl

1 Introduction

This lecture concerns limit cycles in renormalization group (RG) behavior of quantum Hamiltonians. Cyclic behavior is perhaps more common in quantum mechanics than the fixed-point behavior which is well-known from critical phenomena in classical statistical mechanics. We discuss a simple Hamiltonian model that exhibits limit cycle behavior.

The value of the model is its simplicity. The model can serve as a basis for further mathematical studies of the cycle for the purpose of solving complex physical theories. For example, the limit cycle may be related to existence of a set of bound states with binding energies forming a geometric sequence converging to zero. The model offers a possibility of studying in detail how Hamiltonians with bound states behave as operators under the RG transformation in the vicinity of the cycle. This lecture reports on results concerning behavior of marginal and irrelevant operators in the model.

The possibility of limit cycle behavior in RG calculations was originally pointed out by Wilson [1] in the context of theory of strong interactions in particle physics. The classification of operators near a fixed point as relevant, marginal, and irrelevant, was introduced by Wegner [2] in the context of critical phenomena and corrections to scaling in condensed matter physics.

Recent interest in the RG limit cycle stems from the structure of three-nucleon bound-state spectrum in the case of two-body nucleon-nucleon interactions that have a very short range in comparison to the scattering length. This structure was first noticed by Thomas [3] and subsequently discussed by Efimov [4]. It was only recently associated with ultraviolet regularization dependence of parameters in the nuclear potentials by Bedaque, Hammer, and van Kolck [5]. Interactions of atoms, especially in Helium trimers, may exhibit a similar cycle structure even more transparently than effective nuclear forces because more bound states may exist for atoms than for nuclei. In addition, atomic interactions are easier to tune close to the cycle structure by changing experimental conditions than nuclear forces. The cyclic few-body atomic interactions also contribute to many-body dynamics. Braaten, Hammer, and Kusunoki discussed some effects in a Bose-Einstein condensate [6]. LeClair, Roman, and Sierra discussed a theoretical possibility of variation in the superconductivity mechanism [7]. On the other hand, the

S.D. Głazek: *Limit Cycles in Quantum Mechanics*, Lect. Notes Phys. **690**, 65–78 (2006)
www.springerlink.com   © Springer-Verlag Berlin Heidelberg 2006

range of phenomena that may exhibit cycle structure extends also to subnuclear domain. Namely, Braaten and Hammer observed that masses of the up and down quarks appear to be close to special values at which an infrared cyclic behavior may develop in quantum chromodynamics [8]. From a mathematical point of view, it is also interesting that the cyclic behavior occurs in the elementary case of two-body quantum mechanics with the well-known potential r^{-2}, as recently described by Braaten and Phillips [9]. Readers interested in the literature that discusses quantum mechanics with short-range interactions without direct reference to the RG limit cycle can consult a review article by Nielsen, Fedorov, Jensen and Garrido [10]. Braaten and Hammer recently wrote a review article on universal cyclic properties of physical systems with two-body short range potentials and large scattering length in the context of renormalization of the potentials [11]. See also [12].

It should be mentioned that examples of renormalization group maps, which have critical attractors that are not simple fixed points, have been reported in the literature concerning classical dynamical systems. The author has learned about existence of a number of works during the process of preparation of this report [13–20].

The model Hamiltonian reviewed here has been discovered in our RG studies of Hamiltonians that were focused on hadronic parton dynamics in the infinite momentum frame [21]. We have recognized that the model has a RG cycle in a later article [22, 23]. We have subsequently found an analytic solution for the RG behavior of Hamiltonians in the vicinity of the cycle in the model and this enabled us to discuss the RG universality of quantum Hamiltonians with a limit cycle using the model [24]. The purpose of my talk is to describe the cycle in our model, discuss corrections due to irrelevant operators, and show tuning to criticality, which allows the cycle spectrum to clearly appear even in approximations in which the basis of the space of quantum states is limited to about 70. This is attractive because one can inspect what happens in the model using a computer. A machine can provide numerical facts that help in developing intuition which otherwise is not available. I also address issues that are not fully understood and invite more research on mathematics of the RG limit cycle in quantum mechanics. Since atomic interactions can be strongly influenced by varying external fields (e.g., see the recent work of Roberts, Claussen, Cornish, and Wieman on magnetic field dependence of ultracold inelastic atomic collisions [25]), better understanding of quantum Hamiltonians near a RG limit cycle is a prerequisite to systematic experimental searches for cycles in real systems. The distinct feature to look for is the geometric sequence of bound states converging at threshold.

2 Definition of the Model

Our model can be defined in more than one way. Let us consider a non-relativistic one-particle Hamiltonian in 2 dimensions in the presence of a point-like potential

$$H = \frac{-\Delta_r}{2\mu} - g\delta^2(r) \; , \tag{1}$$

and write its eigen-functions in momentum space as

$$\phi(p) = \int \frac{d^2r}{(2\pi)^2} \, e^{-ipr} \, \psi(r) \; . \tag{2}$$

The momentum-space eigenvalue equation reads

$$\frac{p^2}{2\mu} \phi(p) - \frac{g}{(2\pi)^2} \int d^2q \, \phi(q) = E\phi(p) \; . \tag{3}$$

ϕ can be composed of angular momentum eigen-functions $e^{im\varphi}$ with coefficients ϕ_m that depend on $p = |p|$,

$$\phi(p) = \sum_{m=-\infty}^{\infty} e^{im\varphi} \phi_m(p) \; , \tag{4}$$

and it turns out that wave functions with $m \neq 0$ are the same as in the case with no interaction. The wave function with $m = 0$ satisfies the equation

$$\frac{p^2}{2\mu} \phi_0(p) - \frac{g}{2\pi} \int_0^\infty dq \, q \, \phi_0(q) = E\phi_0(p) \; . \tag{5}$$

Further discussion concerns only the component with $m = 0$. Using notation

$$p^2 \to z, \quad \phi_0(p) \to \phi(z), \quad \mu \to \frac{1}{2}, \quad \frac{g}{4\pi} \to g \; , \tag{6}$$

one obtains the eigen-value equation of the form

$$z\,\phi(z) - g \int_0^\infty dz' \, \phi(z') = E\phi(z). \tag{7}$$

If there is a bound state with $E = -E_B$ and $E_B > 0$, the solution takes the well-known shape for factorizable interactions: $\phi(z) = c/(z - E)$ with some constant c, and the eigenvalue condition for E_B is

$$1 = g \int_0^\infty \frac{dz}{z + E_B} \; , \tag{8}$$

which would be fine if not the fact that the integral diverges. This divergence and a cure for it are well known [26–34]. The resulting eigenvalue equations are studied in mathematical physics [35, 36]. The essence of the cure is to regulate the integral somehow and make the coupling constant g depend on the regularization in such a way that the result for E_B does not depend on the regularization when it is being removed. Then, other observables become also independent of the regularization when it is being removed and instead they depend on E_B. I will focus on a RG calculation for Hamiltonians in Wilson's sense of the word [37, 38] and this point of view will be applied in discussion of the limit cycle.

The point is that every interval $2^{n-1} \leq z \leq 2^n$ contributes the same amount $\int_{2^{n-1}}^{2^n} dz\, z^{-1} = \ln 2$ to the complete integral and there are infinitely many such intervals to add. The coupling constant g must be inversely proportional to the number of such intervals included in the integration. Since each of the intervals contributes the same amount, we can consider every interval as a single degree of freedom. The essence of regularization is to limit the number of degrees of freedom. The problem with the divergence is that all these degrees of freedom are coupled with equal strength and the correlation length among an infinite number of degrees of freedom on the energy scale is infinite.

In order to deal with the large number of degrees of freedom it is enough to consider a discrete model in which the continuous variable z is replaced by a set of values $z_n = b^n$ with some $b > 1$ (b replaces the number 2 in the integration interval discussed above). Instead of $\phi(z)$ we have $\phi(z_n) = \phi_n$, and $dz = z_n dn \ln b$. The regularization replaces the integral $\int_0^\infty$ by $\int_\varepsilon^\Lambda$. In the discrete version, we have $\varepsilon = b^M$ with M large negative and $\Lambda = b^N$ with N large positive. The integral is replaced by a sum $\sum_{n=M}^N$ since $dn = 1$. The dependence of g on Λ means that in the discrete model we expect g to depend on N (dependence on ε or M does not require a separate RG study in the model). We denote $g \ln b$ with appropriate factors including π by g_N. Finally, we obtain a Hamiltonian matrix of the form

$$
H_{kl} = \begin{bmatrix} b^N & 0 & \dots & 0 \\ 0 & b^{N-1} & \dots & 0 \\ . & . & \dots & 0 \\ 0 & 0 & \dots & b^M \end{bmatrix}_{kl} - b^{\frac{k+l}{2}} g_N \begin{bmatrix} 1 & 1 & \dots & 1 \\ 1 & 1 & \dots & 1 \\ . & . & \dots & 1 \\ 1 & 1 & \dots & 1 \end{bmatrix}_{kl} , \tag{9}
$$

that acts on the state vectors with components $\chi_l = b^{l/2} \phi_l$.

Our model with a limit cycle is defined by the replacement

$$
g_N \begin{bmatrix} 1 & 1 & \dots & 1 \\ 1 & 1 & \dots & 1 \\ . & . & \dots & 1 \\ 1 & 1 & \dots & 1 \end{bmatrix} \rightarrow g_N \begin{bmatrix} 1 & 1 & \dots & 1 \\ 1 & 1 & \dots & 1 \\ . & . & \dots & 1 \\ 1 & 1 & \dots & 1 \end{bmatrix} + i h_N \begin{bmatrix} 0 & 1 & \dots & 1 \\ -1 & 0 & \dots & 1 \\ . & . & \dots & 1 \\ -1 & -1 & \dots & 0 \end{bmatrix} . \tag{10}
$$

This means that we complement a real symmetric interaction matrix that requires renormalization due to the large number of equal entries with an imaginary skew-symmetric matrix of similarly equal entries. Such imaginary part is allowed in quantum mechanics. It turns out that the ultraviolet-diverging Hamiltonian structure with non-zero imaginary part leads to a limit cycle, or even more chaotic RG behavior [22].

3 Renormalization Group

Our model Hamiltonian matrix is

$$H_{kl}(N, g_N, h_N) = b^{\frac{k+l}{2}} \left(\delta_{kl} - g_N - ih_N s_{kl} \right) ,\tag{11}$$

where $s_{kl} = 1$ for $k > l$ and $s_{lk} = -s_{kl}$. In order to introduce the RG procedure let us first set $h_N = 0$. Then, the eigenvalue equation for the matrix H_{kl} can be rewritten, introducing $\psi_k = b^{k/2}\chi_k$, $\epsilon_k = Eb^{-k}$, and $\sigma_N = \sum_{l=M}^{N} \psi_l$, as the following set of equations:

$$(1 - \epsilon_N)\,\psi_N = g_N\,\sigma_N ,$$
$$(1 - \epsilon_{N-1})\,\psi_{N-1} = g_N\,\sigma_N ,$$
$$\vdots \tag{12}$$
$$(1 - \epsilon_M)\,\psi_M = g_N\sigma_N .$$

The first equation allows us to express ψ_N by all other components with m smaller than N,

$$\psi_N = \frac{g_N}{1 - \epsilon_N - g_N}\,\sigma_{N-1} ,\tag{13}$$

and this value of ψ_N can be inserted into the remaining equations

$$(1 - \epsilon_{N-1})\,\psi_{N-1} = g_N\,\sigma_{N-1} + g_N\psi_N$$
$$= g_N \left[1 + \frac{g_N}{1 - \epsilon_N - g_N} \right] \sigma_{N-1}$$
$$(1 - \epsilon_{N-2})\,\psi_{N-2} = g_N \left[1 + \frac{g_N}{1 - \epsilon_N - g_N} \right] \sigma_{N-1}$$
$$\vdots \tag{14}$$
$$(1 - \epsilon_M)\,\psi_M = g_N \left[1 + \frac{g_N}{1 - \epsilon_N - g_N} \right] \sigma_{N-1} .$$

This set can be written as

$$(1 - \epsilon_{N-1})\,\psi_{N-1} = g_{N-1}\,\sigma_{N-1} ,$$
$$(1 - \epsilon_{N-2})\,\psi_{N-2} = g_{N-1}\,\sigma_{N-1} ,$$
$$\vdots \tag{15}$$
$$(1 - \epsilon_M)\,\psi_M = g_{N-1}\,\sigma_{N-1},$$

when one defines

$$g_{N-1} = g_N \left[1 + \frac{g_N}{1 - \epsilon_N - g_N} \right] . \tag{16}$$

Thus, the Hamiltonian matrix of (11) is turned into a new matrix, smaller by one row and one column of highest energy b^N, $k, l \leq N - 1$,

$$H_{kl}(N - 1, g_{N-1}) = b^{\frac{k+l}{2}} \left(\delta_{kl} - g_{N-1} \right) . \tag{17}$$

The new ultraviolet cutoff is $\Lambda_{N-1} = b^{N-1} = \Lambda/b$, and there is a new coupling constant g_{N-1}. This process can be repeated and one obtains a recursion

$$g_{n-1} = g_n + \frac{g_n^2}{1 - g_n - \epsilon_n} . \tag{18}$$

Since no other change occurs in the Hamiltonian matrix, (18) fully describes the RG transformation for Hamiltonians in our model.

Suppose that we consider n such that the eigenvalue E can be neglected in comparison to b^n. Then the RG transformation simplifies to

$$g_{n-1} = \frac{g_n}{1 - g_n} . \tag{19}$$

If we denote $g_{n-1} = \text{RG}(g_n)$, then $\text{RG}(f) = f/(1 - f)$. This recursion has a solution $f_n = 1/(n + c)$ with arbitrary constant c, which means that $g_n = 1/(c + \ln \Lambda_n / \ln b)$ and the larger is the cutoff Λ_n, the smaller is g_n. This phenomenon is called asymptotic freedom [39, 40]. It is associated with the fixed point, f^*, of the RG transformation, $\text{RG}(f^*) = f^*$, which is $f^* = 0$. When one reverses the RG transformation to calculate g_N for increasing N in terms of g_{n_0} for some finite n_0, g_N approaches 0 for asymptotically large N. The coupling constant g_N vanishes as an inverse of a logarithm of the ultraviolet cutoff Λ.

Let us now speculate a bit about what may happen when the eigenvalue E is not neglected in the recursion. The RG transformation becomes related to the spectrum of the Hamiltonian. In our discrete model, after repeating the transformation $N - M$ times, which can be a very large number, one obtains one last equation

$$1 - g_M(E) - E/b^M = 0 , \tag{20}$$

which must be satisfied if E is to be an eigenvalue. But instead of a recursion for a sequence of coupling constants, there appears now a recursion for a sequence of functions $g_n(E) = f_n(\varepsilon_n)$ in our model. The recursion is $f_{n-1}(bx) = \text{RG}[f_n(x)]$ with $\text{RG}[f(x)] = (1 - x)f(x)/[1 - x - f(x)]$.

If the repeated application of the RG transformation carried every starting function $f_N(x)$ over to a fixed-point function $f^*(x)$ after many iterations, very many eigenvalues would be given by one equation

$$1 - f^*(\epsilon_M) - \epsilon_M = 0 , \qquad (21)$$

provided that ϵ_M is sufficiently large not to produce results sensitive to the infrared cutoff M. The fixed-point function in our model with $h_N = 0$ satisfies the equation $f^*(bx) = (1 - x)f^*(x)/[1 - x - f^*(x)]$, which has a solution

$$f^*(x) = \left[\ldots + \frac{1}{1 - b^2 x} + \frac{1}{1 - bx} + \frac{1}{1 - x} + y \right]^{-1} , \qquad (22)$$

with arbitrary number y (y is determined by one eigenvalue). Most of the spectrum is determined not by the initial function $f_N(x)$ but by the fixed-point function $f^*(x)$ with some y. The largest eigenvalues may be sensitive to the initial function in the Hamiltonian with ultraviolet cutoff N, which can be just a constant. The lowest eigenvalues are sensitive to the infrared cutoff M because the series that extends to $M = -\infty$ in the fixed-point solution is cut off in a model with a finite M. But a large number of eigenvalues is determined by the behavior of $f^*(x)$ as function of x in the region between b^{M-N} and 1. This behavior depends on the properties of the RG transformation rather than on the details of the initial Hamiltonian. Thus, all Hamiltonians that lead to essentially the same RG recursion produce a universal answer for the spectrum.

One can also ask if the initial Hamiltonian may be tuned so that a small number of initial iterations of the RG transformation brings the Hamiltonian right to the vicinity of the fixed-point structure so that further iterations have universal properties. In such case, the spectrum may more clearly exhibit the universal structure in a larger subset of eigenvalues than in the case where the initial Hamiltonian is far away from the criticality that allows the RG transformation to quickly reach the vicinity of the fixed-point. Since the fixed-point is not changed by the RG transformation, it corresponds to a theory with an infinite cutoff. A theory with finite cutoffs must be tuned to yield the fixed-point structure.

We can now proceed to our model with the limit cycle where such universality scenario exists in an explicit form and produces a geometric sequence of bound states [24]. The sequence is related to the fact that there is not just one fixed point, but a whole cycle of them.

4 Limit Cycle

We now consider the Hamiltonian matrix in (11) with $h_N \neq 0$. By eliminating component after component as in the previous section, we arrive at the RG transformation that transforms a Hamiltonian matrix

$$H_{kl}(n, g_n, h_n) = b^{\frac{k+l}{2}} \left(\delta_{kl} - g_n - ih_n s_{kl} \right) , \qquad (23)$$

into the one with one row and one column less and [24]

$$g_{n-1} = g_n + \frac{g_n^2 + h_n^2}{1 - g_n - \epsilon_n} , \tag{24}$$

$$h_{n-1} = h_n \equiv h_N = h . \tag{25}$$

The eigenvalue condition at the end of iteration takes the form of (20). The exact RG recursion can be re-written as

$$\alpha_{n-1} = \alpha_n + \beta_n , \tag{26}$$

when one introduces angles

$$\alpha_n = \arctan \frac{g_n}{h} , \tag{27}$$

$$\beta_n = \arctan \frac{h}{1 - \epsilon_n} . \tag{28}$$

When the eigenvalue E is set to 0 (there may exist such eigenvalue), we have $\beta_n = \arctan h \equiv \beta$, and the successive coupling constants are given by the formula

$$g_{n-p} = h \tan (\alpha_n + p\,\beta) , \tag{29}$$

which implies that $g_{n-p} = g_n$ when $\beta = \pi/\tilde{p}$ with integer $\tilde{p} \geq 3$ and $p = \tilde{p}$. The coupling constant goes through a cycle and returns to the same value because the function tan is periodic with period π. If $\tilde{p}$ were irrational, one would obtain a chaotic RG behavior [22].

Note that for $E = 0$ we have $g_{n-p} = g_n$ regardless of the value of g_n. Therefore, one can seek a simplified RG relationship between g_{n-p} and g_n for E near 0. For this purpose, we introduce functions $f_n(\epsilon_n) = g_n(\epsilon_n)/h$ and calculate the RG transformation for p successive steps of elimination of rows and columns one after another. This way we obtain

$$f_{n-p}(rx) = R_p [f_n(x)] , \tag{30}$$

where $x = \epsilon_n$, $\epsilon_{n-p} = r\,\epsilon_n$, and $r = b^p$. There are $p - 1$ different intermediate functions in this recursion but we focus on the transformation over the entire cycle because it has a fixed point and we can take advantage of the universality that develops around fixed points.

The transformation R_p takes the form

$$f_{n-p}(x_{n-p}) = \frac{f_n(x_n) + z_p(x_n)}{1 - f_n(x_n)\,z_p(x_n)} , \tag{31}$$

where the function $z_p(0) = 0$. For example, when $p = 3$,

$$z_3(x)/h = \frac{(7/4)x(1 - x)}{1 + x^3 - (7/4)x(1 + x)} \tag{32}$$

$$= \frac{7}{4} x + \frac{21}{16} x^2 + \frac{343}{64} x^3 + O(x^4) . \tag{33}$$

The transformation R_p for $p = 3$ will be discussed in a later section of this lecture. A complete analysis for arbitrary integer $p \geq 3$ and to all orders of expansion in powers of x is given in the literature [24].

5 Marginal and Irrelevant Operators

Let us consider first two terms in the function

$$f(x) = f_0 + f_1 x + O(x^2) \ , \tag{34}$$

and find the fixed-point values of the coefficients f_0 and f_1. The fixed point condition is

$$f^*(x) = R_p[f^*] = f^*(x/r) + (1 + f_0^{*\,2}) z_p^{(1)} x/r \ , \tag{35}$$

and the solution is given by $f_0^* = g^*/h$ and $f_1^* = \left(1 + f_0^{*\,2}\right) z_p^{(1)}/(r - 1)$. $z_p^{(1)}$ denotes the first term in the Taylor expansion of $z_p(x)$ in powers of x. Now we can write the function $f(x)$ as a fixed-point function plus an infinitesimal correction,

$$f(x) = f^*(g^*, x) + df(g^*, x) \ , \tag{36}$$

and find the RG evolution of the corrections $df(g^*, x)$ by inspecting the transformation

$$R_p[f^* + df] = f^*(x/r) + df(x/r) + [1 + f_0^{*\,2} + 2f_0^* df_0] z_p^{(1)} x/r \ , \tag{37}$$

in the linearized approximation, $L_p(df) = w \, df(x)$, where L_p denotes the derivative of R_p at the fixed point and w is the Wegner eigenvalue [2]. This way we find that in $df(x) = c_0 + c_1 x$ the numbers c_0 and c_1 must simultaneously satisfy equations

$$c_0 = w \, c_0 \tag{38}$$

$$\frac{1}{r} c_1 + 2 z_p^{(1)} f_0^* \frac{1}{r} c_0 = w \, c_1 \ . \tag{39}$$

The two solutions for the eigenvalues are $w_0 = 1$ and $w_1 = r^{-1} = 1/b^p < 1$.

The marginal operator corresponds to the eigenvalue 1 (it does not change under the RG transformation). In this case c_0 is arbitrary and c_1 is determined by c_0. All coefficients in the expansion of the marginal operator in powers of x are determined by c_0 through the eigenvalue condition [24]. The critical exponent for the marginal Wegner operator, defined by the scaling relation $w_0 = r^{\lambda_0}$, is $\lambda_0 = 0$.

The first irrelevant operator has c_0 equal to zero and c_1 arbitrary. All coefficients in the first irrelevant operator are determined by c_1 in it through the eigenvalue condition. The critical exponent for the first irrelevant Wegner operator, defined by the relation $w_1 = r^{\lambda_1}$, is $\lambda_1 = -1$.

The linearized RG transformation allows us to find all Wegner's eigen-operators using condition

$$df(x, g^*) + z_p(x)[df(x, g^*)f(rx, g^*) + f(x, g^*)wdf(rx, g^*)] = wdf(rx, g^*) ,$$
(40)

and we find that the operators df whose Taylor expansion starts with terms $\sim x^l$ are irrelevant with eigenvalues $w_l = r^{-l}$. This analysis allows us to understand how one should choose the initial Hamiltonian in order to bring it close to criticality.

6 Tuning to a Cycle

We want to make sure that the initial H transforms under the first few RG transformations to such a matrix that the marginal operator takes some convenient value in a cycle and the first few irrelevant operators vanish. The latter condition can be realized by eliminating the first few terms in the Taylor expansion in powers of x.

We consider tuning of our model Hamiltonian to a limit cycle using example with $p = 3$, $b = 2$, $N = 17$, and $M = -51$. In a physical system in which external conditions define the range of available options for tuning the system to a cycle, one has to proceed by analogy and identify the key factors that can expose the cycle within a limited window of energy scales available experimentally.

The idea we use here in the mathematical model is to modify only the four upper-left (high-energy) entries in the matrices H_{kl} in (9) and (10) according to the rule

$$H_{kl} \to \begin{bmatrix} tb^N & 0 & 0 & . \\ 0 & tb^{N-1} & 0 & . \\ 0 & 0 & b^{N-2} & . \\ . & . & . & . \end{bmatrix}_{kl} - b^{\frac{k+l}{2}} \begin{bmatrix} 0 & t+iv & g+ih & . \\ t-iv & 0 & g+ih & . \\ g-ih & g-ih & g & . \\ . & . & . & . \end{bmatrix}_{kl} .$$
(41)

We are lucky doing this, because for $v = \sqrt{3}$ and $t = 20/3$ the non-linear exact RG transformation produces in two iterations

$$g_{N-2}(x) = -7 + 13\,x - \frac{65}{3}\,x^2 + o(x^3) .$$
(42)

This is precisely the first three terms in the marginal operator solution from the previous section. In our example, here we have $x = E/2^{15}$. The first two irrelevant operators with eigenvalues $w_1 = 1/8$ and $w_2 = 1/64$ are absent. Corrections to the limit cycle spectrum die out at the rate given by the third irrelevant operator with eigenvalue $w_3 = 1/512$. This is illustrated for bound state eigenvalues in Table 1 (they are found with help of a simple numerical routine). Note that in order to get rid of the infrared cutoff M we had to extend the matrix to $M = -60$ to see precisely how the eigenvalue $1/512$ appears in the cyclic spectrum.

Table 1. The bound-state spectrum of a quantum Hamiltonian with a cycle tuned to criticality in marginal and first two irrelevant operators. The ratio of successive corrections to the cycle period $2^3 = 8$ for bound-state energies approaches the inverse of Wegner's eigenvalue for third irrelevant operator with critical exponent -3, $1/8^{-3} = 512$, and the entry ~ 400 turns to 511.6 when the infrared cutoff $M = -51$ is changed to $M = -60$. Without tuning, the first irrelevant operator with critical exponent -1 and eigenvalue 8^{-1} would slow down the approach to the cycle 64 times

E	Ratio	w_3^{-1}
-111890.939577163000	8.223556605927	119
-13606.149375382800	8.001879953885	433
-1700.369094987040	8.000004341325	502
-212.546021531960	8.000000008649	511
-26.568252662770	8.000000000017	~ 400
-3.321031582839		

7 Generic Properties of Limit Cycles

In this section we indicate which features of the model cycle are generic and will occur in other cases and which are specific and cannot be guaranteed to occur elsewhere. See [24] for details.

We cannot guarantee that bound states must appear. But we can say that if bound states do appear, they must form a geometric series converging on zero. Hamiltonians in the cycle vary but critical exponents do not vary on the cycle. This is a generic feature that should be helpful in tuning. A marginal operator is guaranteed in continuum, because the operator that changes the Hamiltonian in the direction of the cycle will be a derivative with respect to a continuous parameter that labels the cycle, as the coupling constant g^* did in the discrete model. But since tuning depends on non-linear mappings from H_Λ to coefficients in Wegner's operators, there is no guarantee that one will be able to eliminate any given set of Wegner's operators.

The generic properties of cyclic behavior can be studied in the model without extensive effort required in realistic cases. These properties can be searched for in real physical systems using exact analytic insight into the cycle structure that the model provides.

8 Conclusion

Perhaps the most important question regarding the model that has no answer so far is the question about the scattering states in the continuum limit, when b tends to 1. Another question concerns the ratios of positive eigenvalues. For example, for $p = 3$ and $b = 2$, the successive positive eigenvalues appear with ratio E_n/E_{n-1} very close to $\sqrt{8}$, but not exactly. It is not clear to the author what this ratio should be as function of b and p beyond the suggestion that it should be $b^{\frac{p}{p-1}}$. Precise conditions required for existence of bound states in a cyclic Hamiltonian spectrum are not known.

The existence of limit cycles in quantum mechanics has a philosophical consequence: in search for an ultimate theory we always have to keep in mind that the "ultimate" may mean not a fixed-point structure with parameters reaching their ultimate values but that we may rather wonder endlessly through a sequence of patterns. This comment is particularly interesting in the context of asymptotic freedom as an example of ultimate understanding of microscopic phenomena. The cyclic model shows that asymptotic freedom may be merely an intermediate part of a larger cycle. Besides, since we know that real symmetric Hamiltonians are of measure 0 in the space of complex Hermitean Hamiltonians, one may immediately suggest that the cyclic or even chaotic RG behavior of quantum Hamiltonians is more common than the fixed-point behavior known in classical theories. Hopefully, the concrete example of such surprising mathematical transparency, combined with a considerable scope of possible interpretations and applications by analogy in various branches of physics, will inspire further interest and the unknowns of the model will be resolved.

Acknowledgments

I am grateful to Ken Wilson for illuminating discussions during the course of our work on the model described in this lecture, and on other subjects. I would like to thank the organizers of Qmath9 for inviting me to speak.

References

1. K.G. Wilson. Renormalization group and strong interactions. *Phys. Rev.*, D3:1818, 1971.
2. F.J. Wegner. Corrections to scaling laws. *Phys. Rev.*, B5:4529, 1972.
3. L.H. Thomas. The interaction between a neutron and a proton and the structure of H3. *Phys. Rev.*, 47:903, 1935.
4. V. Efimov. Energy levels arising from resonant two-body forces in a three-body system. *Phys. Lett.*, 33B:563, 1970.
5. P.F. Bedaque, H.W. Hammer, and U. van Kolck. Renormalization of the three-body system with short-range interactions. *Phys. Rev. Lett.*, 82:463, 1999.

6. E. Braaten, H.-W. Hammer, and M. Kusunoki. Efimov states in a Bose-Einstein condensate near a feshbach resonance. *Phys. Rev. Lett.*, 90:170402, 2003.
7. A. LeClair, J.M. Roman, and G. Sierra. Russian doll renormalization group and superconductivity. *Phys. Rev.*, B69:20505, 2004.
8. E. Braaten and H.W. Hammer. An infrared renormalization group limit cycle in qcd. *Phys. Rev. Lett.*, 91:102002, 2003.
9. E. Braaten and D. Phillips. Renormalization-group limit cycle for the r^{-2} potential. *Phys. Rev.*, A70:052111, 2004.
10. E. Nielsen, D.V. Fedorov, A.S. Jensen, and E. Garrido. The three-body problem with short-range interactions. *Phys. Rept.*, 347:373, 2001.
11. E. Braaten and H.W. Hammer. Universality in few-body systems with large scattering length, 2004, INT-PUB-04-27, cond-math/0410417.
12. R.F. Mohr Jr. Quantum mechanical three body problem with short range interactions. *Ph.D. Thesis, Ohio State U., Advisor: R. J. Perry, UMI-31-09134-mc (microfiche)*, 2003. See also R.F. Mohr, R.J. Furnstahl, R.J. Perry, K.G. Wilson, H.-W. Hammer, Precise numerical results for limit cycles in the quantum three-body problem, Annals Phys. 321:225, 2006.
13. O.E. Landford III. Renormalization group methods for circle mappings, in statistical mechanics and field theory: Mathematical aspects, edited by T. C. Dorlas, N. M. Hugenholtz, M. Winnik, Springer Vg. *Lecture Notes in Physics*, 257:176, 1986.
14. O.E. Landford III. Renormalization group methods for circle mappings, in G. Gallavotti, P.F. Zweifel (eds.). *Nonlinear Evolution and Chaotic Phenomena*, Plenum Press, p. 25, 1998.
15. M. Yampolsky. Hyperbolicity of renormalization of critical circle maps. *Publ. Math. Inst. Hautes Etudes Sci.*, 96:1, 2002.
16. M. Yampolsky. Global renormalization horseshoe for critical circle maps. *Commun. Math. Physics*, 240:75, 2003.
17. R.S. MacKay. Renormalization in area-preserving maps. *World Scientific*, 1993.
18. J.M. Green and J. Mao. Higher-order fixed points of the renormalisation operator for invariant circles. *Nonlinearity*, 3:69, 1990.
19. C. Chandre and H.R. Jauslin. Renormalization-group analysis for the transition to chaos in Hamiltonian systems. *Phys. Rep.*, 365:1, 2002.
20. C. Chandre and H.R. Jauslin. Critical attractor and universality in a renormalization-group scheme for three frequency Hamiltonian systems. *Phys. Rev. Lett.*, 81:5125, 1998.
21. S.D. Głazek and K.G. Wilson. Renormalization of overlapping transverse divergences in a model light front Hamiltonian. *Phys. Rev.*, D47:4657, 1993.
22. S.D. Głazek and K.G. Wilson. Limit cycles in quantum theories. *Phys. Rev. Lett.*, 89:230401, 2002.
23. S.D. Głazek and K.G. Wilson. Erratum: Limit cycles in quantum theories [Phys. Rev. Lett. 89, 230401 (2002)]. *Phys. Rev. Lett.*, 92:139901, 2004.
24. S.D. Głazek and K.G. Wilson. Universality, marginal operators, and limit cycles. *Phys. Rev.*, B69:094304, 2004.
25. J.L. Roberts, N.R. Claussen, S.L. Cornish, and C.E. Wieman. Magnetic field dependence of ultracold inelastic collisions near a Feshbach resonance. *Phys. Rev. Lett.*, 85:728, 2000.
26. E.E. Salpeter. Wave functions in momentum space. *Phys. Rev.*, 84:1226, 1951.
27. T.D. Lee. Some special examples in renormalizable field theory. *Phys. Rev.*, 95:1329, 1954.

28. Y. Yamaguchi. Two-nucleon problem when the potential is nonlocal but separable. *Phys. Rev.*, 95:1628, 1954.
29. W. Heisenberg. Lee model and quantisation of non linear field equations. *Nucl. Phys.*, 4:532, 1957.
30. Ya. B. Zel'dovich. Journal of theoretical and experimental physics. *Soviet Physics JETP*, 11:594, 1960.
31. F.A. Berezin and L.D. Faddeev. A remark on Schroedinger's equation with a singular potential. *Dokl. Akad. Nauk USSR*, 137:1011, 1961. *English translation in:* Sov. Math., Dokl. **2**, 372 (1961).
32. R. Jackiw. Delta function potentials in two-dimensional and three-dimensional quantum mechanics. In A. Ali and P. Hoodbhoy, editors, *M. A. B. Bég Memorial Volume*, Singapore, 1991. World Scientific.
33. R.L. Jaffe and L.R. Williamson. The Casimir energy in a separable potential. *Ann. Phys.*, 282:432, 2000.
34. J. Dereziński and R. Früboes. Renormalization of the Friedrichs Hamiltonian. *Rep. Math. Phys.*, 50:433, 2002.
35. D.R. Yafaev. *General Theory*, volume 105 of *Amer. Math. Soc. Trans.* Amer. Math. Soc., Providence RI, 1992.
36. S. Albeverio and P. Kurasov. *Singular Perturbations of Differential Operators*, volume 271 of *Lond. Math. Soc. Lect. Note Ser.* Cambridge University Press, Cambridge, 2000.
37. K.G. Wilson. Model Hamiltonians for local quantum field theory. *Phys. Rev.*, 140:B445, 1965.
38. K.G. Wilson. Model of coupling-constant renormalization. *Phys. Rev.*, D2:1438, 1970.
39. D.J. Gross and F. Wilczek. Ultraviolet behavior of non-abelian gauge theories. *Phys. Rev. Lett.*, 30:1343, 1973.
40. H.D. Politzer. Reliable perturbative results for strong interactions? *Phys. Rev. Lett.*, 30:1346, 1973.

Cantor Spectrum for Quasi-Periodic Schrödinger Operators

Joaquim Puig

Departament de Matemàtica Aplicada I. Universitat Politècnica de Catalunya
Av. Diagonal 647, 08028 Barcelona, Spain.

Abstract. We present some results concerning the Cantor structure of the spectrum of quasi-periodic Schrödinger operators. These are obtained studying the dynamics of the corresponding eigenvalue equations, specially the notion of reducibility and Floquet theory. We will deal with the Almost Mathieu case, and the solution of the "Ten Martini Problem" for Diophantine frequencies, as well as other models.[1]

In recent years there has been substantial progress in the understanding of the structure of the spectrum of Schrödinger operators with quasi-periodic potential. Here we will concentrate on one-dimensional, real analytic quasi-periodic potentials with one or more Diophantine frequencies.

We will begin with the best studied of such operators: the Almost Mathieu operator and the solution of the "Ten Martini Problem" which asks for the Cantor structure of its spectrum. Secondly we will see how this Cantor structure is generic in the set of quasi-periodic real analytic potentials. We will end introducing a different approach to this problem which is helpful to study the phenomenon of "gap opening".

1 The Almost Mathieu Operator & the Ten Martini Problem

The *Almost Mathieu operator* is probably the best studied model among quasi-periodic Schrödinger operators. It is the following second-order difference operator:

$$(H^{AM}_{b,\omega,\phi}x)_n = x_{n+1} + x_{n-1} + b\cos(2\pi\omega n + \phi)x_n, \qquad n \in \mathbb{Z}, \qquad (1)$$

where b is a real parameter (a *coupling* parameter, since for $b = 0$ the operator is trivial), ω is the *frequency*, which we assume to be an irrational number (in most of what follows, also Diophantine) and $\phi \in \mathbb{T} = \mathbb{R}/(2\pi\mathbb{Z})$ will be called the *phase*.

---

[1] This work was done while the author was at the *Departament de Matemàtica Aplicada i Anàlisi* of the *Universitat de Barcelona*. It has been supported by grants DGICYT BFM2003-09504-C02-01 (Spain) and CIRIT 2001 SGR-70 (Catalonia).

J. Puig: *Cantor Spectrum for Quasi-Periodic Schrödinger Operators*, Lect. Notes Phys. **690**, 79–91 (2006)
www.springerlink.com   © Springer-Verlag Berlin Heidelberg 2006

Considered as an operator on $l^2(\mathbb{Z})$, the Almost Mathieu operator is bounded and self-adjoint. The reason for its name comes from the fact that its eigenvalue equation, namely

$$x_{n+1} + x_{n-1} + b\cos(2\pi\omega n + \phi)x_n = ax_n, \qquad n \in \mathbb{Z},$$

(sometimes called *Harper's equation*), is a discretization of the classical *Mathieu equation*,

$$x'' + (a + b\cos(t))\,x = 0\,.$$

which is a second-order periodic differential equation (see Ince [9]). The analogies between the Harper equation and Mathieu equation are quite striking and their comparison illustrates the differences between periodic and quasi-periodic Schrödinger operators.

1.1 The IDS and the Spectrum

The Integrated Density of States (IDS) is a very convenient object for the description of the spectrum of quasi-periodic Schrödinger operators which can be extended to more general operators. Here we introduce it in the case of the Almost Mathieu operator for the sake of concreteness.

Fix some $b \in \mathbb{R}$ and $\phi \in \mathbb{T}$. For any $L \in \mathbb{N}$ we define $H_{b,\omega,\phi}^{AM,L}$ as the restriction of the Almost Mathieu operator to the interval $\{1,\ldots,L-1\}$ with zero boundary conditions at 0 and L. Let

$$k_{b,\omega,\phi}^{AM,L}(a) = \frac{1}{(L-2)}\#\left\{\text{eigenvalues of } H_{b,\omega,\phi}^{AM,L} \leq a\right\}\,.$$

Then Avron & Simon [2] prove that

$$\lim_{L\to\infty} k_{b,\omega,\phi}^{AM,L}(a) = k_{b,\omega}^{AM}(a)\,,$$

which is called *integrated density of states*, IDS for short. This limit is independent of the value of ϕ and it is a continuously increasing function of a.

The IDS can be used to describe the spectrum of quasi-periodic Schrödinger operators in a very nice way. Indeed, the spectrum of $H_{b,\omega,\phi}^{AM}$ is precisely the set of points of increase of the map

$$a \mapsto k_{b,\omega}^{AM}(a)$$

so that the intervals of constancy belong to the resolvent set of the operator (and are called the *spectral gaps*). In particular, this characterization shows that the spectrum of the Almost Mathieu operator (and in general any quasi-periodic Schrödinger operator with irrational frequencies) does not depend on ϕ. Therefore, we write

$$\sigma_{b,\omega}^{AM} = \mathrm{Spec}\left(H_{b,\omega,\phi}^{AM}\right)\,.$$

Remark 1. The IDS for one-dimensional quasi-periodic Schrödinger operators has many other characterizations. It can be linked to the *rotation number* of the corresponding eigenvalue equation. This object was introduced by Johnson & Moser [12] in the continuous case (see Delyon & Souillard [7] for the adaption to the discrete case and Johnson [11] for a review on these different characterizations).

The IDS can also be used to "label" the spectral gaps of the Almost Mathieu operator. This is the contents of the *Gap Labelling Theorem*, by Johnson & Moser [12]: if I is a spectral gap (an interval of constancy of the IDS) then there is an integer $n \in \mathbb{Z}$ such that

$$k_{b,\omega}^{AM}(a) = n\omega, \qquad (\text{modulus } \mathbb{Z})$$

for all $a \in I$. Figure 1 displays the gap labelling for the Almost Mathieu operator at "critical coupling" $b = 2$.

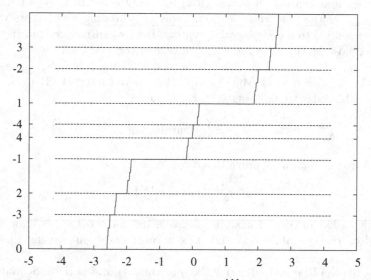

Fig. 1. Schematic view of Gap Labelling for $H_{2,\omega,\phi}^{AM}$ and $b = 2$. The IDS is plot as a function of a. Integers in the vertical direction correspond to values n such that the IDS equals $n\omega$ modulus $\mathbb{Z}$

The Gap Labelling Theorem motivates the following definitions. For any $n \in \mathbb{Z}$ let

$$I(n) = \left\{ a \in \mathbb{R}; k_{b,\omega}^{AM}(a) - n\omega \in \mathbb{Z} \right\} .$$

If $I(n) = [a_-^n, a_+^n]$ for some $a_-^n < a_+^n$ then we will say that (a_-^n, a_+^n) is a *noncollapsed* or *open* spectral gap. If $a_-^n = a_+^n$ then we will call $\{a_-^n\}$ a *collapsed* or *closed* spectral gap. Note that noncollapsed spectral gaps are subsets of the resolvent whereas collapsed spectral gaps belong to the spectrum. In both cases, the endpoints of gaps belong to the spectrum.

In the quasi-periodic case, when ω is an irrational frequency, the possible values of the IDS at gaps define the set of labels

$$\mathcal{M}(\omega) = \{m + n\omega, \quad n, m \in \mathbb{Z}\} \cap [0, 1] \,,$$

which is dense in $[0, 1]$. Since the IDS is a continuously increasing function of a, the spectrum of the Almost Mathieu operator is a Cantor set if all spectral gaps are open. For a general quasi-periodic Schrödinger operator, gaps can be collapsed and, in fact, the spectrum may contain intervals. Figure 2 displays a numerical computation of some of the gaps of the Almost Mathieu operator. None of them appears to be collapsed.

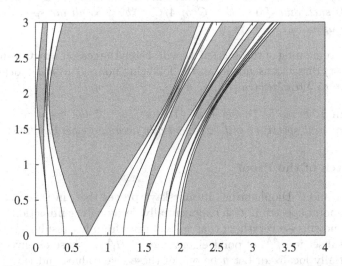

Fig. 2. Numerical computation of the ten biggest spectral gaps for the Almost Mathieu operator. Coupling parameter b is in the vertical direction, whereas the spectral one a is in the horizontal one. Shaded regions correspond to gaps

Due to all this, and to some physical arguments, Simon [18], after an offer by Kac, posed the following problems on the Cantor structure for the spectrum of the Almost Mathieu operator. The first one is the "Ten Martini Problem": for ω irrational and $b \neq 0$ prove that the spectrum of the Almost Mathieu operator is a Cantor set. The second one, which implies the first, is the "Strong (or Dry) Ten Martini Problem" and, under the same hypothesis, asks if all gaps, as predicted by the Gap Labelling Theorem, are open.

Concerning the Ten Martini Problem, we can prove the following [14].

Corollary 1. *Assume that $\omega \in \mathbb{R}$ is Diophantine, that is, there exist positive constants c and $r > 1$ such that*

$$|\sin 2\pi n\omega| > \frac{c}{|n|^r}$$

for all $n \neq 0$. Then, the spectrum of the Almost Mathieu operator, $\sigma_{b,\omega}^{AM}$, is a Cantor set if $b \neq 0, \pm 2$.

Remark 2. Very recently Avila & Jitomirskaya have proved Cantor structure for *all* irrational frequencies. The set of Diophantine frequencies is a total measure subset of the real numbers.

Concerning the dry version of the Ten Martini Problem, using a reducibility theorem by Eliasson [8], one can also say something in the perturbative regime [14]

Corollary 2. *Let $\omega \in \mathbb{R}$ be Diophantine. Then, there is a constant $C = C(\omega) > 0$ such that if $0 < |b| < C$ or $4/C < |b| < \infty$ all the spectral gaps of $\sigma_{b,\omega}^{AM}$ are open.*

In the remaining of this section we will sketch the reason why Corollary 1 is an (almost) direct consequence of the following nonperturbative localization result due to Jitomirskaya [10].

Theorem 1. *Let ω be Diophantine. Then, if $|b| > 2$ the operator $H_{b,\omega,0}^{AM}$ has only pure point spectrum with exponentially decaying eigenfunctions.*

1.2 Sketch of the Proof

Let $b > 2$ and ω Diophantine. Jitomirskaya proves that, in this case, $H_{b,\omega,0}^{AM}$ has pure-point spectrum with exponentially decaying eigenfunctions. In particular (and this is everything that we will need from her result), there exists a dense subset in $\sigma_{b,\omega}^{AM}$ of point eigenvalues of $H_{b,\omega,0}^{AM}$ whose eigenvectors are exponentially localized. Let a be one of these eigenvalues and $\psi = (\psi_n)_{n \in \mathbb{Z}}$ its exponentially localized eigenvector. We are going to see that a is the endpoint of a noncollapsed spectral gap. From this the Cantor structure of the spectrum follows immediately.

By hypothesis $a \in \sigma_{b,\omega}^{AM}$ and $\psi \in l^2(\mathbb{Z})$ satisfy the Harper equation

$$\psi_{n+1} + \psi_{n-1} + b\cos(2\pi\omega n)\psi_n = a\psi_n, \qquad n \in \mathbb{Z},$$

with some constants $A, \beta > 0$ such that

$$|\psi_n| \leq A\exp(-\beta|n|), \qquad n \in \mathbb{Z}.$$

The very special form of the Almost Mathieu operator makes that the Fourier transform of ψ,

$$\tilde{\psi}(\theta) = \sum_{n \in \mathbb{Z}} \psi_n e^{in\theta}, \qquad \theta \in \mathbb{T},$$

which is real analytic in $|\text{Im}\theta| < \beta$, defines the following *quasi-periodic Bloch wave*

$$x_n = \tilde{\psi}\left(2\pi\omega n + \theta\right), \qquad n \in \mathbb{Z},$$

and that this, satisfies the equation

$$(x_{n+1} + x_{n-1}) + \frac{4}{b}\cos(2\pi\omega n + \theta)x_n = \frac{2a}{b}x_n, \qquad n \in \mathbb{Z}. \tag{2}$$

for any $\theta \in \mathbb{T}$. Note that this is again a Harper equation whose parameters have changed

$$(a, b) \mapsto \left(\frac{2a}{b}, \frac{4}{b}\right).$$

This invariance of the Almost Mathieu operator under Fourier transform is known as Aubry duality [1]. Although the argument above requires the existence of a point eigenvalue, Avron & Simon [2] proved the following form of Aubry duality in terms of the IDS

$$k_{b,\omega}^{AM}(a) = k_{4/b,\omega}^{AM}\left(\frac{2a}{b}\right).$$

In particular, if we prove that $2a/b$ is the endpoint of a noncollapsed gap of $\sigma_{4/b,\omega}^{AM}$ we are done.

1.3 Reducibility of Quasi-Periodic Cocycles

Our main tool will be to use the dynamics of the eigenvalue equation (2) to prove that a is an endpoint of a non-collapsed gap. To do so, it is convenient to write down (2) as a first order system

$$\begin{pmatrix} x_{n+1} \\ x_n \end{pmatrix} = \begin{pmatrix} \frac{2a^k}{b} - \frac{4}{b}\cos\theta_n & -1 \\ 1 & 0 \end{pmatrix}\begin{pmatrix} x_n \\ x_{n-1} \end{pmatrix}, \qquad \theta_{n+1} = \theta_n + 2\pi\omega, \tag{3}$$

with $\theta_n \in \mathbb{T}$. Such first-order systems are usually called *quasi-periodic skew-products*. The evolution of the vector $v_n = (x_{n+1}, x_n)^T$ and the angle θ_n can be seen as the iteration of a *quasi-periodic cocycle* on $SL(2, \mathbb{R}) \times \mathbb{T}$

$$(v, \theta) \in \mathbb{R}^2 \times \mathbb{T} \mapsto \left(A_{2a/b,4/b,\omega}^{AM}(\theta), \omega\right)(v, \theta) = \left(A_{2a/b,4/b,\omega}^{AM}(\theta)v, \theta + 2\pi\omega\right),$$

setting

$$A_{2a/b,4/b,\omega}^{AM}(\theta) = \begin{pmatrix} \frac{2a}{b} - \frac{4}{b}\cos\theta & -1 \\ 1 & 0 \end{pmatrix}.$$

That is,

$$v_{n+1} = \begin{pmatrix} \frac{2a}{b} - \frac{4}{b}\cos(2\pi\omega n + \phi) & -1 \\ 1 & 0 \end{pmatrix}\cdots\begin{pmatrix} \frac{2a}{b} - \frac{4}{b}\cos(\phi) & -1 \\ 1 & 0 \end{pmatrix}\cdot v_0$$

and

$$\theta_n = 2\pi\omega n + \theta_0 \, .$$

When the frequency ω is rational the skew-product is periodic and, thanks to Floquet theory, it can be reduced to a skew-product with constant matrix by means of a periodic transformation. Quasi-periodic reducibility tries to extend this theory to the quasi-periodic case. Let us now introduce some basic notions.

Two cocycles (A, ω) and (B, ω) of $SL(2, \mathbb{R}) \times \mathbb{T}$ (not necessarily associated to the Harper equation) are *conjugated* if there exists a continuous $Z : \mathbb{T} \to SL(2, \mathbb{R})$ such that

$$A(\theta)Z(\theta) = Z(\theta + 2\pi\omega)B(\theta), \qquad \theta \in \mathbb{T} \, .$$

In this case the corresponding quasi-periodic skew-products

$$u_{n+1} = A(\theta)u_n, \qquad \theta_{n+1} = \theta_n + 2\pi\omega$$

and

$$v_{n+1} = B(\theta)v_n, \qquad \theta_{n+1} = \theta_n + 2\pi\omega$$

are conjugated through the change $u = Zv$.

Particularly important to our purposes is the case of cocycles which are conjugated to a constant cocycles. A cocycle (A, ω) is *reducible to constant coefficients* if it is conjugated to a cocycle (B, ω) with B not depending on θ.

Remark 3. B is called the *Floquet matrix*. Neither B nor Z are unique.

The fundamental solution of a reducible system $X_n(\phi)$ has the following *Floquet representation*:

$$X_n(\phi) = Z(2\pi n\omega + \phi)B^n Z(\phi)^{-1}X_0(\phi) \, . \tag{4}$$

In particular, and this is an important observation, if $B = I$ then all solutions of the corresponding skew-product are quasi-periodic with frequency ω. If the cocycle comes from a Harper's equation, then all the solutions of this equation are quasi-periodic Bloch waves.

Now let us go back to our dual Harper's equation. In terms of $\tilde\psi$ we have that the relation

$$\begin{pmatrix} \tilde\psi(4\pi\omega + \theta) \\ \tilde\psi(2\pi\omega + \theta) \end{pmatrix} = \begin{pmatrix} \frac{2a}{b} - \frac{4}{b}\cos\theta & -1 \\ 1 & 0 \end{pmatrix} \begin{pmatrix} \tilde\psi(2\pi\omega + \theta) \\ \tilde\psi(\theta) \end{pmatrix}$$

holds for all $\theta \in \mathbb{T}$. The following Lemma shows that, in this situation, the Almost Mathieu cocycle is reducible to constant coefficients.

Lemma 1. *Let $A : \mathbb{T} \to SL(2, \mathbb{R})$ be a real analytic map and ω be Diophantine. Assume that there is a nonzero real analytic map $v : \mathbb{T} \to \mathbb{R}^2$, such that the relation*

$$v(\theta + 2\pi\omega) = A(\theta)v(\theta)$$

holds for all $\theta \in \mathbb{T}$. Then, the quasi-periodic cocycle (A, ω) is reducible to constant coefficients by means of a quasi-periodic transformation which is real analytic. Moreover the Floquet matrix can be chosen to be of the form

$$B = \begin{pmatrix} 1 & c \\ 0 & 1 \end{pmatrix} \tag{5}$$

for some $c \in \mathbb{R}$.

1.4 End of Proof

Classical Floquet theory for periodic Hill's equation relates endpoints of gaps to the corresponding Floquet matrices (which always exist because the system is periodic). It turns out that, if an Almost Mathieu cocycle is reducible to constant coefficients such characterization also holds. In fact, one can prove that if an Almost Mathieu cocycle (or any other quasi-periodic Schrödinger cocycle), for some a, b, ω fixed, is reducible to constant coefficients with Floquet matrix B then a is at the endpoint of a spectral gap of the operator if, and only if, trace $B = \pm 2$. Moreover the gap is collapsed if and, only if, $B = \pm I$.

Therefore, if

$$B = \begin{pmatrix} 1 & c \\ 0 & 1 \end{pmatrix}$$

then the gap is collapsed if, and only if, $c = 0$. Summing up, we have that $2a/b$ is a noncollapsed spectral gap of $\sigma^{AM}_{4/b,\omega}$ if, and only if, the Floquet matrix of the corresponding cocycle, is the identity.

Now we can use an adaption of Ince's argument [9] for the classical Mathieu equation to our case. If B was the identity then, as we learned from Floquet representation (4), there would be two linearly independent real analytic quasi-periodic Bloch waves of Harper's equation

$$x_{n+1} + x_{n-1} + \frac{4}{b}\cos(2\pi\omega n + \phi)x_n = \frac{2a}{b}x_n, \qquad n \in \mathbb{Z}.$$

Passing to the dual, this would tell us that

$$x_{n+1} + x_{n-1} + b\cos(2\pi\omega n)x_n = ax_n, \qquad n \in \mathbb{Z}.$$

has two linearly independent solutions in $l^2(\mathbb{Z})$. This is a contradiction with the limit-point character of the Almost Mathieu operator (or the preservation of the Wronskian for the difference equation).

Therefore $B \neq I$ ($c \neq 0$) so that $2a/b$ is the endpoint of a noncollapsed gap of $\sigma^{AM}_{4/b,\omega}$. Since such endpoints are dense in the spectrum, this must be a Cantor set for all $b \neq 0, \pm 2$ and Diophantine frequencies.

2 Extension to Real Analytic Potentials

In the proof of the Ten Martini Problem that we have presented above, there are some features which are specific of the Almost Mathieu operator. Some other, however, can be extended to more general potentials. Let us try to reproduce the proof for a real analytic potential $V : \mathbb{T} \to \mathbb{R}$ instead of $b \cos \theta$. The corresponding Schrödinger operators are of the form

$$(H_{V,\omega,\phi}x)_n = x_{n+1} + x_{n-1} + V(2\pi\omega n + \phi)x_n \ .$$

The dual model of this operator is the following *long-range operator*,

$$(L_{V,\omega,\phi}x)_n = \sum_{k\in\mathbb{Z}} V_k x_{n+k} + 2\cos(2\pi\omega n + \phi)x_n$$

so that analytic quasi-periodic Bloch waves of $H_{V,\omega,\phi}$ correspond to exponentially localized eigenvectors of $L_{V,\omega,\phi}$. Bourgain & Jitomirskaya [3] proved that, for some $\varepsilon > 0$, $L_{V,\omega,\phi}$ has pure-point spectrum with exponentially localized eigenfunctions for almost all $\phi \in \mathbb{T}$ if

$$|V|_\rho := \sup_{|\mathrm{Im}\theta|<\rho} |V(\theta)| < \varepsilon$$

and ω is Diophantine.

Using this result and some facts on the IDS one can show [15] that for Lebesgue almost every $a \in \mathbb{R}$, the cocycle

$$(A_{a-V}, \omega) = \left(\begin{pmatrix} a - V(\theta) & -1 \\ 1 & 0 \end{pmatrix}, \omega \right)$$

is reducible to constant coefficients if $|V|_\rho < \varepsilon$ and ω is Diophantine. Also, there exists a dense set of values of a in the spectrum such that the corresponding cocycle is reducible to

$$B = \begin{pmatrix} 1 & c \\ 0 & 1 \end{pmatrix} \ .$$

Therefore, these values of a are at endpoints of spectral gaps of $H_{V,\omega,\phi}$. However, we cannot use Ince's argument and it may happen that some of these are collapsed (see Fig. 3). In fact, there are examples of quasi-periodic Schrödinger operators (with V small, ω Diophantine) which do not display Cantor spectrum (see De Concini & Johnson [6]).

Nevertheless, even if c can be zero, Moser & Pöschel [13] showed that, in this reducible setting, a closed gap can be opened by means of an arbitrarily small and generic perturbation of the potential, as it is shown in [15] (the proof by Moser & Pöschel is in the continuous case, although it extends without trouble to the discrete). Here generic is meant in the G_δ-sense, considering the space of real analytic perturbations in some fixed complex strip

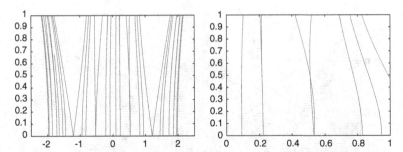

Fig. 3. *Left:* Endpoints of some spectral gaps of $H_{bV,\omega,\phi}$, with $V(\theta) = \cos(\theta) + 0.3\cos(2\theta)$ and several values of b (in the vertical direction) and $\omega = (\sqrt{5} - 1)/2$. Note the spectral gap which is collapsed. *Right:* Magnification of the figure around the collapsed gap

furnished with the supremum norm. Since there is, at most, a countable number of collapsed spectral gaps, we can conclude that, a generic potential with $|V|_\rho < \varepsilon$, for some ρ fixed, has Cantor spectrum for Diophantine frequencies (see again [15]). This generalizes nonperturbatively results obtained by Eliasson [8] on the genericity of Cantor spectrum for quasi-periodic Schrödinger operators.

3 Cantor Spectrum for Specific Models

The results in the previous section on the genericity of Cantor spectrum for quasi-periodic Schrödinger operators have the disadvantage that they cannot be applied to specific examples of Schrödinger operators. In this section we will briefly describe how to get Cantor spectrum, and opening of all gaps, for some prescribed families of quasi-periodic Schrödinger operators. This is joint work with Broer & Simó [4, 16].

Here we will consider continuous Schrödinger operators, and for the sake of definiteness, the following *quasi-periodic Mathieu operator*

$$H_{b,\omega,\phi}^{QPM} x = -x'' + b \sum_{j=1}^{d} \cos(\omega_j t)x \,,$$

where now $x \in L^2(\mathbb{R})$. Let us consider the self-adjoint extension of $H_{b,\omega,\phi}^{QPM}$ to $L^2(\mathbb{R})$ whose spectrum, again, does not depend on ϕ (see Fig. 4). For such operators we can prove the following.

Theorem 2. *Let $d \geq 2$. Then for almost all $\omega = (\omega_1, \dots, \omega_d) \in \mathbb{R}^d$ there is a $C = C(\omega)$ such that for all values of $0 < |b| < C$, except for a countable set, the spectrum of the quasi-periodic Mathieu operator $H_{b,\omega,\phi}^{QPM}$ has all gaps open and, thus, it is a Cantor set.*

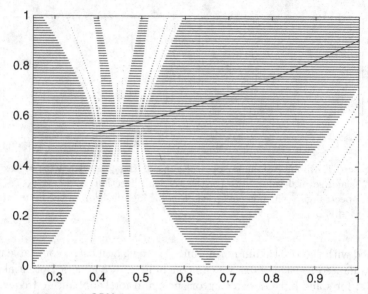

Fig. 4. Spectrum of $H^{QPM}_{b,\omega,\phi}$ for several values of b vertical direction. Shaded regions correspond to gaps. Here $\omega = (1,\gamma)^T$ where $\gamma = (1+\sqrt{5})/2$. Source Broer & Simó [5]

The idea for the proof is based on the study of gap boundaries as functions of the coupling constant b. The set formed by the closure of a certain gap (with fixed label) in the (a,b)-plane will be called a *resonance tongue*. When the boundaries of a certain resonance tongue merge for two different values of b we will speak of an *instability pocket* (see Fig. 5).

In the periodic case it is known (see, e.g. Rellich [17]) that tongue boundaries are real analytic functions. In the quasi-periodic case the same methods cannot be applied, but using KAM techniques it can be seen that tongue boundaries are real analytic if $|b|$ is smaller than a certain constant C which depends on the Diophantine class of ω [16].

Using Birkhoff Normal Form, we show that all these tongue boundaries (which we know are real analytic) have some finite order of contact at $b = 0$ [4]. In particular, each gap can collapse at most a finite number of times. Since the number of gaps is countable we only have to take out a countable subset of $|b| < C$.

Remark 4. This is a perturbative result (the smallness condition on the potential depends on the precise Diophantine conditions on the frequency vector) but it holds irrespectively of the dimension d (contrary to the methods in the first two sections).

Remark 5. The same result holds for any quasi-periodic potential whose Fourier coefficients are all nonzero.

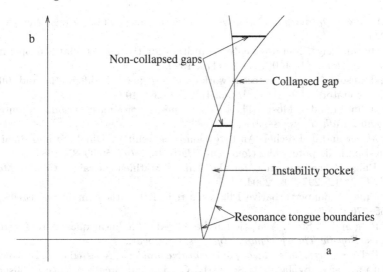

b

Non-collapsed gaps

Collapsed gap

Instability pocket

Resonance tongue boundaries

a

Fig. 5. Resonance tongue with pocket in the (a, b)-plane giving rise to spectral gaps on each horizontal line with constant b. Note how collapse of gaps corresponds to crossings of tongue boundaries at tips of an instability pocket

Remark 6. In [4] it is shown that by means of suitable and arbitrarily small perturbations of the potential of $H^{QPM}_{b,\omega,\phi}$ it is possible to produce pockets at any gap of the operator.

References

1. S. Aubry and G. André. Analyticity breaking and Anderson localization in incommensurate lattices. In *Group theoretical methods in physics (Proc. Eighth Internat. Colloq., Kiryat Anavim, 1979)*, pp. 133–164. Hilger, Bristol, 1980.
2. J. Avron and B. Simon. Almost periodic Schrödinger operators II. The integrated density of states. *Duke Math. J.*, 50:369–391, 1983.
3. J. Bourgain and S. Jitomirskaya. Absolutely continuous spectrum for 1D quasiperiodic operators. *Invent. Math.*, 148(3):453–463, 2002.
4. H.W. Broer, J. Puig, and C. Simó. Resonance tongues and instability pockets in the quasi-periodic Hill-Schrödinger equation. *Comm. Math. Phys*, 241(2–3):467–503, 2003.
5. H.W. Broer and C. Simó. Hill's equation with quasi-periodic forcing: resonance tongues, instability pockets and global phenomena. *Bol. Soc. Brasil. Mat. (N.S.)*, 29(2):253–293, 1998.
6. C. De Concini and R.A. Johnson. The algebraic-geometric AKNS potentials. *Ergodic Theory Dynam. Systems*, 7(1):1–24, 1987.
7. F. Delyon and B. Souillard. The rotation number for finite difference operators and its properties. *Comm. Math. Phys.*, 89(3):415–426, 1983.
8. L.H. Eliasson. Floquet solutions for the one-dimensional quasi-periodic Schrödinger equation. *Comm. Math. Phys.*, 146:447–482, 1992.

9. E.L. Ince. *Ordinary Differential Equations*. Dover Publications, New York, 1944.

10. S. Jitomirskaya. Metal-insulator transition for the almost Mathieu operator. *Ann. of Math. (2)*, 150(3):1159–1175, 1999.

11. R. Johnson. A review of recent work on almost periodic differential and difference operators. *Acta Appl. Math.*, 1(3):241–261, 1983.

12. R. Johnson and J. Moser. The rotation number for almost periodic potentials. *Comm. Math. Phys.*, 84:403–438, 1982.

13. J. Moser and J. Pöschel. An extension of a result by Dinaburg and Sinai on quasi-periodic potentials. *Comment. Math. Helvetici*, 59:39–85, 1984.

14. J. Puig. Cantor spectrum for the Almost Mathieu operator. *Comm. Math. Phys.*, 244(2):297–309, 2004.

15. J. Puig. A nonperturbative Eliasson's reducibility theorem. *Nonlinearity*, 19 355–376 (2006).

16. J. Puig and C. Simó. Analytic families of reducible linear quasi-periodic equations. *Ergodic Theory Dynam. Systems*, to appear.

17. F. Rellich. *Perturbation theory of eigenvalue problems*. Assisted by J. Berkowitz. With a preface by Jacob T. Schwartz. Gordon and Breach Science Publishers, New York, 1969.

18. B. Simon. Almost periodic Schrödinger operators: a review. *Adv. in Appl. Math.*, 3(4):463–490, 1982.

Part II

Quantum Field Theory and Statistical
Mechanics

he next three papers consists in contributions devoted to Quantum Field Theoretic issues or models:

Derezinski and Meissner reconsider the constructions of Quantum bosonic massless fields in 1+1 dimensions, a delicate matter due to the infrared problem. The authors present two constructions of massless fields which avoid the known defects of other constructions, namely, a constraint on smearing functions or an indefinite metric. In the first construction, they obtain a separable Hilbert space but no vacuum vector, and in the second one, a vacuum exists, however the Hilbert space in non-separable.

The contribution of Rivasseau reviews recent progresses obtained by means of field theoretic methods on the understanding of interacting Fermions in two dimensions. It provides an introduction to Fermi liquids, their description by means of Salmhofer's criterion and the models accessible to mathematical analysis in a non-perturbative framework.

The translation invariant Nelson model describing an electron coupled to a massive scalar radiation field is the object of active current studies. Schach-Moller's paper provides a detailed description of the analyticity properties of the bottom of the essential spectrum as a function of the conserved total momentum.

The paper of Abou-Salem and Fröhlich and that of Merkli consider aspects of Quantum Statistical Mechanics in a time dependent setting. Merkli addresses the dynamical notion of return to equilibrium for a small system coupled to a heat bath modelled by a Boson gas. When this gas has a Bose-Einstein condensate and displays coexisting phases, this notion has to be revisited. Abou-Salem and Fröhlich also consider return to equilibrium for small systems in contact with a heat bath when external slowly varying driving forces are present. They define notions of heat flow and work rate which they study in this setup by establishing variants of the adiabatic theorem of Quantum Mechanics.

In their contribution, Nachtergaele and Starr give a pedagogical introduction to recently proven spectral properties of the celebrated Heisenberg models that goes under the name ferromagnetic ordering of energy levels. They also spell out some links and consequences of their results for the analysis of the symmetric simple exclusion process of particles on a graph. It provides in particular a proof of Aldous' Conjecture, in one dimension.

Adiabatic Theorems
and Reversible Isothermal Processes*

Walid K. Abou-Salem and Jürg Fröhlich

Institute for Theoretical Physics, ETH-Hönggerberg
CH-8093 Zürich, Switzerland

Abstract. Isothermal processes of a finitely extended, driven quantum system in contact with an infinite heat bath are studied from the point of view of quantum statistical mechanics. Notions like heat flux, work and entropy are defined for trajectories of states close to, but distinct from states of joint thermal equilibrium. A theorem characterizing *reversible* isothermal processes as *quasi-static* processes (*"isothermal theorem"*) is described. Corollaries concerning the changes of entropy and free energy in reversible isothermal processes and on the 0th law of thermodynamics are outlined.

1 Introduction

The problem to derive the fundamental laws of thermodynamics, the 0th, 1st, 2nd (and 3rd) law, from kinetic theory and non-equilibrium statistical mechanics has been studied since the late 19th century, with contributions by many distinguished theoretical physicists including Maxwell, Boltzmann, Gibbs and Einstein. In this note, we report on some recent results concerning *isothermal processes* that have grown out of our own modest attempts to bring the problem just described a little closer to a satisfactory solution; (see [1] for a synopsis of our results).

In the study of the 0th law and of Carnot processes *isothermal processes* play an important role. Such processes arise when a system with finitely many degrees of freedom, the *"small system"*, is coupled diathermally to an infinitely extended, dispersive system, the *"heat bath"*, e.g., one consisting of black-body radiation, or of a gas of electrons (metals), or of magnons (magnetic materials), at positive temperature. (Diathermal contacts are couplings that preserve all extensive quantities except for the *"internal energy"* of the *"small system"*. The latter will be precisely defined later.) During the past ten years, a particular phenomenon, *"return to equilibrium"*, encountered in the study of isothermal processes of quantum-mechanical systems, has been analyzed for simple (non-interacting) models of heat baths, by several groups of people [2–6]: After coupling the small system to the heat bath, the state of the coupled system approaches an equilibrium state at the temperature

---

*To appear in *Letters in Mathematical Physics*. The authors are thankful for the permission to re-publish it.

W.K. Abou-Salem and J. Fröhlich: *Adiabatic Theorems and Reversible Isothermal Processes*, Lect. Notes Phys. **690**, 95–105 (2006)
www.springerlink.com   © Springer-Verlag Berlin Heidelberg 2006

of the heat bath, as time t tends to ∞. Further, if all contacts of the heat bath to its environment are broken the state of the heat bath returns to its original equilibrium state [7,8]. (The state of an isolated infinite heat bath is thus characterized by a single quantity: its temperature!)

The results in [2–6] are proven using spectral- and resonance theory, starting from the formalism developed in [9]. If $\mathcal{H}^{\mathcal{R}}$ denotes the Hilbert space of state vectors of the infinite heat bath, $\mathcal{R}$, at a *fixed* temperature $(k\beta)^{-1}$, where k denotes Boltzmann's constant, ($\mathcal{H}^{\mathcal{R}}$ is obtained with the help of the GNS construction, see [9,10]), and $\mathcal{H}^{\Sigma}$ denotes the Hilbert space of *pure* state vectors of the "small system" Σ, the Hilbert space of general (in particular mixed) states of the *composed* system, $\mathcal{R} \vee \Sigma$, is given by

$$\mathcal{H} \equiv \mathcal{H}^{\mathcal{R} \vee \Sigma} := \mathcal{H}^{\mathcal{R}} \otimes (\mathcal{H}^{\Sigma} \otimes \mathcal{H}^{\Sigma}) ; \tag{1}$$

see [4, 9]. The equilibrium state of the heat bath, or reservoir, $\mathcal{R}$ at inverse temperature $k\beta$ corresponds to a vector $\Omega^{\mathcal{R}}_{\beta} \in \mathcal{H}^{\mathcal{R}}$, and a general mixed state of the small system Σ can be described as the square root of a density matrix, which is a Hilbert-Schmidt operator on $\mathcal{H}^{\Sigma}$, or, in other terms, as a vector in $\mathcal{H}^{\Sigma} \otimes \mathcal{H}^{\Sigma}$.

The dynamics of the composed system is generated by an (in general, *time-dependent*) thermal Hamiltonian, or *Liouvillian*,

$$L(t) := L_0(t) + g(t)I , \tag{2}$$

(the "standard Liouvillian", see, e.g., [6,9]), where

$$L_0(t) = L^{\mathcal{R}}_{\beta} \otimes (\mathbf{1} \otimes \mathbf{1}) + \mathbf{1} \otimes (H^{\Sigma}_0(t) \otimes \mathbf{1}) - \mathbf{1} \otimes (\mathbf{1} \otimes H^{\Sigma}_0(t)) \tag{3}$$

is the Liouvillian of the uncoupled system, $L^{\mathcal{R}}_{\beta}$ is the Liouvillian of the heat bath, with $L^{\mathcal{R}}_{\beta} \Omega^{\mathcal{R}}_{\beta} = 0$, $H^{\Sigma}_0(t)$ is the (generally time-dependent) Hamiltonian of the small system, and where $g(t)I$ is a spatially localized term describing the interactions between $\mathcal{R}$ and Σ, with a time-dependent coupling constant $g(t)$. Concrete models are analyzed in [2–7].

We will only consider heat baths with a unique equilibrium state at each temperature; (no phase coexistence). If $L_0(t) \equiv L_0$ and $g(t) \equiv g$ are independent of t, for $t \geq t_*$, "return to equilibrium" holds true if we can prove that L has a simple eigenvalue at 0 and that the spectrum, $\sigma(L)$, of L is purely continuous away from 0; see [2–4]. The eigenvector, $\Omega_{\beta} \equiv \Omega^{\mathcal{R} \vee \Sigma}_{\beta}$, of L corresponding to the eigenvalue 0 is the thermal equilibrium state of the coupled system $\mathcal{R} \vee \Sigma$ at temperature $(k\beta)^{-1}$. Since L_0 tends to have a rich spectrum of eigenvalues (embedded in continuous spectrum), it is an a priori surprising consequence of interactions between $\mathcal{R}$ and Σ that the point spectrum of L consists of *only one simple eigenvalue* at 0; (see [2–6,11] for hypotheses on the interaction I and results). Under suitable hypotheses on $\mathcal{R}$ and Σ, see [2–4], one can actually prove that "return to equilibrium" is described by an *exponential law* involving a finite *relaxation time*, $\tau_{\mathcal{R}}$.

If the Liouvillian $L(t)$ of $\mathcal{R} \vee \Sigma$ depends on time t, but with the property that, for all times t, $L(t)$ has a simple eigenvalue at 0 corresponding to an eigenvector $\Omega_\beta(t)$, then $\Omega_\beta(t)$ can be viewed as an *instantaneous equilibrium* (or *reference*) *state*, and $\tau_\mathcal{R}(t)$ is called instantaneous relaxation time of $\mathcal{R} \vee \Sigma$. Let τ be the time scale over which $L(t)$ changes appreciably. Assuming that, at some time t_0, the state , $\Psi(t_0)$, of $\mathcal{R} \vee \Sigma$ is given by $\Omega_\beta(t_0)$, it is natural to compare the state $\Psi(t)$ of $\mathcal{R} \vee \Sigma$ at a *later* time t with the instantaneous equilibrium state $\Omega_\beta(t)$ and to estimate the norm of the difference $\Psi(t) - \Omega_\beta(t)$. One would expect that if $\tau \gg \sup_t \tau_\mathcal{R}(t)$, then

$$\Psi(t) \simeq \Omega_\beta(t) \ .$$

In [12], we prove an "adiabatic theorem", which we call "*isothermal theorem*", saying that

$$\Psi(t) \overset{\tau \to \infty}{\to} \Omega_\beta(t) \ , \tag{4}$$

for all times $t \geq t_0$. The purpose of our note is to carefully state this theorem and various generalizations thereof and to explain some of its consequences; e.g., to show that *quasi-static* ($\tau \to \infty$) isothermal processes are *reversible* and that, in the quasi-static limit, a variant of the 0th *law* holds. We also propose general definitions of heat flux and of entropy for trajectories of states of $\mathcal{R} \vee \Sigma$ sampled in arbitrary isothermal processes and use the "isothermal theorem" to relate these definitions to more common ones. Details will appear in [1, 12].

2 A General "Adiabatic Theorem"

In this section we carefully state a general adiabatic theorem which is a slight improvement of results in [13, 14] concerning adiabatic theorems for Hamiltonians without spectral gaps. Our simplest result follows from those in [13, 14] merely by eliminating the superfluous hypothesis of semiboundedness of the generator of time evolution.

Let $\mathcal{H}$ be a separable Hilbert space, and let $\{L(s)\}_{s \in I}$, with $I \subset \mathbf{R}$ a compact interval, be a family of selfadjoint operators on $\mathcal{H}$ with the following properties:

(A1) The operators $L(s), s \in I$, are selfadjoint on a common domain, $\mathcal{D}$, of definition dense in $\mathcal{H}$.

(A2) The resolvent $R(i, s) := (L(s) - i)^{-1}$ is bounded and differentiable, and $L(s)\dot{R}(i, s)$ is bounded uniformly in $s \in I$, where $(\dot{\ })$ denotes the derivative with respect to s.

Existence of time evolution. If assumptions (A1) and (A2) hold then there exist unitary operators $\{U(s, s')|s, s' \in I\}$ with the properties:
For all s, s', s'' in I,

$$U(s, s) = \mathbf{1} \ , U(s, s')U(s', s'') = U(s, s'') \ ,$$

$U(s, s')$ is strongly continuous in s and s', and

$$i\frac{\partial}{\partial s}U(s, s')\Psi = L(s)U(s, s')\Psi , \tag{5}$$

for arbitrary $\Psi \in \mathcal{D}$, s, s' in I; (U is called a "*propagator*").

This result follows from, e.g., Theorems X.47a and X.70 in [15] in a straightforward way; see also Theorem 2 of Chap. XIV in [16]. (Some sufficient conditions for (A1) and (A2) to hold are discussed in [12].)

In order to prove an adiabatic theorem, one must require some additional assumptions on the operators $L(s)$.

(A3) We assume that $L(s)$ has an eigenvalue $\lambda(s)$, that $\{P(s)\}$ is a family of finite rank projections such that $L(s)P(s) = \lambda(s)P(s)$, $P(s)$ is twice continuously differentiable in s with bounded first and second derivatives, for all $s \in I$, and that $P(s)$ is the spectral projection of $L(s)$ corresponding to the eigenvalue $\lambda(s)$ for almost all $s \in I$.

We consider a quantum system whose time evolution is generated by a family of operators

$$L_\tau(t) := L\left(\frac{t}{\tau}\right), \quad \frac{t}{\tau} =: s \in I , \tag{6}$$

where $\{L(s)\}_{s\in I}$ satisfies assumptions (A1)-(A3). The propagator of the system is denoted by $U_\tau(t, t')$. We define

$$U^{(\tau)}(s, s') := U_\tau(\tau s, \tau s') \tag{7}$$

and note that $U^{(\tau)}(s, s')$ solves the equation

$$i\frac{\partial}{\partial s}U^{(\tau)}(s, s')\Psi = \tau L(s)U^{(\tau)}(s, s')\Psi , \Psi \in \mathcal{D} . \tag{8}$$

Next, we define

$$L_a(s) := L(s) + \frac{i}{\tau}[\dot{P}(s), P(s)] \tag{9}$$

and the corresponding propagator, $U_a^{(\tau)}(s, s')$, which solves the equation

$$i\frac{\partial}{\partial s}U_a^{(\tau)}(s, s')\Psi = \tau L_a(s)U_a^{(\tau)}(s, s')\Psi, \Psi \in \mathcal{D} . \tag{10}$$

The propagator $U_a^{(\tau)}$ describes what one calls the *adiabatic time evolution*. (Note that the operators $L_a(s), s \in I$, satisfy (A1) and (A2), since, by (A3), $\frac{i}{\tau}[\dot{P}(s), P(s)]$ are bounded, selfadjoint operators with bounded derivative in s.)

Adiabatic Theorem. If assumptions (A1)–(A3) hold then
(i) $U_a^{(\tau)}(s', s)P(s)U_a^{(\tau)}(s, s') = P(s')$ (*intertwining property*), for arbitrary s, s' in I, and (ii) $\lim_{\tau\to\infty} \sup_{s,s'\in I} \|U^{(\tau)}(s, s') - U_a^{(\tau)}(s, s')\| = 0$.

A proof of this result can be inferred from [14].

Remarks.

(1) We note that $U^{(\tau)}(s', s) = U^{(\tau)}(s, s')^*$.

(2) With more precise assumptions on the nature of the spectrum of $\{L(s)\}$, one can obtain information about the speed of convergence in (ii), as $\tau \to \infty$; see [12–14]. A powerful strategy to obtain such information is to make use of complex spectral deformation techniques, such as dilatation or spectral-translation analyticity.

These techniques also enable one to prove an

(3) *Adiabatic Theorem for Resonances, [17].* This result resembles the adiabatic theorem described above, but eigenstates of $L(s)$ are replaced by resonance states, and one must require the adiabatic time scale τ to be small as compared to the life time, $\tau_{res}(s)$, of a resonance of $L(s)$, uniformly in $s \in I$. (For shape resonances, the techniques in [18] are useful. Precise statements and proofs can be found in [17].) Similar ideas lead to an adiabatic theorem in non-equilibrium statistical mechanics, [22].

3 The "Isothermal Theorem"

In this section, we turn to the study of isothermal processes of "small" driven quantum systems, Σ, in diathermal contact with a heat bath, $\mathcal{R}$, at a fixed temperature $(k\beta)^{-1}$. Our notations are as in Sect. 1; see, (1), (2), (3).

Let $L_\tau(t) := L(\frac{t}{\tau})$ denote the Liouvillian of the coupled system $\mathcal{R} \vee \Sigma$, where $\{L(s)\}_{s \in I}$ is as in (2) and (1) of Sect. 1 and satisfies assumptions (A1) and (A2) of Sect. 2. The interval

$$I_\tau = \{t \mid \frac{t}{\tau} \in I \subset \mathbf{R}\}$$

is the time interval during which an isothermal process of $\mathcal{R} \vee \Sigma$ is studied.

We assume that Σ is driven "slowly", i.e., that τ is large as compared to the relaxation time $\tau_\mathcal{R} = \max_{s \in I} \tau_\mathcal{R}(s)$ of $\mathcal{R} \vee \Sigma$.

Assumption (A3) of Sect. 2 is supplemented with the following more specific assumption.

(A4) For all $s \in I \equiv [s_0, s_1]$, the operator $L(s)$ has a *single, simple eigenvalue* $\lambda(s) = 0$, the spectrum, $\sigma(L(s)) \backslash \{0\}$, of $L(s)$ being *purely continuous* away from 0. It is also assumed that, for $s \leq s_0$, $L(s) \equiv L$ is independent of s and has spectral properties sufficient to prove return to equilibrium [3, 4].

Let $\Omega_\beta(s) \in \mathcal{H}$ denote the eigenvector of $L(s)$ corresponding to the eigenvalue 0, for $s \leq s_1$. Then $\Omega_\beta(\frac{t}{\tau})$ is the *instantaneous equilibrium state* of $\mathcal{R} \vee \Sigma$ at time t. Let $P(s) = |\Omega_\beta(s)\rangle\langle\Omega_\beta(s)|$ denote the orthogonal projection onto $\Omega_\beta(s)$; $P(s)$ is assumed to satisfy (A3), Sect. 2.

Let $\Psi(t)$ be the "true" state of $\mathcal{R} \vee \Sigma$ at time t; in particular

$$\Psi(t) = U_\tau(t, t')\Psi(t') ,$$

where $U_\tau(t, t')$ is the propagator corresponding to $\{L_\tau(t)\}$; see (5)–(7), Sect. 2. By the property of return to equilibrium and assumption (A4),

$$\Psi(t) = \Omega_\beta \, , t \leq \tau s_0 \, , \tag{11}$$

for an arbitrary initial condition $\Psi(-\infty) \in \mathcal{H}$ at $t = -\infty$.

We set

$$\Psi^{(\tau)}(s) = \Psi(\tau s) \, , \tag{12}$$

and note that, with the notations of (7), (8), Sect. 2, and thanks to (11)

$$\Psi^{(\tau)}(s) = U^{(\tau)}(s, s_0)\Omega_\beta \, , \tag{13}$$

for $s \in I$.

Isothermal Theorem. Suppose that $L(s)$ and $P(s)$ satisfy assumptions (A1)–(A3), Sect. 2, and (A4) above. Then

$$\lim_{\tau \to \infty} \sup_{s \in I} ||\Psi^{(\tau)}(s) - \Omega_\beta(s)|| = 0 \, .$$

Remarks. (1) The isothermal theorem follows readily from the adiabatic theorem of Sect. 2 and from (13) and the definition of $\Omega_\beta(s)$.

(2) We define the expectation values (states)

$$\omega_t^\beta(a) := \left\langle \Omega_\beta\left(\frac{t}{\tau}\right), a\Omega_\beta\left(\frac{t}{\tau}\right) \right\rangle \tag{14}$$

and

$$\rho_t(a) := \langle \Psi(t), a\Psi(t) \rangle \, , \tag{15}$$

where a is an arbitrary bounded operator on $\mathcal{H} = \mathcal{H}^{\mathcal{R} \vee \Sigma}$. Then the isothermal theorem says that

$$\rho_t(a) = \omega_t^\beta(a) + \epsilon_t^{(\tau)}(a) \, , \tag{16}$$

where

$$\lim_{\tau \to \infty} \frac{|\epsilon_t^{(\tau)}(a)|}{||a||} = 0 \, , \tag{17}$$

for all times $t \in I_\tau$.

(3) If the complex spectral deformation techniques of [2,3] are applicable to the analysis of the coupled system $\mathcal{R} \vee \Sigma$ then

$$|\epsilon_t^{(\tau)}(a)| \leq \mathcal{O}(\tau^{-\frac{1}{2}})||a|| \, ; \tag{18}$$

see [11,17].

(4) *All* our assumptions, (A1)–(A4), can be verified for the classes of systems studied in [2–6] for which return to equilibrium has been established therein. They can also be verified for a "quantum dot" coupled to non-interacting electrons in a metal, or for an impurity spin coupled to a reservoir of non-interacting magnons; see [1,11,12].

4 (Reversible) Isothermal Processes

In this section, we study general isothermal processes and use the isothermal theorem to characterize *reversible* isothermal processes.

It will be convenient to view the heat bath $\mathcal{R}$ as the *thermodynamic limit* of an increasing family of quantum systems confined to compact subsets of physical space, as discussed in [10, 19]. The pure states of a quantum mechanical system confined to a bounded region of physical space are unit rays in a separable Hilbert space, while its mixed states are described by density matrices, which are positive trace-class operators with unit trace. Before passing to the thermodynamic limit of the heat bath, the dynamics of the coupled system, $\mathcal{R} \vee \Sigma$, is generated by a family of time-dependent Hamiltonians

$$H(t) \equiv H^{\mathcal{R} \vee \Sigma}(t) := H^{\mathcal{R}} + H^{\Sigma}(t) \,, \tag{19}$$

where

$$H^{\Sigma}(t) = H_0^{\Sigma}(t) + g(t)W \,, \tag{20}$$

$H_0^{\Sigma}(t)$ is as in Sect. 1, and W is the interaction Hamiltonian (as opposed to the interaction Liouvillian, $I = ad_W$, introduced in Sect. 1).

Let $\mathsf{P}(t)$ denote the density matrix describing the state of the coupled system, $\mathcal{R} \vee \Sigma$, at time t, (*before* the thermodynamic limit for $\mathcal{R}$ is taken). Then $\mathsf{P}(t)$ satisfies the Liouville equation

$$\dot{\mathsf{P}}(t) = -i[H(t), \mathsf{P}(t)] \,. \tag{21}$$

The instantaneous equilibrium-, or reference state of the coupled system is given, in the *canonical* ensemble, by the density matrix

$$\mathsf{P}^{\beta}(t) = Z^{\beta}(t)^{-1} e^{-\beta H(t)} \,, \tag{22}$$

where

$$Z^{\beta}(t) = tr(e^{-\beta H(t)}) \tag{23}$$

is the partition function, and tr denotes the trace. We assume that the thermodynamic limits

$$\rho_t(\cdot) = TD \lim_{\mathcal{R}} tr(\mathsf{P}(t)(\cdot)) \tag{24}$$

$$\omega_t^{\beta}(\cdot) = TD \lim_{\mathcal{R}} tr(\mathsf{P}^{\beta}(t)(\cdot)) \tag{25}$$

exist on a suitable kinematical algebra of operators describing $\mathcal{R} \vee \Sigma$; see [4, 10, 19].

The equilibrium state and partition function of a finitely extended heat bath are given by

$$\mathsf{P}_{\mathcal{R}}^{\beta} = (Z_{\mathcal{R}}^{\beta})^{-1} e^{-\beta H^{\mathcal{R}}} \,, \tag{26}$$

$$Z_{\mathcal{R}}^{\beta} = tr(e^{-\beta H^{\mathcal{R}}}) \,, \tag{27}$$

respectively.

Next, we introduce *thermodynamic potentials* for the small system Σ: The *internal energy* of Σ in the "true" state, ρ_t, of $\mathcal{R} \vee \Sigma$ at time t is defined by

$$U^{\Sigma}(t) := \rho_t(H^{\Sigma}(t)) \tag{28}$$

and the *entropy* of Σ in the state ρ_t at time t by

$$S^{\Sigma}(t) := -k\,TD \lim_{\mathcal{R}} tr(\mathsf{P}(t)[ln\mathsf{P}(t) - ln\mathsf{P}_{\mathcal{R}}^{\beta}]) \;. \tag{29}$$

Note that we here define $S^{\Sigma}(t)$ as a *relative* entropy (with the aim of subtracting the divergent contribution of the heat bath to the *total* entropy). A general inequality for traces, see [10], says that

$$S^{\Sigma}(t) \leq 0 \;. \tag{30}$$

The *free energy* of Σ in an *instantaneous equilibrium state*, ω_t^{β}, of $\mathcal{R} \vee \Sigma$ is defined by

$$F^{\Sigma}(t) := -kT\,TD \lim_{\mathcal{R}} ln \left(\frac{Z^{\beta}(t)}{Z_{\mathcal{R}}^{\beta}} \right) \;. \tag{31}$$

Next, we define quantities associated not with states but with the *thermodynamic process* carried out by $\mathcal{R} \vee \Sigma$: the *heat flux* into Σ and the *work rate*, or *power*, of Σ. Let δ denote the so called "imperfect differential". Then

$$\frac{\delta Q^{\Sigma}}{dt}(t) := TD \lim_{\mathcal{R}} -\frac{d}{dt} tr(\mathsf{P}(t)H^{\mathcal{R}}) \;, \tag{32}$$

and

$$\frac{\delta A^{\Sigma}}{dt}(t) := \rho_t(\dot{H}^{\Sigma}(t)) \;; \tag{33}$$

see [1,8,12] for details.

We are now prepared to summarize our main results on isothermal processes. The first two results are general and concern the first law of thermodynamics and the relationship between the rate of change of entropy and the heat flux into Σ. The remaining three results are corollaries pertaining to free energy and changes of entropy in reversible isothermal processes, i.e., processes in which states are sampled at equilibrium, and on the zeroth law of thermodynamics.

(1) From definitions (28), (32) and (33) and the Liouville equation (21) it follows that

$$\dot{U}^{\Sigma}(t) = \frac{\delta Q^{\Sigma}}{dt}(t) + \frac{\delta A^{\Sigma}}{dt}(t) \;, \tag{34}$$

which is the *first law of thermodynamics*; (hardly more than a definition of $\frac{\delta A^{\Sigma}}{dt}$).

(2) Note that, by the unitarity of time evolution and the cyclic invariance of the trace,

$$\frac{d}{dt}tr(\mathsf{P}(t)ln\mathsf{P}(t)) = 0 \,,$$

and

$$\frac{d}{dt}tr(\mathsf{P}(t)lnZ_{\mathcal{R}}^{\beta}) = \frac{d}{dt}lnZ_{\mathcal{R}}^{\beta} = 0 \,.$$

Together with definitions (26), (29) and (32), this implies that

$$\dot{S}^{\Sigma}(t) = \frac{1}{T}\frac{\delta Q^{\Sigma}}{dt}(t) \,, \tag{35}$$

for *arbitrary* isothermal processes at temperature $T = (k\beta)^{-1}$.

(3) Next, we consider an isothermal process of $\mathcal{R} \vee \Sigma$ during a finite time interval $I_{\tau} = [\tau s_0, \tau s_1]$, with s_0 and s_1 fixed. The initial state $\rho_{\tau s_0}$ of $\mathcal{R} \vee \Sigma$ is assumed to be an equilibrium state $w_{\tau s_0}^{\beta}$ of the Liouvillian $L_{\tau}(\tau s_0) = L(s_0)$. We are interested in the properties of such a process when τ becomes large, i.e., when the process is *quasi-static*.

Result. Quasi-static isothermal processes are *reversible* (in the sense that all intermediate states ρ_t of $\mathcal{R} \vee \Sigma$, $t \in I_{\tau}$, converge in norm to *instantaneous equilibrium states* w_t^{β}, as $\tau \to \infty$).

This result is an immediate consequence of the isothermal theorem. It means that, for all practical purposes, an isothermal process with time scale τ is reversible if $\tau \gg \tau_{\mathcal{R}} = \max_{s \in I} \tau_{\mathcal{R}}(s)$.

(4) For reversible isothermal processes, the usual *equilibrium definitions* of internal energy and entropy of the small system Σ can be used:

$$U_{rev}^{\Sigma}(t) := w_t^{\beta}(H^{\Sigma}(t)) \,, \tag{36}$$

$$S_{rev}^{\Sigma}(t) := -k\,TD\lim_{\mathcal{R}} tr(\mathsf{P}^{\beta}(t)[ln\mathsf{P}^{\beta}(t) - ln\mathsf{P}^{\mathcal{R}}]) = \frac{1}{T}(U_{rev}^{\Sigma}(t) - F^{\Sigma}(t)) \,, \tag{37}$$

where the free energy $F^{\Sigma}(t)$ has been defined in (31), and the second equation in (37) follows from (22), (26), (31) and (36). Equations (37) and (31) then imply that

$$\dot{S}_{rev}^{\Sigma}(t) = \frac{1}{T}\left(\frac{d}{dt}w_t^{\beta}(H^{\Sigma}(t)) - w_t^{\beta}(\dot{H}^{\Sigma}(t))\right) \,.$$

Recalling (34) and (35), and applying the isothermal theorem, we find that

$$\dot{S}^{\Sigma}(t) \to \dot{S}_{rev}^{\Sigma}(t) \,, \tag{38}$$

$$\frac{\delta A^{\Sigma}}{dt}(t) \to \dot{F}^{\Sigma}(t) \,, \tag{39}$$

as $\tau \to \infty$.

(5) We conclude this overview by considering a quasi-static isothermal process of $\mathcal{R} \vee \Sigma$ with $H^\Sigma(s) \to H_0^\Sigma$, $g(s) \to 0$, as $s \nearrow s_1$, i.e., the interactions between $\mathcal{R}$ and Σ are switched off at the end of the process. Then the isothermal theorem implies that

$$\lim_{\tau \to \infty} \lim_{s \nearrow s_1} \rho_{\tau s} = \omega_\mathcal{R}^\beta \otimes \omega_\Sigma^\beta, \tag{40}$$

where $\omega_\mathcal{R}^\beta(\cdot) = (Z_\mathcal{R}^\beta)^{-1} tr(e^{-\beta H^\mathcal{R}} \cdot)$, see (26), and

$$\omega_\Sigma^\beta(\cdot) = (Z_\Sigma^\beta)^{-1} tr(e^{-\beta H_0^\Sigma} \cdot) \tag{41}$$

is the Gibbs state of the small system Σ at the temperature $(k\beta)^{-1}$ of the heat bath, *independently* of the properties of the diathermal contact (i.e., of the interaction Hamiltonian W), assuming that (A1)–(A4) hold for $s < s_1$.

This result and the property of return to equilibrium for the heat bath $\mathcal{R}$ yield, in essence, the 0th law of thermodynamics.

Carnot processes and the 2nd law of thermodynamics are discussed in [1]; (see also [10,20,21]). An important variant of the adiabatic theorem for non-equilibrium stationary states will appear in [22]. The analysis in [22] is based on some basic techniques developed in [23].

References

1. Abou-Salem, W.K. and Fröhlich, J.: *Status of the fundamental laws of thermodynamics*, in preparation.
2. Jaksić, V. and Pillet, C.-A.: *On a model of quantum friction II: Fermi's golden rule and dynamics at positive temperature*, Comm. Math. Phys. **176** (1996), 619–644.
3. Jaksić, V. and Pillet, C.-A.: *On a model of quantum friction III: Ergodic properties of the spin-boson system*, Comm. Math. Phys. **178** (1996), 627–651.
4. Bach, V., Fröhlich, J. and Sigal, I.M.: *Return to equilibrium*, J. Math. Phys. **41** (2000), 3985–4060.
5. Merkli, M.: *Positive commutators in non-equilibrium quantum statistical mechanics*, Comm. Math. Phys. **223** (2001), 327–362.
6. Fröhlich, J. and Merkli, M.: *Another return of "return to equilibrium"*, Comm. Math. Phys. **251** (2004), 235–262.
7. Robinson, D.W.: *Return to equilibrium*, Comm. Math. Phys. **31** (1973), 171–189.
8. Fröhlich, J., Merkli, M., Ueltschi, D. and Schwarz, S.: *Statistical mechanics of thermodynamic processes*, in *A garden of quanta*, 345–363, World Sci. Publishing, River Edge, New Jersey, 2003.
9. Haag, R., Hugenholtz, N.M. and Winnink, M.: *On equilibrium states in quantum statistical mechanics*, Comm. Math. Phys. **5** (1967), 215–236.
10. Bratelli, O. and Robinson, D.W.: *Operator algebras and quantum statistical mechanics I,II*, Texts and Monographs in Physics, Springer-Verlag, Berlin, 1987.
11. Abou-Salem, W.K.: *PhD thesis* (2005).

12. Abou-Salem, W.K. and Fröhlich, J., in preparation.
13. Avron, J.E. and Elgart, A.: *Adiabatic theorem without a gap condition*, Comm. Math. Phys. **203** (1999), 445–463.
14. Teufel, S.: *A note on the adiabatic theorem*, Lett. Math. Phys. **58** (2001), 261–266.
15. Reed, M. and Simon, B.: *Methods of modern mathematical physics*, vol. II, Academic Press, New York, 1975.
16. Yosida, K.: *Functional analysis*, 6th ed., Springer-Verlag, Berlin, 1998.
17. Abou-Salem, W.K. and Fröhlich, J.: *Adiabatic theorem for resonances*, in preparation.
18. Fröhlich, J. and Pfeifer, P.: *Generalized time-energy uncertainty relations and bounds on lifetimes of resonances*, Rev. Mod. Phys. **67** (1995), 759–779.
19. Ruelle, D.: *Statistical mechanics: rigorous results*, World Scientific, Singapore, 1999.
20. Ruelle, D.: *Entropy production in quantum spin systems*, Comm. Math. Phys. **224** (2001), 3–16.
21. Fröhlich, J., Merkli, M. and Ueltschi, D.: *Dissipative transport: thermal contacts and tunnelling junctions*, Ann. Henri Poincaré **4** (2003), 897–945.
22. Abou-Salem, W.K.: *An adiabatic theorem for non-equilibrium steady states*, in preparation.
23. Jakšić, V. and Pillet, C.-A.:*Non-equilibrium steady states of finite quantum systems coupled to thermal reservoirs*, Comm. Math. Phys. **226** (2002), 131–162.

Quantum Massless Field in 1+1 Dimensions

Jan Dereziński[1] and Krzysztof A. Meissner[2]

[1] Dept of Math. Methods in Physics, Warsaw University, Hoża 74, 00-682 Warsaw, Poland
Jan.Derezinski@fuw.edu.pl
[2] Institute of Theoretical Physics, Warsaw University, Hoża 69, 00-681 Warsaw Poland
Krzysztof.Meissner@fuw.edu

Abstract. We present a construction of the algebra of operators and the Hilbert space for a quantum massless field in 1+1 dimensions.

1 Introduction

It is usually stated that quantum massless bosonic fields in 1+1 dimensions (with noncompact space dimension) do not exist. With massive fields the correlation function

$$\langle \Omega | \phi(f_1)\phi(f_2)\Omega \rangle = \int \frac{\mathrm{d}p}{2\pi 2E_p} \hat{f}_1^*(E_p, p)\hat{f}_2(E_p, p) \tag{1}$$

(where $E_p = \sqrt{p^2 + m^2}$) is well defined but in the limit $m \to 0$ diverges because of the infrared problem. The limit exists only after adding an additional nonlocal constraint on the smearing functions:

$$\hat{f}(0,0) = \int \mathrm{d}t\mathrm{d}x\, f(t,x) = 0 . \tag{2}$$

Under this constraint it is not difficult to construct massless fields in 1+1 dimension (see eg. [15], where the framework of the Haag-Kastler axioms is used).

Massless fields are extensively used for example in string theory (albeit most often after Wick rotation to the space with Euclidean signature). They also appear as the scaling limit of massive fields [6]. Usually, in these applications, the constraint (2) appears to be present at least implicitly. e.g. in string amplitudes one imposes the condition that sum of all momenta is equal to 0. Nevertheless, it seems desirable to have a formalism for massless 1+1-dimensional fields free of this constraint.

In the literature there are many papers that propose to use an indefinite metric Hilbert space for this purpose [4,9–13]. Clearly, an indefinite metric is not physical and in order to determine physical observables one needs to perform a reduction similar to that of the Gupta-Bleuler formalism used in QED. The outcome of this Gupta-Bleuler-like procedure is essentially equivalent to

J. Dereziński and K.A. Meissner: *Quantum Massless Field in 1+1 Dimensions*, Lect. Notes Phys. **690**, 107–127 (2006)
www.springerlink.com

imposing the constraint (2) [11]. Therefore, we do not find the indefinite metric approach appropriate.

In this paper we present two explicit constructions of (positive definite) Hilbert spaces with representations of the massless Poincaré algebra in 1+1 dimensions and local fields (or at least their exponentials). We allow all test functions f that belong to the Schwartz class on the 1+1 dimensional Minkowski space, without the constraint (2). We try to make sure that as many Wightman axioms as possible are satisfied.

In the first construction we obtain a separable Hilbert space and well defined fields, however we do not have a vacuum vector. In the second construction, the Hilbert space is non-separable, only exponentiated fields are well defined, but there exists a vacuum vector. Thus, neither of them satisfies all Wightman axioms. Nevertheless, we believe that both our constructions are good candidates for a physically correct massless quantum field theory in 1+1 dimensions.

Our constructions have supersymmetric extensions, which we describe at the end of our article.

In the literature known to us the only place where one can find a treatment of massless fields in 1+1 dimension similar to ours is [1,2] by Acerbi, Morchio and Strocchi. Their construction is equivalent to our second (nonseparable) construction. We have never seen our first (separable) construction of massless fields in the literature.

Acerbi, Morchio and Strocchi start from the C^*-algebra associated to the CCR over the symplectic space of solutions of the wave equation parametrized by the initial conditions. Then they apply the GNS construction to the Poincaré invariant quasi-free state obtaining a non-regular representation of CCR.

In our presentation we prefer to use the derivatives of right and left movers to parametrize fields, rather than the initial conditions. We also avoid, as long as possible, to invoke abstract constructions from the theory of C^*-algebras, which may be less transparent to some of the readers. We explain the relationship between our formalism and that of [1, 2]. The symmetry structure of this theory is surprisingly rich. Some of the objects are covariant only under Poincaré group but there are others that are covariant under larger groups: $A_+(1, \mathbb{R}) \times A_+(1, \mathbb{R})$, $SL(2, \mathbb{R}) \times SL(2, \mathbb{R})$, $\mathrm{Diff}_+(\mathbb{R}) \times \mathrm{Diff}_+(\mathbb{R})$, $\mathrm{Diff}_+(S^1) \times \mathrm{Diff}_+(S^1)$.

2 Fields

The action of the 1+1 dimensional free real scalar massless field theory reads

$$S = \frac{1}{2} \int \mathrm{dt}\mathrm{d}x \ \left((\partial_t \phi)^2 - (\partial_x \phi)^2\right) . \tag{3}$$

This leads to the equations of motion

$$(-\partial_t^2 + \partial_x^2)\phi = 0 \ . \tag{4}$$

The solution of (4) is the sum of right and left movers, i.e. functions of $(t-x)$ and $(t+x)$ respectively:

$$\phi(t,x) = \phi_R(t-x) + \phi_L(t+x) \ . \tag{5}$$

We will often used "smeared" fields in the sense

$$\phi(f) = \int \mathrm{dtd}x \, \phi(t,x) f(t,x) \ ,$$

where we assume that f are real Schwartz functions. Because of (5), they can be written in the form

$$\phi(f) = \phi(g_R, g_L) \ ,$$

where

$$\hat{g}_R(k) = \hat{f}(k,k), \qquad \hat{g}_L(k) = \hat{f}(k,-k) \ ,$$

$(k \geq 0)$ and the Fourier transforms of the test function f and g are defined as

$$\hat{f}(E,p) := \int \mathrm{dtd}x \, f(t,x) \, \mathrm{e}^{\mathrm{i}Et - \mathrm{i}px} \ , \tag{6}$$

$$\hat{g}(k) := \int_{-\infty}^{\infty} \mathrm{dt}g(t)\mathrm{e}^{-\mathrm{i}kt} \ . \tag{7}$$

The function g_R corresponds to right movers and g_L to left movers. Note that

$$\hat{g}_R(0) = \hat{g}_L(0) =: \hat{g}(0) \ . \tag{8}$$

$\hat{g}(0)$ is real, because function f is real.

We introduce the notation

$$(g_1|g_2) := \frac{1}{2\pi} \lim_{\epsilon \searrow 0} \left(\int_{\epsilon}^{\infty} \frac{\mathrm{dk}}{k} \hat{g}_1^*(k)\hat{g}_2(k) + \ln(\epsilon/\mu)\hat{g}_1^*(0)\hat{g}_2(0) \right) , \tag{9}$$

where μ is a positive constant having the dimension of mass. For functions that satisfy $\hat{g}(0) = 0$, $(g_1|g_2)$ is a (positive) scalar product – otherwise it is not positive definite and therefore cannot be used directly in the construction of a Hilbert space. Such a scalar product corresponds to quantization of the theory in a constant compensating background.

In view of the infrared divergence we factorize the Hilbert space into two parts – one that is infrared safe and the second that in some sense regularizes the divergent part.

We introduce the creation $a_R^\dagger(k)$, $a_L^\dagger(k)$ and annihilation $a_R(k)$, $a_L(k)$ operators as well as pair of operators (χ, p). They satisfy the commutation relations

$$\left[a_R(k), a_R^\dagger(k')\right] = 2\pi k \delta(k - k') \,,$$

$$\left[a_L(k), a_L^\dagger(k')\right] = 2\pi k \delta(k - k') \,,$$

$$[\chi, p] = i \tag{10}$$

with all other commutators vanishing.

To proceed we choose two real functions $\sigma_R(x)$ and $\sigma_L(x)$ satisfying

$$\hat\sigma_R(0) = \hat\sigma_L(0) = 1 \tag{11}$$

and otherwise arbitrary. To simplify further formulae we define the combinations

$$a_{\sigma R}(k) := a_R(k) - i\hat\sigma_R(k)\chi$$
$$a_{\sigma R}^\dagger(k) := a_R^\dagger(k) + i\hat\sigma_R^*(k)\chi$$
$$a_{\sigma L}(k) := a_L(k) - i\hat\sigma_L(k)\chi$$
$$a_{\sigma L}^\dagger(k) := a_L^\dagger(k) + i\hat\sigma_L^*(k)\chi \tag{12}$$

and therefore

$$\left[a_{\sigma R}(k), a_{\sigma R}^\dagger(k')\right] = 2\pi k \delta(k - k') \,,$$

$$\left[a_{\sigma L}(k), a_{\sigma L}^\dagger(k')\right] = 2\pi k \delta(k - k') \,,$$

$$[a_{\sigma R}(k), p] = \hat\sigma_R(k) \,,$$

$$\left[a_{\sigma R}^\dagger(k), p\right] = -\hat\sigma_R^*(k) \,,$$

$$[a_{\sigma L}(k), p] = \hat\sigma_L(k) \,,$$

$$\left[a_{\sigma L}^\dagger(k), p\right] = -\hat\sigma_L^*(k). \tag{13}$$

Now we are in a position to introduce the field operator $\phi(g_R, g_L)$, depending on a pair of functions g_R, g_L satisfying (8).

$$\phi(g_R, g_L) = \int \frac{dk}{2\pi k} \Big((\hat g_R(k) - \hat g(0)\hat\sigma_R(k))a_{\sigma R}^\dagger(k)$$

$$+ (\hat g_R^*(k) - \hat g(0)\hat\sigma_R^*(k))a_{\sigma R}(k) + (\hat g_L(k) - \hat g(0)\hat\sigma_L(k))a_{\sigma L}^\dagger(k)$$

$$+ (\hat g_L^*(k) - \hat g(0)\hat\sigma_L^*(k))a_{\sigma L}(k) \Big) + \hat g(0)p \,. \tag{14}$$

The field $\phi(g_R, g_L)$ is hermitian and satisfies the commutation relation

$$[\phi(g_{R1}, g_{L1}), \phi(g_{R2}, g_{L2})]$$
$$= (g_{R1}|g_{R2}) - (g_{R2}|g_{R1}) + (g_{L1}|g_{L2}) - (g_{L2}|g_{L1})$$
$$= i2\text{Im}(g_{R1}|g_{R2}) + i2\text{Im}(g_{L1}|g_{L2}) \,. \tag{15}$$

The commutator in (15) does not depend on the functions σ_R, σ_L.

3 Poincaré Covariance

Let $A_+(1, \mathbb{R})$ denote the group of orientation preserving affine transformations of the real line, that is the group of maps $t \mapsto at + b$ with $a > 0$. The proper Poincaré group in 1+1 dimension can be naturally embedded in the direct product of two copies of $A_+(1, \mathbb{R})$, one for the right movers and one for the left movers.

The infinitesimal generators of the right $A_+(1, \mathbb{R})$ group will be denoted H_R (the right Hamiltonian) and D_R (the right generator of dilations) and they satisfy the commutation relations

$$[D_R, H_R] = iH_R .$$

The representation of these operators in terms of the creation and annihilation operators is given by

$$H_R = \int \frac{dk}{2\pi}\, a_{\sigma R}^\dagger(k) a_{\sigma R}(k) ,$$
$$D_R = \frac{i}{2} \int \frac{dk}{2\pi} \left(a_{\sigma R}^\dagger(k) \partial_k a_{\sigma R}(k) - (\partial_k a_{\sigma R}^\dagger(k)) a_{\sigma R}(k) \right) .$$

$$(16)$$

Their action on fields is given by

$$[H_R, \phi(g_R, g_L)] = -i\phi(\partial_t g_R, 0) ,$$
$$[D_R, \phi(g_R, g_L)] = i\phi(\partial_t t g_R, 0) ,$$

and in the exponentiated form by

$$e^{isH_R} \phi(g_R, g_L) e^{-isH_R} = \phi(g_R(\cdot - s), g_L)$$
$$e^{isD_R} \phi(g_R, g_L) e^{-isD_R} = \phi(e^{-s} g_R(e^{-s}\cdot), g_L) .$$

For $(a, b) \in A_+(1, \mathbb{R})$ we set $r_{a,b} g(t) := a^{-1} g(a^{-1}(t - b))$ and

$$R_R(a, b) = e^{i \ln a D_R} e^{ibH_R} .$$

R_R is a unitary representation of $A_+(1, \mathbb{R})$, which acts naturally on the fields:

$$R_R(a, b)\phi(g_R, g_L)R_R(a, b)^\dagger = \phi(r_{a,b} g_R, g_L) .$$

$$(17)$$

Note, however, that $r_{a,b}$ does not preserve the indefinite scalar product (9) unless we impose the constraint $\hat{g}(0)=0$:

$$(r_{a,b} g_1 | r_{a,b} g_2) = (g_1 | g_2) - \ln a \hat{g}_1^*(0) \hat{g}_2(0) .$$

Similarly we introduce the left Hamiltonian H_L and the left generator of dilations D_L satisfying analogous commutation relations and the representation of the left $A_+(1, \mathbb{R})$.

The Poincaré group generators are the Hamiltonian $H = H_R + H_L$, the momentum $P = H_R - H_L$ and the boost operator $\Lambda = D_R - D_L$ (the only Lorentz generator in 1+1 dimensions). The elements of the Poincaré group are of the form

$$(a, b_R), (a^{-1}, b_L) \in A_+(1, \mathbb{R}) \times A_+(1, \mathbb{R}) .$$

The scalar product $(g_{R1}|g_{R2}) + (g_{L1}|g_{L2})$ is invariant wrt the proper Poincaré group.

4 Changing the Compensating Functions

It is important to discuss the dependence of the whole construction on the choice of compensating functions σ_R and σ_L.

Let $\tilde{\sigma}_R$ and $\tilde{\sigma}_L$ be another pair of real functions satisfying (11). Set $\xi_R(x) := \tilde{\sigma}_R(x) - \sigma_R(x)$, $\xi_L(x) := \tilde{\sigma}_L(x) - \sigma_L(x)$. Note that $\hat{\xi}_R(0) = \hat{\xi}_L(0) = 0$. Define

$$U(\xi_R, \xi_L) = \exp\left(\int \frac{dk}{2\pi k}\left(i\chi\hat{\xi}_R^*(k)a_R(k) + i\chi\hat{\xi}_R(k)a_R^\dagger(k)\right.\right.$$
$$\left.\left. + \frac{1}{2}\chi^2(\hat{\xi}_R^*(k)\hat{\sigma}_R(k) - \hat{\xi}_R(k)\hat{\sigma}_R^*(k)) + R \to L\right)\right) . \quad (18)$$

Using the formula

$$e^A B e^{-A} = B + [A, B] + \frac{1}{2}[A, [A, B]] + \cdots \quad (19)$$

we have

$$Ua_R(k)U^{-1} = a_R(k) - i\chi\hat{\xi}_R(k) ,$$
$$Ua_R^\dagger(k)U^{-1} = a_R^\dagger(k) + i\chi\hat{\xi}_R^*(k) ,$$
$$Ua_L(k)U^{-1} = a_L(k) - i\chi\hat{\xi}_L(k) ,$$
$$Ua_L^\dagger(k)U^{-1} = a_L^\dagger(k) + i\chi\hat{\xi}_L^*(k) ,$$
$$U\chi U^{-1} = \chi \quad (20)$$

and

$$UpU^{-1} = p + \int \frac{dk}{2\pi k}\left(-\hat{\xi}_R^*(k)a_R(k) - \hat{\xi}_R(k)a_R^\dagger(k)\right.$$
$$\left. - i\chi\hat{\xi}_R(k)\hat{\sigma}_R^*(k) + i\chi\hat{\xi}_R^*(k)\hat{\sigma}_R(k) + R \to L\right). \quad (21)$$

Using these relations we get for example

$$U\phi_\sigma(g_R, g_L)U^{-1} = \phi_{\tilde{\sigma}}(g_R, g_L) ,$$
$$Ua_{\sigma R}^\dagger U^{-1} = a_{\tilde{\sigma}R}^\dagger ,$$
$$UH_{\sigma R}U^{-1} = H_{\tilde{\sigma}R} , \quad (22)$$

where we made explicit the dependence of ϕ and H_R on σ and $\tilde{\sigma}$. Thus the two constructions – with σ and with $\tilde{\sigma}$ – are unitarily equivalent.

5 Hilbert Space

The Hilbert space of the system is the product of three spaces: $\mathcal{H} = \mathcal{H}_R \otimes \mathcal{H}_L \otimes \mathcal{H}_0$. $\mathcal{H}_R$ is the bosonic Fock space spanned by the creation operators $a_R^\dagger(k)$ acting on the vacuum vector $|\Omega_R\rangle$. Analogously $\mathcal{H}_L$ is the bosonic Fock space spanned by the creation operators $a_L^\dagger(k)$ acting on the vacuum vector $|\Omega_L\rangle$. With the third sector $\mathcal{H}_0$ we have essentially two options. If we take the usual choice $\mathcal{H}_0 = L^2(\mathbb{R}, d\chi)$ then we can define the vacuum state (vacuum expectation value) but there does not exist a vacuum vector. On the other hand, we can take $\mathcal{H}_0 = l^2(\mathbb{R})$, i.e. the space with the scalar product

$$(f|g) = \sum_{\chi \in \mathbb{R}} f^*(\chi)g(\chi) , \tag{23}$$

which is a nonseparable space. It may sound as a nonstandard choice, it has however the advantage of possessing a vacuum vector. The orthonormal basis in the latter space consists of the Kronecker delta functions δ_χ for each $\chi \in \mathbb{R}$. In the nonseparable case, the operator p, and therefore also $\phi(g_R, g_L)$, cannot be defined. But there exist operators e^{isp}, for $s \in \mathbb{R}$, and also $e^{i\phi(g_R,g_L)}$. The commutation relations for these exponential operators follow from the commutation relations for p and $\phi(g_R, g_L)$ described above.

In such a space the vacuum vector is given by

$$|\Omega\rangle = |\Omega_R \otimes \Omega_L \otimes \delta_0\rangle . \tag{24}$$

This vector is invariant under the action of the Poincaré group and the action of the gauge group U. We now prove that it is the unique vector with the lowest energy. Note first that H is diagonal in $\chi \in \mathbb{R}$. Now for an arbitrary $\Phi \in \mathcal{H}_R \otimes \mathcal{H}_L$ and $\chi_1 \in \mathbb{R}$,

$$\langle \Phi \otimes \delta_{\chi_1}|H|\Phi \otimes \delta_{\chi_1}\rangle$$
$$= \int \left(\left\langle \Phi|\left(a_R^\dagger(k) + i\chi_1\hat{\sigma}_R^*(k)\right)\left(a_R(k) - i\chi_1\hat{\sigma}_R(k)\right)\Phi \right\rangle \frac{dk}{2\pi} + R \to L \right) \tag{25}$$

For any χ_1, the expression (25) is nonnegative. If $\chi_1 = 0$, it has a unique ground state $|\Omega_R \otimes \Omega_L\rangle$.

If $\chi_1 \neq 0$, then (25) has no ground state. In fact, it is well known that a ground state of a quadratic Hamiltonian is a coherent state, that is given by a vector of the form

$$|\beta_R, \beta_L\rangle = C \exp \left(\int \frac{dk}{2\pi k} \left(\beta_R(k)a_R^\dagger(k) + \beta_L(k)a_L^\dagger(k) \right) \right) |\Omega_R \otimes \Omega_L\rangle \tag{26}$$

and C is the normalizing constant

$$C = \exp \left(-\frac{1}{2} \int \frac{dk}{2\pi k} \left(|\beta_R(k)|^2 + |\beta_L(k)|^2 \right) \right) . \tag{27}$$

If we set $\Phi = |\beta_R, \beta_L\rangle$ in (25), then we obtain

$$\langle \Phi \otimes \delta_{\chi_1} | H | \Phi \otimes \delta_{\chi_1} \rangle = |\beta_R(k) + i\chi_1 \sigma_R(k)|^2 + R \to L .$$

that takes the minimum for

$$\beta_R(k) = -i\chi_1 \hat{\sigma}_R(k), \quad \beta_L(k) = -i\chi_1 \hat{\sigma}_L(k) .$$

But for $\chi_1 \neq 0$, $|\beta_R, \beta_L\rangle$ is not well defined as a vector in the Hilbert space. To see this we can note that the normalizing constant C equals zero, because then

$$\chi_1^2 \int \frac{dk}{2\pi k} \left(|\hat{\sigma}_R(k)|^2 + |\hat{\sigma}_L(k)|^2 \right) = \infty .$$

(The fact that operators of the form (25) have no ground state is well known in the literature, see eg. [7]).

In the nonseparable case $\mathcal{H}_0 = l^2(\mathbb{R})$, the expectation value

$$\langle \Omega | \cdot | \Omega \rangle =: \omega(\cdot)$$

is a Poincaré-invariant state (positive linear functional) on the algebra of observables. If we take the separable case $\mathcal{H}_0 = L^2(\mathbb{R}, d\chi)$, the state ω can also be given a meaning, even though the vector Ω does not exist (since then δ_0 is not well defined).

Note that in the nonseparable case the state ω can act on an arbitrary bounded operator on $\mathcal{H}$. In the separable case we have to restrict ω to a smaller algebra of operators, say, the algebra (or the C^*-algebra) spanned by the operators of the form $e^{i\phi(g_R, g_L)}$.

The expectation values of the exponentials of the 1+1-dimensional massless field make sense and can be computed, both in the separable and nonseparable case:

$$\omega \left(\exp \left(i\phi(g_R, g_L) \right) \right) = \exp \left(-\frac{1}{2} \int \frac{dk}{2\pi k} \left(|\hat{g}_R(k)|^2 + |\hat{g}_L(k)|^2 \right) \right) . \tag{28}$$

Note that the integral in the exponent of (28) is the usual integral of a positive function, and not its regularization as in (9). Therefore, if $\hat{g}(0) \neq 0$, then this integral equals $+\infty$ and (28) equals zero.

The "two-point functions" of massless fields in 1+1 dimension, even smeared out ones, are not well defined. Formally, they are introduced as

$$\omega \left(\phi(g_{R,1}, g_{L,1}) \phi(g_{R,2}, g_{L,2}) \right) . \tag{29}$$

If we use the nonseparable $\mathcal{H}_0 = l^2(\mathbb{R})$, then field operators $\phi(g_R, g_L)$ is not well defined if $\hat{g}(0) \neq 0$, and thus (29) is not defined. If we use the separable Hilbert space $\mathcal{H}_0 = L^2(\mathbb{R}, d\chi)$, then $\phi(g_{R,1}, g_{L,1}) \phi(g_{R,2}, g_{L,2})$ are unbounded operators and there is no reason why the state ω could act on them. Thus (29) a priori does not make sense. In the usual free quantum field theory, if mass is positive or dimension more than 2, the expectation values the exponentials depend on the smeared fields analytically, and by taking their second derivative at $(g_R, g_L) = (0,0)$, one can introduce the 2-point function. This is not the case for massless field in 1+1 dimension.

The above discussion shows that the problem of the non-positive definiteness of the two-point function, so extensively discussed in the literature [4, 9, 11–13] does not exist in our formalism.

It should be noted that massless fields in 1+1 dimension do not satisfy the Wightman axioms [14]. In the separable case there is no vacuum vector in the Hilbert space; in the nonseparable case there is a vacuum vector, but there are no fields $\phi(f)$, only the "Weyl operators" $e^{i\phi(f)}$.

6 Fields in Position Representation

So far in our discussion we found it convenient to use the momentum representation. The position representation is, however, better suited for many purposes.

Let $\mathcal{W}(t)$ denote the Fourier transform of the appropriately regularized distribution $\frac{\theta(k)}{2\pi k}$, that is

$$\mathcal{W}(t) = \frac{1}{2\pi} \lim_{\epsilon \to 0} \left(\int_{k > \epsilon} \frac{dk}{k} e^{ikt} + \ln(\epsilon/\mu) \right) \tag{30}$$
$$= \frac{1}{2\pi} \left(-\gamma_E - \ln|\mu t| + \frac{i\pi}{2} \mathrm{sgn}(t) \right) = \mathcal{W}^*(-t)$$

where γ_E is the Euler's constant. We can rewrite (9) as

$$(g_1|g_2) = \int_{-\infty}^{\infty} dt ds g_1^*(t) \mathcal{W}(t - s) g_2(s) \tag{31}$$

To describe massless field in the position representation we introduce the operators $\psi_R(t)$ defined as

$$\psi_R(t) = \int \frac{dk}{2\pi k} \left(a_R^\dagger(k) e^{-ikt} + a_R(k) e^{ikt} \right),$$

and similarly for $R \to L$. Note that it is allowed to smear $\psi_R(t)$ and $\psi_L(t)$ only with test functions satisfying

$$\int g(t) dt = 0 .$$

Note that $\mathcal{W}(t-s)$ and $\frac{1}{2}\mathrm{sgn}(t-s)$ are the correlator and the commutator functions for $\psi_R(t)$:

$$\langle \Omega_R | \psi_R(t)\psi_R(s)\Omega_R \rangle = W(t-s) \,,$$

$$[\psi_R(t), \psi_R(s)] = \frac{i}{2}\text{sgn}(t-s) \,,$$

and similarly for $R \to L$.

We introduce also

$$\psi_{\sigma R}(t) = \int \frac{dk}{2\pi k} a^\dagger_{\sigma R}(k)e^{-ikt} + \int \frac{dk}{2\pi k} a_{\sigma R}(k)e^{ikt}$$

$$= \psi_R(t) + \frac{\chi}{2}\int ds\sigma_R(s)\text{sgn}(t-s) \,,$$

as well as $R \to L$.

It is perhaps useful to note that formally we can write

$$\psi_{\sigma R}(t) = Y_R\psi_R(t)Y_R^\dagger \,,$$

where

$$Y_R := \exp\left(i\chi \int dt\sigma_R(t)\psi_R(t)\right)$$

Note that Y_R is not a well defined operator, since $\hat{\sigma}_R(0) \neq 0$.

Expressed in position representation the fields are given by

$$\phi(g_R, g_L) = \int dt(g_R(t) - \hat{g}(0)\sigma_R(t))\psi_{\sigma R}(t) + R \to L + \hat{g}(0)p \,. \tag{32}$$

Since $\int (g_R(t) - \hat{g}(0)\sigma_R(t))dt = 0$ the whole expression is well defined.

The commutator of two fields equals

$$[\phi(f_1), \phi(f_2)] = i \int dt_1 dt_2 dx_1 dx_2 f_1(t_1, x_1) f_2(t_2, x_2)$$

$$\times (\text{sgn}(t_1 - t_2 + x_1 - x_2) + \text{sgn}(t_1 - t_2 - x_1 + x_2))$$

Note that the commutator of fields is causal – it vanishes if the supports of f_1 and f_2 are spatially separated.

7 The $SL(2, \mathbb{R}) \times SL(2, \mathbb{R})$ Covariance

Massless fields in $1+1$ dimension satisfying the constraint (2) actually possess much bigger symmetry than just the $A_+(1, \mathbb{R}) \times A_+(1, \mathbb{R})$ symmetry, they are covariant wrt the action of $SL(2, \mathbb{R}) \times SL(2, \mathbb{R})$ (for right and left movers).

We will restrict ourselves to the action of $SL(2, \mathbb{R})$ for, say, right movers. First we consider it on the level of test functions.

We assume that test functions satisfy $\hat{g}(0) = 0$ and

$$g(t) = O(1/t^2), \quad |t| \to \infty \,. \tag{33}$$

Let

$$C = \begin{bmatrix} a & b \\ c & d \end{bmatrix} \in SL(2, \mathbb{R}) \qquad (34)$$

(i.e. $ad - bc = 1$). We define the action of C on g by

$$(r_C g)(t) = (-ct + a)^{-2} g\left(\frac{dt - b}{-ct + a}\right). \qquad (35)$$

Note that (35) preserves (33) and the scalar product

$$(r_C g_1 | r_C g_2) = (g_1 | g_2),$$

and is a representation, that is $r_{C_1} r_{C_2} = r_{C_1 C_2}$.

We second quantize r_C by introducing the unitary operator $R_R(C)$ on $\mathcal{H}_R$ fixed uniquely by the conditions

$$R_R(C)\Omega_R = \Omega_R,$$

$$R_R(C)\psi_R(t)R_R(C)^\dagger = \psi_R\left(\frac{at + b}{ct + d}\right). \qquad (36)$$

Note that

$$R_R(C)\left(\int dt g_R(t)\psi_R(t)\right) R_R(C)^\dagger = \int dt (r_C g_R)(t)\psi_R(t). \qquad (37)$$

$C \mapsto R_R(C)$ is a representation in $\mathcal{H}_R$. Thus the operators $R_R(C)$ act naturally on fields satisfying (2) (and hence also $\hat{g}_R(0) = 0$).

The fields without the constraint (2) are not covariant with respect to $SL(2, \mathbb{R}) \times SL(2, \mathbb{R})$, since this symmetry fails even at the classical level. What remains is the $A_+(1, \mathbb{R}) \times A_+(1, \mathbb{R})$ symmetry described in (17). Note that $A_+(1, \mathbb{R})$ can be viewed as a subgroup of $SL(2, \mathbb{R})$:

$$A_+(1, \mathbb{R}) \ni (a, b) \mapsto C\begin{bmatrix} a^{1/2} & ba^{-1/2} \\ 0 & a^{-1/2} \end{bmatrix} \in SL(2, \mathbb{R}). \qquad (38)$$

Clearly, on the restricted Hilbert space, under the identification (38), $R_R(a, b)$ coincides with $R_R(C)$.

8 Normal Ordering

In the theory without the compensating sector the normal ordering can be introduced in a standard way. In particular we have

$$: e^{i\phi(g_R, g_L)}: \ = \ e^{\frac{1}{2}(g_R | g_R) + \frac{1}{2}(g_L | g_L)} e^{i\phi(g_R, g_L)} \qquad (39)$$

If the compensating sector is present then the theory does not act in the Fock space any longer and we do not have an invariant particle number operator. It is however possible (and useful) to introduce the notion of the normal ordering. For Weyl operators it is by definition given by (39). For an arbitrary operator, we first decompose it in terms of Weyl operators, and then we apply (39). Note that our definition has an invariant meaning wrt the change of the compensating function: in the notation of (22) we have

$$U : e^{i\phi_\sigma(g_R, g_L)} : U^{-1} =: e^{i\phi_{\tilde\sigma}(g_R, g_L)} : .$$

Normal ordering is Poincaré invariant but suffers anomalies under remaining $A_+(1, \mathbb{R}) \times A_+(1, \mathbb{R})$ transformations (because the prefactor on the rhs of (39) is invariant only under the Poincaré group). If the constraint (2) is satisfied, then normal ordering is $SL(2, \mathbb{R}) \times SL(2, \mathbb{R})$ covariant.

9 Classical Fields

In order to better understand massless quantum fields in $1+1$ dimension it is useful to study the underlying classical system, that is the wave equation in $1+1$ dimension (4).

From the general representation of any classical solution

$$\phi(t, x) = \phi_R(t - x) + \phi_L(t + x) \tag{40}$$

we get (in notation where $f(\pm\infty)$ stands for $\lim_{t \to \pm\infty} f(t)$)

$$\phi(t, \infty) + \phi(t, -\infty) = \phi_R(-\infty) + \phi_L(\infty) + \phi_R(\infty) + \phi_L(-\infty)$$
$$= \phi(\infty, x) + \phi(-\infty, x) \tag{41}$$

It will be convenient to denote by the space of Schwartz functions on $\mathbb{R}$ by $\mathcal{S}$ and by $\partial_0^{-1}\mathcal{S}$ the space of functions whose derivatives belong to $\mathcal{S}$ and satisfy the condition $f(\infty) = -f(-\infty)$.

We are interested only in those solutions that restricted to lines of constant time and lines of constant position belong to $\partial_0^{-1}\mathcal{S}$ (we will denote them as $\mathcal{F}_{11}$). Neglecting a possible global constant shift we therefore assume that they satisfy

$$\phi(t, \infty) + \phi(t, -\infty) = \phi(\infty, x) + \phi(-\infty, x) = 0 \tag{42}$$

$\mathcal{F}_{11}$ is characterized by two numbers

$$\lim_{t \to \infty} \phi(t, x) = -\lim_{t \to -\infty} \phi(t, x) =: c_0 ,$$
$$\lim_{x \to \infty} \phi(t, x) = -\lim_{x \to -\infty} \phi(t, x) =: c_1 .$$

It is natural to distinguish the following subclasses of solutions to (4):

- $\mathcal{F}_{00}$ – solutions that restricted to lines of constant time and to lines of constant position belong to $\mathcal{S}$ i.e. $c_0 = c_1 = 0$.

- $\mathcal{F}_{10}$ – solutions that restricted to lines of constant time belong to $\mathcal{S}$ and restricted to lines of constant position belong to $\partial_0^{-1}\mathcal{S}$ i.e. $c_1 = 0$.
- $\mathcal{F}_{01}$ – solutions that restricted to lines of constant position belong to $\mathcal{S}$ and restricted to lines of constant time belong to $\partial_0^{-1}\mathcal{S}$ i.e. $c_0 = 0$.

There are several useful ways to parametrize elements of $\mathcal{F}_{11}$.

1. **Initial conditions at $t = 0$:**

$$f_0(x) = \phi(0, x) \,,$$
$$f_1(x) := \partial_t \phi(0, x) \,. \tag{43}$$

Here, $f_0 \in \partial_0^{-1}\mathcal{S}$, $f_1 \in \mathcal{S}$. Note that

$$c_0 = \frac{1}{2} \int f_1(x) dx, \qquad c_1 = f_0(\infty) \,.$$

2. **Derivatives of right/left movers:**

$$g_R(t) := -\frac{1}{2} f_0'(-t) + \frac{1}{2} f_1(-t) \,,$$
$$g_L(t) := \frac{1}{2} f_0'(t) + \frac{1}{2} f_1(t) \,.$$

Note that $g_R, g_L \in \mathcal{S}$ and they satisfy

$$\int g_R(t) dt = c_0 - c_1, \qquad \int g_L(t) dt = c_0 + c_1 \,. \tag{44}$$

3. **Right/left movers:**

$$\phi_R(t) = \frac{1}{2} \int g_R(t - u) \operatorname{sgn}(u) du$$
$$\phi_L(t) = \frac{1}{2} \int g_L(t - u) \operatorname{sgn}(u) du \,. \tag{45}$$

Note that $\phi_R, \phi_L \in \partial_0^{-1}\mathcal{S}$ and they satisfy

$$\phi_R(\infty) = -\phi_R(-\infty) = \frac{1}{2}(c_0 - c_1) \,,$$
$$\phi_L(\infty) = -\phi_L(-\infty) = \frac{1}{2}(c_0 + c_1) \,. \tag{46}$$

We can go back from (g_R, g_L) to (f_0, f_1) by

$$f_0(x) = \frac{1}{2} \int g_R(s - x) \operatorname{sgn}(-s) ds + \frac{1}{2} \int g_R(s + x) \operatorname{sgn}(-s) ds \,,$$
$$f_1(x) = g_R(-x) + g_L(x) \,.$$

We can go back from (ϕ_R, ϕ_L) to (g_R, g_L) by

$$g_R = \phi'_R, \quad g_L = \phi'_L .$$

The unique solution of (4) with the initial conditions (43) equals

$$\phi(t, x) = \phi_R(t - x) + \phi_L(t + x) .$$

It will be sometimes denoted by $\phi(g_R, g_L)$.

In the literature, one can find all three parametrizations of solutions of the wave equation. In particular, note that 3. is especially useful in the case of $\mathcal{F}_{00}$, since then $\phi_R, \phi_L \in \mathcal{S}$.

Note that in our paper we use 2. as the standard parametrization of solutions of the wave equation. We are interested primarily in the space $\mathcal{F}_{10}$. Note that $\mathcal{F}_{10}$ are the solutions to the wave equation with $f_0, f_1 \in \mathcal{S}$. Equivalently, for $\mathcal{F}_{10}$, the functions g_R, g_L satisfy

$$\int g_R(t)\mathrm{d}t = \int g_L(t)\mathrm{d}t . \tag{47}$$

We equip the space $\mathcal{F}_{11}$ with the Poisson bracket, which we write for all three parametrizations:

$$\{\phi(g_{R1}, g_{L1}), \phi(g_{R2}, g_{L2})\} = \int f_{01}(x)f_{12}(x)\mathrm{d}x - \int f_{02}(x)f_{11}(x)\mathrm{d}x \tag{48}$$

$$= \int g_{R1}(t)\mathrm{sgn}(s - t)g_{R1}(s)\mathrm{d}t\mathrm{d}s + \int g_{L1}(t)\mathrm{sgn}(s - t)g_{L1}(s)\mathrm{d}t\mathrm{d}s$$

$$= \mathrm{Im}(g_{R1}|g_{R2}) + \mathrm{Im}(g_{L1}|g_{L2}) \tag{49}$$

$$= \frac{1}{2}\int \partial_t\phi_{R1}(t)\phi_{R2}(t)\mathrm{d}t + \frac{1}{2}\int \partial_t\phi_{L1}(t)\phi_{L2}(t)\mathrm{d}t . \tag{50}$$

Above, (f_{0i}, f_{1i}) and (ϕ_{Ri}, ϕ_{Li}) correspond to (g_{Ri}, g_{Li}). The formula in (48) is the usual Poisson bracket for the space of solutions of relativistic 2nd order equations (both wave and Klein-Gordon equations). (49) we have already seen in (15).

The Poisson bracket in $\mathcal{F}_{11}$ is invariant wrt to the conformal group fixing the infinities preserving separately the orientation of right and left movers, that is $\mathrm{Diff}_+(\mathbb{R}) \times \mathrm{Diff}_+(\mathbb{R})$. In the case of $\mathcal{F}_{00}$ we can extend this action to the full orientation preserving conformal group, that is $\mathrm{Diff}_+(S^1) \times \mathrm{Diff}_+(S^1)$, where we identify $\mathbb{R}$ together with the point at infinity with the unit circle.

10 Algebraic Approach

Among mathematical physicists, it is popular to use the formalism of C^*-algebras to describe quantum systems. A description of massless fields in 1+1 dimension within this formalism is sketched in this section.

To quantize the space $\mathcal{F}_{11}$, we consider formal expressions

$$e^{i\phi(g_R, g_L)} \tag{51}$$

equipped with the relations

$$e^{i\phi(g_{R1}, g_{L1})} e^{i\phi(g_{R2}, g_{L2})} = e^{i\mathrm{Im}(g_{R1}|g_{R2}) + i\mathrm{Im}(g_{L1}|g_{L2})} e^{i\phi(g_{R1} + g_{R2}, g_{L1} + g_{L2})},$$
$$\left(e^{i\phi(g_R, g_L)}\right)^\dagger = e^{i\phi(-g_R, -g_L)}.$$

Linear combinations of (51) form a $*$-algebra, which we will denote Weyl $(\mathcal{F}_{11})$. (If we want, we can take its completion in the natural norm and obtain a C^*-algebra).

Note that the group $\mathrm{Diff}_+(\mathbb{R}) \times \mathrm{Diff}_+(\mathbb{R})$ acts on Weyl $(\mathcal{F}_{11})$ by $*$-automorphisms. In other words, we have two actions

$$\mathrm{Diff}_+(\mathbb{R}) \ni F \mapsto \alpha_R(F) \in \mathrm{Aut}(\mathrm{Weyl}(\mathcal{F}_{11})),$$
$$\mathrm{Diff}_+(\mathbb{R}) \ni F \mapsto \alpha_L(F) \in \mathrm{Aut}(\mathrm{Weyl}(\mathcal{F}_{11})),$$

commuting with one another given by

$$\alpha_R(F)\left(e^{i\phi(g_R, g_L)}\right) = e^{i\phi(r_F g_R, g_L)},$$
$$\alpha_L(F)\left(e^{i\phi(g_R, g_L)}\right) = e^{i\phi(g_R, r_F g_L)}. \tag{52}$$

Above, $\mathrm{Aut}(\mathrm{Weyl}(\mathcal{F}_{11}))$ denotes the group of $*$-automorphisms of the algebra $\mathrm{Weyl}(\mathcal{F}))$ and $r_F g(t) := \frac{1}{F'(t)} g(F^{-1}(t))$.

Similarly $\mathrm{Diff}_+(S^1) \times \mathrm{Diff}_+(S^1)$ acts on Weyl $(\mathcal{F}_{00})$ by $*$-automorphisms.

The state ω given by (28) is invariant wrt $A_+(1, \mathbb{R}) \times A_+(1, \mathbb{R})$ on Weyl $(\mathcal{F}_{11})$ and wrt $SL(2, \mathbb{R}) \times SL(2, \mathbb{R})$ on Weyl $(\mathcal{F}_{00})$.

In our paper we restricted ourselves to Weyl $(\mathcal{F}_{10})$.

The constructions presented in this paper give representations of Weyl $(\mathcal{F}_{10})$ in a Hilbert space $\mathcal{H}$ and two commuting with one another strongly continuous unitary representations

$$A_+(1, \mathbb{R}) \ni (a, b) \mapsto R_R(a, b) \in U(\mathcal{H}),$$
$$A_+(1, \mathbb{R}) \ni (a, b) \mapsto R_L(a, b) \in U(\mathcal{H}).$$

implementing the automorphisms (52):

$$\alpha_R(a, b)(A) = R_R(a, b) A R_R(a, b)^\dagger, \tag{53}$$
$$\alpha_L(a, b)(A) = R_L(a, b) A R_L(a, b)^\dagger. \tag{54}$$

In the case of the algebra Weyl $(\mathcal{F}_{00})$ the same is true for $SL(2, \mathbb{R})$.

In Sect. 5 we described two representations that satisfy the above mentioned conditions. The first, call it π_I, represents Weyl $(\mathcal{F}_{10})$ in a separable Hilbert space. Its drawback is the absence of a vacuum vector – a Poincaré invariant vector. The second, call it π_{II}, represents Weyl $(\mathcal{F}_{10})$ in a non-separable Hilbert space. It has an invariant vector $|\Omega\rangle$.

We can perform the GNS construction with ω. As a result we obtain the representation π_{II} together with the cyclic invariant vector Ω. The description of this construction for massless fields in 1+1 dimension can be found in [1], Sect. III D, and [2] Sect. 4. Note, however, that we have not seen the representation π_{I} in the literature, even though one can argue that it is in some ways superior to π_{II}.

Let us make a remark concerning the role played by the functions (σ_R, σ_L). We note that $\mathcal{F}_{00}$ is a subspace of $\mathcal{F}_{10}$ of codimension 1. Fixing (σ_R, σ_L) satisfying (11) allows us to identify $\mathcal{F}_{10}$ with $\mathcal{F}_{00} \oplus \mathbb{R}$. Thus any (g_R, g_L) satisfying (47) is decomposed into the direct sum of $(g_R - \hat{g}(0)\sigma_R, g_L - \hat{g}(0)\sigma_L)$ and $\hat{g}(0)(\sigma_R, \sigma_L)$.

Of course, similar constructions can be performed for the algebra $\mathrm{Weyl}(\mathcal{F}_{11})$ or $\mathrm{Weyl}(\mathcal{F}_{01})$. In the literature, algebras of observables based on $\mathcal{F}_{01}$ appear in the context of "Doplicher-Haag-Roberts charged sectors" in [5, 6, 15].

11 Vertex Operators

Finally, let us make some comments about the so-called vertex operators, often used in string theory [8]. Let δ_y denote the delta function at $y \in \mathbb{R}$.

Let $t_{R1}, \ldots, t_{Rn} \in \mathbb{R}$ correspond to insertions for right movers and $t_{L1}, \ldots, t_{Lm} \in \mathbb{R}$ correspond to insertions for left movers. Suppose that the complex numbers $\beta_{R1}, \ldots, \beta_{Rn}$, and $\beta_{L1}, \ldots, \beta_{Lm}$ denote the corresponding insertion amplitudes and satisfy

$$\sum \beta_{Ri} = \sum \beta_{Lj} \, .$$

Then the corresponding vertex operator is formally defined as

$$V(t_{R1}, \beta_{R1}; \ldots; t_{Rn}, \beta_{Rn}; t_{L1}, \beta_{L1}; \ldots; t_{Lm}, \beta_{Lm}) = \, : \exp\left(i\phi(g_R, g_L)\right) : \, , \tag{55}$$

where

$$g_R = \beta_{R1}\delta_{t_{R1}} + \cdots + \beta_{Rn}\delta_{t_{Rn}}, \tag{56}$$

$$g_L = \beta_{L1}\delta_{t_{L1}} + \cdots + \beta_{Lm}\delta_{t_{Lm}}. \tag{57}$$

Strictly speaking, the rhs of (55) does not make sense as an operator in the Hilbert space. In fact, in order that $e^{i\phi(g_R, g_L)}$ be a well defined operator, we need that

$$\int \frac{dk}{2\pi k} \left|\hat{g}_R(k) - \hat{g}(0)\hat{\sigma}_R(k)\right|^2 + \int \frac{dk}{2\pi k} \left|\hat{g}_L(k) - \hat{g}(0)\hat{\sigma}_L(k)\right|^2 < \infty. \tag{58}$$

This is not satisfied if g_R or g_L are as in (56) and (57).

Nevertheless, proceeding formally, we can deduce various identities. For instance, we have the Poincaré covariance:

$$R_R(a, b_R) R_L(a^{-1}, b_L)$$
$$\times V(t_{R1}, \beta_{R1}; \ldots; t_{Rn}, \beta_{Rn}; t_{L1}, \beta_{L1}; \ldots)$$
$$\times R_L^\dagger(a^{-1}, b_L) R_R^\dagger(a, b_R)$$
$$= V(at_{R1} + b_L, \beta_{R1}; \ldots; a^{-1}t_{Rn} + b_R, \beta_{Rn}; a^{-1}t_{L1} + b_L, \beta_{L1}; \ldots) .$$

If in addition $\sum \beta_{Ri} = 0$, then a similar identity is true for $SL(2, \mathbb{R}) \times SL(2, \mathbb{R})$.

Clearly, we have

$$\omega \left(V(t_{R1}, \beta_{R1}; \ldots; t_{Rn}, \beta_{Rn}; t_{L1}, \beta_{L1}; \ldots; t_{Lm}, \beta_{Lm}) \right)$$

$$= \begin{cases} 1, & \sum \beta_{Ri} = 0; \\ 0, & \sum \beta_{Ri} \neq 0. \end{cases} \tag{59}$$

The following identities are often used in string theory for the calculation of on-shell amplitudes. Suppose that $t_{R1}, \ldots, t_{Rn}$ are distinct, and the same is true for $t_{L1}, \ldots, t_{Ln}$. Then, using (30), we obtain

$$V(t_{R1}, \beta_{R1}; t_{L1}, \beta_{L1}) \cdots V(t_{Rn}, \beta_{Rn}; t_{Ln}, \beta_{Ln})$$

$$= e^{\left(\sum_{i<j} W(t_{Ri} - t_{Rj})) \beta_{Ri}\beta_{Rj} + W(t_{Li} - t_{Lj})\beta_{Li}\beta_{Lj} \right)}$$

$$\times V(t_{R1}, \beta_{R1}; \ldots; t_{Rn}, \beta_{Rn}; t_{L1}, \beta_{L1}; \ldots; t_{Ln}, \beta_{Ln})$$

$$= \prod_{i<j} \left(\frac{t_{Ri} - t_{Rj}}{i\mu e^{\gamma_E}} \right)^{-\beta_{Ri}\beta_{Rj}/2\pi} \left(\frac{t_{Li} - t_{Lj}}{i\mu e^{\gamma_E}} \right)^{-\beta_{Li}\beta_{Lj}/2\pi}$$

$$\times V(t_{R1}, \beta_{R1}; \ldots; t_{Rn}, \beta_{Rn}; t_{L1}, \beta_{L1}; \ldots; t_{Ln}, \beta_{Ln}) .$$

12 Fermions

Massless fermions in 1+1 dimension do not pose such problems as bosons. The fields are spinors, they will be written as $\begin{bmatrix} \lambda_R(t, x) \\ \lambda_L(t, x) \end{bmatrix}$. They satisfy the Dirac equation

$$\begin{bmatrix} \partial_t - \partial_x & 0 \\ 0 & \partial_t + \partial_x \end{bmatrix} \begin{bmatrix} \lambda_R(t, x) \\ \lambda_L(t, x) \end{bmatrix} = 0 .$$

We will also use the fields smeared with real functions f, where the condition (2) is not needed any more:

$$\begin{bmatrix} \lambda_R(f) \\ \lambda_L(f) \end{bmatrix} = \int \begin{bmatrix} \lambda_R(t, x) \\ \lambda_L(t, x) \end{bmatrix} f(t, x) \mathrm{d}t \mathrm{d}x .$$

Because of the Dirac equation, they can be written as

$$\lambda_R(f) = \lambda_R(g_R), \quad \lambda_L(f) = \lambda_L(g_L) \,,$$

where g_R and g_L where introduced when we discussed bosons.

For $k > 0$, we introduce fermionic operators (for right and left sectors) $b_R(k)$ and $b_L(k)$ satisfying the anticommutation relations

$$\{b_R(k), b_R^\dagger(k')\} = 2\pi\delta(k - k') \,,$$
$$\{b_L(k), b_L^\dagger(k')\} = 2\pi\delta(k - k'), \tag{60}$$

with all other anticommutators vanishing. Now

$$\lambda_R(g_R) = \int \frac{dk}{2\pi} \left(g_R^*(k) b_R(k) + g_R(k) b_R^\dagger(k) \right)$$
$$\lambda_L(g_L) = \int \frac{dk}{2\pi} \left(g_L^*(k) b_L(k) + g_L(k) b_L^\dagger(k) \right) \tag{61}$$

The anticommutation relations for the smeared fields read

$$\{\lambda_R^\dagger(g_{R1}), \lambda_R(g_{R2})\} = \int g_{R1}^*(t) g_{R2}(t) dt = \int \frac{dk}{2\pi} \hat{g}_{R1}^*(k) \hat{g}_{R2}(k) \tag{62}$$

and similarly for the left sector. Note the difference of the fermionic scalar product (62) and the bosonic one $(\cdot|\cdot)$.

In terms of space-time smearing functions these anticommutation relations read

$$\{\lambda_R^\dagger(f_1), \lambda_R(f_R)\} = 2 \int dt dx \delta(t + x) f_1(t, x) f_2(t, x) \,,$$
$$\{\lambda_L^\dagger(f_1), \lambda_R(f_L)\} = 2 \int dt dx \delta(t - x) f_1(t, x) f_2(t, x) \,.$$

Fermionic fields are covariant with respect to the group $A_+(1, \mathbb{R}) \times A_+(1, \mathbb{R})$. We will restrict ourselves to discussing the covariance for say, right movers. The right Hamiltonian and the right dilation generator are

$$H_R^f = \int \frac{dk}{2\pi} k b_R^\dagger(k) b_R(k)$$
$$D_R^f = \frac{i}{2} \int \frac{dk}{2\pi} \left(b_R^\dagger(k) k \partial_k b_R(k) - (k \partial_k b_R^\dagger(k)) b_R(k) \right) \,.$$

We have the usual commutation relations for H_R^f and D_R^f and their action on the fields is anomaly-free:

$$[H_R^f, \lambda_R(g_R)] = -i\lambda(\partial_t g_R) \,,$$
$$[D_R^f, \lambda_R(g_R)] = i\lambda((t\partial_t + 1/2)g_R).$$

We have also the covariance with respect to the conformal group $SL(2,\mathbb{R}) \times SL(2,\mathbb{R})$. We need to assume that test functions satisfy

$$g(t) = O(1/t), \quad |t| \to \infty . \tag{63}$$

We define the action of

$$C = \begin{bmatrix} a & b \\ c & d \end{bmatrix} \in SL(2,\mathbb{R}) \tag{64}$$

on g by

$$(r_C^{\mathrm{f}} g)(t) = (-ct + a)^{-1} g \left(\frac{dt - b}{-ct + a} \right) . \tag{65}$$

Note that (65) has a different power than (35). It is a unitary representation for the scalar product $(\cdot | \cdot)_{\mathrm{f}}$.

We second quantize r_C^{f} on the fermionic Fock space by introducing the unitary operator $R_R^{\mathrm{f}}(C)$ fixed uniquely by the conditions

$$R_R^{\mathrm{f}}(C)\Omega_R = \Omega_R ,$$

$$R_R^{\mathrm{f}}(C)\lambda_R(t)R_R^{\mathrm{f}}(C)^\dagger = (ct + d)^{-1} \lambda_R \left(\frac{at + b}{ct + d} \right) . \tag{66}$$

Note that $C \mapsto R_R^{\mathrm{f}}(C)$ is a unitary representation and it acts naturally on fields:

$$R_R^{\mathrm{f}}(C)\lambda_R(g_R)R_R^{\mathrm{f}}(C)^\dagger = \lambda_R(r_C^{\mathrm{f}} g_R), \tag{67}$$

13 Supersymmetry

In this section we consider both bosons and fermions. Thus our Hilbert space is the tensor product of the bosonic and fermionic part. We assume that the bosonic and fermionic operators commute with one another. Clearly, our theory is $A_+(1,\mathbb{R}) \times A_+(1,\mathbb{R})$ covariant. In fact, the right Hamiltonian and the generator of dilations for the combined theory are equal to $H_R + H_R^{\mathrm{f}}$ and $D_R + D_R^{\mathrm{f}}$.

In the case of the theory with the constraint (2), we have also the $SL(2,\mathbb{R}) \times SL(2,\mathbb{R})$. covariance.

On top of that, the combined theory is supersymmetric. The supersymmetry generators Q_R, Q_L are defined as

$$Q_R = \int \frac{dk}{2\pi} \left(a_{\sigma R}^\dagger(k) b_R(k) + a_{\sigma R}(k) b_R^\dagger(k) \right) ,$$

$$Q_L = \int \frac{dk}{2\pi} \left(a_{\sigma L}^\dagger(k) b_L(k) + a_{\sigma L}(k) b_L^\dagger(k) \right) . \tag{68}$$

They satisfy the basic supersymmetry algebra relations without the central charge

$$\{Q_R, Q_R\} = 2(H_R + H_R^f) \,,$$
$$\{Q_L, Q_L\} = 2(H_L + H_L^f) \,,$$
$$\{Q_R, Q_L\} = 0. \tag{69}$$

The action of the supersymmetric charge transforms bosons into fermions and vice versa:

$$[Q_R, \phi(g_R, g_L)] = \lambda_R(g_R) \,,$$
$$[Q_L, \phi(g_R, g_L)] = \lambda_L(g_L) \,,$$
$$[Q_R, \lambda_R(g_R)] = \phi(\partial_t g_R, 0) \,,$$
$$[Q_L, \lambda_L(g_L)] = \phi(0, \partial_t g_L) \,.$$

The pair of operators $\begin{bmatrix} Q_R \\ Q_L \end{bmatrix}$ behaves like a spinor under the Poincaré group. Even more is true: we have the covariance under the group $A_+(1, \mathbb{R}) \times A_+(1, \mathbb{R})$, which for the right movers can be expressed in terms of the following commutation relations:

$$[H_R, Q_R] = 0 \,, \qquad [D_R, Q_R] = -\frac{i}{2} Q_R \,.$$

Acknowledgement

J.D. was partly supported by the European Postdoctoral Training Program HPRN-CT-2002-0277, the Polish KBN grant SPUB127 and 2 P03A 027 25. K.A.M. was partially supported by the Polish KBN grant 2P03B 001 25 and the European Programme HPRN–CT–2000–00152.
J.D. would like to thank S. DeBièvre, C. Gérard and C. Jäkel for useful discussions.

References

1. Acerbi, F., Morchio, G., Strocchi, F.: Infrared singular fields and nonregular representations of canonical commutation relation algebras, Journ. Math. Phys. 34 (1993) 899–914
2. Acerbi, F., Morchio, G., Strocchi, F.: Theta vacua, charge confinement and charged sectors from nonregular representations of CCR algebras, Lett. Math. Phys. 27 (1993) 1–11.
3. Brattelli, O., Robinson D. W.: *Operator Algebras and Quantum Statistical Mechanics, Volume 2*, Springer-Verlag, Berlin, second edition 1996.

4. S. De Bièvre and J. Renaud: *A conformally covariant quantum field in 1+1 dimension*, J. Phys. A34 (2001) 10901–10919.
5. Buchholz, D.: *Quarks, gluons, colour: facts or fiction?*, Nucl. Phys. B469 (1996) 333–356.
6. D. Buchholz and R. Verch, *Scaling algebras and renormalization group in algebraic quantum field theory. II. Instructive examples*, Rev. Math. Phys. 10 (1998) 775–800.
7. J. Dereziński: *Van Hove Hamiltonians – exactly solvable models of the infrared and ultraviolet problem*, Ann. H. Poincaré 4 (2003) 713–738.
8. M.B. Green, J.H. Schwarz and E. Witten: *Superstring theory*, Cambridge Univ. Press, Cambridge 1987.
9. G.W. Greenberg, J.K. Kang and C.H. Woo: *Infrared regularization of the massless scalar free field in two-dimensional space-time via Lorentz expansion*, Phys. Lett. 71B (1977) 363–366.
10. C. Itzykson and J.B. Zuber: *Quantum Field Theory*, McGraw-Hill, 1980, Chap. 11.
11. G. Morchio, D. Pierotti and F. Strocchi: *Infrared and vacuum structure in two-dimensional local quantum field theory models. The massless scalar field*, Journ. Math. Phys. 31 (1990) 1467–1477.
12. N. Nakanishi: *Free massless scalar field in two-dimensional space-time*, Prog. Theor. Phys. 57 (1977) 269–278.
13. N. Nakanishi: *Free massless scalar field in two-dimensional space-time: revisited*, Z. Physik C. Particles and Fields 4 (1980) 17–25.
14. R.F. Streater and A.S. Wightman: *PCT, spin and statistics and all that*, W.A.Benjamin, New York-Amsterdam 1964.
15. R.F. Streater and I.F. Wilde: *Fermion states of a boson field*, Nucl. Phys. B24 (1970) 561–575.

Stability of Multi-Phase Equilibria

Marco Merkli[1,2]

[1] Department of Mathematics and Statistics, McGill University, 805 Sherbrooke
  W., Montreal QC, Canada, H3A 2K6
  merkli@math.mcgill.ca
[2] Centre des Recherches Mathématiques, Université de Montréal, Succursale
  centre-ville, Montréal, QC, Canada, H3C 3J7

1 Stability of a Single-Phase Equilibrium

We expect equilibrium states of very large systems to have a property of
dynamical stability, called the property of return to equilibrium. This means
that if the system is initially in a state that differs only a bit (say locally in
space) from an equilibrium state, then it approaches that equilibrium state in
the large time limit. One may view this irreversible process as a consequence
of the dispersiveness of the dynamics, and the infinite size of the system:
the local spatial disturbance of the equilibrium state, defining the initial
state, propagates out of any bounded region if one waits long enough. Strictly
speaking we expect return to equilibrium only for spatially infinitely extended
systems. If one desires to observe localized events of a large, but finite systems
(a laboratory) then the above description is a good approximation on an
intermediate time scale (time should be large so the system can settle towards
the equilibrium state, but not too large as to avoid recurrences in the finite
system).

A quantum system for which the property of return to equilibrium is easy
to examine is the free Bose gas. If the gas has a Bose-Einstein condensate then
the system has many coexisting phases (many equilibrium states at the same
temperature). We will explain what the property of return to equilibrium
translates into for systems with multiple (multi-phase) equilibria.

If a free Bose gas, modeling an "environment" or a "heat bath", is coupled
to a finite system, e.g. to a spin, then it is not so easy any more to show
the property of return to equilibrium. This question has been examined, for
single-phase equilibria, in a variety of recent publications, [3,6,10,14,15,27].
The case when the gas is in a condensate state is investigated in [18]. We
refer to [11] for a discussion of the above mentioned validity of the infinite
volume approximation of finite systems for intermediate times.

1.1 The Free Bose Gas

Let $\Lambda = [-L/2, L/2]^3 \subset \mathbb{R}^3$ be a box in physical space. Pure states of the
bose gas localized inside Λ are represented by vectors in the bosonic Fock

M. Merkli: *Stability of Multi-Phase Equilibria*, Lect. Notes Phys. **690**, 129–148 (2006)
www.springerlink.com

space

$$\mathcal{F}_\Lambda = \bigoplus_{n=0}^{\infty} P_+ L^2(\Lambda, \mathrm{d}^3 x)^{\otimes n} , \tag{1}$$

where P_+ projects onto functions which are symmetric under permutations of arguments. In order to describe an infinitely extended Bose gas we want to increase the volume $\Lambda \uparrow \mathbb{R}^3$. One may replace Λ by $\mathbb{R}^3$ in expression (1) thereby getting a Hilbert space $\mathcal{F}$ whose vectors represent states (where particles are not constrained to any bounded volume). If the system is in the state $\psi = \{\psi_n\}_{n=0}^{\infty}$ (a normalized vector in $\mathcal{F}$) then the probability of finding n particles is given by $p_n = \|\psi_n\|^2_{L^2(\mathbb{R}^{3n}, \mathrm{d}^{3n} x)}$. Since these probabilities must add up to one we have $p_n \to 0$, as $n \to \infty$. This indicates that if we would like to describe an infinitely extended Bose gas with a fixed (average) density, say one particle per unit volume, then Fock space cannot be the right state space.

In order to describe the infinite system we should take the infinite volume limit of the *expectation functional* ("averages") defined by a vector, or a density matrix on $\mathcal{F}_\Lambda$. To this end it is useful to introduce the *Weyl algebra* $\mathfrak{W}_\Lambda = \mathfrak{W}(L^2(\Lambda, \mathrm{d}^3 x))$. This is the C^*algebra generated by the unitary *Weyl operators*,

$$W(f) = \mathrm{e}^{\mathrm{i}\varphi(f)} , \tag{2}$$

$f \in L^2(\Lambda, \mathrm{d}^3 x)$, where $\varphi(f) = \frac{1}{\sqrt{2}}(a^*(f) + a(f))$, and $a^*(f)$, $a(f)$ are the usual creation- and annihilation operators on $\mathcal{F}_\Lambda$. The Weyl algebra provides us with a rich class of "observables". Namely, $\mathfrak{W}(L^2(\Lambda, \mathrm{d}^3 x))$ is dense (in the weak operator topology) in $\mathcal{B}(\mathcal{F}_\Lambda)$, the set of *all* bounded operators on $\mathcal{F}_\Lambda$. (This is a consequence of the fact that the Weyl algebra acts irreducibly on Fock space.) The Weyl operators satisfy the *Canonical Commutation Relations* (CCR)

$$W(f)W(g) = \mathrm{e}^{-\frac{1}{2}\mathrm{Im}\langle f, g\rangle} W(f + g) . \tag{3}$$

A state of the system is given by a positive linear functional $\omega_\Lambda : \mathfrak{W}_\Lambda \to \mathbb{C}$, normalized as $\omega_\Lambda(\mathbb{1}) = 1$. In view of (3) it is not very surprising that any state on the Weyl algebra is entirely determined by its value on the generators of the algebra, i.e., by the (nonlinear) expectation functional

$$L^2(\Lambda, \mathrm{d}^3 x) \ni f \mapsto E_\Lambda(f) := \omega_\Lambda(W(f)) . \tag{4}$$

The converse is true too: if $E_\Lambda : L^2(\Lambda, \mathrm{d}^3 x) \to \mathbb{C}$ is a functional satisfying certain compatibility conditions, then it defines uniquely a state ω_Λ on $\mathfrak{W}_\Lambda$, via $\omega_\Lambda(W(f)) = E_\Lambda(f)$.

The *dynamics* of Weyl operators is given by

$$t \mapsto \alpha_t^\Lambda(W(f)) = W(\mathrm{e}^{-\mathrm{it}H_\Lambda} f) , \tag{5}$$

where the *one particle Hamiltonian* H_Λ is $-\Delta$ (Laplace operator acting on $L^2(\Lambda, \mathrm{d}^3 x)$ with some classical boundary condition, or a function of this operator).

The procedure of finding (defining!) the infinitely extended system is now reduced to finding the limiting expectation functional $E = \lim_{\Lambda \uparrow \mathbb{R}^3} E_\Lambda$. We are interested in constructing equilibrium states (w.r.t. the dynamics (5)). Those are special states, characterized by an inverse temperature β and a mean density $\bar{\rho}$, which satisfy the so-called KMS condition, [4]. Formally, a state ω on a C^*algebra $\mathfrak{A}$ is a (β, α_t)-KMS state (where α_t is a $*$automorphism group of $\mathfrak{A}$, and $0 < \beta < \infty$) if the KMS condition is satisfied,

$$\omega(A\alpha_t(B)) = \omega(\alpha_{t-i\beta}(B)A) , \tag{6}$$

for suitable observables $A, B \in \mathfrak{A}$.

It is convenient and standard to formulate the theory in Fourier space, where the periodic Laplacian is diagonalized. In the infinite volume limit, the test functions (wave functions of a single particle) f in (2)–(5) are elements of $L^2(\mathbb{R}^3, \mathrm{d}^3k)$. In the following we place ourselves in this setting (all results can be expressed in direct space at the expense of a more cumbersome notation).

The Araki–Woods Construction, [2]

Let $\mathbb{R}^3 \ni k \mapsto \rho(k) > 0$ be a given function (the "continuous momentum-density distribution"), and $\rho_0 \geq 0$ a fixed number (the "condensate density"). Araki and Woods obtain a state of the Bose gas by the following procedure. Put $L^3 \rho_0$ particles in the ground state of the one particle Hamiltonian H_Λ, and a discrete distribution of particles in excited states. Then take the limit $L \to \infty$ while keeping ρ_0 fixed and letting the discrete distribution of excited states tend to $\rho(k)$. In this way one obtains the generating functional

$$E_{\rho, \rho_0}^{\mathrm{AW}}(f) = \exp\left[-\frac{1}{4}\langle f, (1 + 2(2\pi)^3 \rho) f\rangle\right] J_0\left(\sqrt{2(2\pi)^3 \rho_0} |f(0)|\right), \tag{7}$$

where $J_0(\sqrt{\alpha^2 + \beta^2}) = \int_{-\pi}^{\pi} \frac{\mathrm{d}\theta}{2\pi} e^{-i(\alpha \cos \theta + \beta \sin \theta)}$, $\alpha, \beta \in \mathbb{R}$ (Bessel function). As mentioned above, $E_{\rho, \rho_0}^{\mathrm{AW}}$ defines a state of the infinitely extended Bose gas. The physical interpretation is that this state describes a free Bose gas where a sea of particles, all being in the same state (corresponding to the ground state of the finite-volume Hamiltonian), form a condensate with density ρ_0, which is immersed in a gas of particles where $\rho(k)$ particles per unit volume have momentum in the infinitesimal volume d^3k around $k \in \mathbb{R}^3$. If the Hamiltonian in the finite box is taken with periodic boundary conditions the condensate is homogeneous in space (the ground state wave function is a constant in position space). The resulting state is an equilibrium state (satisfies the KMS condition (6) for A, B Weyl operators with test functions $f \in L^2(\mathbb{R}^2, \mathrm{d}^3k)$ and w.r.t. the dynamics $t \mapsto W(e^{-it\omega} f))$ if the momentum density distribution is given by

$$\rho(k) = (2\pi)^{-3} \frac{1}{e^{\beta \omega(k)} - 1} , \tag{8}$$

corresponding to Planck's law of black body radiation, and the condensate density $\rho_0 \geq 0$ is arbitrary. We consider dispersion relations $\omega(k) = |k|^2$ or $\omega(k) = |k|$ (non-relativistic Bosons or massless relativistic ones).

The Grand-Canonical Construction, [16]

The density matrix (acting on Fock space $\mathcal{F}_\Lambda$) for the local grand-canonical equilibrium system is

$$\sigma_{\beta,z_\Lambda}^\Lambda = \frac{e^{-\beta(H_\Lambda - \mu_\Lambda N_\Lambda)}}{\operatorname{Tr} e^{-\beta(H_\Lambda - \mu_\Lambda N_\Lambda)}} \,, \tag{9}$$

where $\mu_\Lambda \in \mathbb{R}$ is the *chemical potential*, $z_\Lambda = e^{\beta\mu_\Lambda}$ is the *fugacity*, and N_Λ is the number operator. For a fixed inverse temperature $0 < \beta < \infty$ define the *critical density* by

$$\rho_{\text{crit}}(\beta) = (2\pi)^{-3} \int \frac{d^3 k}{e^{\beta\omega(k)} - 1} \,, \tag{10}$$

and denote by $\bar{\rho} \geq 0$ the total (mean) density of the gas, whose value we are at liberty to choose,

$$\bar{\rho} = \operatorname{Tr}\left(\sigma_{\beta,z_\Lambda}^\Lambda \frac{N_\Lambda}{L^3}\right) . \tag{11}$$

For each fixed L, this determines the value of z_Λ (as a function of L, $\bar{\rho}$ and β). One then performs the thermodynamic limit of the expectation functional,

$$E_{\beta,\bar{\rho}}^{\text{GC}}(f) = \lim_{L\to\infty} E_{\beta,z_\Lambda}^\Lambda(f) := \lim_{L\to\infty} \operatorname{Tr}\left(\sigma_{\beta,z}^\Lambda W(f)\right) , \tag{12}$$

where the limit $L \to \infty$ is taken under the constraint (11). The limiting generating functional is

$$E_{\beta,\bar{\rho}}^{\text{GC}}(f) = \begin{cases} e^{-\frac{1}{4}\|f\|^2} \exp\left[-\frac{1}{2}\langle f, \frac{z_\infty}{e^{\beta\omega} - z_\infty} f\rangle\right], & \bar{\rho} \leq \rho_{\text{crit}}(\beta) \\ E_{\beta,\bar{\rho}}^{\text{con}}(f), & \bar{\rho} \geq \rho_{\text{crit}}(\beta) \end{cases} \tag{13}$$

where, with $\rho_0 := \bar{\rho} - \rho_{\text{crit}}(\beta) \geq 0$,

$$E_{\beta,\bar{\rho}}^{\text{con}}(f) = \exp\left[-\frac{1}{4}\langle f, (1 + 2(2\pi)^3\rho)f\rangle\right] \exp\left\{-4\pi^3\rho_0|f(0)|^2\right\} , \tag{14}$$

and $\rho = \rho(k)$ is given in (8). For subcritical density, $\bar{\rho} \leq \rho_{\text{crit}}(\beta)$, the number $z_\infty \in [0, 1]$ is determined by the equation

$$\bar{\rho} = (2\pi)^{-3} \int \frac{z_\infty}{e^{\beta\omega(k)} - z_\infty} d^3 k . \tag{15}$$

In the supercritical case, $\bar{\rho} \geq \rho_{\text{crit}}(\beta)$, we have $z_\infty = 1$ which corresponds to a vanishing chemical potential, $\mu_\infty = 0$.

The Canonical Construction, [5]

The density matrix of the canonical local Gibbs state is

$$\mu^\Lambda_{\beta,\bar\rho} = \frac{e^{-\beta H_\Lambda} P_{\bar\rho L^3}}{\mathrm{Tr}\, e^{-\beta H_\Lambda} P_{\bar\rho L^3}} , \tag{16}$$

and $P_{\bar\rho L^3}$ is the projection onto the subspace of Fock space with $\bar\rho L^3$ particles (if $\bar\rho L^3$ is not an integer take a convex combination of canonical states with integer values $\bar\rho_1 L^3$ and $\bar\rho_2 L^3$ extrapolating $\bar\rho L^3$). The limiting generating functional is given by

$$E^C_{\beta,\bar\rho}(f) = \begin{cases} e^{-\frac14\|f\|^2} \exp\left[-\frac12\langle f, \frac{z_\infty}{e^{\beta\omega}-z_\infty} f\rangle\right], & \bar\rho \le \rho_{\mathrm{crit}}(\beta) \\ E^{AW}_{\rho,\rho_0}(f), & \bar\rho \ge \rho_{\mathrm{crit}}(\beta) . \end{cases} \tag{17}$$

It coincides with the grand-canonical generating functional in the subcritical case, and with the Araki-Woods generating functional with ρ given by (8) and $\rho_0 = \bar\rho - \rho_{\mathrm{crit}}$ in the supercritical case.

The grand-canonical and the canonical generating functionals are linked, [5], in the supercritical case, $\rho_0 > 0$, by the Laplace transform

$$E^{con}_{\beta,\bar\rho}(f) = \int_0^\infty K(r;\bar\rho) E^C_{\beta,r}(f)\mathrm{d}r , \tag{18}$$

where the *Kac density* $K(r;\bar\rho)$ is

$$K(r;\bar\rho) = \begin{cases} e^{-(r-\rho_{\mathrm{crit}})/\rho_0}/\rho_0, & r > \rho_{\mathrm{crit}} \\ 0, & r \le \rho_{\mathrm{crit}} . \end{cases} \tag{19}$$

This means that the grand-canonical equilibrium state with supercritical mean density $\bar\rho$ is a superposition of canonical equilibrium states with supercritical densities r, weighted with the Kac density $K(r,\bar\rho)$.

From now on, we focus on the infinite volume equilibrium state $\omega_{\beta,\bar\rho}$ determined by the thermodynamic limit of the canonical expectation functional, (17). (All that follows can be carried out for the grand-canonical expectation functional, (13), see [18], but is notationally less cumbersome in the canonical case.)

1.2 Spontaneous Symmetry Breaking and Multi-Phase Equilibrium

We denote by $E^{AW}_\rho(f)$ the Araki-Woods expectation functional (7) at critical density, where $\rho(k)$ is given by (8), and $\rho_0 = 0$. The corresponding equilibrium state is denoted by ω_β (it is a state on $\mathfrak{W} \equiv \mathfrak{W}(\mathfrak{D})$, where

$$\mathfrak{D} \subset L^2(\mathbb{R}^3, \mathrm{d}^3 k) \tag{20}$$

consists of functions s.t. the right side of (7) is defined).

With the expansion of the Bessel function J_0 given after (7) we can write the canonical expectation functional, for $\bar{\rho} \geq \rho_{\mathrm{crit}}(\beta)$, as a superposition

$$E^{\mathrm{C}}_{\beta,\bar{\rho}}(f) = \int_{S^1} \frac{\mathrm{d}\theta}{2\pi} \mathrm{e}^{-\mathrm{i}\Phi(f,\theta)} E^{\mathrm{AW}}_{\rho}(f) , \tag{21}$$

where the real phase Φ is given by

$$\Phi(f,\theta) = (2\pi)^{3/2} \sqrt{2\rho_0} \left((\mathrm{Re} f(0)) \cos\theta + (\mathrm{Im} f(0)) \sin\theta\right) . \tag{22}$$

Correspondingly, we define the states ω^θ_β on $\mathfrak{W} \equiv \mathfrak{W}(\mathfrak{D})$ by

$$\omega^\theta_\beta(W(f)) = \mathrm{e}^{-\mathrm{i}\Phi(f,\theta)} \omega_\beta(W(f)) . \tag{23}$$

The point of this exercise is to notice that for each θ, (23) defines a β-KMS state w.r.t. the dynamics $\alpha^t(W(f)) = W(\mathrm{e}^{-\mathrm{i}t\omega} f)$. Thus the supercritical equilibrium state $\omega_{\beta,\rho}$ corresponding to (21) is a uniform superposition of the equilibrium states ω^θ_β, $\theta \in S^1$. Of course, we can now take any probability measure $\mathrm{d}\mu$ on S^1 and define a (β, α_t)-KMS state on $\mathfrak{W}$ by

$$\omega^\mu(\cdot) = \int_{S^1} \mathrm{d}\mu(\theta) \omega^\theta_\beta(\cdot) . \tag{24}$$

One easily shows that the states (23) are factor states, so they are extremal. We also point out that they are not invariant under the gauge group $\sigma_s(W(f)) = W(\mathrm{e}^{\mathrm{i}s} f)$, $s \in \mathbb{R}$, which is a symmetry group of the dynamics α_t (meaning that $\alpha_t \circ \sigma_s = \sigma_s \circ \alpha_t$, $s, t \in \mathbb{R}$). The existence of equilibrium states which have "less symmetry" than the dynamics is called *spontaneous symmetry breaking*.

We close this paragraph with some observations on *space mixing properties*. Given a vector $a \in \mathbb{R}^3$ we define $\tau_a(W(f)) := W(f_a)$, where $f_a(x) := f(x - a)$ is the (direct-space) translate of f by a. τ_a defines a (three parameter) group of automorphisms on $\mathfrak{W}$. A state ω on $\mathfrak{W}$ is called *strongly mixing* w.r.t. space translations if $\lim_{|a|\to\infty} \omega(W(f)\tau_a(W(g))) = \omega(W(f))\omega(W(g))$, for any $f, g \in \mathfrak{D}$. This means that if two observables ($W(f)$ and $W(g)$) are spatially separated far from each other then the expectation of the product of the observables is close to the product of the expectation values (independence of random variables). Intuitively, this means that the state ω has a certain property of locality in space: what happens far out in space does not influence events taking place, say, around the origin. For the equilibrium state $\omega_{\beta,\bar{\rho}}$ determined by (17), it is easy to show that

$$\lim_{|a|\to\infty} \omega_{\beta,\bar{\rho}}\big(W(f)\tau_a(W(g))\big)$$

$$= \omega_{\beta,\bar{\rho}}(W(f))\omega_{\beta,\bar{\rho}}(W(g)) \begin{cases} 1, & \bar{\rho} \leq \rho_{\mathrm{crit}}(\beta) \\ \exp\left[-8\pi^3\rho_0 \, \mathrm{Re}(\overline{f(0)}g(0))\right], & \bar{\rho} \geq \rho_{\mathrm{crit}}(\beta) . \end{cases}$$

Consequently, this equilibrium state is strongly mixing w.r.t. space translations if and only if $\rho_0 = 0$, i.e., if and only if there is no condensation. In presence of a condensate, the system exhibits *long range correlations* (what happens far out does influence what happens say at the origin). On the other hand, it is easily verified that each state ω_β^θ is strongly mixing.

1.3 Return to Equilibrium in Absence of a Condensate

Consider the equilibrium state $\omega_{\beta,\rho}$ defined by the expectation functional (17), in the regime $\overline{\rho} \le \rho_{\mathrm{crit}}(\beta)$. We say that $\omega_{\beta,\overline{\rho}}$ has the property of return to equilibrium iff

$$\lim_{t\to\infty} \omega_{\beta,\overline{\rho}}(B^*\alpha_t(A)B) = \omega_{\beta,\overline{\rho}}(B^*B)\omega_{\beta,\overline{\rho}}(A) , \tag{25}$$

for all $A, B \in \mathfrak{W}$. (This means that all states which are normal w.r.t. $\omega_{\beta,\overline{\rho}}$ converge to $\omega_{\beta,\overline{\rho}}$ in the long time limit.) Let us show that

$$\lim_{t\to\infty} \omega_{\beta,\overline{\rho}}(W(g)\alpha_t(W(f))W(h)) = \omega_{\beta,\overline{\rho}}(W(g)W(h))\omega_{\beta,\overline{\rho}}(W(f)) , \tag{26}$$

for all $f, g, h \in \mathfrak{D}$. (26) implies (25). Using the CCR, (3), we obtain

$$W(g)W(e^{i\omega t}f)W(h) = e^{-\frac{1}{2}\mathrm{Im}[\langle g,e^{i\omega t}f\rangle + \langle g+e^{i\omega t}f,h\rangle]}W(e^{i\omega t}f + g + h) . \tag{27}$$

Using the Riemann-Lebesgue lemma, the first factor on the r.h.s. of (27) is seen to have the limit $e^{-\frac{1}{2}\mathrm{Im}\langle g,h\rangle}$, as $t \to \infty$. From (17) we obtain

$$\omega_{\beta,\overline{\rho}}(W(e^{i\omega t}f + g + h))$$
$$= e^{-\frac{1}{4}\|e^{i\omega t}f+g+h\|^2} \exp\left[-\frac{1}{2}\left\|\sqrt{\frac{z_\infty}{e^{\beta\omega(k)} - z_\infty}}\,(e^{i\omega t}f + g + h)\right\|^2\right] , \tag{28}$$

and another application of the Riemann-Lebesgue lemma shows that the r.h.s. of (28) tends to $\omega_{\beta,\overline{\rho}}(W(f))\omega_{\beta,\overline{\rho}}(W(g+h))$, as $t \to \infty$. That's it as for return to equilibrium for $\overline{\rho} \le \rho_{\mathrm{crit}}(\beta)$!

1.4 Return to Equilibrium in Presence of a Condensate

Of course one can do the calculation of the previous section in the case $\overline{\rho} > \rho_{\mathrm{crit}}$ by using explicitly (17) again. One will then notice that due to the presence of the Bessel function J_0 in (7), expressions do not split so nicely into products any longer. A better way of doing things is to realize that the extremal states (23) do have the property of return to equilibrium, as is obvious from the calculation in the previous paragraph and the fact that $\Phi(e^{i\omega t}f, \theta) = \Phi(f, \theta)$. This leads immediately to the following expression for the asymptotic state of an initial condition which is a local perturbation of (normal w.r.t.) a general mixture, (24), of the extremal equilibria:

$$\lim_{t\to\infty} \omega^\mu(B^*\alpha_t(A)B) = \int_{S^1} \mathrm{d}\mu(\theta)\omega^\theta_\beta(B^*B)\omega^\theta_\beta(A) . \tag{29}$$

Hence the time asymptotic state of the system depends on the state it was initially in. We may interpret this "memory effect" as a consequence of the fact that the system has long-range correlations. Although the perturbation propagates away from any bounded set (the dynamics is dispersive), correlations survive in the limit of large times. If B is such that $\omega^\theta_\beta(B^*B) = 1$ for all θ in the support of $\mathrm{d}\mu$ (e.g. if B is unitary) then the asymptotic state is just ω^μ again. In general however, the limit state is a different mixture of the extremal equilibria, with a distribution given by the measure

$$\mathrm{d}\mu_B(\theta) = \omega^\theta_\beta(B^*B)\mathrm{d}\mu(\theta) , \tag{30}$$

another probability measure, which is *absolutely continuous* w.r.t. $\mathrm{d}\mu(\theta)$. In particular, the time asymptotic state (29) is normal w.r.t. ω^μ.

1.5 Spectral Approach

Take a heat reservoir and bring it into contact with a small system, say an N-level system (e.g. a spin, or an idealized atom in a cavity). The interaction gives rise to emission and absorption processes, where particles in the reservoir (Bosons) are swallowed by the small system, which thereby increases its energy, or where the small system releases a particle by lowering its energy.

It is a well know fact that the coupled system has an equilibrium state (Araki's structural stability of equilibria, see also [7]). We want to show return to equilibrium for the coupled system.

Let us outline here a strategy for doing so, introduced in [14, 15], and further developed in [3, 6, 9, 10, 17, 18].

Assume ω is a (β, α_t)-equilibrium state on a C^*algebra $\mathfrak{A}$, where α_t is a dynamics on $\mathfrak{A}$ (a group of *automorphisms). The GNS construction gives a Hilbert space representation $(\mathcal{H}, \pi, \Omega)$ of the pair $(\mathfrak{A}, \omega)$,

$$\omega(A) = \langle \Omega, \pi(A)\Omega \rangle , \tag{31}$$

where $\mathcal{H}$ is the representation Hilbert space, $\Omega \in \mathcal{H}$, and π maps $\mathfrak{A}$ into bounded operators on $\mathcal{H}$. There is a unique selfadjoint operator L on $\mathcal{H}$ satisfying

$$\pi(\alpha_t(A)) = \mathrm{e}^{\mathrm{i}tL}\pi(A)\mathrm{e}^{-\mathrm{i}tL}, \quad L\Omega = 0 , \tag{32}$$

for all $A \in \mathfrak{A}$, $t \in \mathbb{R}$. L is called the (standard) *Liouvillian*. We have

$$\omega(B^*\alpha_t(A)B) = \omega(\alpha_{-\mathrm{i}\beta}(B)B^*\alpha_t(A))$$
$$= \langle \Omega, \pi\left(\alpha_{-\mathrm{i}\beta}(B)B^*\right)\mathrm{e}^{\mathrm{i}tL}\pi(A)\Omega \rangle . \tag{33}$$

We use here that ω satisfies the KMS condition (6), and (31), (32). It becomes now apparent how we can link the long-time behaviour of (33) to the spectral properties of L. If the kernel of L has dimension one (its dimension is at least one due to (32)), and if L has purely absolutely continuous spectrum on $\mathbb{R}\backslash\{0\}$, then the r.h.s. of (33) converges to

$$\langle \Omega, \pi\left(\alpha_{-i\beta}(B)B^*\right)\Omega\rangle\langle\Omega, \pi(A)\Omega\rangle = \omega(\alpha_{-i\beta}(B)B^*)\omega(A)\,, \qquad (34)$$

as $t \to \infty$. The first term on the r.h.s. is just $\omega(B^*B)$ (use again the KMS condition (6) with $t = 0$). The combination of (34) and (33) shows that

$$\lim_{t\to\infty} \omega(B^*\alpha_t(A)B) = \omega(B^*B)\omega(A)\,, \qquad (35)$$

provided $\mathrm{Ker}L = \mathbb{C}\Omega$ and the spectrum of L on $\mathbb{R}\backslash\{0\}$ is purely absolutely continuous. This shows return to equilibrium (c.f. (25))!

A more modest version is return to equilibrium in the ergodic average sense, where the limit in (35) is understood in the ergodic mean,

$$\lim_{T\to\infty} \frac{1}{T}\int_0^T \omega(B^*\alpha_t(A)B)\mathrm{d}t = \omega(B^*B)\omega(A)\,. \qquad (36)$$

Relation (36) follows from (33) and the von Neumann ergodic theorem, provided $\mathrm{Ker}L = \mathbb{C}\Omega$.

2 Stability of Multi-Phase Equilibria

The discussion of Sect. 1.4 motivates the following more abstract

Definitions. 1. Let ω be a state on a C^*algebra $\mathfrak{A}$, invariant w.r.t. a $*$automorphism group α_t of $\mathfrak{A}$. We say that ω *is asymptotically stable (w.r.t. α_t)* if

$$\lim_{t\to\infty} \omega(B^*\alpha_t(A)B) = \omega(B^*B)\omega(A)\,, \qquad (37)$$

for any $A, B \in \mathfrak{A}$.

2. Let ω_ξ, $\xi \in X$ (a measurable space), be a measurable family of states on a C^*algebra $\mathfrak{A}$ (in the sense that $\xi \mapsto \omega_\xi(A)$ is measurable for all $A \in \mathfrak{A}$) and let α_t be a $*$automorphism group of $\mathfrak{A}$. Given any probability measure μ on X we define the state

$$\omega^\mu = \int_X \mathrm{d}\mu(\xi)\, \omega_\xi\,. \qquad (38)$$

We say that *the family ω_ξ is asymptotically stable (w.r.t. α_t)* if, for any μ, A, B, we have

$$\lim_{t\to\infty} \omega^\mu(B^*\alpha_t(A)B) = \int_X \mathrm{d}\mu(\xi)\, \omega_\xi(B^*B)\, \omega_\xi(A)\,. \qquad (39)$$

3. If ω in 1. is a (β, α_t)-KMS state then we say ω *has the property of Return to Equilibrium*. Similarly, if the ω_ξ in 2. are (β, α_t)-KMS states (then so is ω^μ) we say *the family ω_ξ has the property of Return to Equilibrium*.

In the above definitions the dynamics of the system is given by a
∗automorphism group α_t of a C^*algebra $\mathfrak{A}$. While this description applies to
free Fermionic or Bosonic heat reservoirs it does not in case a *Bosonic* reservoir is *coupled* to a small system. The problem is that one does not know how
to define the dynamics for the coupled system as a ∗automorphism group of
the C^*algebra of observables (unless the algebra is modified, see [9]). One circumvents this issue by defining the interacting dynamics as a ∗automorphism
group of the *von Neumann algebra* associated with a reference state (e.g. the
uncoupled equilibrium state). We shall therefore adapt the above definitions
to a setting where the dynamics is not defined on the level of the C^*algebra of
observables, but is rather expressed as a ("Schrödinger") dynamics of states.

Definitions.   **1'.** Let ω be a state on a C^*algebra $\mathfrak{A}$ and denote by
$(\mathcal{H}_\omega, \pi_\omega, \Omega_\omega)$ its GNS representation, $\omega(A) = \langle \Omega_\omega, \pi_\omega(A)\Omega_\omega \rangle$. Suppose σ_t is
a ∗automorphism group of the von Neumann algebra $\pi_\omega(\mathfrak{A})''$. We say that ω
is asymptotically stable (w.r.t. σ_t) if

$$\lim_{t\to\infty} \langle \Omega_\omega, \pi_\omega(B^*)\sigma_t(\pi_\omega(A))\pi_\omega(B)\Omega_\omega \rangle = \omega(B^*B)\omega(A) , \qquad (40)$$

for all $A, B \in \mathfrak{A}$.
   2'. Let ω_ξ, $\xi \in X$ (a measurable space), be a measurable family of states
on a C^*algebra $\mathfrak{A}$ and denote their GNS representations by $(\mathcal{H}_\xi, \pi_\xi, \Omega_\xi)$.
Suppose that, for each ξ, σ_t^ξ is a ∗automorphism group of the von Neumann
algebra $\pi_\xi(\mathfrak{A})''$, s.t. $\xi \mapsto \langle \sigma_t^\xi(A) \rangle_{B\Omega_\xi}$ is measurable, for all $A, B \in \pi_\xi(\mathfrak{A})''$,
$t \in \mathbb{R}$. ($\langle A \rangle_\Omega = \langle \Omega, A\Omega \rangle$.) We say that *the family ω_ξ is asymptotically stable
(w.r.t. σ_t^ξ)* if, for any $A, B \in \mathfrak{A}$, we have

$$\lim_{t\to\infty} \int_X d\mu(\xi)\langle \sigma_t^\xi(\pi_\xi(A)) \rangle_{\pi_\xi(B)\Omega_\xi} = \int_X d\mu(\xi)\, \omega_\xi(B^*B)\, \omega_\xi(A) , \qquad (41)$$

where μ is an arbitrary probability measure on X.
   3'. If ω in 1'. is a (β, σ_t)-KMS state of $\pi_\omega(\mathfrak{A})''$ then we say ω *has the
property of Return to Equilibrium*. Similarly, if the ω_ξ in 2'. are (β, σ_t^ξ)-KMS
states of $\pi_\xi(\mathfrak{A})''$ we say *the family ω_ξ has the property of Return to Equilibrium*.

3 Quantum Tweezers

We investigate a Bose gas in a state with Bose-Einstein condensate coupled
to a small system.

Description of Model and Stability Result

The small system with which the Bose gas with condensate interacts can trap
finitely many Bosons – we call it therefore a quantum dot. One can imagine

the use of such a trap to remove single (uncharged) particles from a reservoir, hence the name quantum tweezers (see also [8] and the references therein).

The pure states of the small system are given by normalized vectors in $\mathbb{C}^d$. We interpret $[1, 0, \ldots, 0]$ as the *ground state* (or "vacuum state"), $[0, 1, 0, \ldots, 0]$ as the first excited state, e.t.c. The Hamiltonian is given by the diagonal matrix

$$H_1 = \operatorname{diag}(0, 1, 2, \ldots, d-1) . \tag{42}$$

Our method applies to any selfadjoint diagonal matrix with non-degenerate spectrum. We introduce the raising and lowering operators, G_+ and G_-,

$$G_+ = \begin{bmatrix} 0 & 0 & \cdots & 0 \\ 1 & 0 & \ddots & \vdots \\ \vdots & \ddots & \ddots & 0 \\ 0 & \cdots & 1 & 0 \end{bmatrix}, \quad G_- = (G_+)^* , \tag{43}$$

(G_+ has ones on its subdiagonal) which satisfy $H_1 G_\pm = G_\pm(H_1 \pm 1)$. The action of G_+ (G_-) increases (decreases) the excitation level of the quantum dot by one. The dynamics of an observable $A \in \mathcal{B}(\mathbb{C}^d)$ (bounded operators on $\mathbb{C}^d$) is given by

$$\alpha_1^t(A) = e^{itH_1} A e^{-itH_1}, \quad t \in \mathbb{R} . \tag{44}$$

The observable algebra of the combined system is the C^*-algebra

$$\mathfrak{A} = \mathcal{B}(\mathbb{C}^d) \otimes \mathfrak{W}(\mathfrak{D}) , \tag{45}$$

where $\mathfrak{D} \subset L^2(\mathbb{R}^3, \mathrm{d}^3 k)$ (Fourier space) consists of $f \in L^2(\mathbb{R}^3, (1+\rho)\mathrm{d}^3 k)$ which are continuous at zero. The non-interacting dynamics is

$$\alpha_0^t = \alpha_1^t \otimes \alpha_2^t , \tag{46}$$

where

$$\alpha_2^t(W(f)) = W(e^{it\omega} f) , \tag{47}$$

$\omega(k) = |k|^2$ or $|k|$, is the free field dynamics. Denote by $\omega_{1,\beta}$ the Gibbs state of the quantum dot, and let ω_β^θ be given as in (23). Then

$$\omega_{\beta,0}^\theta := \omega_{1,\beta} \otimes \omega_\beta^\theta \tag{48}$$

is a (β, α_0^t)-KMS state. We can form different equilibrium states by mixing such states according to any probability measure μ on S^1. Let us now introduce an interaction operator, formally given by the expression

$$\lambda \big(G_+ \otimes a(g) + G_- \otimes a^*(g) \big) , \tag{49}$$

where $\lambda \in \mathbb{R}$ is a coupling constant, the $G_\pm$ are the raising and lowering operators, (43), and $a^\#(g)$ are creation ($\# = *$) and annihilation operators of the heat bath, smeared out with a function $g \in \mathfrak{D}$, called a *form factor*. The operator $G_+ \otimes a(g)$ destroys a Boson and traps it in the quantum dot (whose excitation level is thereby increased by one) and similarly, the effect of $G_- \otimes a^*(g)$ is to release a Boson from the quantum dot. The total number of particles, measured by the "observable" $H_1 + \int_{\mathbb{R}^3} a^*(k)a(k)\mathrm{d}^3 k$, is preserved by the interaction (meaning that (49) commutes with this operator). Since the quantum dot can absorb only finitely many Bosons, the interacting equilibrium state is a (local) perturbation of the non-interacting one. A physically different situation occurs when the condensate is coupled to another reservoir. Then the time-asymptotic states are *non-equilibrium stationary states*.

As we show in Sect. 3.2, the system has equilibrium states w.r.t. the interacting dynamics, and there is again a special family among them (extremal factorial ones), $\omega^\theta_{\beta,\lambda}$, labelled by $\theta \in S^1$, compare with (48).

Let μ be a probability measure on S^1 and set

$$\omega^\mu = \int_{S^1} \mathrm{d}\mu(\theta)\, \omega^\theta_{\beta,\lambda} \,. \tag{50}$$

Our *weak coupling* result on Return of Equilibrium (Theorem 1 below) says that for all μ, A, B

$$\lim_{\lambda \to 0} \lim_{T \to \infty} \frac{1}{T} \int_0^T \mathrm{d}t\, \omega^\mu(B^* \sigma^t_\lambda(A)B) = \int_{S^1} \mathrm{d}\mu(\theta)\, \omega^\theta_{\beta,0}(B^*B)\omega^\theta_{\beta,0}(A) \,, \tag{51}$$

where σ^t_λ is the interacting dynamics. (The expression $\sigma^t_\lambda(A)$ has to be understood *cum grano salis*, in the sense of Definition 2', as σ^t_λ can only be defined on the von Neumann algebra of observables, c.f. Sect. 3.2).

We prove (51) under a condition of regularity and "effectiveness" of the interaction. Let us close this section by discussing the physical meaning of the latter condition. Consider first the Bose gas at critical density $\rho_{\mathrm{crit}}(\beta)$ for some fixed temperature $1/\beta$ (so that there is no condensate, $\rho_0 = 0$). Heuristically, the probability of trapping a Boson in a state f in the quantum dot is given by

$$\left| \langle (G_+ \otimes a(f))(\varphi \otimes \widetilde{\Omega}), \mathrm{e}^{-\mathrm{i}t H_\lambda}(\varphi \otimes \widetilde{\Omega}) \rangle \right|^2 \,, \tag{52}$$

where φ an eigenstate of the quantum dot Hamiltonian and the Bose gas is in the equilibrium state $\widetilde{\Omega}$ (for the calculation, we put the system in a box and $\widetilde{\Omega}$ is a vector in Fock space with Bosons distributed according to a discrete distribution approaching the Planck distribution as the box size increases). The interacting Hamiltonian is $H_\lambda = H_0 + \lambda(G_+ \otimes a(g) + G_- \otimes a^*(g))$. The second order contribution in λ to (52), for large values of t, is

$$P_2 = C\frac{\lambda^2}{(\mathrm{e}^{\beta\omega(1)} - 1)^2}|f(1)g(1)|^2 \,, \tag{53}$$

where we assume that $f(r), g(r)$ are radially symmetric, and where C is a constant independent of β, f, g. P_2 gives the probability of the second order process where a Boson gets trapped in the quantum dot; the excitation energy is 1 (the quantum dot Hamiltonian (42) has equidistant eigenvalues) and the probability density of finding a Boson with energy $\omega(1) = 1$ per unit volume is $\propto (e^\beta - 1)^{-1}$, according to (8). In order not to suppress this trapping process at second order in the coupling constant we assume that $g(1) \neq 0$ ("effective coupling").

Next let us investigate the influence of the condensate. For this we fix a density ρ_0 of the Bose gas and consider very low temperatures ($\beta \to \infty$), so that most particles are in the condensate. For an explicit calculation we take a pure condensate, i.e., $\widetilde{\Omega}$ in (52) is taken to consist only of particles in the ground state (with the prescribed density ρ_0). We calculate the second order in λ of (52) to be

$$Q_2(t) = C(1 - \cos t)\lambda^2 \rho_0^2 |f(0)g(0)|^2 . \tag{54}$$

We see from (54) that if $g(0) = 0$ then there is no coupling to the modes of the condensate: a physically trivial situation where the condensate evolves freely and the small system coupled to the "excited modes" undergoes return to equilibrium. We outline in this note results established in [18], which include the case $g(0) \neq 0$, a situation which could not be handled by approaches developped so far.

3.1 Non-Interacting System

The *states* of the small system are determined by density matrices ρ on the finite dimensional Hilbert space $\mathbb{C}^d$. A density matrix is a positive trace-class operator, normalized as $\operatorname{Tr} \rho = 1$, and the corresponding state

$$\omega_\rho(A) = \operatorname{Tr}(\rho A), \quad A \in \mathcal{B}(\mathbb{C}^d) \tag{55}$$

is a normalized positive linear functional on the C^*-algebra $\mathcal{B}(\mathbb{C}^d)$ of all bounded operators on $\mathbb{C}^d$, which we call the algebra of *observables*. The (Heisenberg-) *dynamics* of the small system is given by (44). Denote the normalized eigenvector of H_1 corresponding to $E_j = j$ by φ_j. Given any inverse temperature $0 < \beta < \infty$ the *Gibbs state* $\omega_{1,\beta}$ is the unique β-KMS state on $\mathcal{B}(\mathbb{C}^d)$ associated to the dynamics (44). The corresponding density matrix is

$$\rho_\beta = \frac{e^{-\beta H_1}}{\operatorname{Tr} e^{-\beta H_1}} . \tag{56}$$

Let ρ be a density matrix of rank d (equivalently, $\rho > 0$) and let $\{\varphi_j\}_{j=0}^{d-1}$ be an orthonormal basis of eigenvectors of ρ, corresponding to eigenvalues $0 < p_j < 1$, $\sum_j p_j = 1$. The GNS representation of the pair $(\mathcal{B}(\mathbb{C}^d), \omega_\rho)$ is given by $(\mathcal{H}_1, \pi_1, \Omega_1)$, where the Hilbert space $\mathcal{H}_1$ and the cyclic (and separating) vector Ω_1 are

$$\mathcal{H}_1 = \mathbb{C}^d \otimes \mathbb{C}^d \,, \tag{57}$$

$$\Omega_1 = \sum_j \sqrt{p_j}\, \varphi_j \otimes \varphi_j \in \mathbb{C}^d \otimes \mathbb{C}^d \,, \tag{58}$$

and the representation map $\pi_1 : \mathcal{B}(\mathbb{C}^d) \to \mathcal{B}(\mathcal{H}_1)$ is

$$\pi_1(A) = A \otimes \mathbb{1} \,. \tag{59}$$

We introduce the von Neumann algebra

$$\mathfrak{M}_1 = \mathcal{B}(\mathbb{C}^d) \otimes \mathbb{1}_{\mathbb{C}^d} \subset \mathcal{B}(\mathcal{H}_1) \,. \tag{60}$$

The modular conjugation operator J_1 associated to the pair $(\mathfrak{M}_1, \Omega_1)$ is given by

$$J_1 \psi_\ell \otimes \psi_r = C_1 \psi_r \otimes C_1 \psi_\ell \,, \tag{61}$$

where C_1 is the antilinear involution $C_1 \sum_j z_j \varphi_j = \sum_j \overline{z_j} \varphi_j$ (complex conjugate). According to (58) and (56) the vector $\Omega_{1,\beta}$ representing the Gibbs state $\omega_{1,\beta}$ is given by

$$\Omega_{1,\beta} = \frac{1}{\sqrt{\mathrm{Tr}\, e^{-\beta H_1}}} \sum_j e^{-\beta E_j/2} \varphi_j \otimes \varphi_j \in \mathcal{H}_1 \,. \tag{62}$$

Denote by ω_{ρ,ρ_0} the state on $\mathfrak{W}(\mathfrak{D})$ whose generating functional is (7), where $\rho(k)$ is given in (8), and $\rho_0 \geq 0$. The GNS representation of the pair $(\mathfrak{W}(\mathfrak{D}), \omega_{\rho,\rho_0})$ has been given in [2] as the triple $(\mathcal{H}_2, \pi_2, \Omega_2)$, where the representation Hilbert space is

$$\mathcal{H}_2 = \mathcal{F} \otimes \mathcal{F} \otimes L^2(S^1, d\sigma) \,, \tag{63}$$

$\mathcal{F} = \mathcal{F}(L^2(\mathbb{R}^3, d^3k))$ is the Bosonic Fock space over $L^2(\mathbb{R}^3, d^3k)$ and $L^2(S^1, d\sigma)$ is the space of L^2-functions on the circle, with uniform normalized measure $d\sigma$ $(= (2\pi)^{-1} d\theta$, when viewed as the space of periodic functions of $\theta \in [-\pi, \pi])$. The cyclic vector is

$$\Omega_2 = \Omega_{\mathcal{F}} \otimes \Omega_{\mathcal{F}} \otimes 1 \tag{64}$$

where $\Omega_{\mathcal{F}}$ is the vacuum in $\mathcal{F}$ and 1 is the constant function in $L^2(S^1, d\sigma)$. The representation map $\pi_2 : \mathfrak{W}(\mathfrak{D}) \to \mathcal{B}(\mathcal{H}_2)$ is given by

$$\pi_2(W(f)) = W_{\mathcal{F}}(\sqrt{1+\rho}f) \otimes W_{\mathcal{F}}(\sqrt{\rho}\overline{f}) \otimes e^{-i\Phi(f,\theta)} \,, \tag{65}$$

where $W_{\mathcal{F}}(f) = e^{i\varphi_{\mathcal{F}}(f)}$ is a Weyl operator in Fock representation and the field operator $\varphi_{\mathcal{F}}(f)$ is

$$\varphi_{\mathcal{F}}(f) = \frac{1}{\sqrt{2}}(a_{\mathcal{F}}^*(f) + a_{\mathcal{F}}(f)) \tag{66}$$

and $a_{\mathcal{F}}^*(f)$, $a_{\mathcal{F}}(f)$ are the smeared out creation, annihilation operators satisfying the commutation relations

$$[a_{\mathcal{F}}(f), a_{\mathcal{F}}^*(g)] = \langle f, g \rangle, \ [a_{\mathcal{F}}(f), a_{\mathcal{F}}(g)] = [a_{\mathcal{F}}^*(f), a_{\mathcal{F}}^*(g)] = 0. \tag{67}$$

Our convention is that $f \mapsto a_{\mathcal{F}}(f)$ is an antilinear map. The phase $\Phi \in \mathbb{R}$ is given in (22). In the absence of a condensate ($\rho_0 = 0 \Rightarrow \Phi = 0$) the third factor in (63)–(65) disappears and the representation reduces to the "Araki-Woods representation" in the form it has appeared in a variety of recent papers. We denote this representation by π_0. More precisely, the GNS representation of $(\mathfrak{W}(\mathfrak{D}), \omega_{\rho,\rho_0=0})$ is given by $(\mathcal{F} \otimes \mathcal{F}, \pi_0, \Omega_0)$, where

$$\pi_0(W(f)) = W_{\mathcal{F}}(\sqrt{1 + \rho} f) \otimes W_{\mathcal{F}}(\sqrt{\rho} \bar{f}), \tag{68}$$

$$\Omega_0 = \Omega_{\mathcal{F}} \otimes \Omega_{\mathcal{F}}. \tag{69}$$

Let us introduce the von Neumann algebras

$$\mathfrak{M}_0 = \pi_0(\mathfrak{W}(\mathfrak{D}))'' \subset \mathcal{B}(\mathcal{F} \otimes \mathcal{F}) \tag{70}$$

$$\mathfrak{M}_2 = \pi_2(\mathfrak{W}(\mathfrak{D}))'' \subset \mathcal{B}(\mathcal{H}_2) \tag{71}$$

which are the weak closures (double commutants) of the Weyl algebra represented as operators on the respective Hilbert spaces. $\mathfrak{M}_2$ splits into a product

$$\mathfrak{M}_2 = \mathfrak{M}_0 \otimes \mathcal{M} \subset \mathcal{B}(\mathcal{F} \otimes \mathcal{F}) \otimes \mathcal{B}(L^2(S^1, d\sigma)), \tag{72}$$

where $\mathcal{M}$ is the abelian von Neumann algebra of all multiplication operators on $L^2(S^1, d\sigma)$. It satisfies $\mathcal{M}' = \mathcal{M}$. Relation (72) follows from this: clearly we have $\mathfrak{M}_0' \otimes \mathcal{M} \subset \mathfrak{M}_2'$, so taking the commutant gives

$$\mathfrak{M}_0 \otimes \mathcal{M} \supset \mathfrak{M}_2. \tag{73}$$

The reverse inclusion is obtained from $\mathbb{1}_{\mathcal{F} \otimes \mathcal{F}} \otimes \mathcal{M} \subset \mathfrak{M}_2$ and $\mathfrak{M}_0 \otimes \mathbb{1}_{L^2(S^1)} \subset \mathfrak{M}_2$ (see [2]).

It is well known that $\mathfrak{M}_0$, the von Neumann algebra corresponding to the situation without condensate, is a factor. That means that its center is trivial, $\mathfrak{Z}(\mathfrak{M}_0) = \mathfrak{M}_0 \cap \mathfrak{M}_0' \cong \mathbb{C}$. However, we have $\mathfrak{Z}(\mathfrak{M}_2) = (\mathfrak{M}_0 \otimes \mathcal{M}) \cap (\mathfrak{M}_0' \otimes \mathcal{M})$,

$$\mathfrak{Z}(\mathfrak{M}_2) = \mathbb{1}_{\mathcal{F} \otimes \mathcal{F}} \otimes \mathcal{M}, \tag{74}$$

so the von Neumann algebra $\mathfrak{M}_2$ is *not a factor*. One can decompose $\mathfrak{M}_2$ into a direct integral of factors, or equivalently, one can decompose ω_{ρ,ρ_0} into an integral over factor states. The Hilbert space (63) is the direct integral

$$\mathcal{H}_2 = \int_{[-\pi,\pi]}^{\oplus} \frac{d\theta}{2\pi} \mathcal{F} \otimes \mathcal{F}, \tag{75}$$

and the formula (see (64), (65), (68), (69))

$$\omega_{\rho,\rho_0}(W(f)) = \langle \Omega_2, \pi_2(W(f)) \Omega_2 \rangle = \int_{-\pi}^{\pi} \frac{d\theta}{2\pi} e^{-i\Phi(f,\theta)} \langle \Omega_0, \pi_0(W(f)) \Omega_0 \rangle \tag{76}$$

shows that π_2 is decomposed as

$$\pi_2 = \int_{[-\pi,\pi]}^{\oplus} \frac{\mathrm{d}\theta}{2\pi}\, \pi_\theta \, , \tag{77}$$

where $\pi_\theta : \mathfrak{W}(\mathfrak{D}) \to \mathcal{B}(\mathcal{F} \otimes \mathcal{F})$ is the representation defined by

$$\pi_\theta(W(f)) = \mathrm{e}^{-\mathrm{i}\Phi(f,\theta)}\pi_0(W(f)) \, . \tag{78}$$

For each fixed θ,

$$\pi_\theta(\mathfrak{W}(\mathfrak{D}))'' = \mathfrak{M}_0 \tag{79}$$

is a factor. Accordingly we have the factor decomposition $\mathfrak{M}_2 = \int_{[-\pi,\pi]}^{\oplus} \frac{\mathrm{d}\theta}{2\pi}\, \mathfrak{M}_0$.

Consider the equilibrium state of the uncoupled system

$$\omega_{\beta,0}^{\mathrm{con}} = \omega_{1,\beta} \otimes \omega_{2,\beta} \, , \tag{80}$$

where $\omega_{1,\beta}$ is the Gibbs state of the quantum dot (see (62)), and where $\omega_{2,\beta} = \omega_{\rho,\rho_0}$ is the equilibrium state of the heat bath at inverse temperature β and above critcal density, introduced after (62). The index 0 in (80) indicates the absence of an interaction between the two systems. The GNS representation of $(\mathfrak{A}, \omega_{\beta,0}^{\mathrm{con}})$ is just $(\mathcal{H}, \pi, \Omega)$, where

$$\mathcal{H} = \mathcal{H}_1 \otimes \mathcal{H}_2$$
$$\pi = \pi_1 \otimes \pi_2 \tag{81}$$
$$\Omega_{\beta,0}^{\mathrm{con}} = \Omega_{1,\beta} \otimes \Omega_2 \, . \tag{82}$$

The free dynamics is α_0^t, (46). Let

$$\mathfrak{M}_\beta^{\mathrm{con}} := \pi(\mathfrak{A})'' = \mathfrak{M}_1 \otimes \mathfrak{M}_2 = \int_{[-\pi,\pi]}^{\oplus} \frac{\mathrm{d}\theta}{2\pi}\, \mathfrak{M}_1 \otimes \mathfrak{M}_0 \subset \mathcal{B}(\mathcal{H}) \tag{83}$$

be the von Neumann algebra obtained by taking the weak closure of all observables of the combined system, when represented on $\mathcal{H}$. We have

$$\pi_2(\alpha_2^t(W(f))) = \int_{[-\pi,\pi]}^{\oplus} \frac{\mathrm{d}\theta}{2\pi} \mathrm{e}^{-\mathrm{i}\Phi(f,\theta)}\pi_0(W(\mathrm{e}^{\mathrm{i}\omega t}f)) \, . \tag{84}$$

It is well known and easy to verify that for $A \in \mathfrak{A}$,

$$(\pi_1 \otimes \pi_0)(\alpha_0^t(A)) = \mathrm{e}^{\mathrm{i}tL_0}(\pi_1 \otimes \pi_0)(A)\mathrm{e}^{-\mathrm{i}tL_0} \, , \tag{85}$$

where the selfadjoint L_0 on $\mathcal{H}_1 \otimes \mathcal{F} \otimes \mathcal{F}$ is given by

$$L_0 = L_1 + L_2 \, , \tag{86}$$
$$L_1 = H_1 \otimes \mathbb{1}_{\mathbb{C}^d} - \mathbb{1}_{\mathbb{C}^d} \otimes H_1 \, , \tag{87}$$
$$L_2 = \mathrm{d}\Gamma(\omega) \otimes \mathbb{1}_{\mathcal{F}} - \mathbb{1}_{\mathcal{F}} \otimes \mathrm{d}\Gamma(\omega) \, . \tag{88}$$

Here $d\Gamma(\omega)$ is the second quantization of the operator of multiplication by ω on $L^2(\mathbb{R}^3, d^3k)$. We will omit trivial factors $\mathbb{1}$ or indices $\mathbb{C}^d$, $\mathcal{F}$ whenever we have the reasonable hope that no confusion can arise. It follows from (83)–(88) that the uncoupled dynamics α_0^t is unitarily implemented in $\mathcal{H}$ by

$$\pi(\alpha_0^t(A)) = e^{it\mathcal{L}_0}\pi(A)e^{-it\mathcal{L}_0} , \tag{89}$$

where the *standard, non-interacting Liouvillian* $\mathcal{L}_0$ is the selfadjoint operator on $\mathcal{H}$ with constant (θ-independent) fiber L_0,

$$\mathcal{L}_0 = \int_{[-\pi,\pi]}^{\oplus} \frac{d\theta}{2\pi} \, L_0 . \tag{90}$$

The r.h.s. of (89) extends to a ∗automorphism group σ_0^t of $\mathfrak{M}_\beta^{\mathrm{con}}$, which we write

$$\sigma_0^t = \int_{[-\pi,\pi]}^{\oplus} \frac{d\theta}{2\pi} \, \sigma_{0,\theta}^t , \tag{91}$$

where $\sigma_{0,\theta}^t$ is the ∗automorphism group of $\mathfrak{M}_1 \otimes \mathfrak{M}_0$ generated by L_0. As is well known,

$$\Omega_{\beta,0} = \Omega_{1,\beta} \otimes \Omega_0 \tag{92}$$

is a $(\beta, \sigma_{0,\theta}^t)$-KMS state of $\mathfrak{M}_1 \otimes \mathfrak{M}_0$. The modular conjugation operator J associated to $(\mathfrak{M}_0, \Omega_{1,\beta} \otimes \Omega_0)$ is

$$J = J_1 \otimes J_0 , \tag{93}$$

where J_1 is given by (61) and where the action of J_0 on $\mathcal{F} \otimes \mathcal{F}$ is determined by antilinearly extending the relation $J_0\pi_0(W(f))\Omega_0 = W_{\mathcal{F}}(\sqrt{\rho}f) \otimes W_{\mathcal{F}}(\sqrt{1+\rho}\,\overline{f})\Omega_0$. J_0 defines an antilinear representation of the Weyl algebra according to $W(f) \mapsto J_0\pi_0(W(f))J_0$, which commutes with the representation π_0 given in (68). We view this as a consequence of the Tomita-Takesaki theory which asserts that $\mathfrak{M}_0{}' = J_0\mathfrak{M}_0J_0$.

It follows from (82), (83), (91) that

$$\Omega_{\beta,0}^{\mathrm{con}} = \int_{[-\pi,\pi]}^{\oplus} \frac{d\theta}{2\pi} \, \Omega_{\beta,0} \tag{94}$$

is a (β, σ_0^t)-KMS state on $\mathfrak{M}_\beta^{\mathrm{con}}$, and that the modular conjugation operator $\mathcal{J}$ associated to $(\mathfrak{M}_\beta^{\mathrm{con}}, \Omega_{\beta,0}^{\mathrm{con}})$ is given by

$$\mathcal{J} = \int_{[-\pi,\pi]}^{\oplus} \frac{d\theta}{2\pi} \, J_1 \otimes J_0 . \tag{95}$$

The standard Liouvillian $\mathcal{L}_0$, (90), satisfies the relation $\mathcal{J}\mathcal{L}_0 = -\mathcal{L}_0\mathcal{J}$ and

$$\mathcal{L}_0\Omega_{\beta,0}^{\mathrm{con}} = 0 . \tag{96}$$

3.2 Interacting System

The field operator $\varphi(f) = \frac{1}{i}\partial_t|_{t=0}\pi(W(tf))$ in the representation π, (81), is easily calculated to be $\varphi(f) = \int_{[-\pi,\pi]}^{\oplus} \frac{d\theta}{2\pi} \varphi_\theta(f)$, where

$$\varphi_\theta(f) = \varphi_{\mathcal{F}}(\sqrt{1+\rho}f) \otimes \mathbb{1} + \mathbb{1} \otimes \varphi_{\mathcal{F}}(\sqrt{\rho}\bar{f}) - \Phi(f,\theta) , \qquad (97)$$

where $\Phi(f,\theta)$ is given in (22), and where $\varphi_{\mathcal{F}}(f)$ is given in (66). The creation operator $a_\theta^*(f) := \frac{1}{\sqrt{2}}(\varphi_\theta(f) - i\varphi_\theta(if))$ has the expression

$$a_\theta^*(f) = a_{\mathcal{F}}^*(\sqrt{1+\rho}f) \otimes \mathbb{1}_{\mathcal{F}} + \mathbb{1}_{\mathcal{F}} \otimes a_{\mathcal{F}}(\sqrt{\rho}\bar{f}) - (2\pi)^{-3/2}\sqrt{\rho_0}f(0)e^{-i\theta} . \qquad (98)$$

Using these expressions it is not difficult to evaluate $\pi(G_+ \otimes a(g)) + $ adjoint (apply (81) to (49)), and to see that the standard interacting Liouvillian is given by

$$\mathcal{L}_\lambda = \int_{[-\pi,\pi]}^{\oplus} \frac{d\theta}{2\pi} L_{\lambda,\theta} , \qquad (99)$$

where the selfadjoint operator $L_{\lambda,\theta}$ is

$$L_{\lambda,\theta} = L_0 + \lambda I_\theta . \qquad (100)$$

Here L_0 is given in (86) and we define

$$I_\theta = I + K_\theta , \qquad (101)$$

$$I = G_+ \otimes \mathbb{1}_{\mathbb{C}^d} \otimes \left\{ a_{\mathcal{F}}(\sqrt{1+\rho}\,g) \otimes \mathbb{1}_{\mathcal{F}} + \mathbb{1}_{\mathcal{F}} \otimes a_{\mathcal{F}}^*(\sqrt{\rho}\,\bar{g}) \right\} + \text{adj.} \qquad (102)$$

$$- \mathbb{1}_{\mathbb{C}^d} \otimes C_1 G_+ C_1 \otimes \left\{ a_{\mathcal{F}}^*(\sqrt{\rho}g) \otimes \mathbb{1}_{\mathcal{F}} + \mathbb{1}_{\mathcal{F}} \otimes a_{\mathcal{F}}(\sqrt{1+\rho}\,\bar{g}) \right\} + \text{adj.}$$

$$K_\theta = K_\theta^1 \otimes \mathbb{1}_{\mathbb{C}^d} \otimes \mathbb{1}_{\mathcal{F}\otimes\mathcal{F}} - \mathbb{1}_{\mathbb{C}^d} \otimes C_1 K_\theta^1 C_1 \otimes \mathbb{1}_{\mathcal{F}\otimes\mathcal{F}} \qquad (103)$$

$$K_\theta^1 = -(2\pi)^{-3/2}\sqrt{\rho_0} \left(G_+ \overline{g(0)}e^{i\theta} + G_- g(0)e^{-i\theta} \right)$$

with C_1, Φ defined in (61), (22) and where the creation and annihilation operators $a_{\mathcal{F}}^*$, $a_{\mathcal{F}}$ are defined by (67). The operator $L_{\lambda,\theta}$ generates a Heisenberg dynamics $\sigma_{\lambda,\theta}^t$ on the von Neumann algebra $\mathfrak{M}_1 \otimes \mathfrak{M}_0$. It is convenient to write (compare with (91))

$$\sigma_\lambda^t = \int_{[-\pi,\pi]}^{\oplus} \frac{d\theta}{2\pi} \sigma_{\lambda,\theta}^t . \qquad (104)$$

To the interacting dynamics (104) corresponds a β-KMS state on $\mathfrak{M}_\beta^{\text{con}}$, the equilibrium state of the interacting system. It is given by the vector

$$\Omega_{\beta,\lambda}^{\text{con}} = (Z_{\beta,\lambda}^{\text{con}})^{-1} \int_{[-\pi,\pi]}^{\oplus} \frac{d\theta}{2\pi} \Omega_{\beta,\lambda}^\theta , \qquad (105)$$

where $Z_{\beta,\lambda}^{\text{con}}$ is a normalization factor ensuring that $\|\Omega_{\beta,\lambda}^{\text{con}}\| = 1$, and where

$$\Omega^\theta_{\beta,\lambda} = (Z^\theta_{\beta,\lambda})^{-1} e^{-\beta(L_0 + \lambda I_{\theta,\ell})/2} \Omega_{\beta,0} \in \mathcal{H}_1 \otimes \mathcal{F} \otimes \mathcal{F} . \qquad (106)$$

$Z^\theta_{\beta,\lambda}$ is again a normalization factor, and $I_{\theta,\ell}$ is obtained by dropping the terms coming with a minus sign in the r.h.s. of both (102) and (103). The fact that $\Omega_{\beta,0}$, (92), is in the domain of the unbounded operator $e^{-\beta(L_0 + \lambda I_{\theta,\ell})/2}$, provided $\|g/\sqrt{\omega}\|_{L^2(\mathbb{R}^3)} < \infty$, can be seen by expanding the exponential in a Dyson series and verifying that the series applied to $\Omega_{\beta,0}$ converges, see e.g. [3]. It then follows from the generalization of Araki's perturbation theory of KMS states, given in [7], that $\Omega^\theta_{\beta,\lambda}$ is a $(\beta, \sigma^t_{\lambda,\theta})$-KMS state on $\mathfrak{M}_1 \otimes \mathfrak{M}_0$, denoted by

$$\omega^\theta_{\beta,\lambda}(\cdot) = \langle \Omega^\theta_{\beta,\lambda}, \cdot \, \Omega^\theta_{\beta,\lambda} \rangle , \qquad (107)$$

and that

$$L_{\lambda,\theta} \, \Omega^\theta_{\beta,\lambda} = 0 . \qquad (108)$$

We conclude that $\Omega^{con}_{\beta,\lambda}$ is a $(\beta, \sigma^t_\lambda)$-KMS state on $\mathfrak{M}^{con}_\beta$, and that $\mathcal{L}_\lambda \Omega^{con}_{\beta,\lambda} = 0$.

3.3 Stability of the Quantum Tweezers, Main Results

We make two assumptions on the form factor g determining the interaction.

(A1) *Regularity.* The form factor g is a function in $C^4(\mathbb{R}^3)$ and satisfies

$$\|(1 + 1/\sqrt{\omega})(k \cdot \nabla_k)^j \sqrt{1 + \rho} \, g\|_{L^2(\mathbb{R}^3, d^3k)} < \infty , \qquad (109)$$

for $j = 0, \ldots, 4$, and $\| (1 + \omega)^2 \sqrt{1 + \rho} \, g\|_{L^2(\mathbb{R}^3, d^3k)} < \infty$.

(A2) *Effective coupling.* We have $\int_{S^2} d\sigma |g(1, \sigma)|^2 \neq 0$, where g is represented in spherical coordinates.

Remarks. 1) The operator $k \cdot \nabla_k$ emerges because we apply the positive commutator method with conjugate operator $\frac{1}{2}(k \cdot \nabla_k + \nabla_k \cdot k)$ (dilation generator). It is important to notice that in order to couple the particles of the Bose-Einstein condensate to the quantum dot, we must treat the case $g(0) \neq 0$. If the dispersion relation is given by $\omega(k) \sim |k|^s$, as $|k| \sim 0$, then (109) is satisfied for $s < 3/2$. (This does not include non-relativistic Bosons, for which $s = 2$.)

2) Condition (A2) is often called the *Fermi Golden Rule Condition*. It guarantees that the processes of absorption and emission of field quanta by the small system, which are the origin of the stability of the equilibrium, are effective (see the discussion at the beginning of Sect. 3). We integrate over a sphere of radius one since the gap between neighbouring eigenvalues of H_1 is equal to one (Bohr frequency).

Let $B \in \mathfrak{A}$, μ a probability measure on S^1, and define a state on $\mathfrak{A}$ by $\omega^\mu_B(A) = \int_{S^1} d\mu(\theta) \omega^\theta_{\beta,\lambda}(B^* A B)$, where $\omega^\theta_{\beta,\lambda}(A) = \langle \Omega^\theta_{\beta,\lambda}, A_\theta \Omega^\theta_{\beta,\lambda} \rangle$, with $A_\theta := (\pi_1 \otimes \pi_\theta)(A)$ (see (107), (59), (78)). We introduce the suggestive notation $\omega^\mu_B(\sigma^t_\lambda(A)) := \int_{S^1} d\mu(\theta) \langle \sigma^t_{\lambda,\theta}(A_\theta) \rangle_{B_\theta \Omega^\theta_{\beta,\lambda}}$, where $\langle A \rangle_\psi = \langle \psi, A\psi \rangle$, and where $\sigma^t_{\lambda,\theta}$ is given in (104).

Theorem 1 (Stability of equilibrium with condensate, [18]). *Assume conditions (A1) and (A2). Let $A, B \in \mathfrak{A}$, and let μ be a probability measure on S^1. Then*

$$\lim_{\lambda \to 0} \lim_{T \to \infty} \frac{1}{T} \int_0^T \omega_B^\mu(\sigma_\lambda^t(A)) \mathrm{d}t = \int_{S^1} \mathrm{d}\mu(\theta) \omega_{\beta,0}^\theta(B^*B) \omega_{\beta,0}^\theta(A) , \qquad (110)$$

where $\omega_{\beta,0}^\theta$ is given in (48).

The proof of Theorem 1 follows the ideas of the spectral method outlined in Subsection 1.5. It is based on the following result, which in turn is proved using positive commutator techniques.

Theorem 2 (Structure of kernel, [18]). *Assume Conditions (A1) and (A2) and let $P_{\beta,\lambda}^\theta$ the projection onto the subspace spanned by the interacting KMS state $\Omega_{\beta,\lambda}^\theta$, (106). Let $\theta \in [-\pi, \pi]$ be fixed. Any normalized element $\psi_\lambda \in \mathrm{Ker}(L_{\lambda,\theta}) \cap \left(\mathrm{Ran}P_{\beta,\lambda}^\theta\right)^\perp$ converges weakly to zero, as $\lambda \to 0$. The convergence is uniform in $\theta \in [-\pi, \pi]$ and in $\beta \geq \beta_0$, for any $\beta_0 > 0$ fixed.*

I would like to express my **thanks** to Joachim Asch and to Alain Joye for having organized *Qmath9*, and for having invited me to make a contribution. My thanks also go to the referee for his careful reading of the manuscript.

References

1. W. Amrein, A. Boutet de Monvel, V. Georgescu: C_0-Groups, Commutator Methods and Spectral Theory of N-body Hamiltonians. Basel-Boston-Berlin: Birkhäuser, 1996.
2. H. Araki, E. Woods: J. Math. Phys. **4**, 637 (1963).
3. V. Bach, J. Fröhlich, I.M. Sigal, I.M.: J. Math. Phys. **41**, no. 6, 3985 (2000).
4. Bratteli, O., Robinson, D.W., Operator Algebras and Quantum Statistical Mechanics I, II. Texts and Monographs in Physics, Springer-Verlag, 1987.
5. J.T. Cannon: Commun. Math. Phys., **29**, 89 (1973).
6. J. Dereziński, J., V. Jakšić: Ann. Henri Poincaré **4**, no. 4, 739 (2003).
7. J. Dereziński, V. Jakšić, C.-A. Pillet: Rev. Math. Phys. **15**, no. 5, 447 (2003).
8. R.B. Diener, B. Wu, M.G. Raizen, Q. Niu: Phys. Rev. Lett. **89**, 070401 (2002).
9. J. Fröhlich, M. Merkli: Math. Phys. Anal. Geom. **7**, no. 3, 239 (2004).
10. J. Fröhlich, M. Merkli: Commun. Math. Phys. **251**, no. 2, 235 (2004).
11. J. Fröhlich, M. Merkli, D. Ueltschi: Ann. Henri Poincaré **4**, no. 5, 897 (2003).
12. J. Fröhlich, M. Merkli, I.M. Sigal: J. Statist. Phys. **116**, no. 1–4, 311 (2004).
13. V. Georgescu, C. Gérard: Commun. Math. Phys. **208**, 275 (1999).
14. V. Jakšić, C.-A. Pillet: Commun. Math. Phys. **176**, 619 (1996).
15. V. Jakšić, C.-A. Pillet: Commun. Math. Phys. **178**, 627 (1996).
16. J.T. Lewis, J.V. Pulé: Commun. Math. Phys. **36**, 1 (1974).
17. M. Merkli: Commun. Math. Phys. **223**, 327 (2001).
18. M. Merkli: Commun. Math. Phys. **257**, 621 (2005)

Ordering of Energy Levels
in Heisenberg Models and Applications

Bruno Nachtergaele[1] and Shannon Starr[2]

[1] Department of Mathematics, University of California, Davis, One Shields
Avenue, Davis, CA 95616-8366, USA
`bxn@math.ucdavis.edu`
[2] Department of Mathematics, University of California, Los Angeles, Box 951555,
Los Angeles, CA 90095-1555, USA
`sstarr@math.ucla.edu`

Abstract. In a recent paper [17] we conjectured that for ferromagnetic
Heisenberg models the smallest eigenvalues in the invariant subspaces of fixed total spin are monotone decreasing as a function of the total spin and called this
property *ferromagnetic ordering of energy levels* (FOEL). We have proved this conjecture for the Heisenberg model with arbitrary spins and coupling constants on a
chain [17, 20]. In this paper we give a pedagogical introduction to this result and
also discuss some extensions and implications. The latter include the property that
the relaxation time of symmetric simple exclusion processes on a graph for which
FOEL can be proved, equals the relaxation time of a random walk on the same
graph with jump rates given by the coupling constants, i.e., the relaxation time is
independent of the number of particles. Therefore, our results also provide a proof
of Aldous' Conjecture in one dimension.

1 Introduction

The ferromagnetic Heisenberg model is the primordial quantum spin model.
It has been studied almost continuously since it was introduced by Heisenberg in 1926. In the course of its long history, this model has inspired an
amazing variety of new developments in both mathematics and physics. The
Heisenberg Hamiltonian is one of the basic, non-trivial quantum many-body
operators, and understanding its spectrum has been a guiding problem of
mathematical physics for generations.

A lot of attention has been given to the Bethe-Ansatz solvable one-
dimensional spin-1/2 model, which has an infinite-dimensional algebra of
symmetries [9]. The results we will discuss here are not related to exact solutions but there is an essential connection with the $SU(2)$ symmetry of the
model, much in the spirit of the famous result by Lieb and Mattis ([15], see
also [14, footnote 6]). The Lieb-Mattis Theorem proves "ordering of energy
levels" for a large class of antiferromagnetic Heisenberg models on bipartite

B. Nachtergaele and S. Starr: *Ordering of Energy Levels in Heisenberg Models and Applications*, Lect. Notes Phys. **690**, 149–170 (2006)
`www.springerlink.com`

lattices. Namely, if the two sublattices are A and B, and all interactions within A and B are ferromagnetic while interactions in between A and B are antiferromagnetic, then the unique ground state multiplet has total spin equal to $|\mathcal{S}_A - \mathcal{S}_B|$, where $\mathcal{S}_A$ and $\mathcal{S}_B$ are the maximum total spins on the two sublattices. Moreover, the minimum energy in the invariant subspace of total spin S, for $S \geq |\mathcal{S}_A - \mathcal{S}_B|$, is monotone increasing as a function of S. The most important example where this theorem provides useful information is the usual antiferromagnet on a bipartite lattice with equal-size sublattices. Then the ground state is a unique spin singlet, and the minimum energy levels for each possible total spin S, are monotone increasing in S. Our aim is a similar result for ferromagnets. To be able to state the ferromagnetic ordering of energy levels (FOEL) property precisely, we first give some definitions.

Let Λ be a finite connected graph with a set of vertices or sites, x, that we will also denote by Λ and a set E of unoriented edges, or bonds, (xy). We will often write $x \sim y \in \Lambda$ to signify that the edge (xy) is present in Λ. In many physical examples one has $\Lambda \subset \mathbb{Z}^d$.

Each site $x \in \Lambda$ has a quantum spin of magnitude $s_x \in \{1/2, 1, 3/2, \ldots\}$, associated with it. The state space at x is $2s_x + 1$-dimensional and we denote by S_x^i, $i = 1, 2, 3$, the standard spin-s_x matrices acting on the xth tensor factor in the Hilbert space $\mathcal{H} = \bigotimes_{x \in \Lambda} \mathbb{C}^{2s_x + 1}$. The isotropic (also called XXX) ferromagnetic Heisenberg Hamiltonian on Λ is given by

$$H_\Lambda = - \sum_{x \sim y \in \Lambda} J_{xy} \boldsymbol{S}_x \cdot \boldsymbol{S}_y , \qquad (1)$$

where the real numbers J_{xy} are the coupling constants, which we will always assume to be strictly positive (that they are positive is what it means to have the *ferromagnetic* Heisenberg model). This model is widely used to describe ferromagnetism at the microscopic level whenever itinerant electron effects can be ignored. Examples are magnetic domain walls and their properties and a variety of dynamical phenomena.

The spin matrices generate an irreducible representation of $SU(2)$ at each vertex. This representation is conventionally denoted by $D^{(s_x)}$. An important feature of the Hamiltonian (1) is that it commutes with $SU(2)$ via the representation

$$\bigotimes_{x \in \Lambda} D^{(s_x)} \qquad (2)$$

or, equivalently, with the total spin matrices defined by

$$S_\Lambda^i = \sum_{x \in \Lambda} S_x^i, \quad i = 1, 2, 3 .$$

and hence also with the Casimir operator given by

$$C = \boldsymbol{S}_\Lambda \cdot \boldsymbol{S}_\Lambda .$$

The eigenvalues of C are $S(S+1)$, $S = S_{\min}, S_{\min}+1, \ldots, S_{\max} \equiv \sum_{x \in \Lambda} s_x$, which are the spin labels of the irreducible representations that occur in the direct sum decomposition of the tensor product representation (2) into irreducible components. The value of $S_{\min}$ is usually 0 or 1/2, but may be larger if one of the s_x is greater than $S_{\max}/2$. The decomposition into irreducible components can be obtained by repeated application of the Clebsch-Gordan series:

$$D^{(s_1)} \otimes D^{(s_2)} \cong D^{(|s_1-s_2|)} \oplus D^{(|s_1-s_2|+1)} \cdots \oplus D^{(s_1+s_2)} . \qquad (3)$$

The label S is called the *total spin*, and the eigenvectors of the eigenvalue $S(S+1)$ of C, are said to have total spin S. Let $\mathcal{H}^{(S)}$ denote the corresponding eigenspace. Since C commutes with H_Λ, the spaces $\mathcal{H}^{(S)}$ are invariant subspaces for H_Λ. For any hermitian matrix H leaving the spaces $\mathcal{H}^{(S)}$ invariant we define

$$E(H, S) = \min \operatorname{spec} H|_{\mathcal{H}^{(S)}} .$$

By Ferromagnetic Ordering of Energy Levels (FOEL) we mean the property

$$E(H, S) \leq E(H, S'), \text{ if } S' < S .$$

for all S and S' in the range $[S_{\min}, S_{\max}]$, and we will speak of *strict* FOEL if the inequality is strict.

In particular, if H_Λ has the FOEL property it follows that its ground state energy is $E(H_\Lambda, S_{\max})$, which is indeed well-known to be the case for the Heisenberg ferromagnets. Moreover, since the multiplet of maximal spin is unique, FOEL also implies that the gap above the ground state is $E(H_\Lambda, S_{\max} - 1) - E(H_\Lambda, S_{\max})$, which is well-known for translation invariant Heisenberg ferromagnets on Euclidean lattices.

Conjecture 1. All ferromagnetic Heisenberg models have the FOEL property.

The FOEL property and the Lieb-Mattis theorem applied to a spin-1 chain of 5 sites is illustrated in Fig. 1.

Our main result is a proof of this conjecture for the special case of arbitrary ferromagnetic Heisenberg models on chains, i.e., one-dimensional model [17, 18, 20].

Theorem 1. *Strict FOEL holds for ferromagnetic XXX spin chains, i.e., for all*

$$H = -\sum_{x=1}^{L-1} J_{x,x+1} \left(\frac{1}{s_x s_{x+1}} \mathbf{S}_x \cdot \mathbf{S}_{x+1} - 1 \right) , \qquad (4)$$

for any choice of $s_x \in \{1/2, 1, 3/2, \ldots\}$ and $J_{x,x+1} > 0$.

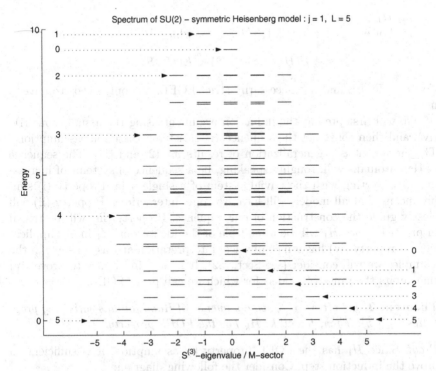

Fig. 1. The spectrum of a ferromagnetic Heisenberg chain consisting of 5 spin-1 spins, and with constant couplings. On the horizontal axis we have plottted the eigenvalue of the third component of the total spin. The spectrum is off-set so that the ground state energy vanishes. The arrows on the right, with label S, indicate the multiplets of eigenvalues $E(H, S)$, i.e., the smallest eigenvalue in the subspace of total spin S. The monotone ordering of the spin labels is the FOEL property. On the left, we have indicated the largest eigenvalues for each value of the total spin. The monotone ordering of their labels in the range $1, \ldots, 5$, is the content of the Lieb-Mattis theorem applied to this system

2 Proof of the Main Result

Our proof of Theorem 1 proceeds by a finite induction argument for a sequence of models with Hamiltonians $H_k = H_k^*$, $1 \leq k \leq N$, on Hilbert spaces $\mathcal{H}_k$, with the following properties:

(i) There is a unitary representation of $SU(2)$, U_k, on $\mathcal{H}_k$, that commutes with H_k.

(ii) There are isometries $V_k : \mathcal{H}_{k+1} \to \mathcal{H}_k \otimes \mathbb{C}^2$, intwining the representations U_{k+1} and $U_k \otimes D^{(1/2)}$, i.e., $V_k U_{k+1}(g) = (U_k(g) \otimes D^{(1/2)}(g))V_k$, for all $g \in SU(2)$, and such that

$$H_{k+1} \geq V_k^*(H_k \otimes \mathbb{1})V_k$$

(iii) H_1 has the FOEL property.

(iv) For every $S = 0, 1/2, 1, 3/2, \ldots$, for which $\mathcal{H}^{(S)} \neq \{0\}$, we have

$$E(H_{k+1}, S + 1/2) < E(H_k, S) \,.$$

To prove FOEL and not necessarily strict FOEL, one only needs to have $\leq$ here.

We will first present the induction argument using the assumptions (i)–(iv), and then construct the sequence H_k satisfying these four assumptions. This argument is a generalization of results in [12] and [17]. The sequence of Hamiltonians will, roughly speaking, be a sequence of systems of increasing size, starting with the trivial system of a single spin. Property (i) simply means that all models will have isotropic interactions. Property (ii) will closely guide the construction of our sequence. Property (iii) will be trivial in practice, since H_1 will be a multiple of the identity on $\mathcal{H}_1$ in our applications. Property (iv) has a nice physical interpretation at least in some of the examples we will consider (see Sect. 5.2). It is our (in) ability to prove (iv) that limits the range of models for which we can prove FOEL.

Theorem 2. *Let $(H_k)_{1 \leq k \leq N}$, be a sequence of Hamiltonians satisfying properties (i)–(iv). Then, for all k, H_k has the FOEL property.*

Proof. Since H_1 has the FOEL property by assumption, it is sufficient to prove the induction step. Consider the following diagram:

$$
\begin{array}{ccccc}
E(H_k, S) & >_1 & E(H_k, S+1) & > & E(H_k, S+2) \\
\vee_2 & {}^{\nearrow}{}_3 & \vee_2 & {}^{\nearrow} & \vee \\
E\left(H_{k+1}, S + \tfrac{1}{2}\right) & >_4 & E\left(H_{k+1}, S + \tfrac{3}{2}\right) & > & E\left(H_{k+1}, S + \tfrac{5}{2}\right)
\end{array}
$$

The inequality labeled 1 is FOEL for H_k, and inequality 2 is property (iv) assumed in the theorem. We will prove inequality 3 (using inequality 1) and, combined with inequality 2 this implies inequality 4, which is the induction step.

As before, we use superscripts to Hilbert spaces to denote their subspaces of fixed total spin. To prove inequality 3, we start from the variational principle:

$$
\begin{aligned}
E(H_{k+1}, S + 1/2) &= \inf_{\phi \in \mathcal{H}_{k+1}^{(S+1/2)}, \|\phi\|=1} \langle \phi, H_{k+1} \phi \rangle \\
&\geq \inf_{\phi \in \mathcal{H}_{k+1}^{(S+1/2)}, \|V_k \phi\|=1} \langle \phi, V_k^*(H_k \otimes \mathbb{1}_2) V_k \phi \rangle \\
&\geq \inf_{\psi \in (\mathcal{H}_k \otimes \mathbb{C}^2)^{(S+1/2)}, \|\psi\|=1} \langle \psi, (H_k \otimes \mathbb{1}_2) \psi \rangle
\end{aligned}
$$

The first inequality uses the fact that V_k is an isometry and property (ii). For the second inequality we enlarged the subspace over which the infimum is taken.

Now, we use the Clebsch-Gordan series (3) to see that $(\mathcal{H}_k \otimes \mathbb{C}^2)^{(S+1/2)} \subset (\mathcal{H}_k^{(S)} \oplus \mathcal{H}_k^{(S+1)}) \otimes \mathbb{C}^2$. Therefore

$$E(H_{k+1}, S + 1/2) \geq \min\{E(H_k, S), E(H_k, S + 1)\} = E(H_k, S + 1) .$$

Clearly, $H_k \otimes \mathbb{1}_2$ restricted to $(\mathcal{H}_k^{(S)} \oplus \mathcal{H}_k^{(S+1)}) \otimes \mathbb{C}^2$ has the same spectrum as H_k restricted to $\mathcal{H}_k^{(S)} \oplus \mathcal{H}_k^{(S+1)}$. The last equality then follows from inequality 1, i.e., the induction hypothesis. This concludes the proof of Theorem 2.

For the proof of Theorem 1 we will apply Theorem 2 to the sequence $(H_k)_{1 \leq k \leq N}$, with $N = 2S_{\max}$, $H_1 = 0$, and $H_N = H$, constructed as follows: for each $k = 1, \ldots, N - 1$, the model with Hamiltonian H_{k+1} is obtained from H_k in one of two ways: either a new spin $1/2$ is added to right of the chain, or the magnitude of the rightmost spin is increased by $1/2$. In both cases, $S_{\max}$ goes up by $1/2$ at each step, hence $N = 2 \sum_{x=1}^{L} s_x$. Each H_k is of the form (4), and we have written the interactions in such a way that the coupling constants $J_{x,x+1}$ can be taken to be independent of k, although this is not crucial since all arguments work for any choice of positive coupling constants at each step. The parameters that change with k are thus L and the set of spin magnitudes $(s_x)_{x=1}^{L}$. To be explicit, the two possible ways of deriving H_{k+1} from H_k are summarized in Table 1.

Table 1. Summary of the k-dependence of the sequence of models used in the proof by induction of Theorem 2

parameter	Case I	Case II
L	$L_{k+1} = L_k + 1$	$L_{k+1} = L_k$
$\{s_x\}$	$s_{L_k+1}(k) = 0, s_{L_k+1}(k+1) = 1/2$	$s_{L_k}(k+1) = s_{L_k}(k) + 1/2$
$\mathcal{H}$	$\mathcal{H}_{k+1} = \mathcal{H}_k \otimes \mathbb{C}^2$	$\mathcal{H}_{k+1} = V(\mathcal{H}_k \otimes \mathbb{C}^2)$

The Hamiltonians are of the form

$$H_k = -\sum_{x=1}^{L_k-1} J_{x,x+1} \left(\frac{1}{s_x(k)s_{x+1}(k)} S_x \cdot S_{x+1} - 1 \right) , \qquad (5)$$

where S_x^i, $i = 1, 2, 3$, are the $2s_x(k) + 1$ dimensional spin matrices. To simplify the notation, the dependence on k will often be omitted further on.

We now have a uniquely defined sequence of Hamiltonians $(H_k)_{1 \leq l \leq N}$, with $H_1 = 0$ and $H_N = H$. Next, we proceed to proving the properties (i)–(iv). Property (i) is obvious by construction. Property (iii) is trivial since $H_1 = 0$. To verify property (ii), we need to distinguish the two cases for the relation between H_k and H_{k+1}, as given in Table 1.

For Case I, $U_{k+1} = U_k \otimes D^{(1/2)}$ and we can take the identity map for V. Property (ii) follows from the positivity of the additional interaction term in H_{k+1}:

$$H_{k+1} = H_k + J_{L_k, L_k+1} \left(\frac{1}{s_{L_k} \cdot (1/2)} \boldsymbol{S}_{L_k} \cdot \boldsymbol{S}_{L_k+1} - 1 \right) .$$

For Case II, we have $\mathcal{H}_k = \mathcal{H}_l \otimes \mathbb{C}^{2s_{L_k}+1}$ and $\mathcal{H}_{k+1} = \mathcal{H}_l \otimes \mathbb{C}^{2s_{L_k+1}+1}$, for some $l < k$, possibly $l = 0, \mathcal{H}_0 = \mathbb{C}$. Since $s_{L_{k+1}} = s_{L_k} + 1/2$, there is a (up to a phase) unique $SU(2)$ intertwining isometry $W : \mathbb{C}^{2s_{L_k+1}} \to \mathbb{C}^{2s_{L_k}+1} \otimes \mathbb{C}^2$, namely the W that identifies the spin $s_{L_{k+1}}$ subrepresentation in $D^{(s_{L_k})} \otimes D^{(1/2)}$. From the intertwining property, the irreducibility of the spin representations, and the $SU(2)$ commutation relations one deduces that there is a constant c such that

$$W^*(S_{L_k}^i(k) \otimes \mathbb{1})W = c S_{L_k}^i(k+1), \quad i = 1, 2, 3 .$$

The constant c is most easily determined by calculating the left and right hand sides on a highest weight vector (a simultaneous eigenvector of C and S^3 with eigenvalues $S(S+1)$ and S, respectively). One finds

$$c = \frac{s_{L_k}(k)}{s_{L_k}(k+1)} .$$

Now, take $V = \mathbb{1}_{\mathcal{H}_l} \otimes W$. It is then straightforward to check that

$$V^* \left(\frac{1}{s_{L_{k-1}} s_{L_k}(k)} \boldsymbol{S}_{L_{k-1}} \cdot \boldsymbol{S}_{L_k} \otimes \mathbb{1}_2 \right) V = \frac{1}{s_{L_{k-1}} s_{L_k}(k+1)} \boldsymbol{S}_{L_{k-1}} \cdot \boldsymbol{S}_{L_k} ,$$

where the spin matrices on the left hand side are of the magnitude determined by $s_{L_{k-1}}$ and $s_{L_k}(k)$, while on the right hand side they are the magnitudes of the spins are $s_{L_{k-1}}$ and $s_{L_k}(k+1)$.

To prove Property (iv), we start by observing that

$$\mathrm{spec}(H_k|_{\mathcal{H}_k^{(S)}}) = \mathrm{spec}(H_k|_{\mathcal{V}_k^{(S)}})$$

where $\mathcal{V}_k^{(S)}$ is the subspace of $\mathcal{H}_k$ of all highest weight vectors of weight S. This is an invariant subspace for H_k and for every eigenvalue of $H_k|_{\mathcal{H}_k^{(S)}}$ there is at least one eigenvector in $\mathcal{V}_k^{(S)}$. Let $d(k,S)$ denote the dimension of $\mathcal{V}_k^{(S)}$.

Property (iv) will be obtained as a consequence of the following proposition and a version of the Perron-Frobenius Theorem.

Proposition 1. *We have $d(k+1, S+1/2) \geq d(k, S)$ and there are bases $\mathcal{B}_k^{(S)}$ for $\mathcal{V}_k^{(S)}$ such that the matrices $A^{(k,S)}$ of $H_k|_{\mathcal{V}_k^{(S)}}$ with respect to these bases have the following properties:*

$$A_{ij}^{(k,S)} \leq 0, \text{ for } 1 \leq i \neq j \leq d(k,S), 1 \leq k \leq N$$
$$A_{ij}^{(k+1,S+1/2)} \leq A_{ij}^{(k,S)}, \text{ for } 1 \leq i,j \leq d(k,S), 1 \leq k \leq N-1 .$$

For reasons of pedagogy and length, we will give the complete proof of this proposition only for the spin 1/2 chain. The proposition provides the assumptions needed to apply a slightly extended Perron-Frobenius theorem (see, e.g., [23]), which we state below.

The standard Perron-Frobenius Theorem makes several statements about square matrices with all entries non-negative, which we will call a non-negative matrix for short. Recall that a non-negative matrix A is called irreducible if there exists an integer $n \geq 1$ such that the matrix elements of A^n are all strictly positive. The standard results are the following: (i) every non-negative matrix has a non-negative eigenvalue equal to its spectral radius (hence it has maximal absolute value among all eigenvalues), and there is a corresponding non-negative eigenvector (i.e., with all components non-negative); (ii) if A is an irreducible non-negative matrix there is a unique eigenvalue with absolute value equal to the spectral radius of A, which is strictly positive and has algebraic (and hence geometric) multiplicity 1. Its corresponding eigenvector can be chosen to have all strictly positive components.

If A is a square matrix with all off-diagonal matrix elements non-positive, we will call A irreducible if there exists a constant c such that $c\mathbb{1} - A$ is irreducible according to the previous definition. From the standard Perron-Frobenius Theorem it immediately follows that the eigenvalue with smallest real part of an irreducible matrix in the last sense is real, has algebraic (and hence geometric) multiplicity 1, and that the corresponding eigenvector can be chosen to have all components strictly positive. In the following, we will repeatedly use the information provided by the standard Perron-Frobenius Theorem as described above without further reference. Let specrad(A) denote the spectral radius of a square matrix A.

Lemma 1. *Let $A = (a_{ij})$ and $B = (b_{ij})$ be non-negative $n \times n$ matrices, and assume that $a_{ij} \leq b_{ij}$, for all $1 \leq i, j \leq n$. Then*

$$\mathrm{specrad}(A) \leq \mathrm{specrad}(B) . \tag{6}$$

If B is irreducible and there is at least one pair ij such that $a_{ij} < b_{ij}$, then

$$\mathrm{specrad}(A) < \mathrm{specrad}(B) . \tag{7}$$

Since the spectral radii are also the eigenvalues of maximal absolute value, the same relations holds for these eigenvalues.

Proof. Let $r = \mathrm{specrad}(A)$. Then A has a non-negative eigenvector, say v, with eigenvalue r. If $a_{ij} \leq b_{ij}$, for all $1 \leq i, j \leq n$, it is clear that there is a non-negative vector w such that

$$Bv = rv + w . \tag{8}$$

This relation implies that $\|B^k\| \geq r^k$, for all positive integers k and, hence, specrad(B) $\geq r$. This proves (6).

To prove (7) for irreducible B such that $a_{ij} \le b_{ij}$ for at least one pair of indices, let k be a positive integer such that B^k is strictly positive. This implies that $B^k v$ has all strictly positive components. From this it is easy to see that the non-negative w such that

$$B^k v = r^k v + w$$

cannot be the zero vector. Therefore there is $z \in \mathbb{R}$ with all strictly positive components such that

$$B^{k+1} v = r^{k+1} v + z \ .$$

Since z is strictly positive, there exists $\varepsilon > 0$ such that, $\varepsilon v \le z$ componentwise, and therefore we can find $\delta > 0$ such that

$$B^{k+1} v = (r + \delta)^{k+1} v + z' \ ,$$

with z' non-negative. We conclude that $\mathrm{specrad}(B) \ge r + \delta > \mathrm{specrad}(A)$.

Note that the argument that proves this lemma could also be used to give a lower bound for the difference of the spectral radii. Since we do not need it, we will not pursue this here. The next theorem is an extension of Lemma 1.

Theorem 3. *Let $A = (a_{ij})$ and $B = (b_{ij})$ be two square matrices of size n and m, respectively, with $n \le m$, both with all off-diagonal matrix elements non-positive, and such that $b_{ij} \le a_{ij}$, for $1 \le i, j \le n$. Then*

$$\inf \mathrm{spec}(B) \le \inf \mathrm{spec}(A) \ . \tag{9}$$

If B is irreducible and either (i) there exists at least one pair ij, $1 \le i, j \le n$, such that $b_{ij} < a_{ij}$; or (ii) $b_{ij} < 0$, for at least one pair ij with at least one of the indices i or $j > n$, then

$$\inf \mathrm{spec}(B) < \inf \mathrm{spec}(A) \ . \tag{10}$$

Proof. Let $c \ge 0$ be a constant such that the matrices $A' = (a'_{ij}) = c\mathbb{1}_n - A$ and $B' = (b'_{ij}) = c\mathbb{1}_m - B$ are non-negative. Define A'' to be the $m \times m$ matrix obtained by extending A' with zeros:

$$(a''_{ij}) = A'' = \begin{bmatrix} A' & 0 \\ 0 & 0 \end{bmatrix} \ .$$

It is easy to see that $a''_{ij} \le b'_{ij}$, for $1 \le i, j \le m$. Therefore, we can apply Lemma 1 with A'' playing the role of A, and B' playing the role of B. Clearly, $\mathrm{specrad}(A'') = c - \inf \mathrm{spec}(A)$ and $\mathrm{specrad}(B') = c - \inf \mathrm{spec}(B)$. Therefore, this proves (9).

Similarly, (10) follows from the additional assumptions and (7).

Proof of Theorem 1: The remaining point was to prove property (iv) needed in the assumptions of Theorem 2. We use Proposition 1, which we will prove in the next section, and apply Theorem 3 with $A = A^{(k,S)}$ and $B = A^{(k+1,S+1/2)}$. This completes the proof.

3 The Temperley-Lieb Basis. Proof of Proposition 1

In the proof of Theorem 1 in the previous section we used the matrix representation of the Hamiltonians restricted to the highest weight spaces given by Proposition 1. We now give the complete proof of that proposition for the spin $1/2$ chain and sketch the proof in the general case.

The main issue is to find a basis of the highest weight spaces with the desired properties. Fortunately for us, such a basis has already been constructed and we only need to show that it indeed had the properties claimed in Proposition 1. For the spin $1/2$ chain we will use the Temperley-Lieb basis [22], and for the general case its generalization to arbitrary spin representations introduced by Frenkel and Khovanov [6].

3.1 The Basis for Spin 1/2

We start with the spin $1/2$ chain, i.e., $s_x = 1/2$, for all x. In this case $S_{\max} = k/2$ and $V_k^{(S)}$ is the subspace of $(\mathbb{C}^2)^{\otimes k}$ consisting of all vectors ψ such that $S^3\psi = S\psi$ and $S^+\psi = 0$. Let n be the "spin-deviation" defined as $S = k/2 - n$. Then, n is a non-negative integer. The case $n = 0$ is trivial since $\dim V_k^{(k/2)} = 1$, namely just the mutiples of the vector $|+\rangle \otimes |+\rangle \otimes \cdots \otimes |+\rangle$, where $|\pm\rangle$ is the basis of $\mathbb{C}^2$ that diagonalizes S^3. For $n \geq 1$, the basis vectors are a tensor product of n singlet vectors $\xi = |+\rangle \otimes |-\rangle - |-\rangle \otimes |+\rangle$, accounting for two sites each, and $k - 2n$ factors equal to $|+\rangle$. Such vectors are sometimes called Hulthén brackets. It is clear that any such factor is a highest weight vector of weight $k/2 - n$, just calculate the action of S^3 and S^+ on such a vector. They are not linearly independent however, except in the trivial case $k = 2$. The contribution of Temperley and Lieb was to show how to select a complete and linearly independent subset, i.e., a basis. How to select the Temperley-Lieb basis, is most easily explained by representing the vectors by configurations of n arcs on the k vertices $1, \ldots, n$. The arcs are drawn above the line of vertices as shown in Fig. 2. Each arc represents a spin singlet ξ, and each unpaired vertex represents a factor $|+\rangle$. The vectors (configurations of arcs) selected for the basis are those that satisfy two properties: (i) the arcs are non-crossing, (ii) no arc spans an unpaired vertex. The resulting set is a (non-orthogonal) basis. E.g., the basis for $k = 5$ and $n = 2$ is shown in Fig. 2. We will use, $\alpha, \beta, \ldots$, to denote arc configurations that obey these rules, and by the corresponding basis vectors will be denoted by $|\alpha\rangle, |\beta\rangle, \ldots$. We will use the notation $[xy] \in \alpha$ to denote that the arc connecting x and y is present in α.

Proof of Proposition 1 for the spin 1/2 chain. The action of the Hamiltonian on the basis vectors has an appealing graphical representation. We can write the Hamiltonian as

$$H_k = -2 \sum_{x=1}^{k-1} J_{x,x+1} U_{x,x+1}$$

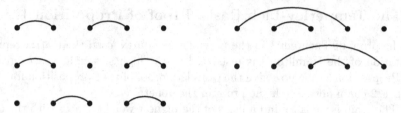

Fig. 2. The possible configurations of 2 arcs on 5 vertices

where $U_{x,x+1} = -\xi \otimes \xi^*$ which, up to a factor -2, is the orthogonal projection onto the singlet vector acting on the xth and $x + 1$st factor in the tensor product. The $U_{x,x+1}$ form a representation of the Temperley-Lieb algebra with parameter $q = 1$ (see, e.g, [10]). It is a straightforward calculation to verify the action of $U_{x,x+1}$ on a basis vector $|\alpha\rangle$: (i) if both x and $x + 1$ are unpaired vertices in α, $U_{x,x+1} |\alpha\rangle = 0$; (ii) if $[x, x + 1] \in \alpha$, we have $U_{x,x+1} |\alpha\rangle = -2 |\alpha\rangle$; (iii) if $[uv] \in \alpha$, with exactly one of the vertices u and v equal to x or $x + 1$, we have $U_{x,x+1} |\alpha\rangle = |\beta\rangle$, where β is obtained form α by removing $[uv]$ and adding $[x, x+1]$; (iv) if $[ux]$ and $[x+1, v]$ are both present in α, we have $U_{x,x+1} |\alpha\rangle = |\beta\rangle$, where β is obtained form α by removing $[ux]$ and $x + 1, v]$, and adding $[uv]$ and $[x, x + 1]$.

The action of $U_{x,x+1}$ on the vector $|\alpha\rangle$ can be graphically represented by placing the diagram shown in Fig. 3 under the diagram for α, and read off the result using the graphical representation of the rules (i)–(iv) shown in Fig. 4. The action of the Hamiltonian is then obtained by summing over x as shown in Figs. 5 and 6.

$$x \qquad x+1$$

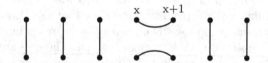

Fig. 3. The graphical representation of $U_{x,x+1}$

The important observation is the action of the Hamiltonian on a basis vector $|\alpha\rangle$ yields a linear combination of basis vectors with non-positive coefficients except possibly for the coefficient of $|\alpha\rangle$ itself, which has the opposite sign resulting from the "bubble" in the graphical representation. This means that all off-diagonal matrix elements are non-positive as claimed for the matrices $A^{k,S}$ in the proposition.

The second will follow from the observation that $A^{k,S}$ is a submatrix of $A^{k+1,S+1/2}$. Note that the spin deviation for $V_k^{(S)}$ and $V_{k+1}^{(S+1/2)}$ is the same, say n. Let us order the basis elements of $V_{k+1}^{(S+1/2)}$ so that all α where the last vertex, $k + 1$, is unpaired, are listed first, and consider the $\alpha\beta$ matrix element of H_{k+1} for such α and β. Then, it is easy to see that there are no

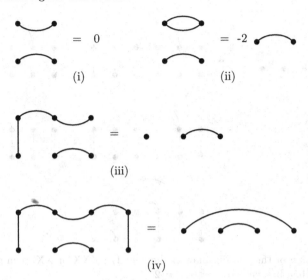

Fig. 4. The graphical rules (i)–(iv) for the action of $U_{x,x+1}$ on a Temperley-Lieb basis vector

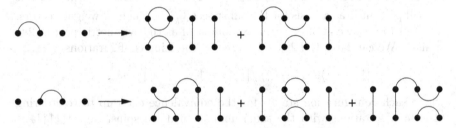

Fig. 5. Action of the Hamiltonian of the spin-1/2 XXX or XXZ chain on a generalized Hulthén bracket, for $L = 4$, $k = 1$

contributions from the $k, k+1$ term in Hamiltonian, since its action results in non-zero coefficients only for configurations where $k+1$ belongs to an arc. This means that these matrix elements are identical to those computed for H_k for basis vectors labeled α' and β' obtained from α and β by dropping the last vertex, $k+1$ which is unpaired.

This completes the proof of Proposition 1 in the case of the pure spin 1/2 chain. Q.E.D.

3.2 The Basis for Higher Spin

We are looking for a basis of the space of highest weight vectors of weight S of the spin chain with Hilbert space $\mathcal{H}_k$. Equivalently, we may look for a basis of the $SU(2)$ intertwiners $D^{(S)} \to \mathcal{H}_k$. There is a graphical algebra of such intertwiners with a very convenient basis, the dual canonical basis,

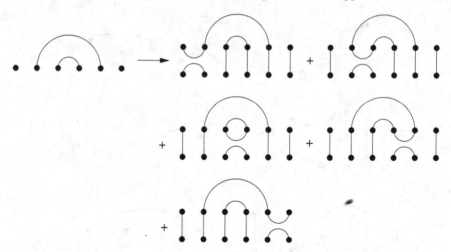

Fig. 6. Action of the Hamiltonian of the spin-1/2 XXX or XXZ chain on a generalized Hulthén bracket, for $k = 6$, $n = 2$

introduced by Frenkel and Khovanov [6]. This is the basis we will use, but we will present it as a basis for the subspaces $V_k^{(S)}$ of highest weight vectors.

The state space at site x can be thought of as the symmetric part of $2s_x$ spins-$\frac{1}{2}$. We can label the $2s_x + 1$ states by the Ising configurations

$$|\uparrow\uparrow \cdots \uparrow\rangle, |\downarrow\uparrow \cdots \uparrow\rangle, |\downarrow\downarrow\uparrow \cdots \uparrow\rangle, \ldots, |\downarrow\downarrow\downarrow \cdots \downarrow\rangle .$$

where each configuration stands for the equivalence class up to re-ordering of all configurations with the same number of down spins. E.g., $|\downarrow\downarrow\uparrow\uparrow\uparrow\rangle$ is the vector normally labelled as $|j, m\rangle = |5/2, 1/2\rangle$, and *not* the tensor $|\downarrow\rangle \otimes |\downarrow\rangle \otimes |\uparrow\rangle \otimes |\uparrow\rangle \otimes |\uparrow\rangle$. The states for a chain of L spins of magnitudes $s_1, \ldots, s_L$ are then tensor products of these configurations. We shall call such vectors *ordered Ising configurations*. These tensor product vectors, in general, are not eigenvectors of the Casimir operator S, i.e., they are not of definite total spin. Suitable linear combinations that do have definite total spin are obtained by extending the Hulthén bracket idea to arbitrary spin as follows. Start from any ordered Ising configuration such that $2M = \# \uparrow - \# \downarrow$. Then, look for the leftmost $\downarrow$ that has a $\uparrow$ to its left, and draw an arc connecting this $\downarrow$ to the rightmost $\uparrow$, left of it. At this point, one may ignore the paired spins, and repeat the procedure until there is no remaining unpaired $\downarrow$ with an unpaired $\uparrow$ to its left. This procedure guarantees that no arcs will cross and no arc will span an unpaired spin. The result, when ignoring all paired spins, is an ordered Ising configuration of a single spin. See Fig. 7 for an example of this procedure. The result is a basis for the spin chain consisting entirely of simultaneous eigenvectors of the total spin and its third component, with eigenvalues S and M, respectively. The value of M is 1/2 times the difference between the number of up spins and the number of down spins in the ordered

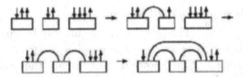

Fig. 7. Construction of a basis vector from an ordered Ising configuration

Ising configuration. The total-spin S is equal to S minus the number of pairs. Clearly, the highest weight vectors are then those that have no unpaired $\downarrow$, i.e., the ordered Ising configuration consists exclusively of up spins.

The vectors can be expanded in the tensor product basis by the following procedure: each arc is replaced by the spin singlet $|\uparrow\rangle \otimes |\downarrow\rangle - |\downarrow\rangle \otimes |\uparrow\rangle$, and the unpaired spins are replaced by their tensor products. Finally, one symmetrizes in each block.

Next, we briefly sketch how the properties claimed in Proposition 1 can be verified. To do this we have to calculate the action of the Hamiltonian on the highest weight vector constructed in the previous paragraph. This is most easily accomplished by deriving a graphical representation for the action of each term in the Hamiltonian as we did in the case of the pure spin 1/2 chain. The Heisenberg interaction for arbitrary spins of magnitude s_x and s_{x+1} can be realized as an interaction between spin $\frac{1}{2}$'s making up the spin s_x and s_{x+1}, conjugated with the projections onto the symmetric vectors. The result is the following:

$$-h_{x,x+1} = \frac{1}{2}\left(\frac{1}{s_x s_{x+1}} \boldsymbol{S}_x \cdot \boldsymbol{S}_{x+1} - 1\right) = \begin{array}{|c|c|} \hline 2s_x & 2s_{x+1} \\ \hline \cdots & \cdots \\ \hline 2s_x & 2s_{x+1} \\ \hline \end{array}.$$

Here, the rectangles with label $2s$ represent the symmetrizing projections on the space of $2s$ spin $\frac{1}{2}$ variables. The fundamental algebraic property that allows us to calculate the matrix elements of H_k graphically is the Jones-Wenzl relation (c.f., [10] and references therein):

$$\begin{array}{|c|c|} \hline 2s & 1 \\ \hline \cdots & \\ \hline 2s+1 & \\ \hline \cdots & \\ \hline 2s & 1 \\ \hline \end{array} = \begin{array}{|c|c|} \hline 2s & 1 \\ \hline \cdots & \\ \hline 2s & 1 \\ \hline \cdots & \\ \hline 2s & 1 \\ \hline \end{array} + \frac{2s}{2s+1}\begin{array}{|c|c|} \hline 2s & 1 \\ \hline \cdots & \\ \hline 2s-1 & \\ \hline \cdots & \\ \hline 2s & 1 \\ \hline \end{array}$$

For any element of the basis introduced above one can now compute the action of the Hamiltonian and write it as a linear combination of the same basis vectors. From the grahical rules it is easy to observe that all off-diagonal matrix elements are non-positive.

As before, it is straighforward to identify the basis for $\mathcal{V}_k^{(S)}$ with a subset of the basis for $\mathcal{V}_{k+1}^{(S+1/2)}$. The label of the rightmost box in any basis vector for the system k is raised by one but the number of arcs remains unchanged.

The crucial property that allows us to compare the two Hamiltonians is the following. When H_{k+1} acts on a basis vector obtained from a corresponding H_k vector as we have just described, the only possible new terms that are generated are off-diagonal terms, which do not contain a bubble and, hence, are negative. The details of the calculation of these matrix elements and further applications will appear elsewhere [18].

4 Extensions

A highly desirable extension of our main result, of course, would be the proof of Conjecture 1 for all ferromagnetic isotropic Heisenberg models on an arbitrary graph! While we have been able to prove some partial results for the spin-1/2 model on an arbitrary tree and a few other graphs, we do not have an argument that works for arbitrary graphs [18]. But there are a few other directions in which one might extend the ordering of energy levels property. The aim of this section is to discus two such generalizations. In the first, the group $SU(2)$ is replaced by the quantum group $SU_q(2)$, $0 < q < 1$. This only works on the chain and, as far as we are aware, leads to information about physically interesting models in the case of the spin 1/2 chain, namely the XXZ chain. The second generalization we consider is isotropic ferromagnetic models with higher order nearest neighbor interaction terms, such as $(\boldsymbol{S}_x \cdot \boldsymbol{S}_{x+1})^2$. For this to be relevant, the spins have to be of magnitude ≥ 1.

4.1 The Spin 1/2 $SU_q(2)$-symmetric XXZ Chain

It is well-known that the translation invariant spin-1/2 XXZ chain with a particular choice of boundary fields is $SU_q(2)$ invariant [21]. This $SU_q(2)$ symmetry can be exploited in much the same way as the $SU(2)$ symmetry of the isotropic model [12, 19]. Here we will show how it leads to a natural $SU_q(2)$ analogue of the FOEL property.

The Hamiltonian of the $SU_q(2)$-invariant ferromagnetic spin-1/2 chain of length $L \geq 2$ is given by

$$H_L = -\sum_{x=1}^{L-1}[\Delta^{-1}(S_x^1 S_{x+1}^1 + S_x^2 S_{x+1}^2) + (S_x^3 S_{x+1}^3 - 1/4)] \qquad (11)$$
$$-A(\Delta)(S_L^3 - S_1^3).$$

where $\Delta > 1$, and

$$A(\Delta) = \frac{1}{2}\sqrt{1 - 1/\Delta^2}$$

This model commutes with one of the two natural representation of $SU_q(2)$ on $(\mathbb{C}^2)^L$, with $q \in (0,1)$, such that $\Delta = (q + q^{-1})/2$. Concretely, this means that H_L commutes with the three generators of this representation defined as follows:

$$S^3 = \sum_{x=1}^{L} \mathbb{1}_1 \otimes \cdots \otimes S_x^3 \otimes \mathbb{1}_{x+1} \otimes \cdots \mathbb{1}_L$$

$$S^+ = \sum_{x=1}^{L} t_1 \otimes \cdots \otimes t_{x-1} \otimes S_x^+ \otimes \mathbb{1}_{x+1} \otimes \cdots \mathbb{1}_L$$

$$S^- = \sum_{x=1}^{L} \mathbb{1}_1 \otimes \cdots \otimes S_x^- \otimes t_{x+1}^{-1} \otimes \cdots t_L^{-1}$$

where

$$t = q^{-2S^3} = \begin{pmatrix} q^{-1} & 0 \\ 0 & q \end{pmatrix}.$$

The $SU_q(2)$ commutation relations are

$$[S^3, S^\pm] = \pm S^\pm, \quad [S^+, S^-] = \frac{q^{2S^3} - q^{-2S^3}}{q - q^{-1}}.$$

Note that one recovers the $SU(2)$ definitions and commutation relations in the limit $q \to 1$. H_L also commutes with the Casimir opeator for $SU_q(2)$, given by

$$C = S^+ S^- + \frac{(qT)^{-1} + qT}{(q^{-1} - q)^2}, \quad T = t \otimes t \otimes \cdots \otimes t.$$

The eigenvalues of C are

$$\frac{q^{-(2S+1)} + q^{2S+1}}{(q^{-1} - q)^2}, \quad S = 0, 1/2, 1, 3/2, \ldots$$

and play the same role as S for the XXX model, e.g., they label the irreducible representations of $SU_q(2)$. The eigenspaces of C are invariant subspaces of H_L and, as before, we denote the smallest eigenvalues of H_L restricted to these invariant subspaces by $E(H_L, S)$. Note that the subspaces depend on q, but their dimensions are constant for $0 < q \leq 1$.

Theorem 4.

$$E(H_L, S+1) < E(H_L, S), \quad \text{for all } S \leq L/2 - 1.$$

The proof of this theorem is identical to the one for the isotropic spin-1/2 chain up to substitution of the singlet vector ξ by the $SU_q(2)$ singlet $\xi_q = q|+\rangle \otimes |-\rangle - |-\rangle \otimes |+\rangle$, and changing the scalar value of the "bubble" to $-(q + q^{-1})$. The details are given in [17].

4.2 Higher Order Interactions

For spins of magnitude greater than 1/2 the Heisenberg interaction is not the only $SU(2)$ invariant nearest neighbor interactions. It is easy to show that the most general $SU(2)$ invariant interaction of two spins of magnitudes s_1 and s_2, i.e., any hermitian matrix commuting with the representation $D^{(s_1)} \otimes D^{(s_2)}$, is an arbitrary polynomial of degree $\leq 2\min\{s_1, s_2\}$ in the Heisenberg interaction with real coefficients:

$$h_{12} = \sum_{m=0}^{2\min\{s_1,s_2\}} J^{(m)}(\boldsymbol{S}_1 \cdot \boldsymbol{S}_2)^m . \tag{12}$$

The definition of the FOEL property only uses $SU(2)$-invariance and therefore applies directly to any Hamiltonian for a quantum spin system on a graph with at each edge an interaction of the form (12). We believe it is possible to determine the exact range of coupling constants $J^{(m)}$ such that FOEL holds for spin s chains with translation invariant interactions. So far, we have carried this out only for the spin-1 chain.

Theorem 5. *FOEL holds for the spin-1 chains with Hamiltonian*

$$H_L = \sum_{x=1}^{L-1} (1 - \boldsymbol{S}_x \cdot \boldsymbol{S}_{x+1}) + \beta(1 - \boldsymbol{S}_x \cdot \boldsymbol{S}_{x+1})^2)$$

with $0 \leq \beta \leq 1/3$. Level crossings occur at $\beta = 1/3$ and FOEL does not hold, in general, for $\beta > 1/3$.

The overall method of proof is the same as for the standard Heisenberg model. Theorem 2 applies directly since only the $SU(2)$ symmetry is used in its proof. The only difference is in the proof of Proposition 1. The same basis for the highest weight spaces is used but verifying the signs of the matrix elements is more involved.

5 Applications

In this section we discuss a number of results that are either consequences of the FOEL property, or other applications of the properties of the Heisenberg Hamiltonian that allowed us to prove FOEL.

5.1 Diagonalization at Low Energy

The most direct applications of the FOEL property are its implications for the low-lying spectrum of the Hamiltonian. FOEL with strict inequality implies that the ground states are the multiplet of maximal spin which, of course, is not a new result. Since the maximal spin multiplet is unique, the first excited state must belong to less than maximal spin and therefore, by FOEL, to $S_{\max} - 1$. In the case of models for which this second eigenvalue can be computed, such as translation invariant models on a lattice, this is again consistent with a well-known fact, namely that the lowest excitation are simple spinwaves. But in the case of arbitrary coupling constants and spin magnitudes in one dimension it proves that the first excited state is represented in the subspace of "one overturned spin" (with respect to the fully polarized ground state), i.e., $S^3 = S_{\max}^3 - 1$, which is a new result.

More generally, the FOEL property can help with determining the spectrum of the Heisenberg model at low energies, whether by numerical or other means, in the following way. Suppose H is a Hamiltonian with the FOEL property. Diagonalize H in the subspaces $\mathcal{H}^{(S_{\max}-n)}$, for $n = 0, 1, \ldots N$, and select those eigenvalues that are less or equal than $E(H, S_{\max} - N)$. It is easy to see that the FOEL property implies that this way you have obtained *all* eigenvalues of the full H below $\leq E(H, S_{\max} - N)$.

This is interesting because you only had to diagonalize the Hamiltonian in an invariant subspace that is explicitly known (by the representation theory of $SU(2)$) and of relatively low dimension: $\dim(\mathcal{H}^{(S_{\max}-n)})$ is $O(L^n)$, while the full Hilbert space has dimension $(2J + 1)^L$ for L spin J variables.

5.2 The Ground States of Fixed Magnetization for the XXZ Chain

The spin $1/2$ XXZ ferromagnetic chain with suitable boundary conditions, or defined on the appropriate infinite-chain Hilbert space has low-energy states that can be interpreted as well-defined magnetic domains in a background of opposite magnetization [11, 19]. Using the techniques we have used for proving FOEL, we can rigorously determine the dispersion relation of a finite droplet of arbitrary size.

The spin $1/2$ XXZ chain can, in principle, be diagonalized using the Bethe Ansatz [13]. There are two complications that may prevent one from obtaining the desired information about its spectrum. The first is that a complete proof of completeness of the Bethe Ansatz eigenstates has been obtained and published only for the XXX chain ($q = \Delta = 1$), although the corresponding result for the XXZ chain has been announced quite some time ago [7]. The second problem is that the eigenvalues are the solutions of complicated sets of equations, such that proving statements as the one we discuss here, may be very hard.

For brevity, let us consider the XXZ Hamiltonian for the inifinite chain defined on the Hilbert space generated by the orthonormal set of vectors representing n down spins in an infinite "sea" of up spins, and let us denote this space by $\mathcal{H}_n$. Define

$$E(n) = \inf \operatorname{spec}(H|_{\mathcal{H}_n}) .$$

As before, the relation between $q \in (0,1)$ and the anisotropy parameter Δ in the XXZ Hamiltonian (11) is given by $\Delta = (q + q^{-1})/2$.

Theorem 6. *For $n \geq 1$, we have*

$$E(n) = \frac{(1 - q^2)(1 - q^n)}{(1 + q^2)(1 + q^n)}.$$

Moreover, $E(n)$ belongs to the continuous spectrum and is the bottom of a band of width

$$4q^n \frac{1 - q^2}{(1 + q^n)(1 - q^n)}.$$

The states corresponding to this band can be interpreted as a droplet of size n with a definite momentum. The formula for the width indicates that the "mass" of a droplet diverges as $n \to \infty$. The proof of this result will appear in a separate paper [16]. If one looks back at the finite-volume eigenvalues $E(H_k, k/2 - n)$ that converge to the bottom of the band in the infinite-volume limit ($k \to \infty$, n fixed), the property (iv) amounts to property that the ground state energy of a droplet of fixed size n is strictly monotone decreasing in the volume. Moreover the finite-volume eigenvalues can be related to E(n), in the above limit, by using the generalization of the Perron-Frobenius result stated in Theorem 3.

5.3 Aldous' Conjecture for the Symmetric Simple Exclusion Process

The Symmetric Simple Exclusion Process (SSEP) is a Markov process defined on particle configurations on a finite graph Λ. For our purposes it is convenient to define the process as a semigroup on $\mathcal{H}_\Lambda \cong l^2(\Omega_\Lambda)$, where Ω_Λ is the space of configurations $\eta : \Lambda \to \{0, 1\}$. One thinks of $\eta(x) = 1$ to indicate the presence of a particle at the vertex x. Let L be defined in $\mathcal{H}_\Lambda$ by the formula

$$(Lf)(\eta) = \sum_{x \sim y} \sum_\eta J_{xy}(f(\eta) - f(\eta^{xy})) \tag{13}$$

where η^{xy} denotes the configuration obtained form η by interchanging the values of $\eta(x)$ and $\eta(y)$. The parameters J_{xy} are positive numbers representing the jump rates at the edges $x \sim y \in \Lambda$.

Clearly, the number of particles is a conserved quantity of the process. Concretely, this means that H_Λ decomposes into a direct sum of invariant subspaces $H_\Lambda^{(n)}$, $n = 0, \ldots, |\Lambda|$, where $H_\Lambda^{(n)}$ consists of all functions supported on configurations η that have exactly n particles, i.e., $\sum_x \eta(x) = n$. 0 is a simple eigenvalue of each of the restrictions $L|_{H_\Lambda^{(n)}}$, and L is non-negative definite. For each n, the corresponding invariant measure is the uniform distribution on n-particle configurations.

Let $\lambda(n)$ denote the smallest positive eigenvalue of $L|_{H_\Lambda^{(n)}}$. Since the dynamics of the SSEP is given by the semigroup $\{e^{-tL}\}_{t\geq 0}$, $\lambda(n)$, determines the speed of relaxation to the invariant measure.

The following conjecture is known as Aldous' Conjecture but on his website [1] Aldous states that it arose in a conversation with Diaconis. So, maybe it should be called the Aldous-Diaconis Conjecture.

Conjecture 2.

$$\lambda(n) = \lambda(1), \quad \text{for all } 1 \leq n \leq |\Lambda| - 1 .$$

Apart from being a striking property, namely that the relaxation rate should be independent of the number of particles, it could also be very useful. The SSEP for one particle is just a random walk on the graph Λ, and many powerful techniques are available to study the relaxation rate of random walks. In physical terms one would say that the conjectured property reduces the many-body problem of finding the relaxation rate for n particles to a single one-particle problem.

Proposition 2. *If the ferromagnetic spin-1/2 Heisenberg model with coupling constants J_{xy} on a graph Λ satisfies FOEL, then Conjecture 2 holds for the SSEP on Λ with jump rates $J_{xy}/2$.*

Proof. The proof is based on the unitary equivalence of L and the ferromagnetic spin 1/2 Heisenberg Hamiltonian H given by

$$H = \sum_{x \sim y \in \Lambda} J_{xy} \left(\frac{1}{4}\mathbb{1} - \boldsymbol{S}_x \cdot \boldsymbol{S}_y \right) .$$

The unitary transformation $U : L^2(\Omega_\Lambda) \to \mathcal{H}_\Lambda = (\mathbb{C}^2)^{\otimes |\Lambda|}$, that relates L and H is explicitly given by

$$L^2(\Omega_\Lambda) \ni f \mapsto Uf = \psi = \sum_\eta f(\eta) |\eta\rangle , \quad \text{where } S_x^3 |\eta\rangle = (\eta_x - 1/2) |\eta\rangle .$$

To see this note that

$$1/4 - \boldsymbol{S}_x \cdot \boldsymbol{S}_y = (1 - t_{xy})/2 ,$$

where t_{xy} interchanges the states at x and y in any tensor product vector. Then

$$H\psi = \frac{1}{2}\sum_{x \sim y}\sum_\eta f(\eta) J_{xy}(1 - t_{xy}) |\eta\rangle$$

$$= \frac{1}{2}\sum_{x \sim y}\sum_\eta J_{xy}(f(\eta) - f(\eta^{xy})) |\eta\rangle$$

$$= \frac{1}{2}\sum_\eta (Lf)(\eta) |\eta\rangle .$$

Therefore, $HUf = ULf$, for all $f \in L^2(\Omega_\Lambda)$.

Under this unitary transformation, the particle number becomes the third component of the total spin:

$$S^3_{\text{tot}} = -|\Lambda|/2 + n \,.$$

The unique invariant measure of SSEP for n particles is the uniform measure on $\{\eta \in \Omega_\Lambda \mid \sum_x \eta_x = n\}$. The corresponding state for the spin model belongs to the unique multiplet of maximal total spin, i.e., is a ground state. $\lambda(n)$ is the next eigenvalue of H is the same value of total S^3. Since the first excited state of H, by FOEL, is a multiplet of total spin $S_{\max} - 1$, this eigenvalue has an eigenvalue with any value of S^3 in the range $-S_{\max} + 1, \ldots, S_{\max} - 1$. We have $S_{\max} = |\Lambda|/2$. Therefore, this corresponds to the range $1 \le n \le |\Lambda| - 1$. Hence, $\lambda(n)$ is independent of n in this range.

In combination with Theorem 1, this proposition has the following corollary.

Corollary 1. *Conjecture 2 holds for chains.*

Our partial result for trees (not discussed here) also implies Conjecture 2 for arbitrary finite trees as well as some graphs derived from trees. These cases of the Aldous-Diaconis conjecture were previously know [2, 8], as well as some other examples where one can compute $\lambda(n)$ exactly [3–5]. Needless to say, a full proof of FOEL, the Aldous-Diaconis Conjecture, or even a proof for additionial special cases, would be of great interest. An interesting direction for generalization considered by Aldous is to also establish the analogous formula for the spectral gap for a card-shuffling model with full $SU(n)$ symmetry, which restricts to the SSEP when one considers cards of only two colors.

Acknowledgement

This work was supported in part by the National Science Foundation under Grant # DMS-0303316. B.N. also thanks the Erwin Schrödinger Institute, Vienna, where part of this work was done, for financial support and the warm and efficient hospitality it offers.

References

1. D. Aldous, http://stat-www.berkeley.edu/users/aldous/problems.ps.
2. R. Bacher, *Valeur propre minimale du laplacien de Coxeter pour le groupe symétrique*, J. Algebra **167** (1994), 460–472.
3. P. Diaconis and L. Saloff-Coste, *Comparison techniques for random walk on finite groups*, Ann. Prob. **21** (1993), 2131–2156.

4. P. Diaconis and M. Shahshahani, *Generating a random permutation with random transpositions*, Z. Wahrsch. Verw. Geb. **57** (1981), 159–179.
5. L. Flatto, A.M. Odlyzko, and D.B. Wales, *Random shuffles and group representations*, Ann. Prob. **13** (1985), 154–178.
6. I.B. Frenkel and M.G. Khovanov, *Canonical bases in tensor products and graphical calculus for $u_q(sl_2)$*, Duke Math. J. **87** (1997), 409–480.
7. E. Gutkin, *Plancherel formula and critical spectral behaviour of the infinite XXZ chain*, Quantum symmetries (Clausthal, 1991), World Scientific, River Edge, NJ, 1993, pp. 84–98.
8. S. Handjani and D. Jungreis, *Rate of convergence for shuffling cards by transpositions*, J. Theor. Prob. **9** (1996), 983–993.
9. M. Jimbo and T. Miwa, *Algebraic analysis of solvable lattice models*, Regional Conference Series in Mathematics, American Mathematical Society, Providence, RI, 1995.
10. L.H. Kauffman and S.L. Lins, *Temperley-Lieb recoupling theory and invariants of 3-manifolds*, Princeton University Press, 1994.
11. T. Kennedy, *Expansions for droplet states in the ferromagnetic XXZ Heisenberg chain*, Markov Processes and Rel. Fields **11** (2005) 223–236
12. T. Koma and B. Nachtergaele, *The spectral gap of the ferromagnetic XXZ chain*, Lett. Math. Phys. **40** (1997), 1–16.
13. V.E. Korepin, N.M. Bogoliubov, and A.G. Izergin, *Quantum inverse scattering method and correlation functions*, Cambridge Monographs on Mathematical Physics, Cambridge University Press, Cambridge, England, 1993.
14. E.H. Lieb, *Two theorems on the Hubbard model*, Phys. Rev. Lett. **62** (1989), 1201–1204.
15. E.H. Lieb and D. Mattis, *Ordering energy levels of interacting spin systems*, J. Math. Phys. **3** (1962), 749–751.
16. B. Nachtergaele, W. Spitzer, and S. Starr, Droplet Excitations for the Spin-1/2 XXZ Chain with Kink Boundary Conditions, arXiv::math-ph/0508049.
17. _____, *Ferromagnetic ordering of energy levels*, J. Stat. Phys. **116** (2004), 719–738.
18. B. Nachtergaele and S. Starr, in preparation.
19. _____, *Droplet states in the XXZ Heisenberg model*, Commun. Math. Phys. **218** (2001), 569–607, math-ph/0009002.
20. B. Nachtergaele and S. Starr, *Ferromagnetic Lieb-Mattis theorem*, Phys. Rev. Lett. **94** (2005), 057206, arXiv:math-ph/0408020.
21. V. Pasquier and H. Saleur, *Common structures between finite systems and conformal field theories through quantum groups*, Nucl. Phys. **B330** (1990), 523–556.
22. H.N.V. Temperley and E.H. Lieb, *Relations between the "percolation" and "colouring" problem and other graph-theoretical problems associated with regular planar lattices: some exact results for the "percolation" problem*, Proc. Roy. Soc. **A322** (1971), 252–280.
23. H. Wielandt, *Unzerlegbare, nicht negative Matrizen*, Math. Z. **52** (1950), 642–648.

Interacting Fermions in 2 Dimensions

Vincent Rivasseau

Laboratoire de Physique Théorique, CNRS, UMR 8627, Université de Paris-Sud, 91405 Orsay
rivass@th.u-psud.fr

Abstract. We provide an introduction to the constructive results on interacting Fermions in two dimensions at thermal equilibrium above the critical temperature of pair condensation.

1 Introduction

Do interacting Fermions in 2 dimensions (above any low-temperature phase) resemble more three dimensional Fermions, i.e. the Fermi liquid, or one dimensional Fermions, i.e. the Luttinger liquid? The short answer to this controversial question is that it depends on the shape of the Fermi surface. Interacting Fermions with a round Fermi surface behave more like three dimensional Fermi liquids, whether interacting Fermions with the square Fermi surface of the Hubbard model at half-filling behave more like a one-dimensional Luttinger liquid.

This statement has been now proved in full mathematical rigor, beyond perturbation theory, using the mathematically precise criterion of Salmhofer. In this lecture we give an introduction to this result, obtained in the series of papers [1]– [5]. I also take this occasion to thank my colleagues and friends M. Disertori, S. Afchain and J. Magnen for their very pleasant collaborations on this subject.

2 Fermi Liquids and Salmhofer's Criterion

The textbooks definition of condensed matter physicists [6] for Fermi liquid simply does not work for interacting models which are parity invariant i.e. in absence of magnetic field. Indeed the textbook definition states that at zero temperature an interacting Fermi liquid should exhibit a discontinuity in its density of states at a certain Fermi surface, just like the free Fermi liquid. A curve in the style of Fig. 1 is usually shown as typical of such a liquid.

Many condensed matter theorists know that this definition is at best a "figment of imagination" to use Professor Anderson's words. Indeed Kohn-Luttinger singularities are known to be generic. This means that for any fixed bare generic interaction, going towards zero temperature at fixed interaction

V. Rivasseau: *Interacting Fermions in 2 Dimensions*, Lect. Notes Phys. **690**, 171–178 (2006)
www.springerlink.com   © Springer-Verlag Berlin Heidelberg 2006

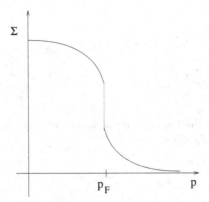

Fig. 1. The wrong picture, still found in many textbooks, for the alleged discontinuity of the selfenergy Σ at the Fermi surface p_F

cannot probe the famous discontinuity supposed to define the Fermi liquid
phase: the system instead undergoes a transition into a different condensed
phase (such as the BCS phase). To conclude from this fact that interacting
Fermi liquids simply do not exist would be wrong because condensed matter
physicists tell us that they know "deep in their bones" that Fermi liquids do
"exist" physically...So a better solution is to work out a correct definition,
that will allow interacting Fermi liquids to exist mathematically as well. Such
a mathematically correct formulation is ultimately needed for the theory
of Fermi liquids just like for thermodynamic limits, Gibbs states or phase
transitions.

It was found by M. Salmhofer a few years ago [7]– [8]. He proposed to
characterize the Fermi liquid behavior by moving a system simultaneously
towards $\lambda = T = 0$, so as to remain above any low temperature phase Kohn
Luttinger singularities. This can be accomplished for instance by sending λ
and T both to 0 along a curve $\lambda |\log T|^p = c$, where c is constant and p is an
integer, typically 1 or 2 (see Fig. 2). Of course no singularity will be met, but
Salmhofer proposed to call the system a Fermi liquid if the second derivative
of the self-energy, as function of the momentum, remains bounded along such
a curve [7].

This criterion works at thermal equilibrium and it has been chosen so
that it is not obeyed by the one dimensional Luttinger liquid, for which
indeed this second derivative along such a curve blows up. To discuss the
physical relevance of Salmhofer's criterion would be too long, and I do not
feel competent at all, in particular for the experimental aspects (is it possible
for example to turn knobs to move at will both temperature and interaction
in as Fermion system? may be yes, by varying temperature and pressure?...)
I just remark that this rigorous criterion is certainly up to now the best for
parity invariant interacting systems, at least because it is the only one.

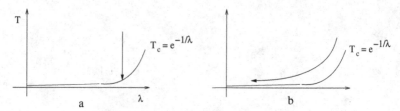

Fig. 2. (a): Lowering the temperature at fixed coupling; (b): Lowering the temperature according to Salmhofer

3 The Models

We consider a gas of Fermions in thermal equilibrium at temperature T, with coupling constant λ. The free propagator for these models is

$$\hat{C}_{a,b}(k) = \delta_{a,b}\frac{1}{ik_0 - e(\overrightarrow{k})} \ , \tag{1}$$

with $e(\overrightarrow{k}) = \epsilon(\overrightarrow{k}) - \mu$, $\epsilon(\overrightarrow{k})$ being the kinetic energy and μ the chemical potential. The relevant values for k_0 are discrete and called the Matsubara frequencies :

$$k_0 = \left(\frac{\pi}{\beta}\right)(2n+1), \ n \in \mathbb{Z} \ , \tag{2}$$

where $\beta = (kT)^{-1}$.

When $T \to 0^+$ (which means $\beta \to +\infty$), k_0 becomes a continuous variable, the corresponding discrete sum becomes an integral, and the corresponding propagator $C_0(x)$ becomes singular on the Fermi surface defined by $k_0 = 0$ and $e(\boldsymbol{k}) = 0$. This Fermi surface depends on the kinetic energy $\epsilon(\overrightarrow{k})$ of the model. For rotation invariant models, $\epsilon(\overrightarrow{k}) = \overrightarrow{k}^2/2m$ and the Fermi surface is a circle in two dimensions and a sphere in three dimensions, with radius $\sqrt{2m\mu}$. These two rotation invariant models, or *jellium models* are respectively nicknamed J_2 and J_3. For the half-filled Hubbard model, nicknamed H_2, $\boldsymbol{x}$ lives on the lattice $\mathbb{Z}^2$, and $\epsilon(\overrightarrow{k}) = \cos k_1 + \cos k_2$ so that at $\mu = 0$ the Fermi surface is a square of side size $\sqrt{2}\pi$, joining the points $(\pm\pi, 0), (0, \pm\pi)$ in the first Brillouin zone.

It is also possible to interpolate continuously between H_2 and J_2 by varying the filling factor of the Hubbard model. Lattice models with next-nearest neighbor hopping are also interesting.

In contrast with the propagator, the interaction is almost unique. Indeed we are interested in long-range physics, so we should start from a quasi-local bare interaction. But there is a unique exactly local such interaction, namely

$$S_V = \lambda \int_V d^3x \left(\sum_{a \in \{\uparrow,\downarrow\}} \bar{\psi}_a(x)\psi_a(x)\right)^2 \ , \tag{3}$$

where $V := [-\beta, \beta[\times V'$ and V' is an auxiliary volume cutoff in two dimensional space, that will be sent to infinity in the thermodynamic limit. Indeed any local polynomial of higher degree is zero since Fermionic fields anticommute.

4 A Brief Review of Rigorous Results

What did the programs of rigorous mathematical study of interacting Fermi systems accomplish until now? Recall that in dimension 1 there is neither superconductivity nor extended Fermi surface, and Fermion systems have been proved to exhibit Luttinger liquid behavior [9]. The initial goal of the studies in two or three dimensions was to understand the low temperature phase of these systems, and in particular to build a rigorous constructive BCS theory of superconductivity. The mechanism for the formation of Cooper pairs and the main technical tool to use (namely the corresponding $1/N$ expansion, where N is the number of sectors which proliferate near the Fermi surface at low temperatures) have been identified [10]. But the goal of building a completely rigorous BCS theory ab initio remains elusive because of the technicalities involved with the constructive control of continuous symmetry breaking.

So the initial goal was replaced with a more modest one, still important in view of the controversies over the nature of two dimensional "Fermi liquids" [11], namely the rigorous control of what occurs before pair formation. The last decade has seen excellent progress in this direction.

As is well known, sufficiently high magnetic field or temperature are the two different ways to break the Cooper pairs and prevent superconductivity. Accordingly two approaches were devised for the construction of "Fermi liquids". One is based on the use of non-parity invariant Fermi surfaces to prevent pair formation. These surfaces occur physically when generic magnetic fields are applied to two dimensional Fermi systems. The other is based on Salmhofer's criterion [7], in which temperature is the cutoff which prevents pair formation.

In a large series of papers [12], the construction of two dimensional Fermi liquids for a wide class of non-parity invariant Fermi surfaces has been completed in great detail by Feldman, Knörrer and Trubowitz. These papers establish Fermi liquid behavior in the traditional sense of physics textbooks, namely as a jump of the density of states at the Fermi surface at zero temperature, but they do not apply to the simplest Fermi surfaces, such as circles or squares, which are parity invariant.

An other program in recent years was to explore which models satisfy Salmhofer's criterion. Of particular interest to us are the three most "canonical" models in more than one dimension namely:

- the jellium model in two dimensions, with circular Fermi surface, nicknamed J_2,

- the half-filled Hubbard model in two dimensions, with square Fermi surface, nicknamed H_2,
- and the jellium model in three dimensions, with spherical Fermi surface, nicknamed J_3.

The study of each model has been divided into two main steps of roughly equal difficulty, the control of convergent contributions and the renormalization of the two point functions. In this sense, five of the six steps of our program are now completed. J_2 is a Fermi liquid in the sense of Salmhofer [1]–[2], H_2 is not, and is a Luttinger liquid with logarithmic corrections, according to [3,4] and [5].

Results similar to [1] - [2] have been also obtained for more general convex curves not necessarily rotation invariant such as those of the Hubbard model at low filling, where the Fermi surface becomes more and more circular, including an improved treatment of the four point functions leading to better constants [13]. Therefore as the filling factor of the Hubbard model is moved from half-filling to low filling, we conclude that there must be a crossover from Luttinger liquid behavior to Fermi liquid behavior. This solves the controversy [11] over the Luttinger or Fermi nature of two-dimensional many-Fermion systems above their critical temperature.

Up to now only the convergent contributions of J_3, which is almost certainly a Fermi liquid, have been controlled [14]. The renormalization of the two point functions for J_3, the last sixth of our program, remains still to be done. This last part is difficult since the cutoffs used in [14] do not conserve momentum. This means that the two point functions that have to be renormalized in this formalism are not automatically one particle irreducible, as is the case both in [2] and in this paper. This complicates their analysis.

5 Multiscale Analysis, Angular Sectors

For any two-dimensional model built until now in the constructive sense, the strategy is the same. It is based on some kind of multiscale expansion, which keeps a large fraction of the theory in unexpanded determinants. The global bound on these determinant (using determinant inequalities such as Gram inequality) is much better than if the determinant was expanded into Feynman graphs which would then be bounded one by one, and the bounds summed. The bound obtained in this way would simply diverge at large order (i.e. not prove any analyticity at all in the coupling constant) simply because there are too many Feynman graphs at large order. But the divergence of a bound does not mean the divergence of the true quantity if the bound is bad. Constructive analysis, which "keeps loops unexpanded" is the correct way to obtain better bounds, which do prove that the true series in fact does not diverge, i.e. has a finite convergence radius in the coupling constant. This

radius howerver shrink when the temperature goes to 0, and a good constructive analysis should establish the correct shrinking rate, which is logarithmic.
This is were *multiscale* rather than *single scale* constructive analysis becomes
necessary.

The basic idea of the multiscale analysis is to slice the propagator according to the size of its denominator so that the slice number i corresponds to
$|k_0 - e(\overrightarrow{k})| \simeq M^{-i}$, where M is some fixed constant.

This multiscale analysis is supplemented within each scale by an angular
"sector analysis". The number of sectors should be kept as small as possible,
so each sector should be as large as possible in the directions tangent to the
Fermi surface in three dimensions, or to the Fermi curve in two dimensions.
What limits however the size of these sectors is the curvature of the surface,
so that stationary phase method could still relate the spatial decay of a
propagator within a sector to its dual size in momentum space. In the case of
a circle, the number of sectors at distance ε of the singularity grows therefore
at least like $\varepsilon^{-1/2}$, hence like a power of ε. However for the half-filled Hubbard
model, since the curvature is "concentrated at the corners" the number of
sectors grows only like $|\log \varepsilon|$. In one dimension there are really only two
sectors since the Fermi singularity is made of two points. A logarithm is
closer to a constant than to a power; this observation is the main reason for
which the half-filled Hubbard model is closer to the one-dimensional Luttinger
liquid than to the three dimensional Fermi liquid.

Momentum conservation rules for sectors which meet at a given vertex
in general are needed to fix the correct *power counting* of the subgraphs of
the model. In the Hubbard case at half filling, these rules are needed only
to fix the correct *logarithmic power counting*, since the growth of sectors
near the singularity is only logarithmic. In both cases the net effect in two
dimensions of these conservation rules is to roughly identify two pairs of
"conserved" sectors at any vertex, so that in each slice the model resembles
an N-component vector model, where N is the number of sectors in the slice.

The multiscale renormalization group analysis of the model then consists
essentially in selecting, for any graph, a tree which is a subtree in each of the
"quasi-local" connected components of the graph according to the momentum
slicing. These connected components are those for which all internal lines are
farther from the Fermi surface than all external lines. The selection of this
tree can be performed in a constructive manner, keeping the remaining loop
fields in a determinant. The combinatoric difficulty related to the fact that a
graph contains many trees has been tackled once and for all thanks to forest
formulas such as those of [15].

6 One and Two Particle Irreducible Expansions

Salmhofer's criterion is stated for the self-energy, i.e. the sum of all one-
particle irreducible graphs for the two point function. Its study requires the

correct renormalization of these contributions. Since angular sectors in a graph may vary from one propagator to the next in a graph, and since different sectors have different decays in different directions, we are in a delicate situation. In order to prove that renormalization indeed does the good that it is supposed to do, one cannot simply rely on the connectedness of these self-energy graphs, but one must use their particle irreducibility explicitly.

So the proof requires a constructive particle irreducible analysis of the self-energy. The following theorem summarizes the results of [1]-DR2

Theorem 1. *The radius of convergence of the jellium two-dimensional model perturbative series for any thermodynamic function is at least $c/|\log T|$, where T is the temperature and c some numerical constant. As T and λ jointly tend to 0 in this domain, the self-energy and its first two momentum derivatives remain uniformly bounded so that the model is a Fermi liquid in the sense of Salmhofer.*

In the case of the jellium model $J2$, this analysis can be performed at the level of one-particle irreducible graphs [2]. The half-filled Hubbard model, however, is more difficult. It came as a surprise to us that although there is no real divergence of the self-energy (the associated counterterm is zero thanks to the particle hole symmetry of the model at half-filling) one really needs a two-particle and one-vertex irreducible constructive analysis to establish the necessary constructive bounds on the self-energy and its derivatives [4]. For parity reasons, the self-energy graphs of the model are in fact not only one-particle irreducible but also two particle and one vertex irreducible, so that this analysis is possible.

This analysis according to the *line form* of Menger's theorem ([16]) leads to the explicit construction of three line-disjoint paths for every self-energy contribution, in a way compatible with constructive bounds. On top of that analysis, another one which is scale-dependent is performed: after reduction of some maximal subsets provided by the scale analysis, two vertex-disjoint paths are selected in every self-energy contribution. This requires a second use of Menger's theorem, now in the *vertex form*. This construction allows to improve the power counting for two point subgraphs, exploiting the particle-hole symmetry of the theory at half-filling, and leads to the desired analyticity result.

Finally an upper bound for the self energy second derivative is combined with a lower bound for the explicit leading self energy Feynman graph [5]. This completes the proof that the Hubbard model violates Salmhofer's criterion, hence is *not* a Fermi liquid, in contrast with the jellium two dimensional model. More precisely the following theorem summarizes the results of [3–5].

Theorem 2. *The radius of convergence of the Hubbard model perturbative series at half-filling is at least $c/\log^2 T$, where T is the temperature and c some numerical constant. As T and λ jointly tend to 0 in this domain, the* *self-energy of the model does not display the properties of a Fermi liquid in the sense of Salmhofer, since the second derivative is not uniformly bounded.*

References

1. M. Disertori and V. Rivasseau, Interacting Fermi liquid in two dimensions at finite temperature, Part I: Convergent Attributions, Commun. Math. Phys. **215**, 251 (2000).
2. M. Disertori and V. Rivasseau, Interacting Fermi liquid in two dimensions at finite temperature, Part II: Renormalization, Commun. Math. Phys. **215**, 291 (2000).
3. V. Rivasseau, The two dimensional Hubbard Model at half-filling: I. Convergent Contributions, Journ. Stat. Phys. **106**, 693–722, (2002).
4. S. Afchain, J. Magnen and V. Rivasseau, Renormalization of the 2-point function of the Hubbard Model at half-filling, cond-mat/0409231 (2004), to appear in Annales Henri Poincaré.
5. S. Afchain, J. Magnen and V. Rivasseau, The Hubbard Model at half-filling, part III: the lower bound on the self-energy, cond-mat/0412401 (2004), to appear in Annales Henri Poincaré.
6. See e.g. E. Fradkin, "Field Theories of Condensed Matter Systems", Frontiers in Physics, 1991
7. M. Salmhofer, Continuous renormalization for Fermions and Fermi liquid theory, Commun. Math. Phys. **194**, 249 (1998).
8. M. Salmhofer, Renormalization, an introduction, Springer Verlag, 1999.
9. G. Benfatto and G. Gallavotti, Renormalization Group, Physics Notes, Chap. 11 and references therein, Vol. 1 Princeton University Press, 1995
10. J. Feldman, J. Magnen, V. Rivasseau and E. Trubowitz, An Intrinsic 1/N Expansion for Many Fermion System, Europhys. Letters **24**, 437 (1993).
11. P.W. Anderson, Luttinger liquid behavior of the normal metallic state of the 2D Hubbard model, Phys Rev Lett. **64** 1839–1841 (1990).
12. Joel Feldman, H. Knörrer and E. Trubowitz, A two dimensional Fermi Liquid, series of papers in Commun. Math. Phys. **247**, 1–319, 2004 and Reviews in Math. Physics, **15**, 9, 949–1169, (2003).
13. G. Benfatto, A. Giuliani and V. Mastropietro, Low temperature Analysis of Two-Dimensional Fermi Systems with Symmetric Fermi surface, Ann. Henri Poincaré, **4** 137, (2003).
14. M. Disertori, J. Magnen and V. Rivasseau, Interacting Fermi liquid in three dimensions at finite temperature, part I: Convergent Contributions, Ann. Henri Poincaré, **2** 733–806 (2001).
15. A. Abdesselam and V. Rivasseau, Trees, Forests and Jungles: A Botanical Garden for Cluster Expansions, in "Constructive Physics", LNP 446, Springer Verlag, 1995.
16. J.A. Bondy and U.S.R. Murty, Graph theory with applications, North-Holland Editions, 1979.

On the Essential Spectrum of the Translation Invariant Nelson Model

Jacob Schach-Møller

Aarhus University, Department of Mathematical Sciences, 8000 Aarhus C,
Denmark
jacob@imf.au.dk

Abstract. Let $\mathbb{R}^\nu \ni \xi \to \Sigma_{\mathrm{ess}}(\xi)$ denote the bottom of the essential spectrum for the fiber Hamiltonians of the translation invariant massive Nelson model, which describes a ν-dimensional electron linearly coupled to a scalar massive radiation field. We prove that, away from a locally finite set, Σ_{ess} is an analytic function of total momentum.

1 The Model and the Result

Let $\mathfrak{h}_{\mathrm{ph}} := L^2(\mathbb{R}^\nu_k)$ and $\mathcal{F} = \Gamma(\mathfrak{h}_{\mathrm{ph}})$ denote the bosonic Fock space constructed from $\mathfrak{h}_{\mathrm{ph}}$. We write $p = -i\nabla_{\mathrm{x}}$ for the momentum operator in $\mathcal{K} := L^2(\mathbb{R}^\nu_{\mathrm{x}})$. The translation invariant Nelson Hamiltonian describing a ν-dimensional electron (or positron) linearly coupled to a massive scalar radiation field has the form

$$H := \Omega(p) \otimes \mathbb{1}_\mathcal{F} + \mathbb{1}_\mathcal{K} \otimes d\Gamma(\omega) + V, \quad \text{on } \mathcal{K} \otimes \mathcal{F},$$

where

$$V := \int_{\mathbb{R}^\nu} \left\{ e^{-i k \cdot \mathrm{x}} v(k)\, \mathbb{1}_\mathcal{K} \otimes \mathbf{a}^*(k) + e^{i k \cdot \mathrm{x}}\, v(k)\, \mathbb{1}_\mathcal{K} \otimes \mathbf{a}(k) \right\} dk.$$

We assume that the form factor v satisfies

$$
\begin{aligned}
& v \in L^2(\mathbb{R}^\nu_k), \ v \text{ real valued}, \ v \neq 0 \text{ a.e.} \\
& \text{and } \forall O \in \mathcal{O}(\nu) : v(Ok) = v(k) \text{ a.e.},
\end{aligned}
\tag{1}
$$

which implies a UV-cutoff. Here $\mathcal{O}(\nu)$ denotes the orthogonal group. The physically interesting choices for the dispersion relations Ω and ω are $\Omega(\eta) = \eta^2/2M$, $\Omega(\eta) = \sqrt{\eta^2 + M^2}$ and $\omega(k) = \sqrt{k^2 + m^2}$, where $M, m > 0$ are the electron and boson masses. We will however work with general forms of both Ω and ω. As for ω, this is partly motivated by the similarity with the Polaron model, cf. [5,11], where ω is not explicitly known. We make no attempt here to say anything about the Polaron model.

The operator H commutes with the total momentum $p \otimes \mathbb{1}_\mathcal{F} + \mathbb{1}_\mathcal{K} \otimes d\Gamma(k)$ and hence fibers as $H \sim \int^\oplus_{\mathbb{R}^\nu} H(\xi) d\xi$, where the fiber Hamiltonians $H(\xi)$, $\xi \in \mathbb{R}^\nu$, are operators on $\mathcal{F}$ given by

J.S. Møller: *On the Essential Spectrum of the Translation Invariant Nelson Model*, Lect. Notes Phys. **690**, 179–195 (2006)
www.springerlink.com

$$H(\xi) = H_0(\xi) + \Phi(v) \quad \text{where} \quad H_0(\xi) = d\Gamma(\omega) + \Omega(\xi - d\Gamma(k)) \qquad (2)$$

and the interaction is

$$\Phi(v) = \int_{\mathbb{R}^\nu} \{v(k)\mathbf{a}^*(k) + v(k)\mathbf{a}(k)\} dk. \qquad (3)$$

We formulate precise assumptions on Ω and ω, which are satisfied by the examples mentioned above. We use the standard notation $\langle t \rangle := (1 + t^2)^{1/2}$.

Condition 1 (The particle dispersion relation) *Let $\Omega \in C^\infty(\mathbb{R}^\nu)$. There exists $s_\Omega \in \{0, 1, 2\}$ such that*

(i)   There exists $C > 0$ such that $\Omega(\eta) \geq C^{-1}\langle \eta \rangle^{s_\Omega} - C$.
(ii) For any multi-index α there exists C_α such that $|\partial^\alpha \Omega(\eta)| \leq C_\alpha \langle \eta \rangle^{s_\Omega - |\alpha|}$.
(iii) Ω is rotation invariant, i.e., $\Omega(O\eta) = \Omega(\eta)$, for all $O \in \mathcal{O}(\nu)$.
(iv) The function $\eta \rightarrow \Omega(\eta)$ is real analytic.

Condition 2 (The photon dispersion relation) *Let $\omega \in C^\infty(\mathbb{R}^\nu)$ satisfy*

(i)   There exists $m > 0$, the photon mass, such that $\inf_{k \in \mathbb{R}^\nu} \omega(k) = \omega(0) = m$.
(ii) $\omega(k) \rightarrow \infty$, in the limit $|k| \rightarrow \infty$.
(iii) There exist $s_\omega \geq 0$, $C_\omega > 0$, and for any multi-index α with $|\alpha| \geq 1$, a C_α such that $\omega(k) \geq C_\omega^{-1}\langle k \rangle^{s_\omega} - C_\omega$ and $|\partial_k^\alpha \omega(k)| \leq C_\alpha \langle k \rangle^{s_\omega - |\alpha|}$.
(iv) ω is rotation invariant, i.e., $\omega(O\eta) = \omega(\eta)$, for all $O \in \mathcal{O}(\nu)$.
(v) ω is real analytic.
(vi) ω is strictly subadditive, i.e. $\omega(k_1) + \omega(k_2) > \omega(k_1 + k_2)$ for all $k_1, k_2 \in \mathbb{R}^\nu$.

Remarks: (1) The assumption that the photons are massive is essential.
(2) One could weaken the assumption $v \in L^2(\mathbb{R}^\nu)$ by taking instead $v/\sqrt{\omega} \in L^2(\mathbb{R}^\nu)$. This is a weaker ultraviolet condition, which still allows for the construction of the Hamiltonian. See [2].
(3) The subadditivity assumption is discussed at the end of this section.
(4) Condition 2 vi) follows from subadditivity $\omega(k_1) + \omega(k_2) \geq \omega(k_1 + k_2)$ together with Condition 2 (i), (iv), and (v).
(5) The assumptions (1), Condition 1 (ii), Condition 2 (i), (ii), and iii) can be relaxed, cf. [10].
  We introduce the bottom of the spectrum and essential spectrum as functions of total momentum

$$\Sigma_0(\xi) := \inf \sigma(H(\xi)) \quad \text{and} \quad \Sigma_{\text{ess}}(\xi) := \inf \sigma_{\text{ess}}(H(\xi)).$$

The energy of a system of n non-interacting bosons, with momenta $\underline{k} \in \mathbb{R}^{n\nu}$, $\underline{k} = (k_1, \ldots, k_n)$, and one interacting electron with momentum $\xi - k^{(n)}$, where $k^{(n)} = k_1 + \cdots + k_n$, is

$$\Sigma_0^{(n)}(\xi; \underline{k}) := \Sigma_0(\xi - k^{(n)}) + \sum_{j=1}^{n} \omega(k_j) \tag{4}$$

and the smallest of such energies

$$\Sigma_0^{(n)}(\xi) := \inf_{\underline{k} \in \mathbb{R}^{n\nu}} \Sigma_0^{(n)}(\xi; \underline{k}), \tag{5}$$

which is a threshold energy for the model. Due to the assumption of strict subadditivity of ω, Condition 2 (vi), we have

$$\Sigma_0^{(n)}(\xi) < \Sigma_0^{(n')}(\xi), \text{ for } n < n'. \tag{6}$$

The function Σ_{ess} can be expressed in terms of Σ_0

$$\Sigma_{\text{ess}}(\xi) = \Sigma_0^{(1)}(\xi). \tag{7}$$

This is the content of the HVZ theorem, see [9, Theorem 1.2 and Corollary 1.4] and [11, Sect. 4]. Write $\mathcal{I}_0 := \{\xi \in \mathbb{R}^\nu | \Sigma_0(\xi) < \Sigma_{\text{ess}}(\xi)\}$ (ξ's with an isolated groundstate) and for $\xi \in \mathbb{R}^\nu$: $\mathcal{I}_0^{(1)}(\xi) := \{k \in \mathbb{R}^\nu : \xi - k \in \mathcal{I}_0\}$.

We recall that $H(\xi)$ is self-adjoint on $\mathcal{D} = \mathcal{D}(H_0(\xi))$, which is independent of ξ. The functions $\xi \to \Sigma_0(\xi), \Sigma_{\text{ess}}(\xi), \Sigma_0^{(n)}(\xi)$ are Lipschitz continuous, rotation invariant, and go to infinity at infinity. For a treatment of the second quantization formalism used in the formulation of the model see [1] or the brief overviews given in [4] and [9]. See also the recent monograph [12] by Spohn, for up to date material on models of non-relativistic QED.

The authors talk at the QMATH9 meeting, was devoted to an overview of results for the spectral functions introduced above. Drawing mostly on work of Fröhlich [6, 7], Spohn [11], and the author [9]. One of these results, [9, Theorem 1.9], states that $\mathbb{R} \ni t \to \Sigma_{\text{ess}}(t\boldsymbol{u})$ is a real analytic function away from a closed countable set, under the additional assumption that ω is also convex. Here $\boldsymbol{u}$ is an arbitrary unit vector. This prompted the following question from Heinz Siedentop: "Is this optimal?". Here is the answer:

Theorem 1. *Fix a unit vector $\boldsymbol{u} \in \mathbb{R}^\nu$. Suppose (1) and Conditions 1 and 2. Then there exists a locally finite set $\mathcal{S} \subset \mathbb{R}$ such that $\mathbb{R}\backslash\mathcal{S} \ni t \to \Sigma_{\text{ess}}(t\boldsymbol{u})$ is analytic. For any connected component $I = (a,b)$ of $\mathbb{R}\backslash\mathcal{S}$ we have either: Σ_{ess} is constant on I, or there exists an analytic function $I \ni t \to \theta(t) \in \mathcal{I}_0^{(1)}(t\boldsymbol{u})$ such that for $t \in I$*

$$\Sigma_{\text{ess}}(t\boldsymbol{u}) = \Sigma_0^{(1)}(t\boldsymbol{u}; \theta(t)\boldsymbol{u}) \text{ and } \nabla\Sigma_{\text{ess}}(t\boldsymbol{u}) = \nabla\omega(\theta(t)\boldsymbol{u}).$$

In the latter case, there furthermore exist integers $1 \leq p, q < \infty$ such that the functions $(a, a + \delta) \ni t \to \theta(a + (t-a)^p))$ and $(b-\delta, b) \ni t \to \theta(b - (b-t)^q)$ extend analytically through a respectively b. (Here δ is chosen such that $a + \delta^p, b - \delta^q \in I$.)

Remarks: (1) The part of (1) requiring v to be real is an input to a Perron-Frobenius argument, see [7, Sect. 2.4] and [9, Sect. 3.3], which ensures that the groundstate of each $H(\xi)$ is non-degenerate. This, together with analytic perturbation theory, implies that $\xi \to \Sigma_0(\xi)$ is analytic in $\mathcal{I}_0$, see [7, Theorem 3.6]. This is in fact the information we need to make the proof of Theorem 1 work. Hence the conclusion of the theorem remains true also for the uncoupled system ($v \equiv 0$) although (1) is not satisfied in this case.

(2) If subadditivity of ω is not assumed we are faced with two problems: (I) We would need to understand the breakup of degenerate critical points of $\underline{k} \to \Sigma_0^{(n)}(\xi; \underline{k})$ for any n, not just $n = 1$. This is a much more difficult problem (but probably doable). (II) The crossing of thresholds $\Sigma_0^{(n)}(\xi)$ may be associated with the disappearance of the groundstate $\Sigma_0(\xi)$ into the essential spectrum. The strategy of the proof below would require that $\Sigma_0(\xi)$ (suitably reparameterized as in Theorem (1) continues analytically into the essential spectrum. This is beyond current technology.

In Sect. 2 below, we study analytic functions of two complex variables which are of the form of $f(x - y) + g(y)$, cf. the definition (4) of $\Sigma_0^{(1)}(\cdot; \cdot)$. In Sect. 3 we apply the results of Sect. 2 to prove Theorem 1. In Appendix A we recall basic properties of Riemannian covering spaces.

2 A Complex Function of Two Variables

We write $(\mathcal{R}_p, \pi_p)$ for the p'th Riemannian cover over $\mathbb{C}\backslash\{0\}$. See Appendix A.

For $z \in \mathbb{C}$ and $r > 0$ we introduce the notation

$$D(z, r) := \{z' \in \mathbb{C} : |z - z'| < r\}, \quad D'(z, r) := D(z, r)\backslash\{z\},$$
$$D'_p(r) := \pi_p^{-1}(D'(0, r)) \subset \mathcal{R}_p.$$

We furthermore use the abbreviations $D(r) \equiv D(0, r)$ and $D'(r) = D'(0, r)$.

Fix $x_0, y_0 \in \mathbb{C}$ and $r_0 > 0$. Let f, g be analytic in $D(x_0 - y_0, 2r_0)$ and $D(y_0, r_0)$ respectively. We define:

$$H(x, y) := f(x - y) + g(y) \text{ for } (x, y) \in \mathcal{D} := D(x_0, r_0) \times D(y_0, r_0).$$

The function H is an analytic function of two variables in the polydisc $\mathcal{D}$.

We suppose that $y \to H(x_0, y)$ has a critical point at $y = y_0$, that is:

$$(\partial_y H)(x_0, y_0) = 0. \tag{8}$$

The aim of this section is to catalogue the breakup of the critical point, counting multiplicity, when x_0 is replaced by an x near x_0. We wish to determine the sets

$$\Theta(x) := \{y \in D(y_0, r_y) : (\partial_y H)(x, y) = 0\}, \tag{9}$$

for $x \in D'(x_0, r_x)$ and r_x, r_y small enough.

We write f and g as convergent power series in the discs $D(x_0 - y_0, 2r_0)$ and $D(y_0, r_0)$ respectively

$$f(z) = \sum_{k=0}^{\infty} f_k(z - (x_0 - y_0))^k \quad \text{and} \quad g(z) = \sum_{k=0}^{\infty} g_k(y - y_0)^k. \tag{10}$$

We can without loss of generality assume $f_0 = g_0 = 0$, and (8) implies $f_1 = g_1$, and hence we can in addition assume $f_1 = g_1 = 0$. (The critical points are independent of f_0, f_1, g_0 and g_1). To summarize

$$f_0 = f_1 = g_0 = g_1 = 0. \tag{11}$$

For a function h, analytic in an open set $\mathcal{U} \subset \mathbb{C}$, we recall that h has a zero of order k at z_0 if $(z - z_0)^{-k} h$ is analytic near z_0, and $(z - z_0)^{-k-1} h$ is singular at z_0. Equivalently h has a zero of order k at z_0 if $\frac{\partial^\ell h}{\partial z^\ell}(z_0) = 0$, for $0 \le \ell < k$, and $\frac{\partial^k h}{\partial z^k}(z_0) \ne 0$. We will use the following notation for roots of unity. For $k \ge 1$, we write the k solutions of $\alpha^k = 1$ as

$$\alpha_\ell^k := e^{i2\pi\ell/k}, \quad \text{for } \ell \in \{1, \dots, k\}. \tag{12}$$

By the p'th root of a complex non-zero constant $C \sim (|C|, \arg(C))$, where $0 \le \arg(C) < 2\pi$, we understand

$$C^{1/p} := |C|^{1/p} e^{i \arg C/p}. \tag{13}$$

In the following we write κ_H for the order of the zero y_0 for the analytic function $(\partial_y H)(x_0, \cdot)$, κ_g for the order of the zero y_0 of $y \to g(y)$, and κ_f for the order of the zero $x_0 - y_0$ of $z \to f(z)$ (recall (11)). We furthermore abbreviate

$$F := \frac{\frac{\partial^{\kappa_f} f}{\partial z^{\kappa_f}}(x_0 - y_0)}{(\kappa_f - 1)!} = \kappa_f f_{\kappa_f}, \quad G := \frac{\frac{\partial^{\kappa_g} g}{\partial z^{\kappa_g}}(y_0)}{(\kappa_g - 1)!} = \kappa_g g_{\kappa_g}, \tag{14}$$

$$M := \frac{(\partial_y^{\kappa_H+1} H)(x_0, y_0)}{\kappa_H!} = (\kappa_H + 1)((-1)^{\kappa_H+1} f_{\kappa_H+1} + g_{\kappa_H+1}) \ne 0.$$

Proposition 1. *Let f, g, x_0, y_0 and r_0 be as above, with $\kappa_g, \kappa_f, \kappa_H < \infty$. Then there exist $0 < r_x, r_y \le r_0$, such that for $x \in D'(x_0, r_x)$ the set of solutions $\Theta(x) \subset D'(y_0, r_y)$ consists of precisely κ_H distinct points, all of which are zeroes of order 1 for $(\partial_y H)(x, \cdot)$. We have the following description of $\Theta(x)$:*

I The case $\kappa_g \le \kappa_H$: There are analytic functions $\theta_\ell : D(r_x) \to D(r_y)$, $\ell \in \{1, \dots, \kappa_g - 2\}$, and $\theta : D'_{\kappa_H - \kappa_g + 2}(r_x) \to D'(r_y)$, such that $\Theta(x) = y_0 + (\cup_{\ell=1}^{\kappa_g - 2}\{\theta_\ell(x - x_0)\}) \cup \theta(\pi_{\kappa_H - \kappa_g + 2}^{-1}(\{x - x_0\}))$. We have the asymptotics

$$\theta_\ell(x) = \frac{\alpha_\ell^{\kappa_g-1}}{\alpha_\ell^{\kappa_g-1} - 1} x + O(|x|^2), \quad \ell \in \{1,\dots,\kappa_g - 2\},$$

$$\theta(\mathbf{x}) = C_I \mathbf{x}^{1/(\kappa_H - \kappa_g + 2)} + O(|\mathbf{x}|^{-2/(\kappa_H - \kappa_g + 2)})$$

where $C_I = [-(\kappa_g - 1)G/M]^{1/(\kappa_H - \kappa_g + 2)}$.

II The case $\kappa_g = \kappa_H + 1$: Here $\kappa_f \geq \kappa_g$ and we write $\kappa_f - 1 = p\kappa_H + q$ and $d = (q, \kappa_H)$ (the greatest common divisor), where $p \geq 1$, $0 \leq q < \kappa_H$. There exists d analytic functions $\theta_\ell : D'_{\kappa_H/d}(r_x) \to D'(r_y)$ such that $\Theta(x) = y_0 + \cup_{\ell=1}^d \theta_\ell(\pi^{-1}_{\kappa_H/d}(\{x - x_0\}))$. We have two possible asymptotics: If $\kappa_f = \kappa_g$ (and hence $q = 0$, $d = \kappa_H$, and $D'_1(r_x) \equiv D'(r_x)$) then

$$\theta_\ell(x) = \frac{C_{II}\alpha_\ell^{\kappa_H}}{C_{II}\alpha_\ell^{\kappa_H} - 1} x + O(|x|^2), \quad \ell \in \{1,\dots,\kappa_H\},$$

where $C_{II} = ((-1)^{\kappa_H} F/G)^{1/\kappa_H} \neq 1$. If $\kappa_f > \kappa_g$ then

$$\theta_\ell(\mathbf{x}) = C'_{II} \alpha_\ell^d \pi_{\kappa_H/d}(\mathbf{x})^p \left(\mathbf{x}^{\frac{1}{\kappa_H/d}}\right)^{\frac{q}{d}} + O(|\mathbf{x}|^{\kappa_f/\kappa_H}), \quad \ell \in \{1,\dots,d\},$$

where $C'_{II} = (F/G)^{1/\kappa_H}$.

III The case $\kappa_g > \kappa_H + 1$: Write $\kappa_g - 1 = p\kappa_H + q$ and $d = (\kappa_H, q)$, where $p \geq 1$ and $0 \leq q < \kappa_H$. There are d analytic functions $\theta_\ell : D'_{\kappa_H/d}(r_x) \to D'(r_y)$ such that $\Theta(x) = y_0 + \cup_{\ell=1}^d \theta_\ell(\pi^{-1}_{\kappa_H/d,x_0}(\{x - x_0\}))$. We have the asymptotics, with $C_{III} = (-G/M)^{1/\kappa_H}$ and $\ell \in \{1,\dots,d\}$,

$$\theta_\ell(\mathbf{x}) = \pi_{\kappa_H/d}(\mathbf{x}) + C_{III} \alpha_\ell^d \pi_{\kappa_H/d}(\mathbf{x})^p \left(\mathbf{x}^{\frac{1}{\kappa_H/d}}\right)^{\frac{q}{d}} + O(|\mathbf{x}|^{\kappa_g/\kappa_H}).$$

If $q = 0$ and hence $d = \kappa_H$ in II and III, then the maps θ_ℓ, a priori defined on $D'_1(r_x) \equiv D'(0, r_x)$, extend to analytic maps from $D(0, r_x)$ by the prescription $\theta_\ell(0) := y_0$. (Note the convention $(0, p) = p$ for $p \neq 0$.)

Remarks: (1) If $\kappa_f = \infty$, then $H(x, y) = g(y)$, and $\theta(x) \equiv y_0$ is the solution to (8). If $\kappa_g = \infty$ then $g = 0$ and $\theta(x) = x$ is the solution to (8). If $\kappa_H = \infty$ then $H(x_0, y) \equiv H(x_0, y_0)$ and hence $g(y) = H(x_0, y_0) - f(x - y)$.
(2) If $\kappa_H = 1$ We get an analytic solution $x \to \theta(x)$ of (8) from the implicit function theorem, cf. [8, Theorem I.B.4]. Proposition 1 II and III then states the possible asymptotics.
(3) A particular consequence is that degenerate critical points are isolated.
(4) In the proof we handle the error term by a fixed point argument. This implies the following important observation. If $x_0, y_0 \in \mathbb{R}$ and f and g are real analytic. Then a branch $\mathbb{R}\setminus\{x_0\} \ni x \to \theta(x) \in \Theta(x)$ is real valued if and only if $\theta(x)$ is real to leading order (the order needed to uniquely determine θ.)

Proof. The plan of the proof is as follows. First we identify enough terms in an asymptotic expansion $\theta(x) = \tilde\theta(x) + z(x)$ of the critical points so that we can separate them. Secondly we use a fixed point argument to show that the remainder, $x \to z(x)$, vanishes at a faster rate than the leading order term $\tilde{\theta}$. Note that it is a general result that for x close to x_0, $\partial_y H(x, \cdot)$ has precisely κ_H zeroes counting their orders. See [8, Lemma 1.B.3]. Our task is to account for those κ_H zeroes. We remark that we could have simply postulated the form of the leading order terms $\tilde{\theta}$, but at the cost of transparency.

We can assume without loss of generality that $x_0 = y_0 = 0$. We wish to solve, for a fixed x in a neighbourhood of 0,

$$(\partial_y H)(x, y) = 0. \tag{15}$$

We begin by collecting some facts. Compute

$$\forall \ell \; : \; (\partial_y^\ell H)(0,0) = (-1)^\ell \frac{\partial^\ell f}{\partial y^\ell}(0) + \frac{\partial^\ell g}{\partial y^\ell}(0). \tag{16}$$

We thus find from (8) and (10), recall (11), that

$$\forall \ell \leq \kappa_H \; : \; (-1)^{\ell+1} f_\ell = g_\ell, \tag{17}$$

and from the definition of κ_g that

$$\forall \ell < \min\{\kappa_H + 1, \kappa_g\} \; : \; f_\ell = g_\ell = 0. \tag{18}$$

Below we will use the following notation for remainders in expansions. Let h be an analytic function in a disc $D(z_0, r_h)$ with expansion $h(z) = \sum_{k=0}^{\infty} h_k (z - z_0)^k$. We write for $z \in D(z_0, r_h)$

$$R_h^\ell(z) := \sum_{k=\ell+1}^{\infty} k\, h_k\, (z - z_0)^{k-\ell-1}. \tag{19}$$

Note that R_h^ℓ are bounded analytic functions in $D(z_0, r_h/2)$.

We separate into the three cases I, II, and III.

Case I ($\kappa_g \leq \kappa_H$): Expand the left-hand side of (15), using (17), (18), and the notation (19):

$$
\begin{aligned}
(\partial_y H)(x, y) = & \sum_{\ell=\kappa_g}^{\kappa_H+1} \ell \left[- f_\ell (x - y)^{\ell-1} + g_\ell y^{\ell-1} \right] \\
& - R_f^{\kappa_H+1}(x - y)(x - y)^{\kappa_H+1} + R_g^{\kappa_H+1}(y) y^{\kappa_H+1} \\
= & \sum_{\ell=\kappa_g}^{\kappa_H} \ell\, g_\ell \left[y^{\ell-1} - (y - x)^{\ell-1} \right] \\
& + M\, y^{\kappa_H} + (\kappa_H + 1) f_{\kappa_H+1}((x - y)^{\kappa_H} - y^{\kappa_H}) \\
& - R_f^{\kappa_H+1}(x - y)(x - y)^{\kappa_H+1} + R_g^{\kappa_H+1}(y) y^{\kappa_H+1}.
\end{aligned}
\tag{20}
$$

First we look for solutions to (15) with asymptotics $\theta(x) \sim |x|$. That is, the leading order term should solve $y^{\kappa_g-1} - (y-x)^{\kappa_g-1} = 0$, i.e. if we put $y = \beta x$ then β should solve $(\beta/(\beta-1))^{\kappa_g-1} = 1$. This gives the following $\kappa_g - 2$ solutions for β

$$\beta_\ell = \frac{\alpha_\ell^{\kappa_g-1}}{\alpha_\ell^{\kappa_g-1} - 1}, \quad \text{for } \ell \in \{1, \kappa_g - 2\}. \tag{21}$$

Note that $\alpha_{\kappa_g-1}^{\kappa_g-1} = 1$ does not give rise to a solution. We thus find in this case $\tilde{\theta}_\ell(x) = \beta_\ell x$. Recall the notation α_ℓ^k from (12).

Secondly we look for solutions to (15) with asymptotics $\theta(x) \sim |x|^\rho$ for some $0 < \rho < 1$. Expanding the terms $(y-x)^{\ell-1}$ in binomial series we identify the highest order terms and are led to require $(\kappa_g - 1)Gxy^{\kappa_g-2} + My^{\kappa_H} = 0$. This gives the equation $y^{\kappa_H-\kappa_g+2} = -[(\kappa_g - 1)G/M]x$. (We note that the $\kappa_g - 2$ zero solutions are the ones we identified in the first step above.) We use the map $\mathcal{R}_{p_I} \ni \mathbf{x} \to \mathbf{x}^{1/p_I}$ introduced in (32), $p_I = \kappa_H - \kappa_g + 2$, to express the solution

$$\theta(\mathbf{x}) = C_I \mathbf{x}^{1/p_I}, \quad \text{where } C_I = [-(\kappa_g - 1)G/M]^{1/p_I}. \tag{22}$$

Case II ($\kappa_g = \kappa_H + 1$): We expand again

$$(\partial_y H)(x,y) = -F(x-y)^{\kappa_f-1} + G y^{\kappa_g-1}$$
$$- R_f^{\kappa_f+1}(x-y)(x-y)^{\kappa_f} + R_g^{\kappa_g+1}(y) y^{\kappa_g}. \tag{23}$$

First we consider the case $\kappa_f = \kappa_g = \kappa_H + 1$. This is similar to the first step above. We look for solutions with the asymptotics $\theta(x) \sim |x|$, and put $\tilde{\theta}(x) = \beta x$. We get the equation $(-1)^{\kappa_H+1}F(\beta - 1)^{\kappa_H} + G\beta^{\kappa_H} = 0$. This leads us to consider the equation $(\beta/(\beta-1))^{\kappa_H} = (-1)^{\kappa_H}F/G$, which has κ_H solutions

$$\beta_\ell = \frac{C_{II}\alpha_\ell^{\kappa_H}}{C_{II}\alpha_\ell^{\kappa_H} - 1}, \quad \text{for } \ell \in \{1, \ldots, \kappa_H\}. \tag{24}$$

Here $C_{II} := ((-1)^{\kappa_H}F/G)^{1/\kappa_H}$. We note that since $(-1)^{\kappa_H+1}F+G = M \neq 0$, cf. (14), we must have $0 < \arg(C_{II}) < 2\pi/\kappa_H$. This observation ensures that we avoid any singularity in the case $|F| = |G|$.

Secondly we assume $\kappa_f > \kappa_g$ and look for solutions to (15) with asymptotics $\theta(x) \sim |x|^\rho$, for some $\rho > 1$. This leads to the equation

$$-F x^{\kappa_f-1} + G y^{\kappa_g-1} = 0.$$

(Here $M = G$.) Let $\kappa_f - 1 = p\kappa_H + q$ and $d = (q, \kappa_H)$ as in the statement of the proposition. We express the solutions as d analytic maps from $\mathcal{R}_{\kappa_H/d}$ to $\mathbb{C}\backslash\{0\}$

$$\tilde{\theta}_\ell(\mathbf{x}) = C'_{II} \, \alpha_\ell^d \, \pi_{\kappa_H/d}(\mathbf{x})^p \, (\mathbf{x}^{1/(\kappa_H/d)})^{q/d} , \quad C'_{II} = (F/G)^{1/\kappa_H} . \tag{25}$$

Case III ($\kappa_g > \kappa_H + 1$): Here we must have $\kappa_f = \kappa_H + 1$. We write down the expansion

$$
\begin{aligned}
(\partial_y H)(x,y) = {} & M \, (y - x)^{\kappa_H} + G \, y^{\kappa_g - 1} \\
& - R_f^{\kappa_H + 2}(x - y)(x - y)^{\kappa_H + 1} + R_g^{\kappa_g + 1}(y) \, y^{\kappa_g} .
\end{aligned}
$$

Here the asymptotics is the same to leading order, namely $y \sim x$. Write $\tilde{\theta}(x) = x + \hat{\theta}$, and look for $\hat{\theta}$ with the asymptotics $\hat{\theta} \sim |x|^\rho$, $\rho > 1$. This gives the equation for $\hat{\theta}$

$$M \, \hat{\theta}^{\kappa_H} + G \, x^{\kappa_g - 1} = 0 ,$$

As above let $\kappa_g - 1 = p\kappa_H + q$ and $d = (q, \kappa_H)$. We express the solutions as d analytic maps from $\mathcal{R}_{\kappa_H/d}$ to $\mathbb{C}\backslash\{0\}$, with $C_{III} = (-G/M)^{1/\kappa_H}$,

$$\tilde{\theta}_\ell(\mathbf{x}) = \pi_{\kappa_H/d}(\mathbf{x}) + C_{III} \, \alpha_\ell^d \, \pi_{\kappa_H/d}(\mathbf{x})^p \, (\mathbf{x}^{1/(\kappa_H/d)})^{q/d} . \tag{26}$$

We have now determined the leading order term in all cases. We proceed to show by a fixed point argument that indeed there is a zero of order 1 near each of the terms identified above. We introduce function spaces

$$\mathcal{Z}_p(\rho, C) := \{ z \in C(D'_p(r_x) \, ; \, \mathbb{C}) \mid |z(\mathbf{x})| \le C \, |\mathbf{x}|^\rho \} ,$$

equipped with sup-norm. If $p = 1$ we identify $D'_1(r_x) \equiv D'(r_x)$. We now describe the procedure which we follow below, so as to cut short the individual arguments. First we write the actual branch of critical points as a sum $\theta(x) = \tilde{\theta}(x) + z(x)$, where $\tilde{\theta}$ is the leading order term as derived above and z is an element of a suitable $\mathcal{Z}_p$. We plug this into an expansion of $(\partial_y H)(x, y)$ and identify leading order terms. These are of two types. One is linear in z and the others are independent of z. The term linear in z is used to construct maps T on $\mathcal{Z}_p$, by $(Tz)(x) = z(x) - (\partial_y H)(x, \tilde{\theta}(x) + z(x))/h(x)$, if hz is the term linear in z. The remaining leading order terms now become leading order terms for Tz and their decay determine the decay of the remainder and hence ρ. The constant C is chosen such that T maps $\mathcal{Z}_p$ into itself. Finally, since terms in Tz depending on z are of higher order we can choose r_x small enough such that T becomes a contraction. Its unique fixed point z_0 is the desired correction to the leading order contribution found above. Note that a fixed point satisfies $(\partial_y H)(x, \tilde{\theta}(x) + z_0(x)) = 0$.

Case I: Write $\theta_\ell(x) = \beta_\ell x + z(x)$, cf. (21), and look for z vanishing faster than $|x|$ at 0. The leading order term linear in z in (20) is

$$(\kappa_g - 1) \, G \left[(\beta_\ell x)^{\kappa_g - 2} - ((\beta_\ell - 1)x)^{\kappa_g - 2} \right] z = (\kappa_g - 1) \, G \, \gamma_\ell \, x^{\kappa_g - 2} z$$

A computation yields $\gamma_\ell := \beta_\ell^{\kappa_g - 2} - (\beta_\ell - 1)^{\kappa_g - 2} \ne 0$. Define maps on $\mathcal{Z}_1(2, C)$

$$(T_\ell z)(x) := z(x) - \frac{(\partial_y H)(x, \beta_\ell x + z(x))}{(\kappa_g - 1) G \gamma_\ell x^{\kappa_g - 2}} .$$

The contributions to $T_\ell z$ which scale as $|x|^2$ (the slowest appearing rate) are

$$-\frac{(\kappa_g + 1)g_{\kappa_g+1}\big[\beta_\ell^{\kappa_g} - (\beta_\ell - 1)^{\kappa_g}\big]}{(\kappa_g - 1)G\gamma_\ell}\, x^2\,, \quad \text{for } \kappa_g < \kappa_H\,,$$

$$-\frac{M\beta_\ell^{\kappa_H} + (\kappa_H + 1)f_{\kappa_H+1}\big[(1 - \beta_\ell)^{\kappa_H} - \beta_\ell^{\kappa_H}\big]}{(\kappa_g - 1)G\gamma_\ell}\, x^2\,, \quad \text{for } \kappa_g = \kappa_H\,.$$

We now choose C large enough such that the norm of the coefficients above are less than C. Choosing r_x sufficiently small turns T_ℓ into contractions on $\mathcal{Z}_1(2, C)$.

Now write $\theta(\mathbf{x}) = C_I \mathbf{x}^{1/p_I} + z(\mathbf{x})$ where $z \in \mathcal{Z}_{p_I}(\kappa_g/p_I, C)$. The term in (20) which is linear in z and of leading order is

$$\big[(\kappa_g - 2)(\kappa_g - 1) G \pi_{p_I}(\mathbf{x}) (C_I \mathbf{x}^{1/p_I})^{\kappa_g - 3} + \kappa_H M (C_I \mathbf{x}^{1/p_I})^{\kappa_H - 1}\big] z$$
$$= p_I M C_I^{\kappa_H - 1} (\mathbf{x}^{1/p_I})^{\kappa_H - 1} z\,,$$

where we used (33). Note that the coefficient is non-zero. Define a map

$$(Tz)(\mathbf{x}) := z(\mathbf{x}) - \frac{(\partial_y H)(\pi_{p_I}(\mathbf{x}), C_I \mathbf{x}^{1/p_I} + z(\mathbf{x}))}{p_I M (C_I \mathbf{x}^{1/p_I})^{\kappa_H - 1}}\,.$$

We wish to show that $T : \mathcal{Z}_{p_I}(\kappa_g/p_I, C) \to \mathcal{Z}_{p_I}(\kappa_g/p_I, C)$ if C is large enough. Let $C(\kappa_g) = f_{\kappa_g+1}$ if $\kappa_g < \kappa_H$ and $C(\kappa_g) = g_{\kappa_H+1}$ if $\kappa_g = \kappa_H$. The term in Tz which vanish to lowest order is

$$\frac{(\kappa_g + 1) C(\kappa_g) \pi_{p_I}(\mathbf{x}) (C_I \mathbf{x}^{1/p_I})^{\kappa_g - 1}}{p_I M (C_I \mathbf{x}^{1/p_I})^{\kappa_H - 1}} = O\Big(|\mathbf{x}|^{\frac{2}{p_I}}\Big)\,,$$

Choosing C and r_x as above finishes case I.

Case II: Consider the case $\kappa_f = \kappa_g = \kappa_H + 1$. Let $\tilde{\theta}_\ell(x) = \beta_\ell x$, where β_ℓ is as in (24). We define maps on $\mathcal{Z}_1(2, C)$

$$(T_\ell z)(x) := z - \frac{(\partial_y H)(x, \beta_\ell x + z(x))}{(\kappa_g - 1) A_\ell x^{\kappa_g - 2}}\,,$$

which is well defined since

$$A_\ell := (-1)^{\kappa_H + 1} F (\beta_\ell - 1)^{\kappa_H - 1} + G \beta_\ell^{\kappa_H - 1}$$
$$= -G (\beta_\ell - 1)^{\kappa_H} \beta_\ell^{-1} C_{II}^{\kappa_H - 1} \big(C_{II} - (\alpha_\ell^{\kappa_H})^{\kappa_H - 1}\big) \neq 0\,.$$

Here we used (24), the definition of C_{II}, and that $0 < \arg(C_{II}) < 2\pi/\kappa_H$. The leading order contributions to $T_\ell z$ are

$$\big[(\kappa_g - 1) A_\ell\big]^{-1}\big(R_f^{\kappa_f+1} (1 - \beta_\ell)^{\kappa_g} + R_g^{\kappa_g+1} \beta_\ell^{\kappa_g}\big) x^2 = O(|x|^2)\,.$$

As above this estimate suffice.

Next we turn to the case $\kappa_f > \kappa_g$. Let $\tilde{\theta}_\ell$ be as in (25) and define maps on $\mathcal{Z}_{\kappa_H/d}(\kappa_f/\kappa_H, C)$ by

$$(T_\ell z)(\mathbf{x}) := z - \frac{(\partial_y H)(\pi_{\kappa_H/d}(\mathbf{x}), \tilde{\theta}_\ell(\mathbf{x}) + z(\mathbf{x}))}{(\kappa_g - 1)G\tilde{\theta}_\ell(\mathbf{x})^{\kappa_g - 2}}.$$

Let $\rho_{\mathrm{II}} := \min\{2\kappa_f - \kappa_g - 1, \kappa_f + \kappa_g - 2\}$. The leading order terms in $T_\ell z$ are

$$-\frac{(\kappa_f - 1)F\pi_{\kappa_H/d}(\mathbf{x})^{\kappa_f - 2}\tilde{\theta}_\ell(\mathbf{x}) - R_f^{\kappa_f + 1}\pi_{\kappa_H/d}(\mathbf{x})^{\kappa_f}}{(\kappa_g - 1)G\tilde{\theta}_\ell(\mathbf{x})^{\kappa_g - 2}} = O(|\mathbf{x}|^{\rho_{\mathrm{II}}}).$$

Since $2\kappa_f - \kappa_g - 1 \geq \kappa_f$ and $\kappa_f + \kappa_g - 2 \geq \kappa_f$, we have $\rho_{\mathrm{II}} \geq \kappa_f$ and conclude, as above, case II.

Case III: Let $\tilde{\theta}_\ell$ be as in (26) and write $\theta(\mathbf{x}) = \tilde{\theta}_\ell(\mathbf{x}) + z(\mathbf{x})$, where we take z from $\mathcal{Z}_{\kappa_H/d}(\kappa_g/\kappa_H, C)$. We define maps

$$(T_\ell z)(\mathbf{x}) := z - \frac{(\partial_y H)(\pi_{\kappa_H/d}(\mathbf{x}), \tilde{\theta}_\ell(\mathbf{x}) + z(\mathbf{x}))}{\kappa_H M[\tilde{\theta}_\ell(\mathbf{x}) - \pi_{\kappa_H/d}(\mathbf{x})]}.$$

Let $\rho_{\mathrm{III}} := \min\{2\kappa_g - \kappa_H - 2, \kappa_H + \kappa_g - 1\}$. The leading order terms in $T_\ell z$ are

$$\frac{(\kappa_g - 1)G\pi_{\kappa_H/d}(\mathbf{x})^{\kappa_g - 2}\tilde{\theta}_\ell(\mathbf{x}) + R_g^{\kappa_g + 1}\pi_{\kappa_H/d}(\mathbf{x})^{\kappa_g}}{\kappa_H M[\tilde{\theta}_\ell(\mathbf{x}) - \pi_{\kappa_H/d}(\mathbf{x})]} = O(|\mathbf{x}|^{\rho_{\mathrm{III}}}).$$

Since $2\kappa_g - \kappa_H - 2 \geq \kappa_g$ and $\kappa_g + \kappa_H - 1 \geq \kappa_g$, we conclude, as above, case III.

Finally we address analyticity of the a priori continuous solutions found above. Maps from $D'(r_x)$ are analytic by the analytic implicit function theorem [8, Theorem I.B.4], and maps from Riemanian covers are locally analytic by the same argument, and hence analytic. That the maps above defined on $D'(r_x)$ extend to analytic functions on the whole disc $D(r_x)$ follows from [3, Theorem V.1.2]. □

3 The Essential Spectrum

In this section we use Proposition 1 to prove Theorem 1.

Let $\mathbf{u} \in \mathbb{R}^\nu$ be a unit vector. We introduce

$$\sigma(t) := \Sigma_0(t\mathbf{u}) \quad \text{and} \quad \sigma^{(1)}(t; s) := \sigma(t - s) + \omega(s),$$

where we abuse notation and identify $\omega(s) \equiv \omega(s\mathbf{u})$. We furthermore write

$$\sigma^{(1)}(t) := \inf_{s \in \mathbb{R}} \sigma^{(1)}(t; s)$$

and $\mathcal{I}_0^{(1)}(t) := \{s \in \mathbb{R} | s\mathbf{u} \in \mathcal{I}_0^{(1)}(t\mathbf{u})\}$.

We begin with three lemmata and a proposition. The first lemma is a special case of [9, Sect. 3.2].

Lemma 1. *Assume $v \in L^2(\mathbb{R}^\nu)$ and Conditions 1 (i), (ii) and 2 (i), (ii), (vi). Let $\xi \in \mathbb{R}^\nu$ and $k \in \mathbb{R}^\nu$. If $\Sigma_0^{(1)}(\xi;k) < \Sigma_0^{(2)}(\xi)$, then $k \in \mathcal{I}_0^{(1)}(\xi)$.*

Using this lemma, cf. (6), we find that

$$\Sigma_{\mathrm{ess}}(\xi) \;=\; \min\{E \mid (\xi, E) \in \mathcal{T}_0^{(1)}\}\,, \tag{27}$$

where $\mathcal{T}_0^{(1)}$ is the set of thresholds coming from one-photon excitations of the ground state. It is defined by

$$\mathcal{T}_0^{(1)} := \{(\xi, E) \in \mathbb{R}^{\nu+1} \mid \exists k \in \mathbb{R}^\nu : E = \Sigma_0^{(1)}(\xi;k) \text{ and } (\xi;k) \in \mathrm{Crit}_0^{(1)}\}\,,$$

$$\mathrm{Crit}_0^{(1)} := \{(\xi, k) \in \mathbb{R}^{2\nu} \mid k \in \mathcal{I}_0^{(1)}(\xi) \text{ and } \nabla_k \Sigma_0^{(1)}(\xi;k) = 0\}\,.$$

There are obvious extensions to higher photon number, which must be included in (27), if ω is not subadditive.

The next lemma is the key to the applicability of Proposition 1.

Lemma 2. *Let $(\xi, k) \in \mathrm{Crit}_0^{(1)}$, such that $\xi \neq 0$ and $\nabla\omega(k) \neq 0$. Then there exists $\theta \in \mathbb{R}$ such that $k = \theta\xi$.*

Proof. Let $\xi \neq 0$ and $k \in \mathbb{R}^\nu$ be a critical point $\nabla_k \Sigma_0^{(1)}(\xi;k) = 0$. Then

$$\nabla\Sigma_0(\xi - k) \;=\; \nabla\omega(k) \tag{28}$$

Write $\nabla\omega(k) = c_1 k$ and $\nabla\Sigma_0(\xi - k) = c_2(\xi - k)$, using rotation invariance. Here $c_1 \neq 0$. From (28) we find $c\xi = (1 + c)k$, where $c = c_2/c_1$. Since $\xi \neq 0$, we conclude the result. $\qquad\square$

We write in the following, for $r \geq 0$, $B(r) = \{k \in \mathbb{R}^\nu \mid \|k\| = r\}$. For a unit vector $\boldsymbol{u}$ and radii $r_1, r_2 \geq 0$ we write for $t \in \mathbb{R}$

$$\mathcal{C}_t \equiv \mathcal{C}_t(r_1, r_2; \boldsymbol{u}) := \{k \in B(r_1) \mid t\,\boldsymbol{u} - k \in B(r_2)\}\,.$$

We leave the proof of the following lemma to the reader. (Draw a picture.) It deals with the stability of critical points which are not covered by Lemma 2.

Lemma 3. *Let $r_1, r_2 \geq 0$, $\boldsymbol{u} \in \mathbb{R}^\nu$ be a unit vector, and assume $\nu \geq 2$. Suppose $t_0 \in \mathbb{R}$ and $\mathcal{C}_{t_0} \neq \emptyset$. There exists a neighbourhood $\mathcal{U}$ of t_0, such that:*

i) If $\mathcal{C}_{t_0} \not\subset \{-r_1\boldsymbol{u}, +r_1\boldsymbol{u}\}$ and $t \in \mathcal{U}$, then $\mathcal{C}_t \neq \emptyset$.

ii) If $\mathcal{C}_{t_0} \subset \{-r_1\boldsymbol{u}, +r_1\boldsymbol{u}\}$, then $\mathcal{C}_{t_0} = \{k_0\}$. Let $\sigma = \boldsymbol{u} \cdot k_0(t_0 - \boldsymbol{u} \cdot k_0) \in \{-r_1 r_2, 0, +r_1 r_2\}$ ($\sigma = 0$ iff either r_1 or r_2 equals 0) and $t \in \mathcal{U}\backslash\{t_0\}$. If $\sigma(t - t_0) > 0$ then $\mathcal{C}_t \neq \emptyset$, and if $\sigma(t - t_0) \leq 0$, then $\mathcal{C}_t = \emptyset$.

Proposition 2. *Let $t_0 \in \mathbb{R}$ be such that $\sigma^{(1)}(t_0) < \Sigma_0^{(2)}(t_0\boldsymbol{u})$. There exist $0 < \delta \leq 1$ and an analytic function $\mathcal{U}_\delta\backslash\{t_0\} \ni t \to \theta(t)$, where $\mathcal{U}_\delta = (t_0 - \delta, t_0 + \delta)$, such that $\sigma^{(1)}(t) = \sigma^{(1)}(t; \theta(t))$, for $t \in \mathcal{U}_\delta\backslash\{t_0\}$. Furthermore, there exist integers $1 \leq p_\ell, p_r < \infty$, such that the functions $(t_0 - \delta, t_0) \ni t \to \theta(t_0 - (t_0 - t)^{p_\ell})$ and $(t_0, t_0 + \delta) \ni t \to \theta(t_0 + (t - t_0)^{p_r})$ extend analytically through t_0.*

Remark: The reason for only studying $\sigma^{(1)}$ where it is smaller than $\Sigma_0^{(2)}$, is the need for having global minima of $s \to \sigma^{(1)}(t; s)$ in $\mathcal{I}_0^{(1)}(t)$, cf. Lemma 1. This may not be true in general. If $\nu = 1, 2$ or ω is convex, this consideration is unnecessary. See [9, Theorem 1.5 i)] and Lemma 2.

Proof. Pick $\tilde{\delta} > 0$ such that $\sigma^{(1)}(t) < \Sigma_0^{(2)}(tu)$, for $|t - t_0| \leq \tilde{\delta}$. For $t \in \mathcal{U}_{\tilde{\delta}}$, let $\mathcal{G}_t = \{s \in \mathbb{R} | \sigma^{(1)}(t; s) = \sigma^{(1)}(t)\}$ be the set of global minima for $s \to \sigma^{(1)}(t; s)$. We recall from [9, Proof of Theorem 1.9] (an application of Lemma 1) that the sets $\mathcal{G}_t$ are finite, and all $s \in \mathcal{G}_t$ are zeros of finite order for the analytic function $\mathcal{I}_0^{(1)}(t) \ni s \to \partial_s \sigma^{(1)}(t; s)$.

Secondly we remark that for any $\tilde{t} > 0$ and $\bar{\sigma} \in \mathbb{R}$, the set $\{(t, s) \in \mathbb{R}^2 | |t| \leq \tilde{t}$ and $\sigma^{(1)}(t; s) \leq \bar{\sigma}\}$ is compact. From this remark and the finiteness of the $\mathcal{G}_t$'s we conclude from Proposition 1 and a compactness argument that the set

$$\widetilde{S} := \{t \in \mathcal{U}_{\tilde{\delta}} | \exists s \in \mathcal{G}_t, n \in \mathbb{N} \text{ s.t. } \partial_s^2 \sigma^{(1)}(t; s) = 0 \text{ and } \partial^n \sigma(t - s) \neq 0\}$$

is locally finite. In particular, for $t \in \mathcal{U}_{\tilde{\delta}} \backslash \widetilde{S}$ the global minima $s \in \mathcal{G}_t$ are all either simple zeroes of $s \to \partial_s \sigma^{(1)}(t; s)$, or zeroes of infinite order for $s \to \partial \sigma(t - s)$.

Suppose first that $t_0 \notin \widetilde{S}$. For $s_0 \in \mathcal{G}_{t_0}$ which are simple zeroes of $s \to \partial_s \sigma^{(1)}(t; s)$ we obtain from the analytic implicit function theorem analytic solutions θ to $\partial_s \sigma^{(1)}(t; \theta(t)) = 0$, defined in a neighbourhood of t_0. For $s_0 \in \mathcal{G}_{t_0}$ which are zeroes of infinite order of $s \to \partial \sigma(t_0 - s)$ (and not already included in the first case), we take $\theta(t) \equiv s_0$ which solves $\partial_s \sigma^{(1)}(t; \theta(t)) = 0$ near t_0. See Remark 1), with $\kappa_f = \infty$, after Proposition 1. We have thus for some $0 < \delta' < \tilde{\delta}$ constructed $|\mathcal{G}_{t_0}|$ analytic functions θ_ℓ defined in $\mathcal{U}_{\delta'}$, such that $\mathcal{G}_{t_0} = \{\theta_\ell(t_0)\}$ and $\mathcal{G}_t \subset \{\theta_\ell(t)\}$ (by continuity) for $t \in \mathcal{U}_{\delta'}$. Hence $\sigma^{(1)}(t) = \min_{1 \leq \ell \leq |\mathcal{G}_{t_0}|} \sigma^{(1)}(t; \theta_\ell(t))$, for $t \in \mathcal{U}_{\delta'}$.

If $|\mathcal{G}_{t_0}| = 1$ take $\delta = \delta'$ and $\theta = \theta_1$. If $|\mathcal{G}_{t_0}| > 1$ choose $0 < \delta < \delta'$, ℓ_1, and ℓ_2 such that the following choice works: $\theta(t) = \theta_{\ell_1}(t)$, for $t_0 - \delta < t < t_0$, and $\theta(t) = \theta_{\ell_2}(t)$, for $t_0 < t < t_0 + \delta$. This proves the result if $t_0 \notin \widetilde{S}$.

It remains to treat $t_0 \in \widetilde{S}$. Here we get from Proposition 1 a $0 < \delta' < \tilde{\delta}$ and two families of analytic functions $\{\theta_\ell^{\text{left}}\}$ and $\{\theta_\ell^{\text{right}}\}$ defined in $(t_0 - \delta', t_0)$ and $(t_0, t_0 + \delta')$ respectively, which parameterize the critical points for t near t_0, which comes from $\mathcal{G}_{t_0}$. Furthermore $\mathcal{G}_t \subset \{\theta_\ell^{\text{left}}(t)\}$, $t_0 - \delta' < t < t_0$ and $\mathcal{G}_t \subset \{\theta_\ell^{\text{right}}(t)\}$, $t_0 < t < t_0 + \delta'$. Note that the number of branches to the left and to the right need not be the same, but both are finite.

We are finished if we can prove that, for $\ell \neq \ell'$, the function $\sigma^{(1)}(t; \theta_\ell^{\text{left}}(t)) - \sigma^{(1)}(t; \theta_{\ell'}^{\text{left}}(t))$ is either identically zero, or it does not vanish on a sequence of t's converging to t_0 from the left. Similarly for the right of t_0. (If there is only one branch, then it continues analytically through t_0 and we are done).

In the following we work to the left of t_0 and drop the superscript "left". The region to the right of t_0 can be treated similarly. There exists $1 \leq p, p' < \infty$ and analytic functions $\theta : \mathcal{R}_p \to \mathbb{C}$ and $\theta' : \mathcal{R}_{p'} \to \mathbb{C}$ such that θ_ℓ and $\theta_{\ell'}$ are branches of θ and θ' respectively. See Proposition 1. That is, there exist $0 \leq q < p$ and $0 \leq q' < p'$ such that

$$\theta_\ell(t) = \theta(R_p^\rho(t_0 - t)) \text{ and } \theta_{\ell'}(t) = \theta'(R_{p'}^{\rho'}(t_0 - t)),$$

where $\rho = \pi + 2\pi q$ and $\rho' = \pi + 2\pi q'$. Recall notation from (30) and (31). Here we used the canonical embedding $\mathbb{R}\backslash\{0\} \ni r \to (t, 0) \in \mathcal{R}_p$, for any p. Since $z \to \theta(R_p^\rho(P_p(z)))$ and $z \to \theta'(R_{p'}^{\rho'}(P_{p'}(z)))$ are analytic and bounded functions from $D'(r)$ (for some $r > 0$), they extend to analytic functions on $D(r)$. We can hence define an analytic function

$$h(z) = \sigma^{(1)}(z; \theta(R_p^\rho(P_p((t_0 - z)^{p'})))) - \sigma^{(1)}(z; \theta'(R_{p'}^{\rho'}(P_{p'}((t_0 - z)^p))))$$

in $D(r^{1/pp'})$. The function h is either identically zero or has only isolated zeroes (one is at t_0). This now implies that $\sigma^{(1)}(t; \theta_\ell(t)) - \sigma^{(1)}(t; \theta_{\ell'}(t)) = h(t_0 - (t_0 - t)^{1/pp'})$ is either identically zero or has finitely many zeroes near t_0. We can now choose $0 < \delta < \delta'$ and θ as above. This concludes the proof. $\square$

Proof of Theorem 1: Proposition 2 covers the case $\nu = 1$. In the following we assume $\nu \geq 2$. It suffices to prove the theorem locally near any $t_0 \in \mathbb{R}$. For the global minima at $t \in \mathbb{R}$ we write

$$\mathcal{M}_t := \{ k \in \mathbb{R}^\nu \mid \Sigma_0^{(1)}(t\boldsymbol{u}; k) = \Sigma_0^{(1)}(t\boldsymbol{u}) \}$$

and we introduce two subsets

$$\mathcal{M}_t^{\|} := \{ k \in \mathcal{M}_t \mid k \parallel \boldsymbol{u} \} \text{ and } \mathcal{M}_t^0 := \{ k \in \mathcal{M}_t \mid \nabla\omega(k) = 0 \}.$$

We begin with the following note. Let $t_0 \in \mathbb{R}$. If $\mathcal{M}_{t_0}^{\|} = \emptyset$ ($\mathcal{M}_{t_0}^0 = \emptyset$) then there exists a neighbourhood $\mathcal{U} \ni t_0$ such that for $t \in \mathcal{U}$ we have $\mathcal{M}_t^{\|} = \emptyset$ ($\mathcal{M}_t^0 = \emptyset$).

Let $t_0 \in \mathbb{R}$. First consider the case $\mathcal{M}_{t_0}^0 = \emptyset$. For $t \in \mathcal{U}$, chosen as above, we have $\Sigma_{\mathrm{ess}}(t\boldsymbol{u}) = \sigma^{(1)}(t)$, which by Proposition 2 concludes the proof.

We can now assume that $\mathcal{M}_{t_0}^0 \neq \emptyset$. By analyticity and rotation invariance, the set of k's such that $\nabla\omega(k) = 0$ is a set of concentric balls, with a locally finite set of radii. If $k \in \mathcal{M}_{t_0}^0$ then $Ok \in \mathcal{M}_{t_0}^0$ for any $O \in \mathcal{O}(\nu; \boldsymbol{u})$, where $\mathcal{O}(\nu; \boldsymbol{u}) := \{O \in \mathcal{O}(\nu) \mid O\boldsymbol{u} = \boldsymbol{u}\}$.

Let

$$\widetilde{\mathcal{M}}_t^0 = \{ k \in \mathcal{I}_0^{(1)}(t) \mid (t\boldsymbol{u}, k) \in \mathrm{Crit}_0^{(1)} \text{ and } \nabla\omega(k) = 0 \},$$

$$\tilde{\sigma}^{(1)}(t) := \min_{k \in \widetilde{\mathcal{M}}_t^0} \Sigma_0^{(1)}(t\boldsymbol{u}; k),$$

with the convention that $\tilde{\sigma}^{(1)}(t) = +\infty$ if $\widetilde{\mathcal{M}}_t^0 = \emptyset$. Then by Lemma 2

$$\Sigma_0^{(1)}(t\boldsymbol{u}) = \min\{\sigma^{(1)}(t), \tilde{\sigma}^{(1)}(t)\}.$$

We work only to the left of t_0, i.e. we take $t \leq t_0$. The case $t \geq t_0$ can be treated similarly.

We proceed to find a $\delta > 0$ such that $\sigma^{(1)}$ is either constant equal to $\Sigma_0^{(1)}(t_0 \boldsymbol{u})$, for $t_0 - \delta < t \leq t_0$ or it satisfies $\tilde{\sigma}^{(1)}(t) \geq \sigma^{(1)}(t) = \Sigma_0^{(1)}(t\boldsymbol{u})$, for $t_0 - \delta < t \leq t_0$. This concludes the result, since both $\sigma^{(1)}$ and a constant function, suitably reparameterized, continues analytically though t_0. See Proposition 2.

First we consider the case where: (**A**) Σ_0 is not constant on the connected component of $\mathcal{I}_0$ containing $t_0 \boldsymbol{u} - k_0$, for any $k_0 \in \mathcal{M}_{t_0}^0$. (**B**) For any $k_0 \in \mathcal{M}_{t_0}^0$ we have (with $r_1 = |k_0|$ and $r_2 = |t_0 \boldsymbol{u} - k_0|$): $\mathcal{C}_{t_0}(r_1, r_2; \boldsymbol{u}) \subset \{-r_1 \boldsymbol{u}, r_1 \boldsymbol{u}\}$ and $\boldsymbol{u} \cdot k_0(t_0 - \boldsymbol{u} \cdot k_0) \leq 0$. Assuming (**A**) and (**B**) we have by Lemma 3 ii) a $\delta > 0$, such that

$$\mathcal{C}_t(r_1, r_2; \boldsymbol{u}) = \emptyset, \quad \text{for } t_0 - \delta < t < t_0. \tag{29}$$

We proceed to argue that (**A**) and (**B**) implies $\Sigma_0^{(1)}(t\boldsymbol{u}) = \sigma^{(1)}(t)$, for $t < t_0$. It suffices to find a $\delta > 0$ such that $\mathcal{M}_t^0 = \emptyset$, $t_0 - \delta < t < t_0$. Suppose to the contrary that there exists a sequence $t_n \to t_0$, with $t_n < t_0$ and $\mathcal{M}_{t_n}^0 \neq \emptyset$. Let $k_n \in \mathcal{M}_{t_n}^0$. We can assume, by possibly passing to a subsequence, that $k_n \to k_\infty$. Here we used that $\omega(k) \to \infty$ as $|k| \to \infty$. Clearly $k_\infty \in \mathcal{M}_{t_0}^0$, and hence $k_\infty \in \mathcal{C}_{t_0}(r_1, r_2; \boldsymbol{u})$ for some r_1, r_2. Since the possible r_1's and r_2's are isolated, we must have a $\bar{n}$ such that $\forall n > \bar{n}: k_n \in \mathcal{C}_{t_n}(r_1, r_2; \boldsymbol{u})$. This contradicts (29).

For the remaining case we assume one of the following: (**C**) There exists $k_0 \in \mathcal{M}_{t_0}^0$ such that Σ_0 is constant on the connected component of $\mathcal{I}_0$ containing $t_0 \boldsymbol{u} - k_0$. (The converse of (**A**) above.) (**D**) There exists $k_0 \in \mathcal{M}_{t_0}^0$ such that either $\mathcal{C}_{t_0}(r_1, r_2; \boldsymbol{u}) \not\subset \{-r_1 \boldsymbol{u}, r_1 \boldsymbol{u}\}$ or $\mathcal{C}_{t_0}(r_1, r_2; \boldsymbol{u}) \subset \{-r_1 \boldsymbol{u}, r_1 \boldsymbol{u}\}$ and $\boldsymbol{u} \cdot k_0(t_0 - \boldsymbol{u} \cdot k_0) > 0$. Again $r_1 = |k_0|$ and $r_2 = |t_0 \boldsymbol{u} - k_0|$. (The converse of **B**) above.)

There exists $\delta > 0$ such that: In the case (**C**), for $t_0 - \delta < t < t_0$, there exists $k \in \widetilde{\mathcal{M}}_t^0$ with $\Sigma_0^{(1)}(t\boldsymbol{u}; k) = \Sigma_0^{(1)}(t_0 \boldsymbol{u})$. In case (**D**) we have by Lemma 3 (i) and (ii), for $t_0 - \delta < t < t_0$, likewise $k \in \widetilde{\mathcal{M}}_t^0$ with $\Sigma_0^{(1)}(t\boldsymbol{u}; k) = \Sigma_0^{(1)}(t_0 \boldsymbol{u})$. Hence, if either (**C**) or (**D**) are satisfied we have $\tilde{\sigma}^{(1)}(t) \leq \Sigma_0^{(1)}(t_0 \boldsymbol{u})$, for $t_0 - \delta < t < t_0$.

In order to show the converse inequality $\tilde{\sigma}^{(1)}(t) \geq \Sigma_0^{(1)}(t_0 \boldsymbol{u})$, we assume to the contrary that there exists a sequence $t_n \to t_0$ and $k_n \in \widetilde{\mathcal{M}}_{t_n}^0$ such that $\Sigma_0^{(1)}(t_n \boldsymbol{u}; k_n) < \Sigma_0^{(1)}(t_0 \boldsymbol{u})$. As above we can assume $k_n \to k_\infty \in \mathcal{M}_{t_0}^0$. If Σ_0 is constant on the connected component of $t_0 \boldsymbol{u} - k_\infty$, then $\Sigma_0^{(1)}(t_n \boldsymbol{u}; k_n) = \Sigma_0^{(1)}(t_0 \boldsymbol{u})$ is a constant sequence for n large enough, which is a contradiction. If Σ_0 is not constant on the connected component of $t_0 \boldsymbol{u} - k_\infty$, then

$|k_n| = |k_\infty|$ and $|t_n \mathbf{u} - k_n| = |t_0 \mathbf{u} - k_\infty|$, for n large enough, and again we conclude $\Sigma_0^{(1)}(t_n \mathbf{u}; k_n) = \Sigma_0^{(1)}(t_0 \mathbf{u})$ is a constant sequence, which is a contradiction. □

A Riemannian Covers

Let $\mathcal{R}_p = (0, \infty) \times \mathbb{R}/2\pi p \mathbb{Z}$, equipped with the product topology. We write $\mathbf{z} = (|\mathbf{z}|, \arg(\mathbf{z}))$ for elements of $\mathcal{R}_p$ and introduce the p-cover $(\mathcal{R}_p, \pi_p)$ of $\mathbb{C}\backslash\{0\}$ by

$$\pi_p : \mathcal{R}_p \to \mathbb{C}\backslash\{0\} \text{ where } \pi_p(\mathbf{z}) := |\mathbf{z}| \, e^{i \arg(\mathbf{z})}.$$

Note that π_p is locally a homeomorphism and thus provides a chart which turns $\mathcal{R}_p$ into an analytic surface. If $\mathcal{U} \subset \mathcal{R}_p$ is such that π_p is 1–1 on $\mathcal{U}$, write $\pi_\mathcal{U}^{-1}$ for the inverse homeomorphism from $\pi_p(\mathcal{U})$ to $\mathcal{U}$.

We will use the concept of analytic functions to, from, and between cover spaces. This is just special cases of what it means to be an analytic map between two analytic surfaces, cf. [3, Sects. IX.6 and IX.7]

Let $\mathcal{V} \subset \mathbb{C}$, $\mathcal{V}_p \subset \mathcal{R}_p$, and $\mathcal{V}_q \subset \mathcal{R}_q$ be open sets, and $f_1 : \mathcal{V} \to \mathcal{R}_p$, $f_2 : \mathcal{V}_p \to \mathcal{R}_q$, and $f_3 : \mathcal{V}_q \to \mathbb{C}$ continuous maps.

We say f_1 is analytic if for any $z_0 \in \mathcal{V}$ there exists an open set $\mathcal{U} \subset \mathcal{V}$ with $z_0 \in \mathcal{U}$, such that the map $\mathcal{U} \ni z \to \pi_p(f_1(z))$ is analytic in the usual sense.

We say f_2 is analytic if for any $\mathbf{z}_0 \in \mathcal{V}_p$ there exists an open set $\mathcal{U}_p \subset \mathcal{V}_p$ with $\mathbf{z}_0 \in \mathcal{U}_p$ and $\pi_p : \mathcal{U}_p \to \mathbb{C}$ 1–1, such that $\pi_p(\mathcal{U}_p) \ni z \to \pi_q(f_2(\pi_{\mathcal{U}_p}^{-1}(z)))$ is analytic in the usual sense.

We say f_3 is analytic if for any $\mathbf{z}_0 \in \mathcal{V}_q$ there exists an open set $\mathcal{U}_q \subset \mathcal{V}_q$ with $\mathbf{z}_0 \in \mathcal{U}_q$ and $\pi_q : \mathcal{U}_q \to \mathbb{C}$ 1-1, such that the map $\pi_q(\mathcal{U}_q) \ni z \to f_3(\pi_{\mathcal{U}_q}^{-1}(z))$ is analytic in the usual sense.

With this definition it is easy to check that $f_2 \circ f_1$, $f_3 \circ f_2$ and $f_3 \circ f_2 \circ f_1$ are analytic maps. We give three examples which we use in Sect. 2. The first example is $P_p : \mathbb{C}\backslash\{0\} \to \mathcal{R}_p$, defined by

$$P_p(z) := (|z|^p, p \arg(z)). \tag{30}$$

Second example: Let $\rho \in \mathbb{R}$. We define a map $R_p^\rho : \mathcal{R}_p \to \mathcal{R}_p$ by

$$R_p^\rho(\mathbf{z}) := (|\mathbf{z}|, \arg(\mathbf{z}) + \rho \mod 2\pi p). \tag{31}$$

The third example is the map $\mathcal{R}_p \ni \mathbf{z} \to \mathbf{z}^{1/p} \in \mathbb{C}\backslash\{0\}$ defined by

$$\mathbf{z}^{1/p} := |\mathbf{z}|^{1/p} \, e^{i \arg(\mathbf{z})/p}. \tag{32}$$

The three examples above are all analytic and in addition bijections. We have

$$P_p(\mathbf{z}^{1/p}) = (P_p(\mathbf{z}))^{1/p} = \mathbf{z} \text{ and } (\mathbf{z}^{1/p})^p = \pi_p(\mathbf{z}). \tag{33}$$

References

1. F.A. Berezin, *The method of second quantization*, 1 ed., Academic Press, New York, San Francisco, London, 1966.
2. L. Bruneau and J. Derezinski, *Pauli-Fierz Hamiltonians defined as quadratic forms*, Rep. Math. Phys., **54** (2004), 169–199.
3. J.B. Conway, *Functions of one complex variable*, 2 ed., Graduate texts in mathematics, vol. 11, Springer-Verlag, New York, 1978.
4. J. Derezinski and C. Gérard, *Asymptotic completeness in quantum field theory. massive Pauli-Fierz Hamiltonians*, Rev. Math. Phys. **11** (1999), 383–450.
5. R.P. Feynman, *Statistical mechanics. A set of lectures*, Frontiers in physics, W.A. Benjamin, Inc., Reading, Massechusets, 1972.
6. J. Fröhlich, *On the infrared problem in a model of scalar electrons and massless scalar bosons*, Ann. Inst. Henri Poincaré **19** (1973), 1–103.
7. _____, *Existence of dressed one-electron states in a class of persistent models*, Fortschr. Phys. **22** (1974), 159–198.
8. R.C. Gunning and H. Rossi, *Analytic functions of several complex variables*, Series in modern analysis, Prentice-Hall, Inc., Englewood Cliffs, N. J., 1965.
9. J.S. Møller, *The translation invariant massive Nelson model: I. The bottom of the spectrum*, Ann. Henri Poincaré **6** (2005) 1091–1135.
10. "The Fröhlich polaron revisited". Submitted.
11. H. Spohn, *The polaron at large total momentum*, J. Phys. A **21** (1988), 1199–1211.
12. _____, *Dynamics of charged particles and their radiation field*, Cambridge University Press, 2004.

Part III

Quantum Kinetics
and Bose-Einstein Condensation

he Bose-Einstein Condensation (BEC) is the object of intense current investigations in mathematical physics. Three contributions are devoted to this fascinating phenomenon. Aizenman, Lieb, Seiringer, Solovej, and Yngvason address the rigorous analysis of a mathematical model, the Bose-Hubbard model, describing the transition from Bose-Einstein condensate to a state of localized bosons as the strength of a periodic potential is varied. Superradiance describes the collective emission of radiation from a macroscopic ensemble of atoms, such as a Bose-Einstein condensate. Two models of bosons with internal structure interacting with a single mode photon field describing BEC and superradiance are studied by Pulé, Verbeure, V.A. Zagrebnov. They make a rigorous analysis of the equilibrium states, determine the thermodynamical functions and prove occurrence of spontaneous symmetry breaking. Understanding the dynamics of Bose-Einstein condensates is important to make contact with experiments about BEC. Schlein reports about recent progresses on the mathematical derivation of the Gross-Pitaevskii equation describing the dynamics of such condensates, from microscopic equations. These progresses rely on a fine analysis of certain properties of N-bosons systems as the number of bosons becomes large, in a suitable mean-field approximation.

The determination and justification of effective descriptions of phenomena taking place on a macroscopic scale from fundamental microscopic equations are important tasks of mathematical physics. The following contributions are devoted to this activity. Erdős, Salmhofer and Yau address the description of the dynamics of a Quantum particle in a random potential on a macroscopic times scale beyond the weak coupling limit; that is when a diffusive behavior settles in, in dimension greater or equal to three. They prove that the expectation value of the suitably rescaled Wigner transform of the solution to the Schrödinger equation is described by the heat equation. The contribution of Spohn deals with the justification of the phonon Boltzmann equation by considering a periodic harmonic crystal weakly perturbed by a cubic onsite potential with random initial conditions distributed according to certain scaled laws. Conjecturing that only a certain type of Feynman diagrams contribute to the expansion of the corresponding averaged Wigner function in the kinetic limit, the author makes the Boltzmann equation emerge in the limit. In their paper, Bokanowski, López, Sánchez and Soler review the derivation and properties of the so-called X^α method which is used to provide corrections to the Hartree-Fock approach of the description of Quantum transport in semi-conductors. The analysis of this effective equation provides a mathematical description of certain types of solutions experimentally observed in semi-conductors and Quantum Chemistry.

Bose-Einstein Condensation as a Quantum Phase Transition in an Optical Lattice*

M. Aizenman[1], E.H. Lieb[1], R. Seiringer[1], J.P. Solovej[2] and J. Yngvason[3,4]

[1] Departments of Mathematics and Physics, Jadwin Hall, Princeton University, P. O. Box 708, Princeton, New Jersey 08544
[2] Department of Mathematics, University of Copenhagen, Universitetsparken 5, DK-2100 Copenhagen, Denmark
[3] Institut für Theoretische Physik, Universität Wien, Boltzmanngasse 5, A-1090 Vienna, Austria
[4] Erwin Schrödinger Institute for Mathematical Physics, Boltzmanngasse 9, A-1090 Vienna, Austria

Abstract. One of the most remarkable recent developments in the study of ultra-cold Bose gases is the observation of a reversible transition from a Bose Einstein condensate to a state composed of localized atoms as the strength of a periodic, optical trapping potential is varied. In [1] a model of this phenomenon has been analyzed rigorously. The gas is a hard core lattice gas and the optical lattice is modeled by a periodic potential of strength λ. For small λ and temperature Bose-Einstein condensation (BEC) is proved to occur, while at large λ BEC disappears, even in the ground state, which is a Mott-insulator state with a characteristic gap. The inter-particle interaction is essential for this effect. This contribution gives a pedagogical survey of these results.

1 Introduction

One of the most remarkable recent developments in the study of ultracold Bose gases is the observation of a reversible transition from a Bose-Einstein condensate to a state composed of localized atoms as the strength of a periodic, optical trapping potential is varied [2, 3]. This is an example of a quantum phase transition [4] where quantum fluctuations and correlations

---

*Contribution to the proceedings of QMath9, Giens, France, Sept. 12–16, 2004. Talk given by Jakob Yngvason. © 2004 by the authors. This paper may be reproduced, in its entirety, for non-commercial purposes.

Work supported in part by US NSF grants PHY 9971149 (MA), PHY 0139984-A01 (EHL), PHY 0353181 (RS) and DMS-0111298 (JPS); by an A.P. Sloan Fellowship (RS); by EU grant HPRN-CT-2002-00277 (JPS and JY); by FWF grant P17176-N02 (JY); by MaPhySto – A Network in Mathematical Physics and Stochastics funded by The Danish National Research Foundation (JPS), and by grants from the Danish research council (JPS).

rather than energy-entropy competition is the driving force and its theoretical understanding is quite challenging. The model usually considered for describing this phenomenon is the Bose-Hubbard model and the transition is interpreted as a transition between a superfluid and a *Mott insulator* that was studied in [5] with an application to He^4 in porous media in mind. The possibility of applying this scheme to gases of alkali atoms in optical traps was first realized in [6]. The article [7] reviews these developments and many recent papers, e.g., [8–16] are devoted to this topic. These papers contain also further references to earlier work along these lines.

The investigations of the phase transition in the Bose-Hubbard model are mostly based on variational or numerical methods and the signal of the phase transition is usually taken to be that an ansatz with a sharp particle number at each lattice site leads to a lower energy than a delocalized Bogoliubov state. On the other hand, there exists no rigorous proof, so far, that the true ground state of the model has off-diagonal long range order at one end of the parameter regime that disappears at the other end. In this contribution, which is based on the paper [1], we study a slightly different model where just this phenomenon can be rigorously proved and which, at the same time, captures the salient features of the experimental situation.

Physically, we are dealing with a trapped Bose gas with short range interaction. The model we discuss, however, is not a continuum model but rather a lattice gas, i.e., the particles are confined to move on a d-dimensional, hypercubic lattice and the kinetic energy is given by the discrete Laplacian. Moreover, when discusssing BEC, it is convenient not to fix the particle number but to work in a grand-canonical ensemble. The chemical potential is fixed in such a way that the average particle number equals half the number of lattice sites, i.e., we consider *half filling*. (This restriction is dictated by our method of proof.) The optical lattice is modeled by a periodic, one-body potential. In experiments the gas is enclosed in an additional trap potential that is slowly varying on the scale of the optical lattice but we neglect here the inhomogeneity due to such a potential and consider instead the thermodynamic limit.

In terms of bosonic creation and annihilation operators, $a_{\mathbf{x}}^{\dagger}$ and $a_{\mathbf{x}}$, our Hamiltonian is expressed as

$$H = -\frac{1}{2}\sum_{\langle \mathbf{xy} \rangle}(a_{\mathbf{x}}^{\dagger}a_{\mathbf{y}} + a_{\mathbf{x}}a_{\mathbf{y}}^{\dagger}) + \lambda\sum_{\mathbf{x}}(-1)^{\mathbf{x}}a_{\mathbf{x}}^{\dagger}a_{\mathbf{x}} + U\sum_{\mathbf{x}}a_{\mathbf{x}}^{\dagger}a_{\mathbf{x}}(a_{\mathbf{x}}^{\dagger}a_{\mathbf{x}} - 1) . \quad (1)$$

The sites $\mathbf{x}$ are in a cube $\Lambda \subset \mathbb{Z}^d$ with opposite sides identified (i.e., a d-dimensional torus) and $\langle \mathbf{xy} \rangle$ stands for pairs of nearest neighbors. Units are chosen such that $\hbar^2/m = 1$.

The first term in (1) is the discrete Laplacian $\sum_{\langle \mathbf{xy} \rangle}(a_{\mathbf{x}}^{\dagger} - a_{\mathbf{y}}^{\dagger})(a_{\mathbf{x}} - a_{\mathbf{y}})$ minus $2d\sum_{\mathbf{x}} a_{\mathbf{x}}^{\dagger}a_{\mathbf{x}}$, i.e., we have subtracted a chemical potential that equals d.

The optical lattice gives rise to a potential $\lambda(-1)^{\mathbf{x}}$ which alternates in sign between the A and B sublattices of even and odd sites. The inter-atomic on-site repulsion is U, but we consider here only the case of a *hard-core interaction*, i.e., $U = \infty$. If $\lambda = 0$ but $U < \infty$ we have the Bose-Hubbard model. Then all sites are equivalent and the lattice represents the attractive sites of the optical lattice. In our case the adjustable parameter is λ instead of U and for large λ the atoms will try to localize on the B sublattice. The Hamiltonian (1) conserves the particle number N and it is shown in [1], Appendix A, that, for $U = \infty$, the lowest energy is obtained uniquely for $N = \frac{1}{2}|\Lambda|$, i.e., half the number of lattice sites. Because of the periodic potential the unit cell in this model consists of two lattice sites, so that we have on average one particle per unit cell. This corresponds, physically, to filling factor 1 in the Bose-Hubbard model.

For given temperature T, we consider grand-canonical thermal equilibrium states, described by the Gibbs density matrices $Z^{-1} \exp(-\beta H)$ with Z the normalization factor (partition function) and $\beta = 1/T$ the inverse temperature. Units are chosen so that Boltzmann's constant equals 1. The thermal expectation value of some observable $\mathcal{O}$ will be denoted by $\langle \mathcal{O} \rangle = Z^{-1} \mathrm{Tr}\, \mathcal{O} \exp(-\beta H)$. In the proof of BEC we focus on dimensions $d \geq 3$, but, using the technique employed in [17], an extension to the ground state in two dimensions is possible.

Our main results about this model can be summarized as follows:

1. If T and λ are both small, there is Bose-Einstein condensation. In this parameter regime the one-body density matrix $\gamma(\mathbf{x}, \mathbf{y}) = \langle a_\mathbf{x}^\dagger a_\mathbf{y} \rangle$ has exactly one large eigenvalue (in the thermodynamic limit), and the corresponding condensate wave function is $\phi(\mathbf{x}) = $ constant.
2. If either T or λ is big enough, then the one-body density matrix decays exponentially with the distance $|\mathbf{x} - \mathbf{y}|$, and hence there is *no BEC*. In particular, this applies to the ground state $T = 0$ for λ big enough, where the system is in a Mott insulator phase.
3. The Mott insulator phase is characterized by a gap, i.e., a jump in the chemical potential. We are able to prove this, at half-filling, in the region described in item 2 above. More precisely, there is a cusp in the dependence of the ground state energy on the number of particles; adding or removing one particle costs a non-zero amount of energy. We also show that there is no such gap whenever there is BEC.
4. The interparticle interaction is essential for items 2 and 3. Non-interacting bosons *always display BEC* for low, but positive T (depending on λ, of course).
5. For all $T \geq 0$ and all $\lambda > 0$ the diagonal part of the one-body density matrix $\langle a_\mathbf{x}^\dagger a_\mathbf{x} \rangle$ (the one-particle density) is *not constant*. Its value on the A sublattice is constant, but strictly less than its constant value on the B sublattice and this discrepancy survives in the thermodynamic limit. In contrast, in the regime mentioned in item 1, the off-diagonal long-range order is constant, i.e., $\langle a_\mathbf{x}^\dagger a_\mathbf{y} \rangle \approx \phi(\mathbf{x})\phi(\mathbf{y})^*$ for large $|\mathbf{x} - \mathbf{y}|$ with $\phi(\mathbf{x}) = $ constant.

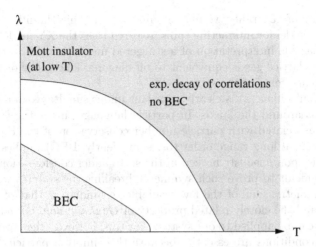

Fig. 1. Schematic phase diagram at half-filling

Because of the hard-core interaction between the particles, there is at most one particle at each site and our Hamiltonian (with $U = \infty$) thus acts on the Hilbert space $\mathcal{H} = \bigotimes_{x \in \Lambda} \mathbb{C}^2$. The creation and annihilation operators can be represented as 2×2 matrices with

$$a_x^\dagger \leftrightarrow \begin{pmatrix} 0 & 1 \\ 0 & 0 \end{pmatrix}, \quad a_x \leftrightarrow \begin{pmatrix} 0 & 0 \\ 1 & 0 \end{pmatrix}, \quad a_x^\dagger a_x \leftrightarrow \begin{pmatrix} 1 & 0 \\ 0 & 0 \end{pmatrix},$$

for each $x \in \Lambda$. More precisely, these matrices act on the tensor factor associated with the site x while $a_x^\dagger$ and a_x act as the identity on the other factors in the Hilbert space $\mathcal{H} = \bigotimes_{x \in \Lambda} \mathbb{C}^2$. Thus the operators at different sites commute as appropriate for Bosons, but on each site they satisfy anticommutation relations, reflecting the hard core condition.

The Hamiltonian can alternatively be written in terms of the spin 1/2 operators

$$S^1 = \frac{1}{2} \begin{pmatrix} 0 & 1 \\ 1 & 0 \end{pmatrix}, \quad S^2 = \frac{1}{2} \begin{pmatrix} 0 & -i \\ i & 0 \end{pmatrix}, \quad S^3 = \frac{1}{2} \begin{pmatrix} 1 & 0 \\ 0 & -1 \end{pmatrix}.$$

The correspondence with the creation and annihilation operators is

$$a_x^\dagger = S_x^1 + i S_x^2 \equiv S_x^+, \quad a_x = S_x^1 - i S_x^2 \equiv S_x^-,$$

and hence $a_x^\dagger a_x = S_x^3 + \frac{1}{2}$. (This is known as the Matsubara-Matsuda correspondence [18].) Adding a convenient constant to make the periodic potential positive, the Hamiltonian (1) for $U = \infty$ is thus equivalent to

$$\begin{aligned} H &= -\frac{1}{2} \sum_{\langle xy \rangle} (S_x^+ S_y^- + S_x^- S_y^+) + \lambda \sum_x \left[\tfrac{1}{2} + (-1)^x S_x^3 \right] \\ &= -\sum_{\langle xy \rangle} (S_x^1 S_y^1 + S_x^2 S_y^2) + \lambda \sum_x \left[\tfrac{1}{2} + (-1)^x S_x^3 \right]. \end{aligned} \tag{2}$$

Without loss of generality we may assume $\lambda \geq 0$. This Hamiltonian is well known as a model for interacting spins, referred to as the XY model [19]. The last term has the interpretation of a staggered magnetic field. We note that BEC for the lattice gas is equivalent to off-diagonal long range order for the 1- and 2-components of the spins.

The Hamiltonian (2) is clearly invariant under simultaneous rotations of all the spins around the 3-axis. In particle language this is the $U(1)$ gauge symmetry associated with particle number conservation of the Hamiltonian (1). Off-diagonal long range order (or, equivalently, BEC) implies that this symmetry is spontaneously broken in the state under consideration. It is notoriously difficult to prove such symmetry breaking for systems with a continuous symmetry. One of the few available techniques is that of *reflection positivity* (and the closely related property of *Gaussian domination*) and fortunately it can be applied to our system. For this, however, the hard core and half-filling conditions are essential because they imply a particle-hole symmetry that is crucial for the proofs to work. Naturally, BEC is expected to occur at other fillings, but no one has so far found a way to prove condensation (or, equivalently, long-range order in an antiferromagnet with continuous symmetry) without using reflection positivity and infrared bounds, and these require the addtional symmetry.

Reflection positivity was first formulated by K. Osterwalder and R. Schrader [20] in the context of relativistic quantum field theory. Later, J. Fröhlich, B. Simon and T. Spencer used the concept to prove the existence of a phase transition for a classical spin model with a continuous symmetry [21], and E. Lieb and J. Fröhlich [22] as well as F. Dyson, E. Lieb and B. Simon [19] applied it for the analysis of quantum spin systems. The proof of off-diagonal long range order for the Hamiltonian (2) (for small λ) given here is based on appropriate modifications of the arguments in [19].

2 Reflection Positivity

In the present context reflection positivity means the following. We divide the torus Λ into two congruent parts, Λ_{L} and Λ_{R}, by cutting it with a hyperplane orthogonal to one of the d directions. (For this we assume that the side length of Λ is even.) This induces a factorization of the Hilbert space, $\mathcal{H} = \mathcal{H}_{\mathrm{L}} \otimes \mathcal{H}_{\mathrm{R}}$, with

$$\mathcal{H}_{\mathrm{L,R}} = \bigotimes_{\mathbf{x} \in \Lambda_{\mathrm{L,R}}} \mathbb{C}^2 \, .$$

There is a natural identification between a site $\mathbf{x} \in \Lambda_{\mathrm{L}}$ and its mirror image $\vartheta\mathbf{x} \in \Lambda_{\mathrm{R}}$. If F is an operator on $\mathcal{H} = \mathcal{H}_{\mathrm{L}}$ we define its reflection θF as an operator on $\mathcal{H}_{\mathrm{R}}$ in the following way. If $F = F_{\mathbf{x}}$ operates non-trivially only on one site, $\mathbf{x} \in \Lambda_{\mathrm{L}}$, we define $\theta F = V F_{\vartheta\mathbf{x}} V^\dagger$ where V denotes the unitary particle-hole transformation or, in the spin language, rotation by π around the 1-axis. This definition extends in an obvious way to products of operators on single sites and then, by linearity, to arbitrary operators on $\mathcal{H}_L$. Reflection positivity of a state $\langle\cdot\rangle$ means that

$$\langle F\theta\overline{F}\rangle \geq 0 \qquad (3)$$

for any F operating on $\mathcal{H}_L$. Here $\overline{F}$ is the complex conjugate of the operator F in the matrix representation defined above, i.e., defined by the basis where the operators S_x^3 are diagonal.

We now show that reflection positivity holds for any thermal equilibrium state of our Hamiltonian. We can write the Hamiltonian (2) as

$$H = H_L + H_R - \frac{1}{2}\sum_{\langle\mathbf{xy}\rangle\in M}(S_\mathbf{x}^+ S_\mathbf{y}^- + S_\mathbf{x}^- S_\mathbf{y}^+)\,, \qquad (4)$$

where H_L and H_R act non-trivially only on $\mathcal{H}_L$ and $\mathcal{H}_R$, respectively. Here, M denotes the set of bonds going from the left sublattice to the right sublattice. (Because of the periodic boundary condition these include the bonds that connect the right boundary with the left boundary.) Note that $H_R = \theta H_L$, and

$$\sum_{\langle\mathbf{xy}\rangle\in M}(S_\mathbf{x}^+ S_\mathbf{y}^- + S_\mathbf{x}^- S_\mathbf{y}^+) = \sum_{\langle\mathbf{xy}\rangle\in M}(S_\mathbf{x}^+\theta S_\mathbf{x}^+ + S_\mathbf{x}^-\theta S_\mathbf{x}^-)\,.$$

For these properties it is essential that we included the unitary particle-hole transformation V in the definition of the reflection θ. For reflection positivity it is also important that all operators appearing in H (4) have a *real* matrix representation. Moreover, the minus sign in (4) is essential.

Using the Trotter product formula, we have

$$\mathrm{Tr}\,F\theta\overline{F}e^{-\beta H} = \lim_{n\to\infty}\mathrm{Tr}\,F\theta\overline{F}\,\mathcal{Z}_n$$

with

$$\mathcal{Z}_n = \left[e^{-\frac{1}{n}\beta H_L}\theta e^{-\frac{1}{n}\beta H_L}\prod_{\langle\mathbf{xy}\rangle\in M}\left(1 + \frac{\beta}{2n}[S_\mathbf{x}^+\theta S_\mathbf{x}^+ + S_\mathbf{x}^-\theta S_\mathbf{x}^-)]\right)\right]^n\,. \qquad (5)$$

Observe that $\mathcal{Z}_n$ is a sum of terms of the form

$$\prod_i A_i\theta A_i\,, \qquad (6)$$

with A_i given by either $e^{-\frac{1}{n}\beta H_L}$ or $\sqrt{\frac{\beta}{2n}}S_\mathbf{x}^+$ or $\sqrt{\frac{\beta}{2n}}S_\mathbf{x}^-$. All the A_i are real matrices, and therefore

$$\mathrm{Tr}_\mathcal{H}\,F\theta\overline{F}\prod_i A_i\theta A_i = \mathrm{Tr}_\mathcal{H}\,F\prod_i A_i\,\theta\left[\overline{F}\prod_j A_j\right] = \left|\mathrm{Tr}_{\mathcal{H}_L}\,F\prod_i A_i\right|^2 \geq 0\,. \qquad (7)$$

Hence $\mathrm{Tr}\,F\theta\overline{F}\,\mathcal{Z}_n$ is a sum of non-negative terms and therefore non-negative. This proves our assertion.

3 Proof of BEC for Small λ and T

The main tool in our proof of BEC are *infrared bounds*. More precisely, for $\mathbf{p} \in \Lambda^*$ (the dual lattice of Λ), let $\widetilde{S}_{\mathbf{p}}^{\#} = |\Lambda|^{-1/2} \sum_{\mathbf{x}} S_{\mathbf{x}}^{\#} \exp(i\mathbf{p} \cdot \mathbf{x})$ denote the Fourier transform of the spin operators. We claim that

$$(\widetilde{S}_{\mathbf{p}}^1, \widetilde{S}_{-\mathbf{p}}^1) \leq \frac{T}{2E_{\mathbf{p}}} , \tag{8}$$

with $E_{\mathbf{p}} = \sum_{i=1}^{d}(1 - \cos(p_i))$. Here, p_i denotes the components of $\mathbf{p}$, and $(,)$ denotes the Duhamel two point function at temperature T, defined by

$$(A, B) = \int_0^1 \mathrm{Tr}\left(Ae^{-s\beta H} B e^{-(1-s)\beta H}\right) ds / \mathrm{Tr}\, e^{-\beta H} \tag{9}$$

for any pair of operators A and B. Because of invariance under rotations around the S^3 axis, (8) is equally true with S^1 replaced by S^2, of course.

The crucial lemma (*Gaussian domination*) is the following. Define, for a complex valued function h on the bonds $\langle \mathbf{xy} \rangle$ in Λ,

$$Z(h) = \mathrm{Tr} \exp\left[-\beta K(h)\right] , \tag{10}$$

with $K(h)$ the modified Hamiltonian

$$K(h) = \frac{1}{4} \sum_{\langle \mathbf{xy} \rangle} \left(\left(S_{\mathbf{x}}^+ - S_{\mathbf{y}}^- - h_{\mathbf{xy}}\right)^2 + \left(S_{\mathbf{x}}^- - S_{\mathbf{y}}^+ - \overline{h_{\mathbf{xy}}}\right)^2\right) + \lambda \sum_{\mathbf{x}} \left[\tfrac{1}{2} + (-1)^{\mathbf{x}} S_{\mathbf{x}}^3\right] . \tag{11}$$

Note that for $h \equiv 0$, $K(h)$ agrees with the Hamiltonian H, because $(S^{\pm})^2 = 0$. We claim that, for any *real valued* h,

$$Z(h) \leq Z(0) . \tag{12}$$

The infrared bound then follows from $d^2 Z(\varepsilon h)/d\varepsilon^2|_{\varepsilon=0} \leq 0$, taking $h_{\mathbf{xy}} = \exp(i\mathbf{p} \cdot \mathbf{x}) - \exp(i\mathbf{p} \cdot \mathbf{y})$. This is not a real function, though, but the negativity of the (real!) quadratic form $d^2 Z(\varepsilon h)/d\varepsilon^2|_{\varepsilon=0}$ for real h implies negativity also for complex-valued h.

The proof of (12) is very similar to the proof of the reflection positivity property (3) given above. It follows along the same lines as in [19], but we repeat it here for convenience of the reader.

The intuition behind (12) is the following. First, in maximizing $Z(h)$ one can restrict to gradients, i.e., $h_{\mathbf{xy}} = \hat{h}_{\mathbf{x}} - \hat{h}_{\mathbf{y}}$ for some function $\hat{h}_{\mathbf{x}}$ on Λ. (This follows from stationarity of $Z(h)$ at a maximizer $h_{\max}$.) Reflection positivity implies that $\langle A\theta\overline{B}\rangle$ defines a scalar product on operators on $\mathcal{H}_{\mathrm{L}}$, and hence there is a corresponding Schwarz inequality. Moreover, since reflection positivity holds for reflections across *any* hyperplane, one arrives at the so-called *chessboard inequality*, which is simply a version of Schwarz's inequality for multiple reflections across different hyperplanes. Such a chessboard estimate implies that in order to maximize $Z(h)$ it is best to choose the function $\hat{h}_{\mathbf{x}}$ to be constant. In the case of classical spin systems [21], this intuition can be turned into a complete proof of (12). Because of non-commutativity of $K(h)$ with $K(0) = H$, this is not possible in the quantum case. However, one can proceed by using the Trotter formula as follows.

Let $h_{\max}$ be a function that maximizes $Z(h)$ for real valued h. If there is more than one maximizer, we choose $h_{\max}$ to be one that vanishes on the largest number of bonds. We then have to show that actually $h_{\max} \equiv 0$. If $h_{\max} \neq 0$, we draw a hyperplane such that $h_{\mathbf{xy}} \neq 0$ for at least one pair $\langle \mathbf{xy} \rangle$ crossing the plane. We can again write

$$K(h) = K_L(h) + K_R(h) + \frac{1}{4} \sum_{\langle \mathbf{xy} \rangle \in M} \left((S_{\mathbf{x}}^+ - S_{\mathbf{y}}^- - h_{\mathbf{xy}})^2 + (S_{\mathbf{x}}^- - S_{\mathbf{y}}^+ - h_{\mathbf{xy}})^2 \right) .$$

(13)

Using the Trotter formula, we have $Z(h) = \lim_{n \to \infty} \alpha_n$, with

$$\alpha_n = \mathrm{Tr} \left[e^{-\beta K_L/n} e^{-\beta K_R/n} \prod_{\langle \mathbf{xy} \rangle \in M} e^{-\beta(S_{\mathbf{x}}^+ - S_{\mathbf{y}}^- - h_{\mathbf{xy}})^2/4n} e^{-\beta(S_{\mathbf{x}}^- - S_{\mathbf{y}}^+ - h_{\mathbf{xy}})^2/4n} \right]^n .$$

(14)

For any matrix, we can write

$$e^{-D^2} = (4\pi)^{-1/2} \int_{\mathbb{R}} dk \, e^{ikD} e^{-k^2/4} .$$

(15)

If we apply this to the last two factors in (14), and note that $S_{\mathbf{y}}^- = \theta S_{\mathbf{x}}^+$ if $\langle \mathbf{xy} \rangle \in M$. Denoting by $\mathbf{x}_1, \dots, \mathbf{x}_l$ the points on the left side of the bonds in M, we have that

$$\alpha_n = (4\pi)^{-nl} \int_{R^{2nl}} d^{2nl} k \, \mathrm{Tr} \left[e^{-\beta K_L/n} e^{-\beta K_R/n} e^{ik_1(S_{\mathbf{x}_1}^+ - \theta S_{\mathbf{x}_1}^+)\beta^{1/2}/2n^{1/2}} \dots \right]$$
$$\times e^{-k^2/4} e^{-ik_1 h_{\mathbf{x}_1 \vartheta \mathbf{x}_1} \beta^{1/2}/2n^{1/2} \dots} .$$

(16)

Here we denote $k^2 = \sum k_i^2$ for short. Since matrices on the right of M commute with matrices on the left, and since all matrices in question are *real*, we see that the trace in the integrand above can be written as

$$\mathrm{Tr} \left[e^{-\beta K_L/n} e^{ik_1 S_{\mathbf{x}_1}^+ \beta^{1/2}/2n^{1/2}} \dots \right] \overline{\mathrm{Tr} \left[e^{-\beta K_R/n} e^{ik_1 \theta S_{\mathbf{x}_1}^+ \beta^{1/2}/2n^{1/2}} \dots \right]} . \quad (17)$$

Using the Schwarz inequality for the k integration, and "undoing" the above step, we see that

$$
\begin{aligned}
|\alpha_n|^2 \leq \Bigg(& (4\pi)^{-nl} \int_{R^{2nl}} d^{2nl}k \, e^{-k^2/4} \\
& \times \operatorname{Tr} \left[e^{-\beta K_L/n} e^{-\beta \theta K_L/n} e^{ik_1(S^+_{x_1} - \theta S^+_{x_1})\beta^{1/2}/2n^{1/2}} \cdots \right] \Bigg) \\
& \times \Bigg((4\pi)^{-nl} \int_{R^{2nl}} d^{2nl}k \, e^{-k^2/4} \\
& \times \operatorname{Tr} \left[e^{-\beta \theta K_R/n} e^{-\beta K_R/n} e^{ik_1(S^+_{x_1} - \theta S^+_{x_1})\beta^{1/2}/2n^{1/2}} \cdots \right] \Bigg).
\end{aligned}
\tag{18}
$$

In terms of the partition function $Z(h)$, this means that

$$
|Z(h_{\max})|^2 \leq Z(h^{(1)})Z(h^{(2)}), \tag{19}
$$

where $h^{(1)}$ and $h^{(2)}$ are obtained from $h_{\max}$ by reflection across M in the following way:

$$
h^{(1)}_{\mathbf{xy}} = \begin{cases} h_{\mathbf{xy}} & \text{if } \mathbf{x,y} \in \Lambda_L \\ h_{\vartheta \mathbf{x}\vartheta \mathbf{y}} & \text{if } \mathbf{x,y} \in \Lambda_R \\ 0 & \text{if } \langle \mathbf{xy} \rangle \in M \end{cases} \tag{20}
$$

and $h^{(2)}$ is given by the same expression, interchanging L and R. Therefore also $h^{(1)}$ and $h^{(2)}$ must be maximizers of $Z(h)$. However, one of them will contain strictly more zeros than $h_{\max}$, since $h_{\max}$ does not vanish identically for bonds crossing M. This contradicts our assumption that $h_{\max}$ contains the maximal number of zeros among all maximizers of $Z(h)$. Hence $h_{\max} \equiv 0$ identically. This completes the proof of (12).

The next step is to transfer the upper bound on the Duhamel two point function (8) into an upper bound on the thermal expectation value. This involves convexity arguments and estimations of double commutators like in Sect. 3 in [19]. For this purpose, we have to evaluate the double commutators

$$
[\widetilde{S}^1_{\mathbf{p}}, [H, \widetilde{S}^1_{-\mathbf{p}}]] + [\widetilde{S}^2_{\mathbf{p}}, [H, \widetilde{S}^2_{-\mathbf{p}}]] = -\frac{2}{|\Lambda|} \left(H - \tfrac{1}{2}\lambda |\Lambda| + 2 \sum_{\langle \mathbf{xy} \rangle} S^3_{\mathbf{x}} S^3_{\mathbf{y}} \cos \mathbf{p} \cdot (\mathbf{x-y}) \right). \tag{21}
$$

Let $C_{\mathbf{p}}$ denote the expectation value of this last expression,

$$
C_{\mathbf{p}} = \langle [\widetilde{S}^1_{\mathbf{p}}, [H, \widetilde{S}^1_{-\mathbf{p}}]] + [\widetilde{S}^2_{\mathbf{p}}, [H, \widetilde{S}^2_{-\mathbf{p}}]] \rangle \geq 0.
$$

The positivity of $C_{\mathbf{p}}$ can be seen from an eigenfunction-expansion of the trace. From [19, Corollary 3.2 and Theorem 3.2] and (8) we infer that

$$
\langle \widetilde{S}^1_{\mathbf{p}} \widetilde{S}^1_{-\mathbf{p}} + \widetilde{S}^2_{\mathbf{p}} \widetilde{S}^2_{-\mathbf{p}} \rangle \leq \frac{1}{2} \sqrt{\frac{C_{\mathbf{p}}}{E_{\mathbf{p}}}} \coth \sqrt{\beta^2 C_{\mathbf{p}} E_{\mathbf{p}}/4}. \tag{22}
$$

Using $\coth x \leq 1 + 1/x$ and Schwarz's inequality, we obtain for the sum over all $\mathbf{p} \neq 0$,

$$\sum_{\mathbf{p}\neq 0}\langle \widetilde{S}_{\mathbf{p}}^1 \widetilde{S}_{-\mathbf{p}}^1 + \widetilde{S}_{\mathbf{p}}^2 \widetilde{S}_{-\mathbf{p}}^2\rangle \leq \frac{1}{\beta}\sum_{\mathbf{p}\neq 0}\frac{1}{E_{\mathbf{p}}} + \frac{1}{2}\left(\sum_{\mathbf{p}\neq 0}\frac{1}{E_{\mathbf{p}}}\right)^{1/2}\left(\sum_{\mathbf{p}\neq 0} C_{\mathbf{p}}\right)^{1/2}. \tag{23}$$

We have $\sum_{\mathbf{p}\in\Lambda^*} C_{\mathbf{p}} = -2\langle H\rangle + \lambda|\Lambda|$, which can be bounded from above using the following lower bound on the Hamiltonian:

$$H \geq -\frac{|\Lambda|}{4}\left[d(d+1) + 4\lambda^2\right]^{1/2} + \frac{1}{2}\lambda|\Lambda|. \tag{24}$$

This inequality follows from the fact that the lowest eigenvalue of

$$-\frac{1}{2}S_{\mathbf{x}}^1 \sum_{i=1}^{2d} S_{\mathbf{y}_i}^1 - \frac{1}{2}S_{\mathbf{x}}^2 \sum_{i=1}^{2d} S_{\mathbf{y}_i}^2 + \lambda S_{\mathbf{x}}^3 \tag{25}$$

is given by $-\frac{1}{4}[d(d+1) + 4\lambda^2]^{1/2}$. This can be shown exactly in the same way as [19, Theorem C.1]. Since the Hamiltonian H can be written as a sum of terms like (25), with $\mathbf{y}_i$ the nearest neighbors of $\mathbf{x}$, we get from this fact the lower bound (24).

With the aid of the sum rule

$$\sum_{\mathbf{p}\in\Lambda^*}\langle \widetilde{S}_{\mathbf{p}}^1 \widetilde{S}_{-\mathbf{p}}^1 + \widetilde{S}_{\mathbf{p}}^2 \widetilde{S}_{-\mathbf{p}}^2\rangle = \frac{|\Lambda|}{2}$$

(which follows from $(S^1)^2 = (S^2)^2 = 1/4$), we obtain from (23) and (24) the following lower bound in the thermodynamic limit:

$$\lim_{\Lambda\to\infty}\frac{1}{|\Lambda|}\langle \widetilde{S}_0^1 \widetilde{S}_0^1 + \widetilde{S}_0^2 \widetilde{S}_0^2\rangle$$

$$\geq \frac{1}{2} - \frac{1}{2}\left(\frac{1}{2}\left[d(d+1) + 4\lambda^2\right]^{1/2} c_d\right)^{1/2} - \frac{1}{\beta}c_d, \tag{26}$$

with c_d given by

$$c_d = \frac{1}{(2\pi)^d}\int_{[-\pi,\pi]^d}d\mathbf{p}\,\frac{1}{E_{\mathbf{p}}}. \tag{27}$$

This is our final result. Note that c_d is finite for $d \geq 3$. Hence the right side of (26) is positive, for large enough β, as long as

$$\lambda^2 < \frac{1}{c_d^2} - \frac{d(d+1)}{4}.$$

In $d = 3$, $c_3 \approx 0.505$ [19], and hence this condition is fulfilled for $\lambda \lesssim 0.960$. In [19] it was also shown that dc_d is monotone decreasing in d, which implies a similar result for all $d > 3$.

The connection with BEC is as follows. Since H is real, also $\gamma(\mathbf{x},\mathbf{y})$ is real and we have

$$\gamma(\mathbf{x}, \mathbf{y}) = \langle S_\mathbf{x}^+ S_\mathbf{y}^- \rangle = \langle S_\mathbf{x}^1 S_\mathbf{y}^1 + S_\mathbf{x}^2 S_\mathbf{y}^2 \rangle \ .$$

Hence, if $\varphi_0 = |\Lambda|^{-1/2}$ denotes the constant function,

$$\langle \varphi_0 | \gamma | \varphi_0 \rangle = \langle \widetilde{S}_0^1 \widetilde{S}_0^1 + \widetilde{S}_0^2 \widetilde{S}_0^2 \rangle \ ,$$

and thus the bound (26) implies that the largest eigenvalue of $\gamma(\mathbf{x}, \mathbf{y})$ is bounded from below by the right side of (26) times $|\Lambda|$. In addition one can show that the infrared bounds imply that there is at most *one* large eigenvalue (of the order $|\Lambda|$), and that the corresponding eigenvector (the "condensate wave function") is strictly constant in the thermodynamic limit [1]. The constancy of the condensate wave function is surprising and is not expected to hold for densities different from $\frac{1}{2}$, where particle-hole symmetry is absent. In contrast to the condensate wave function the particle density shows the staggering of the periodic potential [1, Theorem 3]. It also contrasts with the situation for zero interparticle interaction, as discussed at the end of this paper.

In the BEC phase there is *no gap* for adding particles beyond half filling (in the thermodynamic limit): The ground state energy, E_k, for $\frac{1}{2}|\Lambda| + k$ particles satisfies

$$0 \le E_k - E_0 \le \frac{(\text{const.})}{|\Lambda|} \tag{28}$$

(with a constant that depends on k but not on $|\Lambda|$.) The proof of (28) is by a variational calculation, with a trial state of the form $(\widetilde{S}_0^+)^k |0\rangle$, where $|0\rangle$ denotes the absolute ground state, i.e., the ground state for half filling. (This is the unique ground state of the Hamiltonian, as can be shown using reflection positivity. See Appendix A in [1].) Also, in the thermodynamic limit, the energy per site for a given density, $e(\varrho)$, satisfies

$$e(\varrho) - e(\tfrac{1}{2}) \le \text{const.} \left(\varrho - \tfrac{1}{2}\right)^2 \ . \tag{29}$$

Thus there is no cusp at $\varrho = 1/2$. To show this, one takes a trial state of the form

$$|\psi_\mathbf{y}\rangle = e^{i\varepsilon \sum_\mathbf{x} S_\mathbf{x}^2} (S_\mathbf{y}^1 + \tfrac{1}{2})|0\rangle \ . \tag{30}$$

The motivation is the following: we take the ground state and first project onto a given direction of S^1 on some site $\mathbf{y}$. If there is long-range order, this should imply that essentially all the spins point in this direction now. Then we rotate slightly around the S^2-axis. The particle number should then go up by $\varepsilon|\Lambda|$, but the energy only by $\varepsilon^2|\Lambda|$. We refer to [1, Sect. IV] for the details.

The absence of a gap in the case of BEC is not surprising, since a gap is characteristic for a Mott insulator state. We show the occurrence of a gap, for large enough λ, in the next section.

4 Absence of BEC and Mott Insulator Phase

The main results of this section are the following: If either

- $\lambda \geq 0$ and $T > d/(2\ln 2)$, or
- $T \geq 0$ and $\lambda \geq 0$ such that $\lambda + |e(\lambda)| > d$, with $e(\lambda) =$ ground state energy per site,

then there is exponential decay of correlations:

$$\gamma(\mathbf{x}, \mathbf{y}) \leq (\text{const.}) \exp(-\kappa|\mathbf{x} - \mathbf{y}|) \tag{31}$$

with $\kappa > 0$. Moreover, for $T = 0$, the ground state energy in a sector of fixed particle number $N = \frac{1}{2}|\Lambda| + k$, denoted by E_k, satisfies

$$E_k + E_{-k} - 2E_0 \geq (\lambda + |e(\lambda)| - d)|k| . \tag{32}$$

I.e, for large enough λ the chemical potential has a jump at half filling.

The derivation of these two properties is based on a path integral representation of the equilibrium state at temperature T, and of the ground state which is obtained in the limit $T \to \infty$. The analysis starts from the observation that the density operator $e^{-\beta H}$ has non-negative matrix elements in the basis in which $\{S_\mathbf{x}^3\}$ are diagonal, i.e. of states with specified particle occupation numbers. It is convenient to focus on the dynamics of the 'quasi-particles' which are defined so that the presence of one at a site $\mathbf{x}$ signifies a deviation there from the occupation state which minimizes the potential-energy. Since the Hamiltonian is $H = H_0 + \lambda W$, with H_0 the hopping term in (2) and W the staggered field, we define the quasi-particle number operators $n_\mathbf{x}$ as:

$$n_\mathbf{x} = \frac{1}{2} + (-1)^\mathbf{x} S_\mathbf{x}^3 = \begin{cases} a_\mathbf{x}^\dagger a_\mathbf{x}, & \text{for } \mathbf{x} \text{ even} \\ 1 - a_\mathbf{x}^\dagger a_\mathbf{x}, & \text{for } \mathbf{x} \text{ odd} \end{cases} . \tag{33}$$

Thus $n_\mathbf{x} = 1$ means presence of a particle if $\mathbf{x}$ is on the A sublattice (potential maximum) and absence if $\mathbf{x}$ is on the B sublattice (potential minimum).

The collection of the joint eigenstates of the occupation numbers, $\{|\{n_\mathbf{x}\}\rangle\}$, provides a convenient basis for the Hilbert space. The functional integral representation of $\langle\{n_\mathbf{x}\}| e^{-\beta(H_0 + \lambda W)} |\{n_\mathbf{x}\}\rangle$ involves an integral over configurations of quasi-particle loops in a $space \times time$ for which the (imaginary) "time" corresponds to a variable with period β. The fact that the integral is over a positive measure facilitates the applicability of statistical-mechanics intuition and tools. One finds that the quasi-particles are suppressed by the potential energy, but favored by the entropy, which enters this picture due to the presence of the hopping term in H. At large λ, the potential suppression causes localization: long 'quasi-particle' loops are rare, and the amplitude for long paths decays exponentially in the distance, both for paths which may occur spontaneously and for paths whose presence is forced through the insertion of sources, i.e., particle creation and annihilation operators. Localization is also caused by high temperature, since the requirement of periodicity implies that at any site which participates in a loop there should be be at least two jumps during the short 'time' interval $[0, \beta)$ and the amplitude for even a single jump is small, of order β.

The path integral described above is obtained through the Dyson expansion

$$e^{t(A+B)} = e^{tA} \sum_{m \geq 0} \int_{0 \leq t_1 \leq t_2 \leq \cdots \leq t_m \leq t} B(t_m) \cdots B(t_1) dt_1 \cdots dt_m \qquad (34)$$

for any matrices A and B and $t > 0$, with $B(t) = e^{-tA} B e^{tA}$. (The $m = 0$ term in the sum is interpreted here as 1.)

In evaluating the matrix elements of $e^{-\beta H} = e^{-\beta(H_0 + \lambda W)}$, in the basis $\{|\{n_\mathbf{x}\}\rangle\}$, we note that W is diagonal and $\langle\{n_\mathbf{x}\}|H_0|\{n'_\mathbf{x}\}\rangle$ is non-zero only if the configurations $\{n_\mathbf{x}\}$ and $\{n'_\mathbf{x}\}$ differ at exactly one nearest neighbor pair of sites where the change corresponds to either a creation of a pair of quasi-particles or the annihilation of such a pair. I.e., the matrix elements are zero unless $n_\mathbf{x} = n'_\mathbf{x}$ for all $\mathbf{x}$ except for a nearest neighbor pair $\langle \mathbf{xy} \rangle$, where $n_\mathbf{x} = n_\mathbf{y}$, $n'_\mathbf{x} = n'_\mathbf{y}$, and $n_\mathbf{x} + n'_\mathbf{x} = 1$. In this case, the matrix element equals $-1/2$.

Introducing intermediate states, the partition function can thus be written as follows:

$$\text{Tr } e^{-\beta H} = \sum_{m=0}^{\infty} \int_{0 \leq t_1 \leq t_2 \leq \cdots \leq t_m \leq \beta} \sum_{|\{n_\mathbf{x}^{(i)}\}\rangle, 1 \leq i \leq m}$$

$$\times \exp\left(-\lambda \sum_{i=1}^{m}(t_i - t_{i-1}) \sum_\mathbf{x} n_\mathbf{x}^{(i)}\right) dt_1 \cdots dt_m$$

$$\times (-1)^m \langle\{n_\mathbf{x}^{(1)}\}|H_0|\{n_\mathbf{x}^{(m)}\}\rangle \langle\{n_\mathbf{x}^{(m)}\}|H_0|\{n_\mathbf{x}^{(m-1)}\}\rangle$$

$$\times \langle\{n_\mathbf{x}^{(m-1)}\}|H_0|\{n_\mathbf{x}^{(m-2)}\}\rangle \cdots \langle\{n_\mathbf{x}^{(2)}\}|H_0|\{n_\mathbf{x}^{(1)}\}\rangle \qquad (35)$$

with the interpretation $t_0 = t_m - \beta$. Note that the factor in the last two lines of (35) equals $(1/2)^m$ if adjacent elements in the sequence of configurations $\{n_\mathbf{x}^{(i)}\}$ differ by exactly one quasi-particle pair, otherwise it is zero.

Expansions of this type are explained more fully in [23]. A compact way of writing (35) is:

$$\text{Tr } e^{-\beta H} = \int v(d\omega) e^{-\lambda|\omega|} . \qquad (36)$$

Here the "path" ω stands for a set of disjoint oriented loops in the "space-time" $\Lambda \times [0, \beta]$, with periodic boundary conditions in "time"". Each ω is parametrized by a number of jumps, m, jumping times $0 \leq t_1 \leq t_2 \leq \cdots \leq t_m \leq \beta$, and a sequence of configurations $\{n_\mathbf{x}^{(i)}\}$, which is determined by the initial configuration $\{n_\mathbf{x}^{(1)}\}$ plus a sequence of "rungs" connecting nearest neighbor sites, depicting the creation or annihilation of a pair of neighboring

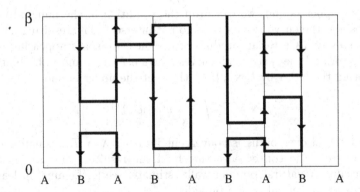

Fig. 2. Loop gas describing paths of quasi-particles for particle number $N = |\Lambda|/2 - 1$. A line on an A site means presence of a particle, while on a B site it means absence. The horizontal rungs correspond to hopping of a particle

quasi-particles (see Fig. 2). As in Feynman's picture of QED, it is convenient to regard such an event as a jump of the quasi-particle, at which its time-orientation is also reversed. The length of ω, denoted by $|\omega|$, is the sum of the vertical lengths of the loops. The measure $v(d\omega)$ is determined by (35); namely, for a given sequence of configurations $\{n_{\mathbf{x}}^{(i)}\}$, $1 \leq i \leq m$, the integration takes places over the times of the jumps, with a measure $(1/2)^m dt_1 \cdots dt_m$.

One may note that the measure $v(d\omega)$ corresponds to a Poisson process of random configurations of oriented "rungs", linking neighboring sites at random times, and signifying either the creation or the annihilation of a pair of quasiparticles. The matrix element $\langle\{n_{\mathbf{x}}\}|e^{-\beta H}|\{n_{\mathbf{x}}'\}\rangle$ gets no contribution from rung configurations that are inconsistent, either internally or with the boundary conditions corresponding to the specified state vectors. A consistent configuration yields a family of non-overlapping loops which describe the motion of the quasi-particles in in the "space-time" $\Lambda \times [0, \beta)$. Each such configuration contributes with weight $e^{-\lambda|\omega|}$ to the above matrix element (another positive factor was absorbed in the measure $v(d\omega)$). One may note that long paths are suppressed in the integral (38) at a rate which increases with λ.

Likewise, for $\mathbf{x} \neq \mathbf{y}$, we can write

$$\mathrm{Tr}\, a_{\mathbf{x}}^\dagger a_{\mathbf{y}} e^{-\beta H} = \int_{\mathcal{A}^{(\mathbf{x},\mathbf{y})}} v(d\omega) e^{-\lambda|\omega|} , \qquad (37)$$

where $\mathcal{A}^{(\mathbf{x},\mathbf{y})}$ denotes the set of all loops that, besides disjoint closed loops, contain one curve which avoids all the loops and connects $\mathbf{x}$ and $\mathbf{y}$ at time zero. The one-particle density matrix can thus be written

$$\gamma(\mathbf{x}, \mathbf{y}) = \frac{\int_{\mathcal{A}^{(\mathbf{x},\mathbf{y})}} v(d\omega) e^{-\lambda|\omega|}}{\int v(d\omega) e^{-\lambda|\omega|}} . \qquad (38)$$

For an upper bound, we can drop the condition in the numerator that the loops and the curve from $\mathbf{x}$ to $\mathbf{y}$ do not intersect. The resulting measure space is simply a Cartesian product of the measure space appearing in the denominator and the space of all curves, ζ, connecting $\mathbf{x}$ and $\mathbf{y}$, both at time 0. Denoting the latter by $\mathcal{B}(\mathbf{x}, \mathbf{y})$, we thus get the upper bound

$$\gamma(\mathbf{x}, \mathbf{y}) \leq \int_{\mathcal{B}(\mathbf{x}, \mathbf{y})} v(d\zeta) e^{-\lambda|\zeta|} . \tag{39}$$

The integral over paths is convergent if either λ or T is small enough, and away from the convergence threshold the resulting amplitude decays exponentially. A natural random walk estimate, see [1, Lemma 4], leads to the claimed exponential bound provided

$$d\left(1 - e^{-\beta\lambda}\right) < \lambda . \tag{40}$$

This includes, in particular, the cases $T > d$ for any λ, and $\lambda > d$ for any T.

Exponential decay actually holds for the larger range of parameters where

$$d\left(1 - e^{-\beta(\lambda-f)}\right) < \lambda - f , \tag{41}$$

where $f = f(\beta, \lambda) = -(\beta|\Lambda|)^{-1}\ln\mathrm{Tr}\, e^{-\beta H}$ is the free energy per site. Note that $f < 0$. This condition can be obtained by a more elaborate estimate than the one used in obtaining (39) from (38), as shown in [1, Lemma 3]. The argument there uses reflection positivity of the measure $v(d\omega)$. Using simple bounds on f one can then obtain from (41) the conditions stated in the beginning of this section.

The proof of the energy gap is based on an estimate for the ratio $\frac{\mathrm{Tr}\,\mathcal{P}_k e^{-\beta H}}{\mathrm{Tr}\,\mathcal{P}_0 e^{-\beta H}}$ where $\mathcal{P}_k$ projects onto states in Fock space with particle number $N = \frac{1}{2}|\Lambda| + k$, expressing numerator and denominator in terms of path integrals. The integral for the numerator is over configurations ω with a non-trivial winding number k. Each such configuration includes a collection of "non-contractible" loops with total length at least $\beta|k|$. An estimate of the relative weight of such loops yields the bound

$$\frac{\mathrm{Tr}\,\mathcal{P}_k e^{-\beta H}}{\mathrm{Tr}\,\mathcal{P}_0 e^{-\beta H}} \leq (\mathrm{const.})(|\Lambda|/|k|)^{|k|} \left(e^{1-(\mathrm{const.})\beta}\right)^{|k|} \tag{42}$$

which gives for $\beta \to \infty$

$$E_k - E_0 \geq (\mathrm{const.})|k| \tag{43}$$

independently of $|\Lambda|$. We refer to [1] for details.

5 The Non-Interacting Gas

The interparticle interaction is essential for the existence of a Mott insulator phase for large λ. In case of absence of the hard-core interaction, there is

BEC for any density and any λ at low enough temperature (for $d \geq 3$). To see this, we have to calculate the spectrum of the one-particle Hamiltonian $-\frac{1}{2}\Delta + V(\mathbf{x})$, where Δ denotes the discrete Laplacian and $V(\mathbf{x}) = \lambda(-1)^{\mathbf{x}}$. The spectrum can be easily obtained by noting that V anticommutes with the off-diagonal part of the Laplacian, i.e., $\{V, \Delta + 2d\} = 0$. Hence

$$\left(-\frac{1}{2}\Delta - d + V(\mathbf{x})\right)^2 = \left(-\frac{1}{2}\Delta - d\right)^2 + \lambda^2 , \tag{44}$$

so the spectrum is given by

$$d \pm \sqrt{\left(\sum_i \cos p_i\right)^2 + \lambda^2} , \tag{45}$$

where $\mathbf{p} \in \Lambda^*$. In particular, $E(\mathbf{p}) - E(0) \sim \frac{1}{2}d(d^2 + \lambda^2)^{-1/2}|\mathbf{p}|^2$ for small $|\mathbf{p}|$, and hence there is BEC for low enough temperature. Note that the condensate wave function is of course *not* constant in this case, but rather given by the eigenfunction corresponding to the lowest eigenvalue of $-\frac{1}{2}\Delta + \lambda(-1)^{\mathbf{x}}$.

6 Conclusion

In this paper a lattice model is studied, which is similar to the usual Bose-Hubbard model and which describes the transition between Bose-Einstein condensation and a Mott insulator state as the strength λ of an optical lattice potential is increased. While the model is not soluble in the usual sense, it is possible to prove rigorously all the essential features that are observed experimentally. These include the existence of BEC for small λ and its suppression for large λ, which is a localization phenomenon depending heavily on the fact that the Bose particles interact with each other. The Mott insulator regime is characterized by a gap in the chemical potential, which does not exist in the BEC phase and for which the interaction is also essential. It is possible to derive bounds on the critical λ as a function of temperature.

References

1. M. Aizenman, E.H. Lieb, R. Seiringer, J.P. Solovej, and J. Yngvason, *Bose-Einstein quantum phase transition in an optical lattice model*, Phys. Rev. A **70**, 023612-1–12 (2004).
2. M. Greiner, O. Mandel, T. Esslinger, T.E. Hänsch, I. Bloch, *Quantum phase transition from a superfluid to a Mott insulator in a gas of ultracold atoms*, Nature **415**, 39 (2002).
3. M. Greiner, O. Mandel, T.E. Hänsch, I. Bloch, *Collapse and revival of the matter wave field of a Bose-Einstein condensate*, Nature **419**, 51 (2002).
4. S. Sachdev, *Quantum Phase Transitions*, Cambridge University Press, 1999.

5. M.P.A. Fisher, P.B. Weichman, G. Grinstein, D.S. Fisher, *Boson localization and the superfluid-insulator transition*, Phys. Rev. B **40**, 546–570 (1989).
6. D. Jaksch, C. Bruder, J.I. Cirac, C.W. Gardiner, P. Zoller, *Cold bosonic atoms in optical lattices*, Phys. Rev. Lett. **81**, 3108–3111 (1998).
7. W. Zwerger, *Mott-Hubbard transition of cold atoms in optical lattices*, Journal of Optics B **5**, 9–16 (2003).
8. J.J. Garcia-Ripoll, J.I. Cirac, P. Zoller, C. Kollath, U. Schollwoeck, J. von Delft, *Variational ansatz for the superfluid Mott-insulator transition in optical lattices*, Optics Express, **12**, 42–54 (2004).
9. K. Ziegler, *Phase Transition of a Bose Gas in an Optical Lattice*, Laser Physics **13**, 587–593 (2003).
10. Z. Nazario, D.I. Santiago, *Quantum States of Matter of Simple Bosonic Systems: BEC's, Superfluids and Quantum Solids*, Arxiv: cond-mat/0308005 (2003).
11. G.M. Genkin, *Manipulating the superfluid – Mott insulator transition of a Bose-Einstein condensate in an amplitude-modulated optical lattice*, Arxiv: cond-mat/ 0311589 (2003).
12. K. Ziegler, *Two-component Bose gas in an optical lattice at single-particle filling*, Phys. Rev. A **68**, 053602 (2003).
13. D.B.M. Dickerscheid, D. van Oosten, P.J.H. Denteneer, H.T.C. Stoof, *Ultracold atoms in optical lattices*, Phys. Rev. A **68**, 043623 (2003).
14. A.M. Rey, K. Burnett, R. Roth, M. Edwards, C.J. Williams, C.W. Clark, *Bogoliubov approach to superfluidity of atoms in an optical lattice*, J. Phys. B **36**, 825–841 (2003).
15. O. Morsch, E. Arimondo, *Ultracold atoms and Bose-Einstein condensates in optical lattices*, Lecture Notes in Physics Vol. 602, Springer (2002).
16. E. Altman, A. Auerbach. *Oscillating Superfluidity of Bosons in Optical Lattices*, Phys. Rev. Lett. **89**, 250404 (2002).
17. T. Kennedy, E.H. Lieb, S. Shastry, *The XY Model has Long-Range Order for all Spins and all Dimensions Greater than One*, Phys. Rev. Lett. **61**, 2582–2584 (1988).
18. T. Matsubara, H. Matsuda, *A lattice model of liquid helium*, Progr. Theor. Phys. **16**, 569–582 (1956).
19. F.J. Dyson, E.H. Lieb, B. Simon, *Phase Transitions in Quantum Spin Systems with Isotropic and Nonisotropic Interactions*, J. Stat. Phys. **18**, 335–383 (1978).
20. K. Osterwalder, R. Schrader, *Axioms for Euclidean Green's Functions*, Commun. Math. Phys. **31**, 83–112 (1973); Commun. Math. Phys. **42**, 281–305 (1975).
21. J. Fröhlich, B. Simon, T. Spencer, *Phase Transitions and Continuous Symmetry Breaking*, Phys. Rev. Lett. **36**, 804 (1976); *Infrared bounds, phase transitions and continuous symmetry breaking*, Commun. Math. Phys. **50**, 79 (1976).
22. J. Fröhlich, E.H. Lieb, *Phase Transitions in Anisotropic Lattice Spin Systems*, Commun. Math. Phys. **60**, 233–267 (1978).
23. M. Aizenman, B. Nachtergaele, *Geometric Aspects of Quantum Spin States*, Commun. Math. Phys. **164**, 17–63 (1994).

Long Time Behaviour
to the Schrödinger–Poisson–X^α Systems

Olivier Bokanowski[1], José L. López[2], Óscar Sánchez[2], and Juan Soler[2]

[1] Lab. Jacques Louis Lions (Paris VI), UFR Mathématique B. P. 7012
& Université Paris VII
boka@math.jusssieu.fr
[2] Departamento de Matemática Aplicada, Facultad de Ciencias, Universidad de
Granada, Campus Fuentenueva, 18071 Granada
jllopez@ugr.es
ossanche@ugr.es
jsoler@ugr.es

1 Introduction

This paper[1] is intended to constitute a review of some mathematical theories
incorporating quantum corrections to the Schrödinger-Poisson (SP) system.
More precisely we shall focus our attention in the electrostatic Poisson potential with corrections of power type.

   The SP system is a simple model used for the study of quantum transport
in semiconductor devices that can be written as a system of infinitely many
coupled equations (mixed-state) as

$$i\hbar\frac{\partial\psi_j}{\partial t} = -\frac{\hbar^2}{2m}\Delta_x\psi_j + \left(\frac{\gamma}{4\pi|x|} * n_\psi\right)\psi_j\,, \quad j \in \mathbb{N}\,, \tag{1}$$

$\psi_j = \psi_j(x,t)$ being the j-th component of the vector wave function and where
$t > 0$ is the time variable, $x \in \mathbb{R}^3$ is the position variable, $\psi_j(x,0) = \phi_j(x)$
is the j-th component of the initial condition and $n_\psi := \sum_{j\in\mathbb{N}}\lambda_j|\psi_j|^2$ holds
for the charge density. Here, $\{\lambda_j\}$ is a nonnegative sequence denoting the
occupation probabilities such that the normalization property $\sum_{j\in\mathbb{N}}\lambda_j = 1$
is fulfilled. In (1), $\hbar$ and m stand for the Planck constant and the particle
mass respectively. The Poisson potential is $V_\psi := \frac{\gamma}{4\pi|x|} * n_\psi$, where $*$ stands
for the x-convolution, $\gamma = 1$ for repulsive interactions and $\gamma = -1$ for attractive interactions. In the sequel we are only concerned with the repulsive
system modeling electrostatic forces. The solutions to the SP system have
been proved to be *strongly* dispersive in the sense

---

[1]This work was partially supported by the European Union, TMR–contract
IHP HPRN–CT–2002–00282. J.L.L., O.S. and J.S. were also partially sponsored by
DGES (Spain), Project MCYT BFM2002–00831.

$$\|\psi(\cdot,t)\|_{L^p(\mathbb{R}^3)} \leq \frac{C(\phi)}{t^{\beta(p)}} , \quad p > 2 , \quad \beta(p) > 0 , \tag{2}$$

first in the single-state case $\lambda = (1,0,0,\dots)$ in [16] and later in the mixed-state case in [7]. These bounds were recently improved in [33]. Here, the key point is the positive character of the total energy

$$\frac{1}{2}\sum_{j\in\mathbb{N}}\lambda_j \int_{\mathbb{R}^3} |\nabla_x\psi_j|^2 \, dx + \frac{1}{2}\int_{\mathbb{R}^3} |\nabla_x V_\psi|^2 \, dx .$$

As consequence, other behaviours like steady-states, breathers,... are not allowed by the SP system in contrast to experimental evidence (see [29]).

The motivation for the X^α approach

$$i\hbar\frac{\partial\psi_j}{\partial t} = -\frac{\hbar^2}{2m}\Delta_x\psi_j + \left(\frac{1}{4\pi|x|} * n_\psi\right)\psi_j - C_S\, n_\psi^\alpha\psi_j , \quad j \in \mathbb{N} , \tag{3}$$

as a correction to the Hartree–Fock system stems from experimentation, which evidences that effects due to the Coulomb charging energy could strongly modify the electron tunneling. In this direction, the Schrödinger–Poisson–Slater (SPS) system incorporates quantum corrections of power type to the SP system with $\alpha = \frac{1}{3}$. The nonlinear Slater term $-C_S\, n_\psi^{\frac{1}{3}}$ comes out as a local correction to the Fock term V_{ex} (see Sect. 2) and should be understood as a quantum effect, following [10,34] (see also [1,4,13] for a mathematical approach), contrary to the Poisson term which has a classical counterpart. Physically, the *Slater constant* C_S is positive for the case of electrons. Notice that if a different normalization $\sum_{j\in\mathbb{N}}\lambda_j = a > 0$ is assumed, then we are led to an analogous problem with modified constants. In particular, the value of C_S is genuinely relevant for the subsequent dynamical stability analysis.

One important feature of the SPS system is that the total energy operator

$$E[\psi] = E_K[\psi] + E_P[\psi] , \tag{4}$$

with

$$E_K[\psi] = \frac{1}{2}\sum_{j\in\mathbb{N}}\lambda_j \int_{\mathbb{R}^3} |\nabla_x\psi_j|^2 \, dx \quad \text{kinetic energy} ,$$

$$E_P[\psi] = \frac{1}{2}\int_{\mathbb{R}^3} |\nabla_x V_\psi|^2 \, dx - \frac{3}{4}C_S \int_{\mathbb{R}^3} n_\psi^{\frac{4}{3}} \, dx \quad \text{potential energy} ,$$

can reach negative values. This implies some relevant consequences:

(a) The minimum of the total energy operator is negative.
(b) There are steady-state (standing wave) solutions, i.e. solutions with constant density.
(c) There are solutions (even with positive energy) which preserve the L^p norm and do not decay with time. Thus, there are solutions that do not exhibit strong dispersive character in the sense of (2).

(d) Otherwise, in certain regime it can be proved that the solutions are dispersive in a statistical sense $(\Delta x)^2 = O(t^2)$ i. e., the variance associated with the solution grows with time. This may aim to the study of other type of weaker dispersion properties implying loss of charge at $t \to \infty$.

To carry on this analysis the preserved quantities and the Galilean invariance of the system are exploited.

These features show important qualitative differences concerning the dynamics of solutions to the SPS system when compared to the behaviour of solutions to Schrödinger–Poisson or Hartree–Fock systems (see [7,15,16,18]). Accordingly, the SPS system seems to be best realistic and fit properly to semiconductor and heterostructure modeling.

Another justification for this approach stems from the ambit of quantum chemistry, especially due to the high number of calculations necessary to evaluate the Fock term (usually of order N^4, N being the number of particles). Several X^α approaches to Fock's correction (for example the Slater approach) have been proved relevant in different contexts. The quantum correction $n_\psi^{\frac{1}{3}}\psi$ is also known as the Dirac exchange term. Another interesting approach comes up from the limit of heavy atoms, i.e. the high–charge–of–nuclei limit. This leads to so–called Thomas–Fermi correction ($\alpha = \frac{2}{3}$) of the kinetic energy (see [19, 20]), which can be alternatively seen as a correction to Fock's interaction showing up as a repulsive potential (see [20]).These local approximations to nonlocal interaction terms yield excellent results when studying stationary states, for example in the frame of quantum chemistry (see [11] and [8,19,23] for details on derivations and analysis of these systems). Then, the number of calculations is reduced from N^4 to N^3, even leaving a margin to improvements. On the contrary, there is no rigorous foundation of X^α models in the time-dependent case. Following the classical ideas of thermodynamical limits in statistical mechanics (see [12]), some progress is being currently done on this subject via continuum and mean–field limits of the N-quantum–particle system (see for instance [2]).

In the single-state case $\lambda = (1, 0, 0, \dots)$ the system is reduced to only one equation (for the wave function $\psi(x,t)$ with normalized constants $\hbar = m = 1$). In this case the model belongs to a wider class of nonlinear Schrödinger equations with power nonlinearities, already analyzed in [9]. We also note that the case of N coupled equations, with $\lambda_j = \frac{1}{N}$ if $1 \leq j \leq N$ and $\lambda_j = 0$ otherwise, corresponds to the X^α correction in Kohn–Sham equations [17]. This model belongs to the density–functional theory approach in Molecular Quantum Chemistry (see [26]) and constitutes a local approximation of the time-dependent Hartree–Fock system (see Sect. 2). Concerning the mixed-state case, an existence and uniqueness theory in L^2 and H^1 has been recently dealt with in [5] for the SPS system, being the basic ingredient a generalization of Strichartz' inequalities to mixed states (already exploited in [7]) and Pazy's fixed–point techniques [27].

Our first aim in these notes is to set up the origins of the mixed SPS system and of other related models, as well as to show that they are well–posed in $\mathbb{R}^3$. The analysis of some of the qualitative differences between X^α models and the Schrödinger–Poisson and Hartree–Fock systems shall also be dealt with. We are mainly interested in the exponent $\alpha = \frac{1}{3}$ studied in semiconductor theory, which is derived from the Fock term by means of a low density limit (see [5]).

Some of our results hold true under precise hypotheses relating the values of the Slater constant, the mass and the energy of the system. It is known that the Slater constant is a characteristic of the component metals in the semiconductor device (see [14]) when interpreting the exchange–correlation potential of Kohn–Sham type. Our analysis covers the whole range of variation for these constants and the relations among them appear in a natural way, not being a restriction from a physical point of view. In this spirit, we shall briefly report on the connections among the physical constants and the minimum of energy in Sect. 5. Attractive interactions are also of interest in applications, for instance to quantum gravity (see [28]). We finally remark that most of our results are valid for other X^α-approaches else than Slater's.

2 On the Derivation of the Slater Approach

We now describe how $-C_S\, n_\psi^{\frac{1}{3}}$ enters as a local approximation to the Hartree–Fock exchange potential, which is itself an approximation of the N-electron problem. In this case, the system evolution is represented by an N-electron state wave function $\Psi = \Psi(x_1, \ldots, x_N, t)$ $(x_i \in \Omega = \mathbb{R}^3,\ 1 \le i \le N)$ that belongs to $L_a^2(\Omega^N) = \bigwedge_{i=1}^{N} L^2(\Omega)$, that is the space of L^2 functions which are antisymmetric with respect to the exchange of any two variables x_i and x_j. The antisymmetry assumption stems from the Pauli exclusion principle, which establishes that two electrons cannot share the same position, so that the probability density $|\Psi|^2$ vanishes as $x_i = x_j$. However, for a rigorous treatment the spin variables of the electrons must be considered. We can then assume that the evolution of the system is described by the Schrödinger equation

$$i\hbar \frac{\partial \Psi}{\partial t} = \mathcal{H}\Psi = \left(-\frac{\hbar^2}{2m}\Delta_x + \sum_{j=1}^{N} V_{ext}(x_j) + V_{ee} \right)\Psi\,,$$

where V_{ext} is an external potential (such as interactions with nuclei) and

$$V_{ee} := \sum_{1 \le i < j \le N} \frac{1}{|x_i - x_j|}$$

is the electron–electron Coulomb potential, which mixes the variables x_i. Then, the total Coulomb energy is given by $U(\Psi) := \langle \Psi, V_{ee}\Psi \rangle_{L^2(\Omega^N)}$.

In the Hartree–Fock system, determinantal wave functions

$$\Psi(x_1, \ldots, x_N, t) = \frac{1}{\sqrt{N!}} \det(\psi_i(x_j, t))$$

are used to represent the electrons and guarantee that the system obeys the Pauli principle. This expression contains the one–electron wave functions ψ_i. Now, we assume $\langle \psi_i, \psi_j \rangle = \delta_{ij}$, $1 \leq i, j \leq N$, and skip t for simplicity. Then, classical quantum calculations (see [26]) give $U(\Psi) = E_{coul}(n_\psi) + E_{ex}(\psi)$, where $\psi = (\psi_1, \psi_2, \ldots, \psi_N)$ and $n_\psi = \sum_{j=1}^{N} |\psi_j|^2$ is the density,

$$E_{coul}(n_\psi) = \frac{1}{2} \int_{\Omega \times \Omega} \frac{n_\psi(x) n_\psi(y)}{|x - y|} \, dx \, dy$$

is the direct Coulomb energy and where the exchange energy is defined by

$$E_{ex}(\psi) = -\frac{1}{2} \int_{\Omega \times \Omega} \frac{|D_\psi(x,y)|^2}{|x - y|} \, dx \, dy, \quad D_\psi(x,y) := \sum_{j=1}^{N} \psi_j(x) \overline{\psi_j}(y).$$

The Hartree–Fock equations can then be obtained as extrema points of the energy functional

$$\langle \Psi, \mathcal{H}\Psi \rangle = \frac{\hbar^2}{2m} \sum_{j=1}^{N} \int_\Omega |\nabla_x \psi_j|^2 \, dx + \int_\Omega n_\psi V_{ext} \, dx + U(\Psi)$$

under the constraint $\langle \psi_i, \psi_j \rangle_{L^2(\Omega)} = \delta_{ij}$, which leads to (see [23,25])

$$-\frac{\hbar^2}{2m} \Delta_x \psi_j + V_{ext} \, \psi_j + \left(\frac{1}{|x|} * n_\psi \right) \psi_j + V_{ex} \bullet \psi_j = \epsilon_j \psi_j, \quad 1 \leq j \leq N,$$

where the term V_{ex} is a nonlocal operator defined by

$$(V_{ex} \bullet \psi_j)(x,t) := -\int_\Omega \frac{1}{|x - y|} D_\psi(x,y) \psi_j(y,t) \, dy \qquad (5)$$

and the ϵ_j are constants. The time-dependent Hartree–Fock equations are

$$i\hbar \frac{\partial \psi_j}{\partial t} = -\frac{\hbar^2}{2m} \Delta_x \psi_j + V_{ext} \psi_j + \left(\frac{1}{|x|} * n_\psi \right) \psi_j + (V_{ex} \bullet \psi_j), \quad 1 \leq j \leq N.$$

The Poisson (multiplicative) operator $\frac{1}{|x|} * n_\psi$ as well as the exchange operator V_{ex} can be obtained as functional derivatives of the Coulomb and exchange energies $E_{coul}(n_\psi)$ and $E_{ex}(\psi)$, respectively. Existence and uniqueness of solutions to this problem can be found in [25]. We also refer to [18] for a general existence result for a system coupling the quantum Hartree–Fock equations for electrons with a classical dynamics for the nuclei. For the semiclassical and long-time asymptotics of the Hartree–Fock system, see [15].

Remark 1. In a simpler model for which antisymmetry might not be assumed, such as $\Psi = \prod_{j=1}^{N} \psi(x_j)$ (with $\int |\psi|^2 \, dx = 1$), we would find

$$U(\Psi) = \frac{N(N-1)}{2} \int_{\Omega \times \Omega} \frac{|\psi(x)|^2 |\psi(y)|^2}{|x-y|} \, dx \, dy = \left(\frac{N-1}{N} \right) E_{coul}(n_\psi)$$

with $n_\psi = N|\psi|^2$, thus no exchange term is expected to appear.

There is no exact local expression for V_{ex}. However, V_{ex} can be well approximated by the multiplicative operator $-C_S \, n_\psi^{\frac{1}{3}}$ (for some constant C_S, that can be adjusted to fit experimental evidence) as proposed by Slater [34]. This is also called the "X^α method". We refer to [26] for a review of such sort of approximations and numerical experiments. Justifications of the Slater approximation can be found in [1] (with $\Omega = \mathbb{R}^3$) and [3,4,6,13] for a periodic model (when Ω is a box), in the limit $N \to \infty$ of a large number of particles.

We give now an idea of Slater's proof. Firstly we search for a local approximation $(V_{ex} \bullet \psi_j)(x) \sim V_{av}(x)\psi_j(x)$. However, $\frac{(V_{ex} \bullet \psi_j)(x)}{\psi_j(x)}$ is not a local operator. Following [34], we can approximate V_{ex} by averaging the Hartree–Fock exchange potential by the weighted densities $|\psi_j|^2$, i.e.

$$V_{av}(x) = \sum_{j=1}^{N} \frac{(V_{ex} \bullet \psi_j)(x)}{\psi_j(x)} \frac{|\psi_j(x)|^2}{n_\psi(x)} = -\int_{\Omega} \frac{|D_\psi(x,y)|^2}{|x-y| \, n_\psi(x)} \, dy \ . \tag{6}$$

Note that the formal approximation $V_{ex} \sim V_{av}$ is an exact integral identity, since when the exchange energy is evaluated one finds

$$E_{ex} = \frac{1}{2} \int_{\Omega} V_{av}(x) n_\psi(x) \, dx = \frac{1}{2} \sum_{i=1}^{N} \int_{\Omega} (V_{ex} \bullet \psi_i)(x) \overline{\psi_i}(x) \, dx \ .$$

The next step is to use a plane-wave approximation (see [13]). We assume now that $\Omega = [-\frac{L}{2}, \frac{L}{2}]^3$ is a box with periodic boundary conditions. Denote $n_0 := \frac{N}{|\Omega|}$ the averaged electron density in the box of volume $|\Omega| = L^3$. Now we take as first-order approximations to the single-particle wave functions the plane-wave states

$$\psi_j(x) = \frac{1}{\sqrt{|\Omega|}} e^{ik_j \cdot x} \ ,$$

with $k_j \in \frac{2\pi}{L}\mathbb{Z}^3$ and $k_j \in B_R$, B_R being the (euclidean) ball centered at 0 with smallest radius R in order to minimize the kinetic energy

$$E_{kin} = \sum_{j=1}^{N} \int_{\Omega} |\nabla_x \psi_j|^2 \, dx = \sum_{j=1}^{N} |k_j|^2 \ .$$

In particular $\frac{4}{3}\pi R^3 \sim \left(\frac{2\pi}{L} \right)^3 N = (2\pi)^3 n_0$, so that $R \sim C n_0^{\frac{1}{3}}$ with $C > 0$. Also $\langle \psi_i, \psi_j \rangle_{L^2(\Omega)} = \delta_{ij}$ and $n_\psi = \sum_{j=1}^{N} |\psi_j|^2 = \frac{N}{|\Omega|} = n_0$. As consequence,

$$D_\psi(x,y) = \frac{1}{\sqrt{|\Omega|}} \sum_{j=1}^{N} e^{ik_j \cdot (x-y)} \sim \frac{1}{(2\pi)^3} \int_{B_R} e^{ih \cdot (x-y)} \, dh \ . \qquad (7)$$

This (nontrivial) continuous approximation is related to number-theory results (see [13]). Actually, the right-hand side of (7) is reduced to

$$D_\psi(x,y) = CR^3 f(R(x-y)) \text{ where } f(t) = \frac{\sin(|t|) - |t|\cos(|t|)}{|t|^3} \ .$$

Finally, by using a change of variables in (6) and the periodic boundary conditions, we get

$$V_{av} \sim -\frac{1}{n_0} \int_\Omega \frac{|D_\psi(x, x+h)|^2}{|h|} \, dh \sim -CR \int_{R\Omega} \frac{|f(t)|^2}{|t|} \, dt \ .$$

If we assume $R \to \infty$, we find that V_{av} can be expressed as the product of $R = C n_0^{\frac{1}{3}}$ times a convergent integral: $V_{av} \sim -C n_0^{\frac{1}{3}}$. This approach depends on the dimension. In dimension d we would have $V_{av} \sim -C_d \, n_0^{\frac{1}{d}}$, although for $d \neq 3$ we should also possibly consider other interaction than $\frac{1}{|x|}$.

To conclude, we note that it is also important to understand if the above approximation still holds for varying densities $n_\psi(x) \neq n_0$. We refer to [4, 6] for a proof of $V_{av} \sim -C_S \, n_\psi^{\frac{1}{3}}$ in a first-order approximation, for densities $n_\psi(x)$ close to n_0 and in the limit $N \to \infty$.

Remark 2. The Slater and Dirac approximations of E_{ex} are related through $E_{ex}^{Dirac}(\psi) = -C_D \int_\Omega n_\psi^{\frac{4}{3}} \, dx$, for some $C_D > 0$ (see [10]) and $n_\psi = \sum_{j=1}^{N} |\psi_j|^2$. Then, $D_{\overline{\psi_i}} E_{ex}^{Dirac}(\psi) = -\left(\frac{4}{3} C_D \, n_\psi^{\frac{1}{3}}\right) \psi_i$.

3 Some Results Concerning Well Posedness and Asymptotic Behaviour

In the sequel we shall assume $\hbar = m = 1$ and denote $U(t)$ the propagator of the free Schrödinger Hamiltonian.

3.1 Existence and Uniqueness of Physically Admissible Solutions

Definition 1. *Let $1 \leq p, q \leq \infty$ and $T > 0$.*

(a) We define $L_T^{q,p}(\lambda) := L^q([0, T]; L^p(\lambda))$, $L_{loc}^{q,p}(\lambda) := L_{loc}^q((0, \infty); L^p(\lambda))$ and $L^{q,p}(\lambda) := L^q((0, \infty); L^p(\lambda))$, where

$$L^p(\lambda) := \left\{ \phi = \{\phi_j\}_{j \in \mathbb{N}}, \ \|\phi\|_{L^p(\lambda)} = \left(\sum_{j \in \mathbb{N}} \lambda_j \|\phi_j\|_{L^p(\mathbb{R}^3)}^2 \right)^{\frac{1}{2}} < \infty \right\},$$

(b) If $\frac{1}{p} + \frac{1}{p'} = \frac{1}{q} + \frac{1}{q'} = 1$, we define

$$X_T^{q,p} := L_T^{\infty,2}(\lambda) \cap L_T^{q,p}(\lambda) , \quad Y_T^{q,p} := L_T^{1,2}(\lambda) + L_T^{q',p'}(\lambda) ,$$

equipped with the norms

$$\|\psi\|_{X_T^{q,p}} := \|\psi\|_{L_T^{\infty,2}(\lambda)} + \|\psi\|_{L_T^{q,p}(\lambda)}$$

and

$$\|\psi\|_{Y_T^{q,p}} := \inf_{\psi_1 + \psi_2 = \psi} \left(\|\psi_1\|_{L_T^{1,2}(\lambda)} + \|\psi_2\|_{L_T^{q',p'}(\lambda)} \right) .$$

In the case $T = \infty$ we omit the index T and denote

$$X^{q,p} := L^{\infty,2}(\lambda) \cap L_{loc}^{q,p}(\lambda) , \quad Y^{q,p} := L^{1,2}(\lambda) + L_{loc}^{q',p'}(\lambda) .$$

Also, in the single-state case we omit the index λ and denote $L_T^{q,p}$, $L_{loc}^{q,p}$ and $L^{q,p}$ instead of $L_T^{q,p}(\lambda)$, $L_{loc}^{q,p}(\lambda)$ and $L^{q,p}(\lambda)$.

We now introduce the concept of solutions we shall deal with.

Definition 2 (Mild solution). *Given* $T > 0$, *we say that* $\psi \in X_T^{q,p}$ *is a mild solution of the SPS system if it solves the integral equation*

$$\psi(x,t) = U(t)\phi(x) - i \int_0^t U(t-s) \left(V_\psi \psi - C_S\, n_\psi^{\frac{1}{3}} \psi \right)(x,s)\, ds , \qquad (8)$$

where $V(\psi) = \frac{1}{4\pi|x|} * n_\psi$ *and* $n_\psi = \sum_{j \in \mathbb{N}} \lambda_j |\psi_j|^2$.

It can be seen that (8) makes sense when $\psi \in X_T^{q,p}$ for some well chosen (q,p) (see [5]). Given $\phi \in L^2(\lambda)$, we first aim to search for a fixed point of

$$\Gamma[\psi](x,t) := U(t)\phi(x) - i \int_0^t U(t-s) \left(V_\psi \psi - C_S\, n_\psi^{\frac{1}{3}} \psi \right)(x,s)\, ds .$$

For that, we use the definition of admissible pair given in [9].

Definition 3. *For* $d = 3$, *we say that a pair* (q,p) *is admissible (and denote it by* $(q,p) \in S$*) if* $2 \le p < 6$ *and* $\frac{2}{q} = 3(\frac{1}{2} - \frac{1}{p})$.

Existence and uniqueness of solutions to the initial value problem for the SPS system (see [5]) is established in the following

Theorem 1. *Let* $(q,p) \in S$ *with* $p > 3$ *and* $\phi \in L^2(\lambda)$. *We have*

(a) *There exists a unique global mild solution* $\psi \in X^{q,p} \cap C^0([0,\infty); L^2(\lambda))$ *of the SPS system. Furthermore,*

(i) $\forall t > 0$, $\forall j, k \in \mathbb{N}$, $\langle \psi_j(\cdot, t), \psi_k(\cdot, t) \rangle_{L^2(\mathbb{R}^3)} = \langle \phi_j, \phi_k \rangle_{L^2(\mathbb{R}^3)}$. In partic-
ular, $\|\psi(\cdot, t)\|_{L^2(\lambda)} = \|\phi\|_{L^2(\lambda)}$ (charge preservation).

(ii) If ψ^1, ψ^2 solve the SPS system with initial data $\phi^1, \phi^2 \in L^2(\lambda)$, then

$$\|\psi^2 - \psi^1\|_{X_T^{q,p}} \le C_T \|\phi^2 - \phi^1\|_{L^2(\lambda)} \quad \forall T > 0,$$

where $C_T > 0$ is a constant that only depends on T and $(\|\phi^j\|_{L^2(\lambda)})_{j=1,2}$.

(b) If $\phi \in H^1(\lambda)$, then $\nabla_x \psi \in X^{q,p}$ and the total energy $E[\psi](t)$ is time preserved

(c) If $C_S = 0$ the above results still hold true for $p > 2$ and $(q, p) \in \mathcal{S}$.

Notice that under the assumption $\phi \in H^1(\lambda)$ there are also other pre-served quantities: the linear momentum $\langle p \rangle(t) := \langle \psi, -i\nabla_x \psi \rangle_{L^2(\lambda)}$, the an-gular momentum $\langle j \rangle(t) := \langle \psi, -i(x \wedge \nabla_x)\psi \rangle_{L^2(\lambda)}$ and the boost operator $\langle b \rangle(t) := \langle \psi, (x - it\nabla_x)\psi \rangle_{L^2(\lambda)}$ (see [9] for the single-state case). These invari-ances result via Noether's theorem in conservation laws of the system and play a relevant role in the analysis of the behaviour of solutions.

3.2 Minimum of Energy

For the sake of simplicity we shall deal with the single-state case from now on. It is not surprising that the Slater term gives rise to some qualitative differences in the behaviour of the solutions to the SPS system when com-pared to solutions to the SP system. While the Schrödinger–Poisson energy operator is positive in the repulsive case, it may become negative when the Slater nonlinearity is also considered. Indeed, an application of the inequality

$$\|\nabla_x V_\psi\|_{L^2}^2 \le C \|\psi\|_{L^2}^{\frac{4}{3}} \|\psi\|_{L^{\frac{8}{3}}}^{\frac{8}{3}} \quad \forall \psi \in L^2(\mathbb{R}^3) \cap L^{\frac{8}{3}}(\mathbb{R}^3), \tag{9}$$

C being the best possible constant, yields negativeness for all times of the potential energy associated with the solution $\psi(x, t)$ under the following as-sumption on the initial data ϕ of the SPS initial value problem:

$$\frac{2C}{3} \le \frac{C_S}{\|\phi\|_{L^2(\mathbb{R}^3)}^{\frac{4}{3}}}.$$

Remark 3. The inequality (9) as well as an upper bound for the sharp con-stant C were found by Lieb and Oxford in [22]. In our context, this bound takes the value $C = \frac{1.092}{2\pi} = 0.1737$.

Furthermore, since the potential energy might be initially negative in the repulsive case, we can find initial data for which the total energy is also negative by scaling arguments $(\psi_\sigma(x) = \sigma^{\frac{3}{2}} \psi(\sigma x))$.

The fact that the total energy can reach negative values allows to establish significant differences in the asymptotic behaviour of solutions to the SP and the SPS systems. For the repulsive SP system it is well known the strong dispersion property (see [7, 16]): given $\phi \in L^2$, $\|\psi(\cdot, t)\|_{L^p}$ tends to zero asymptotically in time for $p \in (2, 6]$. At variance, $\|\psi(\cdot, t)\|_{L^{\frac{8}{3}}}$ cannot go to zero as $t \to \infty$ if ψ is a solution to the repulsive SPS system such that $E[\phi] < 0$. This is because $E[\psi]$ is time preserved and the Slater term is the only nonpositive contribution to the total energy.

Motivated by these estimates, one can proceed to analyze the following minimization problem associated with the total energy of the SPS system

$$I_M = \inf \left\{ E[\psi];\ \psi \in H^1(\mathbb{R}^3),\ \|\psi\|_{L^2} = M \right\} . \tag{10}$$

Existence of a minimizer for (10) implies two interesting consequences.

(a) The existence of stationary profiles, which are periodic-in-time solutions to the SPS system preserving the density.
(b) The derivation of optimal bounds for the kinetic energy linked to solutions ψ for which $E[\psi]$ is well-defined.

Observe that things are sensibly different if we deal with the repulsive SP system, since in this case the minimization problem has no solution because the energy infimum is always set to 0, which is not a minimum except for the case $M = 0$. In particular, stationary solutions do not exist.

The technical difficulties arising in the (nonconvex) minimization problem (10) stem from the invariance of $E[\psi]$ by the noncompact group of translations as well as by the change of sign in the terms contributing to the potential energy, which prevents rearrangement techniques to be used. The possible loss of compactness due to that invariance has to be detected by the techniques used in the proofs. In this way, two methods are commonly used in the literature to solve this kind of problems: the concentration–compactness method [24] and the method of nonzero weak convergence after translations [21]. In fact, it can be proved that every minimizing sequence is *in essence* relatively compact provided that a certain subadditivity property is strict. This condition implies that a minimizing sequence is concentrated in a bounded domain. By considering this sort of arguments (under certain technical assumptions) the results stated in the sequel were all proved in [32].

Theorem 2. *Let $C_S, M > 0$ be such that*

$$M < \left(\frac{7 C_S}{10 C} \right)^{\frac{3}{4}}, \tag{11}$$

where C is given by (9). Then there exists a minimizer $\psi_M \in C^\infty(\mathbb{R}^3)$ of (10) which satisfies the following Euler–Lagrange equation associated with the total energy functional $E[\psi]$:

$$-\frac{1}{2}\Delta\psi_M(x) + \frac{1}{4\pi}\int_{\mathbb{R}^3}\frac{|\psi_M(x')|^2\psi_M(x)}{|x-x'|}\,dx' - C_S|\psi_M|^{\frac{2}{3}}\psi_M(x)$$
$$= \beta\psi_M(x) \qquad (12)$$

in a distributional sense, for some $\beta < 0$.

From Theorem 2 it can be deduced the existence of standing waves $\psi(x,t) = e^{-i\beta t}\psi(x)$ as solutions of the SPS system in the repulsive case. Actually, these are time-periodic solutions which preserve the density. For this kind of solutions, the repulsive SPS system is reduced to the time-independent Schrödinger equation

$$\beta\psi = -\frac{1}{2}\Delta\psi + V_\psi\psi - C_S\,n_\psi^{\frac{1}{3}}\psi, \qquad \lim_{|x|\to\infty}\psi = 0\,, \qquad (13)$$

coupled to the Poisson equation

$$\Delta V_\psi = n_\psi\,, \qquad \lim_{|x|\to\infty}V_\psi = 0\,. \qquad (14)$$

Equations (13)–(14) can be rewritten as the Euler–Lagrange equation associated with (10) (cf. (12)). Then, Theorem 2 implies the existence of solutions ψ_M. Let us also note that these solutions do not exist for the repulsive SP system, for which all solutions are dispersive.

3.3 Optimal Kinetic Energy Bounds

Minimization of $E[\psi]$ implies the minimization of

$$T(\psi) = -\frac{1}{4}\frac{(E_P[\psi])^2}{E_K[\psi]}\,.$$

This yields $E[\psi_M] = \frac{1}{2}E_P[\psi_M] = -E_K[\psi_M]$ which is a virial theorem. In the next result we use this fact to deduce optimal bounds for $E_K[\psi]$ depending on $E[\phi]$ and on the minimum of $E[\psi]$.

Proposition 1. We have $E_K^-[\psi] \le E_K[\psi] \le E_K^+[\psi]$ with

$$E_K^\pm[\psi] = -2\,I_M\left(1 - \frac{E[\psi]}{2I_M} \pm \sqrt{1 - \frac{E[\psi]}{I_M}}\right) \qquad (15)$$

for all $\psi \in H^1(\mathbb{R}^3)$ solution to the SPS system, where I_M is given by (10). Furthermore, if M satisfies (11) the energy bounds $E_K^\pm[\psi]$ are optimal.

4 Long-Time Behaviour

We finally deal with the long-time evolution of solutions to the SPS system. The standard arguments usually leading to the derivation of L^p bounds for the solutions to nonlinear Schrödinger equations are fruitless in our case. This is basically due to the fact that the sign of $E_P[\psi]$ depends on the balance between the Coulombian potential and the Slater correction. We then have to introduce new techniques. By arguing as in [30] we first write an equation modeling the dispersion of solutions to the SPS system. Define

$$(\Delta x)^2 := \langle x^2 \rangle(t) - \langle x \rangle^2(t) , \quad (\Delta p)^2 := \langle p^2 \rangle(t) - \langle p \rangle^2(t) ,$$

where x denotes the position operator, $p = \frac{1}{i}\nabla_x$ is the momentum operator and where the symbol $\langle a \rangle$ stands for the expected value of the function (or functional) $a(x,t)$: $\langle a \rangle = \int_{\mathbb{R}^3} \overline{\psi}(x,t)\,a(x,t)\,\psi(x,t)\,dx$. In terms of Δx and Δp it can be proved (see [32]) that the dispersion equation for a SPS solution $\psi(x,t)$ with initial data in $\Sigma = \{u \in H^2,\ xu \in L^2\}$ reads

$$\frac{d^2}{dt^2}(\Delta x)^2(t) = 2\left(E[\psi] - \frac{1}{2}\langle p \rangle^2(t) \right) + (\Delta p)^2(t) ,$$

or equivalently

$$\frac{d^2}{dt^2}\langle x^2 \rangle = 2\left(\frac{1}{2}\langle p^2 \rangle + E[\psi] \right) . \tag{16}$$

Some simple computations starting from (16) lead us to an extended pseudo-conformal law satisfied by the solutions $\psi(x,t)$ to the SPS system:

$$\frac{d}{dt}\left(\|(x + it\nabla_x)\psi\|_{L^2} + t^2 \int_{\mathbb{R}^3} V_\psi n_\psi\,dx - \frac{3}{2}C_S\,t^2 \int_{\mathbb{R}^3} |\psi|^{\frac{8}{3}}\,dx \right)$$
$$= t \int_{\mathbb{R}^3} V_\psi n_\psi\,dx - \frac{3}{2}C_S\,t \int_{\mathbb{R}^3} |\psi|^{\frac{8}{3}}\,dx . \tag{17}$$

Equation (16) allows to deduce two important properties concerning the long time behaviour of positive-energy solutions.

(a) Positive–energy solutions tend to expand unboundedly.
(b) A decay bound for the potential energy can be established.

Proposition 2. *Let $\phi \in \Sigma$ be the initial data of the SPS system such that $E(\phi) > 0$. Then, the system expands unboundedly for large times and the position dispersion $\langle x^2 \rangle(t)$ grows like $O(t^2)$.*

We can straightforwardly deduce lower bounds for the L^p norm of the solutions, which are either positive constants or coincide with the usual decay rates of the free Schrödinger equation, depending on a relation linking the total energy, the mass and the linear momentum.

Corollary 1. *Let $\psi(x,t)$ be a solution of the SPS system with initial data $\phi \in \Sigma$ such that*

$$E[\phi] < \frac{1}{2} \frac{|\langle p \rangle [\phi]|^2}{\|\phi\|_{L^2}} . \tag{18}$$

Then, there exist positive constants C, C' and C'' depending on $\|\phi\|_{L^2}$, $E[\phi]$, $|\langle p \rangle(0)[\phi]|^2$ and p such that

$$\|\psi(t)\|_{L^p(\mathbb{R}^3)} \geq C, \quad E_P[\psi] \leq -C', \quad \forall t \geq 0, \quad \forall p \in \left[\frac{8}{3}, 6\right] . \tag{19}$$

In the case that

$$E[\phi] \geq \frac{1}{2} \frac{|\langle p \rangle [\phi]|^2}{\|\phi\|_{L^2}} , \tag{20}$$

the following lower bound

$$\|\psi(\cdot,t)\|_{L^p(\mathbb{R}^3)} \geq \frac{C''}{t^{\frac{3p-6}{2p}}} , \quad \forall t > \xi > 0, \quad \forall p \in [2,6] , \tag{21}$$

holds. Furthermore, if $E[\phi] < 0$ and $\|\phi\|_{L^2} = 1$ we have

$$(\Delta x)^2 = O(t^2) .$$

The proof combines the dispersion properties with the Galilean invariance of the system (see [5]). In fact, this property guarantees that if $\psi(x,t)$ is a solution to the SPS system with initial data ϕ, then the solution corresponding to the initial data $\psi_N(x,0) = e^{iNx}\phi(x)$, with $N \in \mathbb{R}^3$, is $\psi_N(x,t) = e^{iNx-itN^2}\psi(x-2tN,t)$. The inequalities (18) and (20) are motivated by this property. We also remark that these solutions verify

$$\|\psi_N(\cdot,t)\|_{L^p} = \|\phi\|_{L^p} , \quad p \geq 2$$

and $E_P(\psi_N) = E_P(\phi)$.

The next result provides a positive rate-of-decay estimate for the potential energy. Recall that the potential energy may be negative as shown before. In the case that the initial condition satisfies (20) we find additional information on the potential energy in the following

Proposition 3. *Let ψ be the (unique) solution to the SPS system with initial data $\phi \in \Sigma$. Then*

$$E_P[\psi](t) \leq \frac{C_\xi}{t} \quad \forall t \geq \xi > 0 , \tag{22}$$

where C_ξ is a positive constant depending on ξ.

Consider now the function

$$f_\psi(t) = \|(x+it\nabla)\psi(\cdot,t)\|_{L^2}^2 + t^2 \int_{\mathbb{R}^3} |\nabla_x V(x,t)|^2 \, dx$$

$$- \frac{3}{2} C_S t^2 \int_{\mathbb{R}^3} |\psi(x,t)|^{\frac{8}{3}} \, dx . \tag{23}$$

From (17) and (22) we get

$$f_\psi(t) \leq C + \int_\xi^t \frac{C_\xi s}{s} ds \leq C_\xi t \quad \forall t \geq \xi > 0 .$$

The evolution of f_ψ (more precisely, the evolution of its sign) implies qualitative differences in the behaviour of ψ. The following result provides either a decay estimate for $E_P[\psi]$ (in the attractive case) either a weak decay property for some $L^{p,q}$ norms of ψ.

Corollary 2. Let f_ψ be as in (23). If there exists $t_0 \in \mathbb{R}^+$ such that $f_\psi(t_0) < 0$, then $f_\psi(t) < 0$ for all $t \geq t_0$ and

$$2E_P[\psi] \leq \left(\frac{f_\psi(t_0)}{t_0}\right)\frac{1}{t} < 0 \quad \forall t \geq t_0 .$$

Otherwise, we have

$$\int_\xi^\infty \|\psi(s)\|_{L^p(\mathbb{R}^3)}^{\frac{4p}{3(p-2)}} ds \leq C \quad \forall p \in (2, 6] ,$$

where C is a positive constant depending on p, $\|\phi\|_{L^2}$, $\|x\phi\|_{L^2}$ and $\xi > 0$.

5 On the General X^α Case

As before, $E[\psi]$ also reaches negative values when the Poisson potential is coupled with power nonlinearities $C_\alpha|\psi|^\alpha\psi$ with $\alpha \in (0, \frac{4}{3}]$. Even combinations of some of these terms (with plus or minus sign) make this property to hold (see [19]), leading eventually to a convex energy functional to be minimized via standard techniques for convex operators.

Let us remark that in the interval $\left(0, \frac{4}{3}\right]$ there are two critical (with respect to the minimization of the energy functional) exponents. The Slater exponent $\alpha = \frac{1}{3}$ is critical because for all $0 < \alpha < \frac{1}{3}$ it is easy to prove that the energy may reach negative values. However, in $\left(\frac{1}{3}, \frac{2}{3}\right)$ it can be proved that the energy also assumes negative values if a positive constant $K(\alpha)$ depending upon α exists such that the interaction constant C_α and the total charge $\|\phi\|_{L^2}$ fulfill the following relation

$$\frac{C_\alpha}{\|\phi\|_{L^2}^{4(1-2\alpha)}} > K(\alpha) > 0 .$$

The other critical case for the energy minimization is $\alpha = \frac{2}{3}$ because the total energy operator is not bounded from below when it reaches negative values. This opens the problem of studying the balance of the kinetic energy against the potential energy [31].

References

1. V. Bach: Accuracy of mean field approximations for atoms and molecules. Commun. Math. Phys. **155**, 295 (1993) .

2. C. Bardos: The weak coupling limit of systems of N quantum particles. In: *Euroconference on Asymptotic Methods and Applications in Kinetic and Quantum-Kinetic Theory'*, ed. by L.L. Bonilla, J. Soler, J.L. Vázquez (Granada 2001).

3. O. Bokanowski, B. Grébert, N. Mauser: Approximations de l'énergie cinétique en fonction de la densité pour un modèle de Coulomb périodique. C.R. Acad. Sci., Math. Phys. **329** 85 (1999).

4. O. Bokanowski, B. Grébert, N. Mauser: Local density approximations for the energy of a periodic Coulomb model. Math. Mod. Meth. Appl. Sci. **13** 1185 (2003).

5. O. Bokanowski, J.L. López, J. Soler: On an exchange interaction model for quantum transport: the Schrödinger–Poisson–Slater system. Math. Mod. Meth. Appl. Sci. **13** 1 (2003).

6. O. Bokanowski, N. Mauser: Local approximation for the Hartree–Fock exchange potential: a deformation approach. Math. Mod. Meth. Appl. Sci. **9(6)** 941 (1999).

7. F. Castella: L^2 solutions to the Schrödinger–Poisson system: existence, uniqueness, time behaviour and smoothing effects. Math. Mod. Meth. Appl. Sci. **8** 1051 (1997).

8. I. Catto, P.L. Lions: Binding of atoms and stability of molecules in Hartree and Thomas–Fermi type theories. Part 1: A necessary and sufficient condition for the stability of general molecular systems. Commun. Partial Diff. Equ. **17** 1051 (1992).

9. T. Cazenave: *An Introduction to Nonlinear Schrödinger Equations*, 2nd edn (Textos de Métodos Matemáticos 26, Universidade Federal do Rio de Janeiro 1993).

10. P.A.M. Dirac: Note on exchange phenomena in the Thomas–Fermi atom. Proc. Cambridge Philos. Soc. **26** 376 (1931).

11. R.M. Dreizler, E.K.U. Gross: *Density Functional Theory* (Springer, Berlin Heidelberg New York 1990).

12. R.S. Ellis: *Entropy, Large Deviations and Statistical Mechanics* (Springer, Berlin Heidelberg New York 1985).

13. G. Friesecke: Pair correlation and exchange phenomena in the free electron gas. Comm. Math. Phys. **184** 143 (1997).

14. M.K. Harbola, V. Sahni: Quantum-mechanical interpretation of the exchange-correlation potential of Kohn–Sham density–functional theory. Phys. Rev. Lett. **62** 448 (1989).

15. I. Gasser, R. Illner, P.A. Markowich et al.: Semiclassical, $t \to \infty$ asymptotics and dispersive effects for Hartree–Fock systems. M^{2}AN **32(6)** 699 (1998).

16. R. Illner, P. Zweifel, H. Lange: Global existence, uniqueness and asymptotic behaviour of solutions of the Wigner–Poisson and Schrödinger–Poisson systems. M^{2}AS **17** 349 (1994).

17. W. Kohn, L.J. Sham: Self-consistent equations including exchange and correlation effects. Phys. Rev. **140 A** 1133 (1965).

18. C. LeBris, E. Cancès: On the time-dependent Hartree–Fock equations coupled with a classical nuclear dynamics. Math. Mod. Meth. Appl. Sci. **9(7)** 963 (1999).

19. E.H. Lieb: Thomas–Fermi and related theories of atoms and molecules. Rev. Mod. Phys. **53** 603 (1981).

20. E.H. Lieb: Thomas–Fermi theory. In: *Kluwer Encyclopedia of Mathematics*, suppl. Vol 2 (2000) pp. 311–313.

21. E.H. Lieb, M. Loss: *Analysis* (Graduate Studies in Mathematics, Vol. 14. Amer. Math. Soc., Providence, Rhode Island 2001).

22. E.H. Lieb, S. Oxford: Improved lower bound on the indirect Coulomb energy. Int. J. Quant. Chem. **19** 427 (1981).

23. E.H. Lieb, B. Simon: The Hartree–Fock theory for Coulomb systems. Comm. Math. Phys. **53** 185 (1977).

24. P.L. Lions: The concentration-compactness principle in the calculus of variations. The locally compact case. Ann. Inst. H. Poincaré **1** 109 & 223 (1984).

25. P.L. Lions: Solutions of Hartree–Fock equations for Coulomb systems. Comm. Math. Phys. **109**, 33 (1987).

26. R.G. Parr, W. Yang: *Density Functional Theory of Atoms and Molecules* (Oxford University Press 1989).

27. A. Pazy: *Semigroups of Linear Operators and Applications to Partial Differential Equations* (Springer, Berlin Heidelberg New York 1983).

28. R. Penrose: On gravity's role in quantum state reduction. Gen. Rel. Grav. **28**, 581 (1996).

29. A. Puente, L. Serra: Oscillation modes of two-dimensional nanostructures within the time-dependent local-spin-density approximation. Phys. Rev. Lett. **83**, 3266 (1999).

30. E. Ruíz Arriola, J. Soler: A variational approach to the Schrödinger–Poisson system: asymptotic behaviour, breathers and stability. J. Stat. Phys. **103**, 1069 (2001).

31. O. Sánchez, in preparation.

32. O. Sánchez, J. Soler: Long-time dynamics of the Schrödinger–Poisson–Slater system. J. Stat. Phys. **114**, 179 (2004).

33. O. Sánchez, J. Soler: Asymptotic decay estimates for the repulsive Schrödinger–Poisson system. Math. Mod. Appl. Sci. **27**, 371 (2004).

34. J.C. Slater: A simplification of the Hartree–Fock method. Phys. Rev. **81(3)**, 385 (1951).

Towards the Quantum Brownian Motion

László Erdős[1], Manfred Salmhofer[2], and Horng-Tzer Yau[3]

[1] Institute of Mathematics, University of Munich, Theresienstr. 39, D-80333 Munich
lerdos@mathematik.uni-muenchen.de

[2] Theoretical Physics, University of Leipzig, Augustusplatz 10, D-04109 Leipzig, and Max–Planck Institute for Mathematics, Inselstr. 22, D-04103 Leipzig
Manfred.Salmhofer@itp.uni-leipzig.de

[3] Department of Mathematics, Stanford University, CA-94305, USA
yau@math.stanford.edu

Abstract. We consider random Schrödinger equations on $\mathbf{R}^d$ or $\mathbf{Z}^d$ for $d \geq 3$ with uncorrelated, identically distributed random potential. Denote by λ the coupling constant and ψ_t the solution with initial data ψ_0. Suppose that the space and time variables scale as $x \sim \lambda^{-2-\kappa/2}, t \sim \lambda^{-2-\kappa}$ with $0 < \kappa \leq \kappa_0$, where κ_0 is a sufficiently small universal constant. We prove that the expectation value of the Wigner distribution of ψ_t, $\mathbf{E}W_{\psi_t}(x, v)$, converges weakly to a solution of a heat equation in the space variable x for arbitrary L^2 initial data in the weak coupling limit $\lambda \to 0$. The diffusion coefficient is uniquely determined by the kinetic energy associated to the momentum v.

1 Introduction

Brown observed almost two centuries ago that the motion of a pollen suspended in water was erratic. This led to the kinetic explanation by Einstein in 1905 that Brownian motion was created by the constant "kicks" on the relatively heavy pollen by the light water molecules. Einstein's theory, based upon Newtonian dynamics of the particles, in fact postulated the emergence of the Brownian motion from a classical non-dissipative reversible dynamics. Einstein's theory became universally accepted after the experimental verification by Perrin in 1908, but it was far from being mathematically rigorous.

The key difficulty is similar to the justification of Boltzmann's molecular chaos assumption (Stoßzahlansatz) standing behind Boltzmann's derivation of the Boltzmann equation. The point is that the dissipative character emerges only in a scaling limit, as the number of degrees of freedom goes to infinity.

The first mathematical definition of the Brownian motion was given in 1923 by Wiener, who constructed the Brownian motion as a scaling limit of random walks. This construction was built upon a stochastic microscopic dynamics which by itself are dissipative.

The derivation of the Brownian motion from a Hamiltonian dynamics was not seriously investigated until the end of the seventies, when several

L. Erdős et al.: *Towards the Quantum Brownian Motion*, Lect. Notes Phys. **690**, 233–257 (2006)
www.springerlink.com

results came out almost simultaneously. Kesten and Papanicolaou [16] proved that the velocity distribution of a particle moving in a random scatterer environment (so-called Lorenz gas with random scatterers) converges to the Brownian motion in a weak coupling limit for $d \geq 3$. The same result was obtained in $d = 2$ dimensions by Dürr, Goldstein and Lebowitz [8]. In this model the bath of light particles is replaced with random static impurities. In a very recent work [18], Komorowski and Ryzhik have controlled the same evolution on a longer time scale and proved the convergence to Brownian motion of the position process as well.

Bunimovich-Sinai [5] proved the convergence of the periodic Lorenz gas with a hard core interaction to a Brownian motion. In this model the only source of randomness is the distribution of the initial condition. Finally, Dürr-Goldstein-Lebowitz [7] proved that the velocity process of a heavy particle in a light ideal gas converges to the Ornstein-Uhlenbeck process that is a version of the Brownian motion. This model is the closest to the one in Einstein's kinetic argument.

An analogous development happened around the same time towards the rigorous derivation of the Boltzmann equation. It was proved by Gallavotti [14], Spohn [26] and Boldrighini, Bunimovich and Sinai [3] that the dynamics of the Lorenz gas with random scatterers converges to the linear Boltzmann equation at low density on the kinetic time scale. Lanford [19] has proved that a truly many-body classical system, a low density gas with hard-core interaction, converges to the nonlinear Boltzmann equation for short macroscopic times.

Brownian motion was discovered and theorized in the context of classical dynamics. Since it postulates a microscopic Newtonian model for atoms and molecules, it is natural to replace the Newtonian dynamics with the Schrödinger dynamics and investigate if Brownian motion correctly describes the motion of a quantum particle in a random environment as well. One may of course take first the semiclassical limit, reduce the problem to the classical dynamics and then consider the scaling limit. This argument, however, does not apply to particles (or Lorenz scatterers) of size comparable with the Planck scale. It is physically more realistic and technically considerably more challenging to investigate the scaling limit of the quantum dynamics directly *without any semiclassical limit*. We shall prove that Brownian motion also describes the motion of a quantum particle in this situation. It is remarkable that the Schrödinger evolution, which is time reversible and describes wave phenomena, converges to a Brownian motion.

The random Schrödinger equation, or the quantum Lorentz model, is given by the evolution equation:

$$i\partial_t \psi_t(x) = H\psi_t(x), \qquad H = H_\omega = -\frac{1}{2}\Delta_x + \lambda V_\omega(x) \tag{1}$$

where $\lambda > 0$ is the coupling constant and V_ω is the random potential.

The first time scale with a non-trivial limiting dynamics is the weak coupling limit, $\lambda \to 0$, where space and time are subject to kinetic scaling and the coupling constant scales as

$$t \to t\varepsilon^{-1}, \quad x \to x\varepsilon^{-1}, \quad \lambda = \sqrt{\varepsilon} \,. \tag{2}$$

Under this limit, the appropriately rescaled phase space density (Wigner distribution, see (10) later) of the solution to the Schrödinger evolution (1) converges weakly to a linear Boltzmann equation. This was first established by Spohn (1977) [25] if the random potential is a Gaussian random field and the macroscopic time is small. This method was extended to study higher order correlations by Ho, Landau and Wilkins [13]. A different method was developed in [10] where the short time restriction was removed. This method was also extended to the phonon case in [9] and to the lattice case in [6].

For longer time scales, one expects a diffusive dynamics since the long time limit of a Boltzmann equation is a heat equation. We shall therefore take a time scale longer than in the weak coupling limit (2), i.e. we set $t \sim \lambda^{-2-\kappa}$, $\kappa > 0$. Our aim is to prove that the limiting dynamics of the Schrödinger evolution in a random potential under this scaling is governed by a heat equation. This problem requires to control the Schrödinger dynamics up to a time scale $\lambda^{-2-\kappa}$. This is a much harder task than first deriving the Boltzmann equation from Schrödinger dynamics on the kinetic scale and then showing that Boltzmann equation converges to a diffusive equation under a different limiting procedure. Quantum correlations that are small on the kinetic scale and are neglected in the first limit, may contribute on the longer time scale.

We consider two models in parallel. In the discrete setup we put the Schrödinger equation (1) on $\mathbf{Z}^d$, i.e. we work with the Anderson model [2]. Thus the kinetic energy operator on $\ell^2(\mathbf{Z}^d)$ is given by

$$(\Delta f)(x) := 2d\, f(x) - \sum_{|e|=1} f(x+e) \tag{3}$$

and the random potential is given by

$$V_\omega(x) = \sum_{\gamma \in \mathbf{Z}^d} V_\gamma(x) \,, \qquad V_\gamma(x) := v_\gamma \delta(x - \gamma) \tag{4}$$

where v_γ are real i.i.d. random variables and δ is the lattice delta function, $\delta(0) = 1$ and $\delta(y) = 0$, $y \neq 0$.

In the continuum model we consider the usual Laplacian, $-\frac{1}{2}\Delta_x$, as the kinetic energy operator on $L^2(\mathbf{R}^d)$. The random potential is given by

$$V_\omega(x) = \int_{\mathbf{R}^d} B(x - y)\mathrm{d}\mu_\omega(y) \,, \tag{5}$$

where μ_ω is a Poisson process $\{y_\gamma \ : \ \gamma = 1, 2, \ldots\}$ on $\mathbf{R}^d$ with unit density and i.i.d. random masses, v_γ, i.e. $\mu_\omega = \sum_\gamma v_\gamma \delta(\cdot - y_\gamma)$, and $B : \mathbf{R}^d \to \mathbf{R}$ is a smooth, radially symmetric function with rapid decay, with 0 in the support of $\widehat{B}$.

Since we investigate large distance phenomena, there should be no physical difference between the continuum and discrete models. On the technical level, the discrete model is more complicated due to the non-convexity of the energy surfaces of the discrete Laplacian in momentum space. However, the continuum model also has an additional technical difficulty: the large momentum regime needs a separate treatment.

Our proof builds upon the method initiated in [10]. In that paper the continuum model with a Gaussian random field was considered. Here we also consider the discrete model and non-Gaussian randomness, in order to demonstrate that these restrictions are not essential. On the Boltzmann scale this extension has also been achieved by Chen [6]. The other reason for working on the lattice as well is to make a connection with the extended state conjecture in the Anderson model.

We recall that the Anderson model was invented to describe the electric conduction properties of disordered metals. It was postulated by Anderson that for localized initial data the wave functions for large time are localized for large coupling constant λ and are extended for small coupling constant (away from the band edges and in dimension $d \geq 3$). The localization conjecture was first established rigorously by Goldsheid, Molchanov and Pastur [15] in one dimension, by Fröhlich-Spencer [12], and later by Aizenman-Molchanov [1] in several dimensions, and many other works have since contributed to this field. The extended state conjecture, however, has remained a difficult open problem and only very limited progress has been made.

Most approaches on extended states focused on the spectral property of the random Hamiltonian. It was proved by Klein [17] that all eigenfuctions are extended on the Bethe lattice. In Euclidean space, Schlag, Shubin and Wolff [24] proved that the eigenfunctions cannot be localized in a region smaller than $\lambda^{-2+\delta}$ for some $\delta > 0$ in $d = 2$. Chen [6], extending the method of [10] to the lattice case, proved that the eigenfunctions cannot be localized in a region smaller than λ^{-2} in any dimension $d \geq 2$ with logarithmic corrections. Lukkarinen and Spohn [21] have employed a similar technique for studying energy transport in a harmonic crystal with weakly perturbed random masses.

A special class of random Schrödinger equation was proposed to understand the dynamics in the extended region. Instead of random potential with i.i.d. random variables, one considers a random potential $V_\omega(x)$ with a power law decay, i.e.,

$$V_\omega(x) = h(x)\omega_x , \qquad h(x) \sim |x|^{-\eta}$$

where ω_x are mean zero i.i.d. random variables and $\eta > 0$ is a fixed parameter.

If $\eta \geq 1$ a standard scattering argument yields that for λ small enough H_ω has absolutely continuous spectrum. Using cancellation properties of the random potential, Rodnianski and Schlag [22] have improved the same result to $\eta > 3/4$ in $d \geq 2$ and recently, J. Bourgain [4] has extended it to $\eta > 1/2$.

For $\eta > 1/2$ the particle becomes essentially ballistic at large distances and there are only finitely many effective collisions.

In summary, in all known results [4,6,22,24] for the Anderson model (or its modification) in Euclidean space the number of effective collisions are finite. In the scaling of the current work (13), the number of effective scatterings goes to infinity in the scaling limit, as it should be the case if we aim to obtain a Brownian motion.

As in [6], our dynamical result also implies that the eigenfunctions cannot be localized in a region smaller than $\lambda^{-2-\delta}$ for some $\delta > 0$ and dimension $d \geq 3$ (one can choose $\delta = \kappa/2$ with κ from Theorem 1). Though this result is the strongest in the direction of eigenfunction delocalization, we do not focus on it here.

Our main result is that the time reversible Schrödinger evolution with random impurities on a time scale $\lambda^{-2-\kappa}$ is described by a dissipative dynamics. In fact, this work is the first rigorous result where a heat equation is established from a time dependent quantum dynamics without first passing through a semiclassical limit.

In this contribution we explain the result and the key ideas in an informal manner. The complete proof is given in [11].

2 Statement of Main Result

We consider the discrete and the continuum models in paralell, therefore we work either on the d-dimensional lattice, $\mathbf{Z}^d$, or on the continuous space, $\mathbf{R}^d$. We always assume $d \geq 3$. Let

$$H_\omega := -\frac{1}{2}\Delta + \lambda V_\omega \qquad (6)$$

denote a random Schrödinger operator acting on $\mathcal{H} = l^2(\mathbf{Z}^d)$, or $\mathcal{H} = L^2(\mathbf{R}^d)$. The kinetic energy operator and the random potential are defined in (3)–(5). We assume that $\mathbf{E}v_\gamma = \mathbf{E}v_\gamma^3 = 0$, $\mathbf{E}v_\gamma^2 = 1$ and $\mathbf{E}v_\gamma^{2d} < \infty$.

In the discrete case, the Fourier transform is given by

$$\widehat{f}(p) \equiv (\mathcal{F}f)(p) := \sum_{x \in \mathbf{Z}^d} e^{-2\pi i p \cdot x} f(x) \,,$$

where $p = (p^{(1)}, \ldots, p^{(d)}) \in \mathbf{T}^d := [-\frac{1}{2}, \frac{1}{2}]^d$. Sometimes an integral notation will be used for the normalized summation over any lattice $(\delta\mathbf{Z})^d$:

$$\int (\cdots) \mathrm{d}x := \delta^d \sum_{x \in (\delta\mathbf{Z})^d} (\cdots) \,.$$

The inverse Fourier transform is given by

$$(\mathcal{F}^{-1}\widehat{g})(x) = \dots \int_{(\mathbf{T}/\delta)^d} \widehat{g}(p)e^{2\pi i p \cdot x}\mathrm{d}p \, .$$

In the continuous case the Fourier transform and its inverse are given by

$$(\mathcal{F}f)(p) := \int_{\mathbf{R}^d} e^{-2\pi i p \cdot x} f(x)\mathrm{d}x \, , \qquad (\mathcal{F}^{-1}\widehat{g})(x) = \int_{\mathbf{R}^d} \widehat{g}(p)e^{2\pi i p \cdot x}\mathrm{d}p \, .$$

We will discuss the two cases in parallel, in particular we will use the unified integral notations $\int(\cdots)\mathrm{d}x$ and $\int(\cdots)\mathrm{d}p$. The letters x, y, z will always be used for position space coordinates (hence elements of $(\delta\mathbf{Z})^d$ or $\mathbf{R}^d$). The letters p, q, r, u, v, w denote for d-dimensional momentum variables (elements of $(\mathbf{T}/\delta)^d$ or $\mathbf{R}^d$).

The Fourier transform of the kinetic energy operator is given by

$$\left(\mathcal{F}\left[-\frac{1}{2}\Delta\right]f\right)(p) = e(p)\widehat{f}(p) \, .$$

The dispersion law, $e(p)$, is given by

$$e(p) := \sum_{i=1}^d (1 - \cos(2\pi p^{(i)})), \qquad \text{and} \qquad e(p) := \frac{1}{2}p^2$$

in the discrete and in the continuous case, respectively.

For $h : \mathbf{T}^d \to \mathbf{C}$ and an energy value $e \in [0, 2d]$ we introduce the notation

$$[h](e) := \int h(v)\delta(e - e(v))\mathrm{d}v := \int_{\Sigma_e} h(q)\frac{\mathrm{d}\nu(q)}{|\nabla e(q)|} \tag{7}$$

where $\mathrm{d}\nu(q) = \mathrm{d}\nu_e(q)$ is the restriction of the d-dimensional Lebesgue measure to the level surface $\Sigma_e := \{q \, : \, e(q) = e\} \subset \mathbf{T}^d$. By the co-area formula it holds that

$$\int_0^{2d} [h](e)\mathrm{d}e = \int h(v)\mathrm{d}v \, . \tag{8}$$

We define the projection onto the energy space of the free Laplacian by

$$\langle h(v) \rangle_e := \frac{[h](e)}{\Phi(e)} \, , \qquad \text{where} \qquad \Phi(e) := [1](e) = \int \delta(e - e(u))\mathrm{d}u \, . \tag{9}$$

In the continuous case we define analogous formulas for any function $h : \mathbf{R}^d \to \mathbf{C}$ and energy value $e \geq 0$.

Define the *Wigner transform* of a function $\psi \in L^2(\mathbf{Z}^d)$ or $\psi \in L^2(\mathbf{R}^d)$ via its Fourier transform by

$$W_\psi(x, v) := \int e^{2\pi i w \cdot x} \overline{\widehat{\psi}\left(v - \frac{w}{2}\right)} \widehat{\psi}\left(v + \frac{w}{2}\right)\mathrm{d}w \, .$$

In the lattice case the integration domain is the double torus $(2\mathbf{T})^d$ and x runs over the refined lattice, $x \in (\mathbf{Z}/2)^d$. For $\varepsilon > 0$ define the rescaled Wigner distribution as

$$W_\psi^\varepsilon(X, V) := \varepsilon^{-d} W_\psi\left(\frac{X}{\varepsilon}, V\right).$$ (10)

(with $X \in (\varepsilon\mathbf{Z}/2)^d$ in the lattice case).

The weak coupling limit is defined by the following scaling:

$$\mathcal{T} := \varepsilon t, \quad \mathcal{X} := \varepsilon x, \quad \varepsilon = \lambda^2.$$ (11)

In the limit $\varepsilon \to 0$ the Wigner distribution $W^\varepsilon_{\psi_{\varepsilon^{-1}\mathcal{T}}}(\mathcal{X}, V)$ converges weakly to the Boltzmann equation ([6, 10])

$$\left(\partial_\mathcal{T} + \frac{1}{2\pi}\nabla e(V) \cdot \nabla_\mathcal{X}\right)F_\mathcal{T}(\mathcal{X}, V) = \int dU\sigma(U, V)\left[F_\mathcal{T}(\mathcal{X}, U) - F_\mathcal{T}(\mathcal{X}, V)\right]$$ (12)

where $\frac{1}{2\pi}\nabla e(V)$ is the velocity. The collision kernel is given by

$$\sigma(U, V) := 2\pi\delta(e(U) - e(V)) \qquad \text{discrete case}$$

$$\sigma(U, V) := 2\pi|\widehat{B}(U - V)|^2\delta(e(U) - e(V)) \qquad \text{continuous case}.$$

Note that the Boltzmann equation can be viewed as the generator of a Markovian semigroup on phase space. In particular, the validity of the Boltzmann equation shows that all correlation effects become negligible in this scaling limit.

Now we consider the long time scaling, i.e. with some $\kappa > 0$,

$$x = \lambda^{-\kappa/2-2}X = \varepsilon^{-1}X, \quad t = \lambda^{-\kappa-2}T = \varepsilon^{-1}\lambda^{-\kappa/2}T, \quad \varepsilon = \lambda^{\kappa/2+2}.$$ (13)

Theorem 1. *[Quantum Diffusion on Lattice] Let $d = 3$ and $\psi_0 \in \ell^2(\mathbf{Z}^d)$ be an initial wave function with $\widehat{\psi}_0 \in C^1(\mathbf{T}^d)$. Let $\psi(t) = \psi^\lambda_{t,\omega}$ solve the Schrödinger equation (1). Let $\widetilde{\mathcal{O}}(x, v)$ be a function on $\mathbf{R}^d \times \mathbf{T}^d$ whose Fourier transform in the first variable, denoted by $\mathcal{O}(\xi, v)$, is a C^1 function on $\mathbf{R}^d \times \mathbf{T}^d$ and*

$$\int_{\mathbf{R}^d} d\xi \int dv |\mathcal{O}(\xi, v)| \|\xi\| \leq C.$$ (14)

Fix $e \in [0, 2d]$. Let f be the solution to the heat equation

$$\partial_T f(T, X, e) = \nabla_X \cdot D(e)\nabla_X f(T, X, e)$$ (15)

with the initial condition

$$f(0, X, e) := \delta(X)\left[|\widehat{\psi}_0(v)|^2\right](e)$$

and the diffusion matrix D

$$D_{ij}(e) := \frac{\left\langle \sin(2\pi v^{(i)}) \cdot \sin(2\pi v^{(j)}) \right\rangle_e}{2\pi \, \Phi(e)} \qquad i,j = 1,2,3 \, . \tag{16}$$

Then for $\kappa < 1/2000$ and ε and λ related by (13), the Wigner distribution satisfies

$$\lim_{\varepsilon \to 0} \int_{(\varepsilon \mathbf{Z})^d} dX \int_{\mathbf{R}^d} dv \tilde{O}(X,v) \mathbf{E} W^\varepsilon_{\psi(\lambda - \kappa - 2T)}(X,v) \tag{17}$$

$$= \int_{\mathbf{R}^d} dX \int_{\mathbf{R}^d} dv \, \tilde{O}(X,v) f(T,X,e(v)) \, .$$

By the symmetry of the measure $\langle \cdot \rangle_e$ under each sign flip $v_j \to -v_j$ we see that $D(e)$ is a constant times the identity matrix:

$$D_{ij}(e) = D_e \, \delta_{ij}, \qquad D_e := \frac{\left\langle \sin^2(2\pi v^{(1)}) \right\rangle_e}{2\pi \, \Phi(e)} \, ,$$

in particular we see that the diffusion is nondegenerate.

The diffusion matrix can also be obtained from the long time limit of the Boltzmann equation (12). For any fixed energy e, let

$$L_e f(v) := \int du \, \sigma(u,v)[f(u) - f(v)], \qquad e(v) = e \, , \tag{18}$$

be the generator of the momentum jump process on Σ_e with the uniform stationary measure $\langle \cdot \rangle_e$. The diffusion matrix in general is given by the velocity autocorrelation function

$$D_{ij}(e) = \int_0^\infty dt \, \left\langle \sin(2\pi v^{(i)}(t)) \cdot \sin(2\pi v^{(j)}(0)) \right\rangle_e \, , \tag{19}$$

where $v(t)$ is the process generated by L_e. Since the collision kernel $\sigma(U,V)$ is uniform, the correlation between $v(t)$ and $v(0)$ vanishes after the first jump and we obtain (16), using

$$\int du \, \sigma(u,v) = 2\pi\Phi(e) \, , \qquad e(v) = e \, .$$

The result in the continuum case is analogous. The diffusion matrix is again a constant times the identity matrix, $D_{ij}(e) = D_e \delta_{ij}$, and D_e is again given by the velocity autocorrelation function

$$D_e := \frac{1}{3(2\pi)^2} \int_0^\infty dt \, \langle v(t) \cdot v(0) \rangle_e \tag{20}$$

using the spatial isotropy. In this case D_e cannot be computed as a simple integral since the outgoing velocity u in the transition kernel $\sigma(u,v)$ of the momentum process depends on the direction of v.

Theorem 2. *[Quantum Diffusion on $\mathbf{R}^d$] Let $d = 3$ and $\psi_0 \in L^2(\mathbf{R}^d)$ be an initial wave function with $|\widehat{\psi}_0(v)|^2|v|^N \in L^2$ for a sufficiently large N.*

Let $\psi(t) = \psi_{t,\omega}^\lambda$ solve the Schrödinger equation (1). Let $\widetilde{\mathcal{O}}(x, v)$ be a function whose Fourier transform in x, denoted by $\mathcal{O}(\xi, v)$, is a C^1 function on $\mathbf{R}^d \times \mathbf{R}^d$ and

$$\iint d\xi dv |\mathcal{O}(\xi, v)| \|\xi\| \leq C . \tag{21}$$

Let $e > 0$ and let f be the solution to the heat equation

$$\partial_T f(T, X, e) = D_e \, \Delta_X f(T, X, e) \tag{22}$$

with diffusion constant D_e given in (20) and with the initial condition

$$f(0, X, e) := \delta(X)\left[|\widehat{\psi}_0(v)|^2\right](e) .$$

Then for $\kappa < 1/500$ and ε and λ related by (13), the Wigner distribution satisfies

$$\lim_{\varepsilon \to 0} \iint_{\mathbf{R}^d \times \mathbf{R}^d} dX dv \widetilde{\mathcal{O}}(X, v) \mathbf{E} W_{\psi(\lambda^{-\kappa-2}T)}^\varepsilon(X, v) \tag{23}$$

$$= \iint_{\mathbf{R}^d \times \mathbf{R}^d} dX dv \, \widetilde{\mathcal{O}}(X, v) f(T, X, e(v)) .$$

The main tool of our proof is to use the Duhamel expansion to decompose the wave function into elementary wavefuctions characterized by their collision histories with the random obstacles. Assume for the moment that the randomness is Gaussian and high order expectations can be computed by Wick pairing. The higher order cumulants arising from a non-Gaussian randomness turn out to be negligible by a separate argument. Therefore, when computing the expectation of a product involving ψ and $\bar{\psi}$ (e.g. $\mathbf{E} W_\psi$), we pair the obstacles in the collision histories of ψ and $\bar{\psi}$ and we thus generate Feynman graphs.

If we take only the Laplacian as the free part in the expansion, even the amplitudes of individual graphs diverge in the limit we consider. However, this can be remedied by a simple resummation of all two-legged insertions caused by the lowest order self-energy contribution. The resummation is performed by choosing an appropriate reference Hamiltonian H_0 for the expansion. After this rearrangement, all graphs have a finite amplitude in our scaling limit, and the so-called ladder graphs give the leading contribution.

Each non-ladder graph has a vanishing amplitude as $\lambda \to 0$ due to oscillatory integrals, in contrast to the ladder graphs where no oscillation is present. However, the number of non-ladder graphs grows as $k!$, where $k \sim \lambda^2 t \sim \lambda^{-\kappa}$ is the typical number of collisions. To beat this factorial growth, we need to give a very sharp bound on the individual graphs.

We give a classification of arbitrary large graphs, based on counting the number of vertices carrying oscillatory effects. The number of these vertices is called the *degree* of the graph. For the ladder graphs, the degree is zero. For general graphs, the degree is roughly the number of vertices after removing all ladder and anti-ladder subgraphs. We thus obtain an extra λ^c power (for some $c > 0$) *per non-(anti) ladder vertex*. This strong improvement is sufficient to beat the growth of the combinatorics in the time scale we consider. To our knowledge, nothing like this has been done in a graphical expansion before.

For a comparison, the unperturbed Green functions in the perturbation expansion for the many-fermion systems for small temperature and for the random Schrödinger equation for large time are given by

$$\frac{1}{ip_0 + p^2 - \mu}, \qquad \frac{1}{p^2 - \alpha + i\eta}.$$

In the many-fermion case, $p_0 \in M_F = \{\frac{\pi}{\beta}(2n+1) : n \in \mathbf{Z}\}$ where $\beta \sim T^{-1}$ is the inverse temperature. In the random Schrödinger case, $\eta \sim t^{-1}$. Their L^2 properties are different:

$$\frac{1}{\beta} \sum_{p_0 \in M_F} \int dp |ip_0 + p^2 - \mu|^{-2} \sim |\log \beta|, \qquad \int dp |p^2 - \alpha + i\eta|^{-2} \sim \eta^{-1}$$

Notice the divergence is more severe for the random Schrödinger equation case.

Finally we note that the threshold $\kappa < 1/2000$ in our theorem can be significantly improved with more detailed arguments. However, one cannot go beyond $\kappa = 2$ with only improvements on estimates of the individual graphs. The Duhamel formula must be expanded at least up to $k = \lambda^2 t = \lambda^{-\kappa}$, which is the typical number of collisions up to time t. Even if one proves for most graphs the best possible estimate, λ^{2k}, it cannot beat the $k!$ combinatorics when $k \gg \lambda^{-2}$, i.e., $\lambda^{2k} k! \gg 1$ for $k \gg \lambda^{-2}$. A different resummation procedure is needed beyond this threshold to exploit cancellations among these graphs.

3 Sketch of the Proof

We present the main ideas of the proof for the lattice case and comment on the modifications for the continuous case.

3.1 Renormalization

Before expanding the solution of the Schrödinger equation (1) via the Duhamel formula, we perform a renormalization of the "one-particle propagator" by splitting the Hamiltonian as $H = H_0 + \tilde{V}$, with H_0 already containing the part of the self-energy produced by immediate recollisions with the same obstacle. This effectively resums all such immediate recollisions.

Let $\theta(p) := \Theta(e(p))$, where $\Theta(\alpha) := \lim_{\varepsilon \to 0+} \Theta_\varepsilon(\alpha)$ and

$$\Theta_\varepsilon(\alpha) := \int \frac{1}{\alpha - e(q) + i\varepsilon} dq . \tag{24}$$

We have

$$\operatorname{Im} \Theta(\alpha) = -\pi \Phi(\alpha) \tag{25}$$

with Φ defined in (9).

We rewrite the Hamiltonian as $H = H_0 + \widetilde{V}$ with

$$H_0 := \omega(p) := e(p) + \lambda^2 \theta(p), \qquad \widetilde{V} := \lambda V - \lambda^2 \theta(p) . \tag{26}$$

Our renormalization includes only the lowest order self-energy. This suffices on the time scales we consider.

3.2 The Expansion and the Stopping Rules

Iterating the Duhamel formula

$$e^{-itH} = e^{-itH_0} - i \int_0^t ds\, e^{-i(t-s)H} \widetilde{V} e^{-isH_0} \tag{27}$$

gives for any fixed integer $N \geq 1$

$$\psi_t := e^{-itH} \psi_0 = \sum_{n=0}^{N-1} \psi_n(t) + \Psi_N(t) , \tag{28}$$

with

$$\psi_n(t) := (-i)^n \int_0^t [ds_j]_1^{n+1}\ e^{-is_{n+1}H_0} \widetilde{V} e^{-is_n H_0} \widetilde{V} \ldots \widetilde{V} e^{-is_1 H_0} \psi_0 \tag{29}$$

being the fully expanded terms and

$$\Psi_N(t) := (-i) \int_0^t ds\, e^{-i(t-s)H} \widetilde{V} \psi_{N-1}(s) \tag{30}$$

is the non-fully expanded or error term. We used the shorthand notation

$$\int_0^t [ds_j]_1^n := \int_0^t \ldots \int_0^t \left(\prod_{j=1}^n ds_j \right) \delta \left(t - \sum_{j=1}^n s_j \right) .$$

Since each potential $\widetilde{V}$ in (29), (30) is a summation itself, $\widetilde{V} = -\lambda^2 \theta(p) + \lambda \sum_\gamma V_\gamma$, both of these terms in (29) and (30) are actually big summations over so-called elementary wavefunctions, which are characterized by their collision history, i.e. by a sequence of obstacles labelled by $\gamma \in \mathbf{Z}^d$ and a label ϑ corresponding to an insertion of $-\lambda^2 \theta(p)$.

Because this expansion is generated by iteration of (27), the sequences defining collision histories can be obtained recursively. This allows us to refine the Duhamel expansion by using stopping rules that depend on the type of collision history. We call a sequence *nonrepetitive* if the only repetitions in potential labels γ occur in gates (immediate recollisions). The iteration of (27) is stopped when adding a new entry to the sequence makes it violate this condition. This can happen because of a recollision, a nested recollision, or a triple collision. The precise definition of these recollision types is given in [11]. The iteration is also stopped when the last entry in the sequence causes the total number of gates and ϑ's to reach 2. If the sequence stays nonrepetitive and the total number of gates and ϑ's stays below 2, the iteration is stopped when the number of non-gate potential labels reaches

$$K = \lambda^{-\delta}(\lambda^2 t) \ . \tag{31}$$

Note that K is much bigger than the expected typical number of collisions, $\lambda^2 t$.

We denote the sum of the truncated elementary non-repetitive wave functions with at most one λ^2 power from the non-skeleton indices or ϑ's and with K skeleton indices by $\psi_{*s,K}^{(\leq 1),nr}$. The superscript (≤ 1) refers to the number of gates and ϑ's, each of which gives a factor λ^2. By this splitting, we arrive at the following modified Duhamel formula, in which all non-error terms are nonrepetitive.

Proposition 1. *[Duhamel formula] For any $K \geq 1$ we have*

$$\psi_t = e^{-itH}\psi_0 = \sum_{k=0}^{K-1} \psi_{t,k}^{(\leq 1),nr}$$

$$\tag{32}$$

$$+ \int_0^t ds \, e^{-i(t-s)H} \left\{ \psi_{*s,K}^{(\leq 1),nr} + \sum_{k=0}^{K} \left(\psi_{*s,k}^{(2),last} + \psi_{*s,k}^{(\leq 1),rec} + \psi_{*s,k}^{(\leq 1),nest} + \psi_{*s,k}^{(\leq 1),tri} \right) \right\}$$

The terms under the integral correspond to the various stopping criteria indicated above. For the precise definition of the corresponding wave functions, see [11].

The main contribution comes from the non-repetitive sequences with $k < K$, i.e. from the first term in (32). The estimate of the terms in the second line (32) first uses the unitarity of the full evolution

$$\left\| \int_0^t ds \, e^{-i(t-s)H} \psi_s^{\#} \right\| \leq t \cdot \sup_{s \leq t} \|\psi_s^{\#}\| \ . \tag{33}$$

For $\# = rec, nest, tri$ we will use the fact the Feynman graphs arising in the expectation $\mathbf{E}\|\psi_s^\#\|^2$ contain an additional oscillatory factor, which renders them smaller than the corresponding non-repetitive term. It turns out that the oscillation effect from one single recollision, nest or triple collision is already sufficient to overcome the additional factor t arising from the crude bound (33). This fact relies on estimates on singular integrals concentrating on the energy level sets Σ_e. It is a well-known fact from harmonic analysis, that such singular integrals can more effectively be estimated for convex level sets. This is why the non-convexity of the energy shells is a major technical complication for the discrete model in comparison with the continuous case, where the level sets are spheres.

Non-skeleton labels also give rise to a smallness effect due to a cancellation between gates and ϑ's, however, one such cancellation would not be sufficient to beat the t–factor. This is why at least two such cancellations are necessary in $\psi_{*s,k}^{(2),last}$. Finally, the term $\psi_{*s,K}^{(\le 1),nr}$ is small because it has unusually many collisions, thanks to the additional factor $\lambda^{-\delta}$ in the definition of K.

In this exposition we focus only on the non-repetitive terms, $\psi_{t,k}^{(\le 1),nr}$, because estimating them involves the main new ideas. The error terms are estimated by laborious technical modifications of these ideas.

3.3 The L^2 Norm of the Non-Repetitive Wavefunction

We first estimate the L^2 norm of the fully expanded wave function with no gates or ϑ, $\psi_{t,k}^{(0),nr}$. This is the core of our analysis.

Feynman Graphs

The wavefunction

$$\psi_{t,k}^{(0),nr} = \sum_\gamma \int_0^t [ds_j]_1^{k+1} \; e^{-is_{k+1}H_0} V_{\gamma_k} e^{-is_k H_0} V_{\gamma_{k-1}} \ldots e^{-is_2 H_0} V_{\gamma_1} e^{-is_1 H_0} \psi_0$$

where the summation is over all sequences for which the potential labels γ_i are all different. Therefore every term in

$$\mathbf{E}\|\psi_{t,k}^{(0),nr}\|^2 = \sum_{\gamma,\gamma'} \mathbf{E} \int \overline{\psi_{t,\gamma}} \psi_{t,\gamma'}$$

has $2k$ potential terms, and their expectation,

$$\mathbf{E} \, \overline{V_{\gamma_1} V_{\gamma_2} \ldots V_{\gamma_k}} V_{\gamma_1'} V_{\gamma_2'} \ldots V_{\gamma_k'} ,$$

is zero, using $\mathbf{E}V_\gamma = 0$, unless the potentials are paired. Since there is no repetition within γ and γ', all these pairings occur between γ and γ', therefore every pairing corresponds to a permutation on $\{1, 2, \ldots, k\}$. The set of such permutations is denoted by $\mathcal{P}_k$ and they can be considered as a map between the indices of the γ and γ' labels.

We recall the following identity from Lemma 3.1 of [10]

$$\int_0^t [ds_j]_1^{k+1} \prod_{j=1}^{k+1} e^{-is_j\omega(p_j)} = \frac{ie^{\eta t}}{2\pi} \int_{\mathbb{R}} d\alpha\, e^{-i\alpha t} \prod_{j=1}^{k+1} \frac{1}{\alpha - \omega(p_j) + i\eta}$$

for any $\eta > 0$. We will choose $\eta := t^{-1}$. Therefore, we have

$$\mathbf{E}\|\psi_{t,k}^{(0),nr}\|^2 = \frac{\lambda^{2k}e^{2t\eta}}{(2\pi)^2} \sum_{\sigma \in \mathcal{P}_k} \sum_{\substack{\gamma_1,\ldots,\gamma_k \\ \gamma_i \neq \gamma_j}} \int d\mathbf{p}d\tilde{\mathbf{p}}\, \delta(p_{k+1} - \tilde{p}_{k+1})$$

(34)

$$\times\, \mathbf{E} \prod_{j=1}^{k} \overline{\widehat{V}_{\gamma_j}(p_{j+1} - p_j)} \widehat{V}_{\gamma_j}(\tilde{p}_{\sigma(j)+1} - \tilde{p}_{\sigma(j)}) M(k,\mathbf{p},\tilde{\mathbf{p}},\eta) \overline{\widehat{\psi}_0(p_1)} \widehat{\psi}_0(\tilde{p}_1)$$

with $\mathbf{p} = (p_1, p_2, \ldots, p_{k+1})$, $\int d\mathbf{p} := \int_{(\mathbf{T}^d)^{k+1}} dp_1 dp_2 \ldots dp_{k+1}$, similarly for $\tilde{\mathbf{p}}$ and $d\tilde{\mathbf{p}}$, and

$$M_\eta(k,\mathbf{p},\tilde{\mathbf{p}}) := \int\!\!\int_{\mathbb{R}} d\alpha d\beta\, e^{i(\alpha-\beta)t} \left(\prod_{j=1}^{k+1} \frac{1}{\alpha - \overline{\omega}(p_j) - i\eta} \frac{1}{\beta - \omega(\tilde{p}_j) + i\eta} \right).$$

(35)

We compute the expectation:

$$\mathbf{E} \prod_{j=1}^{k} \overline{\widehat{V}_{\gamma_j}(p_{j+1} - p_j)} \widehat{V}_{\gamma_j}(\tilde{p}_{\sigma(j)+1} - \tilde{p}_{\sigma(j)})$$

$$= \sum_{\substack{\gamma_1,\ldots,\gamma_k \\ \gamma_i \neq \gamma_j}} \prod_{j=1}^{k} e^{i\gamma_j(p_{j+1}-p_j-(\tilde{p}_{\sigma(j)+1}-\tilde{p}_{\sigma(j)}))}.$$

(36)

In the continuous model this formula also contains a product of $\widehat{B}$-terms, where B was the single site potential function in (5). These factors are included into the definition of M_η. Most importantly, they provide the necessary decay in the momentum variables in the case of non-compact momentum space. A similar idea was used in [10].

Due to the restriction $\gamma_i \neq \gamma_j$, (36) is not a simple product of delta functions in the momenta. We have to use a connected graph expansion that is well known in the polymer expansions of field theory (see, e.g. [23]). We do not give the details here, we only note that the result is a weighted sum over partitions of the index set $\{1, \ldots, k\}$. Each term in the sum is a product of delta functions labelled by the lumps of the partition and each delta funtion imposes the Kirchoff Law for the incoming and outgoing momenta of the lump and its σ-image. The trivial partition, where each lump has a single element, carries the main contribution. Estimating the terms with nontrivial partitions can be reduced to estimates for the trivial partition [11]. We therefore discuss only the contribution from the trivial partition to $\mathbf{E}\|\psi_{t,k}^{(0),nr}\|^2$, given by $\sum_{\sigma \in \mathcal{P}_k} V_\eta(k,\sigma)$, where

$$V_\eta(k,\sigma) := \frac{\lambda^{2k} e^{2t\eta}}{(2\pi)^2} \int \mathrm{dp}\mathrm{d}\tilde{\mathbf{p}} M_\eta(k,\mathbf{p},\tilde{\mathbf{p}}) \delta(\tilde{p}_{k+1} - p_{k+1}) \overline{\widehat{\psi}_0(p_1)} \widehat{\psi}_0(\tilde{p}_1)$$

$$\times \prod_{i=1}^{k} \delta\left(p_{i+1} - p_i - (\tilde{p}_{\sigma(i)+1} - \tilde{p}_{\sigma(i)})\right) \tag{37}$$

This complicated formula can be encoded by a Feynman graph and $V_\eta(k,\sigma)$ is called the value or amplitude of the graph. The Feynman graph for the trivial partition corresponds to the usual Feynman graphs for the Gaussian case discussed in [10] and we briefly describe their construction. A Feynman graph consists of two directed horizontal lines (upper and lower) with k collision vertices on each that represent the collision histories of $\bar{\psi}$ and ψ, respectively. These two lines are joined at the two ends. This corresponds to evaluating the L^2-norm on one end and inserting the initial wavefunction ψ_0 on the other end. Each horizontal segment carries a momentum, $p_1, p_2, \ldots p_{k+1}$ and $\tilde{p}_1, \tilde{p}_2, \ldots \tilde{p}_{k+1}$ and a corresponding (renormalized) propagator, $(\alpha - \overline{\omega}(p_j) - i\eta)^{-1}$ and $(\beta - \omega(\tilde{p}_j) + i\eta)^{-1}$. Here α and β are the dual variables to the time on each line and they will be integrated out, see (35). Finally, the collision vertices are paired. Each pairing line joins an upper and a lower vertex and thus can be encoded with a permutation $\sigma \in \mathcal{P}_k$. It is useful to think of the momenta as flowing through the lines of the graph. The delta function associated to each pairing line in the value of the graph (37) then expresses the Kirchhoff Law for the flow of momenta adjacent to the two vertices.

A typical graph with trivial partition is shown on Fig. 1. For the special case of the identity permutation $\sigma = id$ we obtain the so-called ladder graph (Fig. 2). The following proposition shows that the ladder gives the main contribution.

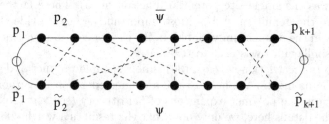

Fig. 1. Typical Feynman graph with no lumps

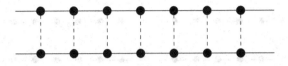

Fig. 2. Ladder graph

The Main Contribution is the Ladder

Proposition 2 (L^2-estimate). *Let $\eta^{-1} := t$, $t = O(\lambda^{2+\kappa})$ and $k \leq K := \lambda^{-\delta}(\lambda^2 t)$. For sufficiently small λ, κ and δ there exists a positive number $c_1(\kappa, \delta)$ such that*

$$\mathbf{E}\|\psi_{t,k}^{(0),nr}\|^2 = V_\eta(k, id) + O_\delta\left(\lambda^{c_1(\kappa,\delta)}\right). \tag{38}$$

The threshold values for κ, δ and the explicit form of $c_1(\kappa, \delta)$ are found in [11].

Sketch of the proof. As mentioned above, we discuss only how to estimate the contributions from the trivial partition, but for an arbitrary permutation σ.

Given a permutation $\sigma \in \mathcal{P}_k$, we define a $(k+1) \times (k+1)$ matrix $M = M(\sigma)$ as follows

$$M_{ij}(\sigma) := \begin{cases} 1 & \text{if} & \tilde{\sigma}(j-1) < i \leq \tilde{\sigma}(j) \\ -1 & \text{if} & \tilde{\sigma}(j) < i \leq \tilde{\sigma}(j-1) \\ 0 & \text{otherwise} \end{cases} \tag{39}$$

where, by definition, $\tilde{\sigma}$ is the extension of σ to a permutation of $\{0, 1, \ldots, k+1\}$ by $\tilde{\sigma}(0) := 0$ and $\tilde{\sigma}(k+1) := k+1$. It is easy to check that

$$V_\eta(k, \sigma) := \frac{\lambda^{2k} e^{2t\eta}}{(2\pi)^2} \int \mathrm{d}\mathbf{p}\mathrm{d}\tilde{\mathbf{p}}\, M_\eta(k, \mathbf{p}, \tilde{\mathbf{p}}) \prod_{i=1}^{k+1} \delta\left(\tilde{p}_i - \sum_{j=1}^{k+1} M_{ij} p_j\right), \tag{40}$$

in other words, the matrix M encodes the dependence of the $\tilde{p}$-momenta on the p-momenta. This rule is transparent in the graphical representation of the Feynman graph: the momentum p_j appears in those $\tilde{p}_i$'s which fall into its "domain of dependence", i.e. the section between the image of the two endpoints of p_j, and the sign depends on the ordering of these images (see Fig. 3).

The matrix $M(\sigma)$ has several properties that follow easily from this structure:

Lemma 1. *For any permutation $\sigma \in \mathcal{P}_k$ the matrix $M(\sigma)$ is*
  (i) invertible;
  (ii) totally unimodular, i.e. any subdeterminant is 0 or ± 1.

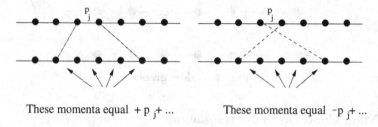

These momenta equal $+ p_j + \dots$ These momenta equal $- p_j + \dots$

Fig. 3. Domain of momenta dependencies

The following definition is crucial. It establishes the necessary concepts to measure the complexity of a permutation.

Definition 1 (Peak, valley and slope). *Given a permutation $\sigma \in \mathcal{P}_k$ let $\tilde{\sigma}$ be its extension. A point $(j, \sigma(j))$, $j \in I_k := \{1, 2, \dots, k\}$, on the graph of σ is called* **peak** *if $\tilde{\sigma}(j-1) > \sigma(j) < \tilde{\sigma}(j+1)$, it is called* **valley** *if $\tilde{\sigma}(j-1) < \sigma(j) > \tilde{\sigma}(j+1)$, otherwise it is called* **slope***. Additionally, the point $(k+1, k+1)$ is also called valley. The set $I = \{1, 2, \dots, k+1\}$ is partitioned into three disjoint subsets, $I = I_p \cup I_v \cup I_s$, such that $i \in I_p, I_v$ or I_s depending on whether $(\tilde{\sigma}^{-1}(i), i)$ is a peak, valley or slope, respectively. Finally, an index $i \in I_v \cup I_s$ is called* **ladder index** *if $|\tilde{\sigma}^{-1}(i) - \tilde{\sigma}^{-1}(i-1)| = 1$. The set of ladder indices is denoted by $I_\ell \subset I$ and their cardinality is denoted by $\ell = \ell(\sigma) := |I_\ell|$. The number of non-ladder indices, $d(\sigma) := k+1 - \ell(\sigma)$ is called the* **degree** *of the permutation σ.*

Remarks: (i) The terminology of peak, valley, slope, ladder comes from the graph of the permutation $\tilde{\sigma}$ viewed as a function on $\{0, 1, \dots, k+1\}$ in a coordinate system where the vertical axis is oriented downward.

(ii) For $\sigma = id$ we have $I_p = \emptyset$, $I_s = \{1, 2, \dots, k\}$, $I_v = \{k+1\}$ and $I_\ell = \{1, 2, \dots, k+1\}$. In particular, $d(id) = 0$ and $d(\sigma) > 0$ for any other permutation $\sigma \neq id$.

The following theorem shows that the degree of the permutation $d(\sigma)$ measures the size of $V_\eta(k, \sigma)$. This is the key theorem in our method and we will sketch its proof separately in Sect. 3.4.

Theorem 3. *Let $\eta^{-1} := t$, $t = O(\lambda^{2+\kappa})$ with a sufficiently small κ. Let $\sigma \in \mathcal{P}_k$ and assume that $k \leq K = \lambda^{-\delta}(\lambda^2 t)$. For sufficiently small κ and δ there exists $c_2(\kappa, \delta) > 0$ such that*

$$|V_\eta(k, \sigma)| \leq \left(C\lambda^{c_2(\kappa,\delta)} \right)^{d(\sigma)}, \qquad \lambda \ll 1. \qquad (41)$$

This theorem is complemented by the following lemma:

Lemma 2. *Let $k = O(\lambda^{-\kappa-\delta})$, $d > 0$ integer and let $\gamma > \kappa + \delta$. Then*

$$\sum_{\substack{\sigma \in \mathcal{P}_k \\ d(\sigma) \geq d}} \lambda^{\gamma d(\sigma)} \leq O\left(\lambda^{d(\gamma-\kappa-\delta)}\right) \tag{42}$$

for all sufficiently small λ.

The proof follows from the combinatorial estimate on the number of permutations with a given degree:

$$\#\{\sigma \in \mathcal{P}_k \ : \ d(\sigma) = d\} \leq (Ck)^d .$$

From Theorem 3 and Lemma 2 we immediately obtain an estimate on the contribution of the trivial lumps to $\mathbf{E}\|\psi_{t,k}^{(0),nr}\|^2$ if κ and δ are sufficiently small:

$$\sum_{\substack{\sigma \in \mathcal{P}_k \\ \sigma \neq id}} |V_\eta(k,\sigma)| \leq O_\delta\left(\lambda^{c_3(\kappa,\delta)}\right) \tag{43}$$

with some appropriate $c_3(\kappa, \delta) > 0$.

3.4 Sketch of the Proof of the Main Technical Theorem

In this section we explain the proof of Theorem 3. We set

$$E_\eta(M) := \lambda^{2k} \iint_{-4d}^{4d} d\alpha d\beta \int d\mathbf{p} \prod_{i=1}^{k+1} \frac{1}{|\alpha - \overline{\omega}(p_i) - i\eta|}$$

$$\times \prod_{j=1}^{k+1} \frac{1}{|\beta - \omega(\sum_{\ell=1}^{k+1} M_{j\ell}p_\ell) + i\eta|} . \tag{44}$$

For the continuous model, the definition includes the $\widehat{B}$ factors to ensure the integrability for the large momentum regime. It is easy to check that $V_\eta(k, \sigma)$ is estimated by $E_\eta(M(\sigma))$ modulo constant factors and negligible additive terms coming from the regime where α or β is big.

The denominators in this multiple integral are almost singular in certain regimes of the high dimensional space of all momenta. The main contribution comes from the overlap of these singularities. The overlap structure is encoded in the matrix M, hence in the permutation σ, in a very complicated entangled way. Each variable p_j may appear in many denominators in (44), so successive integration seems very difficult. We could not find the exact order (as a power of λ) of this multiple integral but we conjecture that true order is essentially $\lambda^{2d(\sigma)}$. Our goal in Theorem 3 is to give a weaker bound of order $\lambda^{cd(\sigma)}$, i.e. that is still a λ-power linear in the degree, but the coefficient considerably smaller than 2.

Notice that the α-denominators in (44) correspond to the columns of M and the β-denominators corresponds to the rows. For this presentation we will use j to label row indices and i to label column indices. We recall the sets I_v, I_p, I_ℓ from Definition 1 and we will view these sets as subsets of the row indices of M.

First we notice that if $j \in (I_\ell \setminus I_v)$, i.e. j is a non-valley ladder row, then there exists a column index $i = c(j)$ such that the momentum p_i appears only in the j-th β-denominator. In other words, the i-th column of M has a single nonzero element (that is actually ± 1) and it is in the j-th row. Therefore the $\mathrm{d}p_i$ integral can be performed independently of the rest of the integrand by using the following elementary but quite involved bound for small κ:

$$\sup_{w,\alpha,\beta} \int_{\mathbf{T}^d} \mathrm{d}p_i \frac{\lambda^2}{|\alpha - \overline{\omega}(p_i) - i\eta|\, |\beta - \omega(\pm p_i + w) + i\eta|} \leq 1 + O(\lambda^{1/4}) \, . \quad (45)$$

Note that the constant of the main term is exactly 1. This fact is important, since in graphs with low degree this estimate has to be raised to a power $|I_\ell \setminus I_v|$ that may be comparable with k. Clearly for $k \leq K \sim \lambda^{-\kappa-\delta}$ and $\kappa + \delta < 1/4$ we have

$$\left(1 + O(\lambda^{1/4})\right)^k \leq const \, , \quad (46)$$

but had 1 been replaced with a bigger constant in (45), we would obtain an exponentially big factor $(const)^k$ that would not be affordable. The precise constant 1 in the estimate (45) is related to the appropriate choice of the renormalization $\theta(p)$ in $\omega(p)$.

After the non-valley ladder rows are integrated out, and the corresponding rows and columns are removed from the matrix M, we obtain a smaller matrix $M^{(1)}$ describing the remaining denominators. In $M^{(1)}$ we keep the original labelling of the rows from M.

Now we estimate some of the β-denominators in (44) by L^∞ norm, i.e. by η^{-1}. This is a major overestimate, but these denominators are chosen in such a way that the entangled structure imposed by M becomes much simpler and many other denominators can be integrated out by L^1-bounds that are only logarithmic in λ.

We start with estimating all β-denominators in rows $j \in I_p$ by the trivial L^∞-norm. The corresponding rows are removed from $M^{(1)}$, in this way we obtain a matrix $M^{(2)}$. Let

$$I^* := I \setminus \left(I_p \cup (I_\ell \setminus I_v)\right)$$

be the remaining row indices after removing the peaks and the non-valley ladders.

Then we inspect the remaining rows $j \in I^*$ of $M^{(2)}$ in increasing order. The key observation is that for each $j \in I^*$ there exists a column index, $i = c(j)$, such that the variable p_i appears **only** in the j-th β-denominator, provided that all β-denominators with $j' < j$ have already been integrated out. In view of the structure of $M^{(1)}$, it means that for any $j \in I^*$ there exists a column $i = c(j)$ such that the only nonzero element among $\{M^{(2)}_{ij'} : j' \geq j\}$ is $M^{(2)}_{ij}$. This fact follows from the structure of $M(\sigma)$ and from the fact that all rows with $j \in I_p$ have been removed.

This property allows us to remove each remaining β-denominator, one by one, by estimating integrals of the type

$$\int_{\mathbf{T}^d} dp_i \, \frac{1}{|\alpha - \overline{\omega}(p_i) - i\eta|} \frac{1}{|\beta - \omega(\pm p_i + w) + i\eta|} \leq \frac{C\eta^{-\tau}}{|w|} , \tag{47}$$

where w is a linear combination of momenta other than p_i. The absolute value $|w|$ is interpreted as the distance of w from the nearest critical point of the dispersion relation $e(p)$. The variable p_i at this stage of the procedure appears only in these two denominators.

The exponent τ can be chosen zero (with logarithmic corrections) for the continuous model and this fact has already been used in [10]. For the discrete model we can prove (47) with $\tau = 3/4 + 2\kappa$ and we know that the exponent cannot be better than $1/2$. The reason for the weaker estimate is the lack of convexity of the level set Σ_e. Replacing $\omega(p)$ with $e(p)$ for a moment, the inequality (47) with $\tau = 0$ essentially states that the level set $\{\alpha = e(p)\}$ and its shifted version $\{\beta = e(p + w)\}$ intersect each other transversally, unless w is close to zero. Indeed, the transversal intersection guarantees that the volume of the p values, where *both* denominators are of order η, is of order η^2. Then a standard argument with dyadic decomposition gives the result with a logarithmic factor. For translates of spheres the transversal intersection property holds, unless $w \sim 0$. However, in certain points of the level sets Σ_e of the discrete dispersion relation the curvature vanishes, in fact Σ_e even contains straight lines for $2 \leq e \leq 4$. The transversal intersection fails in certain regions and results in a weaker bound.

Neglecting the point singularity $|w|^{-1}$ in (47) for a moment (see Sect. 3.5 later), we easily see that with this algorithm one can bound $E_\eta(M(\sigma))$ by $\lambda^{2(k-q)}\eta^{-p}\eta^{-\tau(k-p-q)}$, modulo logarithmic factors, where $p = |I_p|$ is the number of peak indices and $q := |I_\ell \setminus I_v|$ is the number of non-valley ladder indices. From the definitions it follows that the sets I_v, I_p and $I_\ell \setminus I_v$ are disjoint and $|I_v| = p + 1$. Thus we have $2p + 1 + q \leq k + 1$. Therefore

$$\lambda^{2(k-q)}\eta^{-p(1-\tau)-\tau(k-q)} \leq (\lambda^4 t^{\tau+1})^{(k-q)/2} \leq (\lambda^4 t^{\tau+1})^{\frac{1}{2}[d(\sigma)-1]} \tag{48}$$

since $q \leq \ell$. If $\tau < 1$, then with a sufficiently small κ we see that $\lambda^4 t^{\tau+1}$ is a positive power of λ. Thus we obtain a bound where the exponent of λ is linear in $d(\sigma)$. With a more careful estimate one can remove the additional -1 in the exponent. In particular, for the continuous case with $\tau = 0$ this argument works up to $\kappa < 2$.

We end this section with a remark. Apparently the bound $\kappa < 2$ (or, equivalently, $t \ll \lambda^{-4}$) shows up in two different contexts in this argument. To avoid misunderstandings, we explain briefly that neither of these two appearences is the genuine signature of the expected threshold $\kappa = 2$ for our expansion method to work. The true reason is the one mentioned in the introduction: even the best possible bound, $\lambda^{2d(\sigma)}$, on the graph with permutation σ, cannot beat the $k!$ combinatorics of the graphs beyond $\kappa = 2$.

In the argument above, on one hand, $\kappa < 2$ is related to the error term in the ladder calculation (45). This error term can be improved to $\lambda^2 |\log \lambda|$ and it is apparently due to the fact that the renomalization term $\theta(p)$ was solved only up to lowest order. An improvement may be possible by including more than the lowest order of the self-energy.

The second apperance of $\kappa < 2$, or $t \ll \lambda^{-4}$, at least for the continuous model, is in (48) and it is due to the fact that certain β-denominators are overestimated by L^∞. This is again a weakness of our method; we did overestimates in order to simplify the integrand.

3.5 Point Singularities

The argument in the previous section has neglected the point singularity arising from (47). While a point singularity is integrable in $d \geq 3$ dimensions, it may happen that exactly the same linear combinations of the independent variables keep on accumulating by the repeated use of the bound (47). In that case at some point a high negative power of $|w|$ needs to be integrated. While it is possible to improve the estimate (47) by changing the denominator on the right hand side to $|w| + \eta$, this would still yield further negative η-powers.

It is easy to see that this phenomenon does occur. Primarily this would have occurred if we had not treated the ladders separately: if p_i's are ladder variables, then the corresponding w momenta in (47) are indeed the same. Although we have removed the ladders beforehand, the same phenomenon occurs in case of a graph which contains ladder *only as a minor but not as a subgraph*. Our separate ladder integration procedure (45) can be viewed as a very simple renormalization of the ladder subgraphs. The correct procedure should renormalize all ladder minors as well.

To cope with this difficulty, we have to follow more precisely the point singularities. To this end, we define the following generalization of $E_\eta(M)$. For any index set $I' \subset I = \{1, 2, \ldots, k+1\}$, any $|I'| \times (k+1)$ matrix M, any ν integer and any $\nu \times (k+1)$ matrix $\mathcal{E}$ we define

$$E_\eta(I', M, \mathcal{E}) := \lambda^{2k} e^{2t\eta} \iint_{-4d}^{4d} d\alpha d\beta \int d\mathbf{p}$$

$$\times \left(\prod_{i \in I'} \frac{1}{|\alpha - \overline{\omega}(p_i) - i\eta|} \frac{1}{|\beta - \omega(\sum_{j=1}^{k+1} M_{ij} p_j) + i\eta|} \right) \prod_{\mu=1}^{\nu} \frac{1}{|\sum_{j=1}^{k+1} \mathcal{E}_{\mu j} p_j|} \tag{49}$$

We follow the same procedure as described in Sect. 3.4, but we also keep track of the evolution of the point singularity matrix $\mathcal{E}$. At the beginning $I' = I$, $\nu = 0$ and $\mathcal{E}$ is not present. After the first non-ladder type integration, a point singularity will appear from (47). Some of the point singularities may get integrated out later as one of their variables become integration variable. Therefore we will need the following generalization of (47):

Lemma 3. *There exists a constant C such that for any index set A*

$$\sup_{|\alpha|,|\beta|\leq 4d} \int \frac{1}{|\alpha - \overline{\omega}(p) - i\eta|\,|\beta - \omega(r+p) + i\eta|} \prod_{a\in A} \frac{1}{|r_a + p|}\, dp$$

$$\leq C\eta^{-\tau'}|\log\eta|^3 \sum_{a\in A} \left(\prod_{\substack{a'\in A \\ a'\neq a}} \frac{1}{|r_a - r_{a'}|}\right) \frac{1}{|r|}\,. \qquad \Box \qquad (50)$$

For the continuous model $\tau' = 0$ while for the discrete model $\tau' = \frac{7}{8} + 2\kappa$.

Using this lemma, we can keep record of the evolution of the point singularity matrix $\mathcal{E}$ at an intermediate step of our integration algorithm. These matrices change by simple operations reminiscent to the Gaussian elimination.

Three complications occur along this procedure, we briefly describe how we resolve them:

(1) The inequality (50) does not allow higher order point singularities. Although it is possible to generalize it to include such singularities as well, we followed a technically simpler path. In addition to the indices $j \in I_p$, we select further β-denominators that we estimate by the trivial L^∞ bound. These additional indices are chosen in such a way, that (i) the number of remaining rows be at least $\frac{1}{3}d(\sigma)$; (ii) the point singularity matrix be of full rank at every step of the algorithm. This second criterion guarantees that no higher order point singularities occur. Since every $\mathcal{E}$ can be derived from M by a procedure that is close to Gaussian elimination and M is invertible (Lemma 1), the full-rank property is relatively easy to guarantee.

(2) The full-rank property actually needs to be guaranteed in a quantitative way, at least the entries of $\mathcal{E}$ needs to be controlled. These entries appear in the point singularity denominators of (50) and their inverses would appear in the estimate. The key observation is that each entry of every matrix $\mathcal{E}$ along the procedure is always 0, 1 or -1. It is actually easier to prove a stronger statement, namely that every $\mathcal{E}$ is a *totally unimodular matrix*. The proof follows from the fact that every $\mathcal{E}$ can be derived from M by elementary Gaussian elimination steps plus zeroing out certain rows and columns. Such steps preserve total unimodularity and M is totally unimodular by Lemma 1.

(3) After all β-denominators are eliminated, we are left with an integral of the form

$$E_\eta(J,\emptyset,\mathcal{E}) := \int_{-4d}^{4d} d\alpha \int \left(\prod_{i\in J} dp_i\right) \prod_{i\in J} \frac{1}{|\alpha - \overline{\omega}(p_i) - i\eta|} \prod_{\mu=1}^{\nu} \frac{1}{|\sum_{i\in J}\mathcal{E}_{\mu i}p_i|}$$

$$(51)$$

for some index set J and some point singularity matrix obtained along the integration procedure. Without the point singularities, this integral could be estimated by the $|J|$-th power of $|\log \eta|$. Since $\mathcal{E}$ is totally unimodular, a similar estimate can be obtained for (51) as well.

4 Computation of the Main Term and Its Convergence to a Brownian Motion

Our goal is to compute the Wigner distribution $\mathbf{E}W^\varepsilon_{\psi_t}(X, v)$ with $t = \lambda^{-2-\kappa}T$ and $\varepsilon = \lambda^{2+\kappa/2}$. From Proposition 2, and similar bounds on the repetitive terms in (32), we can restrict our attention to the ladder graph. The following lemma is a more precise version of the ladder integration Lemma (45) and it is crucial to this computation. We present it for the more complicated discrete case. The proof is a tedious calculation in [11].

Lemma 4. *Suppose $f(p)$ is a C^1 function on $\mathbf{T}^d$. Recall $0 < \kappa < 1/16$ and define $\gamma := (\alpha + \beta)/2$. Let η satisfy $\lambda^{2+4\kappa} \leq \eta \leq \lambda^{2+\kappa}$. Then for $|r| \leq \lambda^{2+\kappa/4}$ we have,*

$$\int \frac{\lambda^2 f(p)}{(\alpha - \overline{\omega}(p - r) - i\eta)(\beta - \omega(p + r) + i\eta)} \, dp \tag{52}$$

$$= -2\pi i \lambda^2 \int \frac{f(p)\, \delta(e(p) - \gamma)}{(\alpha - \beta) + 2(\nabla e)(p) \cdot r - 2i[\lambda^2 \mathrm{Im}\Theta(\gamma) + \eta]} \, dp + O(\lambda^{1/2 - 8\kappa}|\log \lambda|).$$

Since the Boltzmann collision kernel is uniform on the energy shell, the calculation of $\mathbf{E}W^\varepsilon_{\psi_t}(X, v)$ is more straightforward for the discrete case. We present the sketch of this calculation, the continuous model requires a little more effort at this stage.

Let $\varepsilon = \lambda^{2+\kappa/2}$ be the space scale. After rescaling the Wigner function at time t, we compute $\widehat{W}(\varepsilon\xi, v)$ tested against a smooth, decaying function $\mathcal{O}(\xi, v)$. In particular ξ is of order 1. After the application of Lemma 4 (with $v = v_{k+1}$) and change of variables $a := (\alpha + \beta)/2$ and $b := \lambda^{-2}(\alpha - \beta)$, we obtain

$$\langle \mathcal{O}, \mathbf{E}\widehat{W} \rangle := \int dv d\xi \, \mathcal{O}(\xi, v) \mathbf{E}\widehat{W}(\varepsilon\xi, v) = \sum_{k \leq K} \iint_{\mathbf{R}} \frac{da d\beta}{(2\pi)^2 \lambda^2} \int dv \, e^{it(\alpha - \beta) + 2\eta t}$$

$$\times \mathcal{O}(\xi, v_{k+1}) \widehat{W}_0(\varepsilon\xi, v_1) \prod_{j=1}^{k+1} \frac{\lambda^2}{\left(\alpha - \overline{\omega}(v_i + \frac{\varepsilon\xi}{2}) - i\eta\right)\left(\beta - \omega(v_i - \frac{\varepsilon\xi}{2}) + i\eta\right)}$$

$$\approx \sum_{k \leq K} \iint_{\mathbf{R}} \frac{da db}{(2\pi)^2} \, e^{it\lambda^2 b} \int \left(\prod_j \frac{-2\pi i\, \delta(e(v_j) - a) dv_j}{b + \lambda^{-2}\varepsilon\nabla e(v_j) \cdot \xi - 2i\mathcal{I}(a)} \right)$$

$$\times \widehat{W}_0(\varepsilon\xi, v_1) \mathcal{O}(\xi, v_{k+1}) ,$$

where we defined $\mathcal{I}(\gamma) := \mathrm{Im}\Theta(\gamma)$ for brevity. We used $\eta = \lambda^{2+\kappa}$ to estimate the error terms. The main term (left hand side above) however, is independent of η, so we can choose $\eta = \lambda^{2+4\kappa}$ for the rest of the calculation and we note that Lemma 4 holds for this smaller η as well. This is the reason why the $e^{2\eta t}$ factor is negligible.

We expand the fraction up to second order in ε, we get

$$\frac{-i}{b + \lambda^{-2}\varepsilon\nabla e(v_j) \cdot \xi - 2i\mathcal{I}(a)} \approx \frac{-i}{b - 2i\mathcal{I}(a)}\left[1 - \frac{\lambda^{-2}\varepsilon\nabla e(v_j) \cdot \xi}{b - 2i\mathcal{I}(a)}\right.$$
$$\left. + \frac{\lambda^{-4}\varepsilon^2[\nabla e(v_j) \cdot \xi]^2}{(b - 2i\mathcal{I}(a))^2}\right]$$

By symmetry of the measure $2\pi\delta(e(v) - a)dv$ under the sign flip, $v \to -v$ and using $(\nabla e)(v) = -\nabla e(-v)$, we see that the first order term vanishes after the integration. We also define the matrix

$$D(a) := \frac{1}{2\mathcal{I}(a)}\int d\mu_a(v)\, \frac{\nabla e(v)}{2\pi} \otimes \frac{\nabla e(v)}{2\pi}$$

After integrating out all momenta and changing the b variable we obtain

$$\langle O, E\widehat{W}\rangle \approx \sum_{k \le K}\int d\xi \int_{\mathbf{R}}\frac{2\mathcal{I}(a)da}{2\pi}\langle O(\xi, \cdot)\rangle_a\langle\widehat{W}_0(\varepsilon\xi, \cdot)\rangle_a \int_{\mathbf{R}}\frac{db}{2\pi}\, e^{2i\lambda^2 tb\mathcal{I}(a)}$$

$$\times\left(\frac{-i}{b - i}\right)^{k+1} \times \left[1 + \frac{(2\pi)^2\varepsilon^2\lambda^{-4}\langle\xi, D(a)\xi\rangle}{2\mathcal{I}(a)} \cdot \frac{1}{(b - i)^2}\right]^{k+1}$$

We sum up the geometric series and perform a residue calculation to evaluate the db integral. We obtain that the main contribution comes from $k \sim 2\lambda^2 t\mathcal{I}(e)$, so the truncation $k \le K$ can be neglected and the result of the db integration is

$$\int_{\mathbf{R}}\frac{db}{2\pi}(\cdots) \approx \exp\left(-\frac{(2\pi)^2\varepsilon^2 t\langle\xi, D(e)\xi\rangle}{\lambda^2}\right)$$

To obtain a nontrivial limit, $\varepsilon^2 t/\lambda^2 \sim 1$ is necessary. Noting that $t = \lambda^{-2-\kappa}T$ with $T = O(1)$, we see that indeed the space must be rescaled by $\varepsilon = \lambda^{2+\kappa/2}$. Finally

$$\langle O, E\widehat{W}\rangle \approx \int d\xi \int_{\mathbf{R}}\frac{2\mathcal{I}(a)da}{2\pi}\langle O(\xi, \cdot)\rangle_a\langle\widehat{W}_0(\varepsilon\xi, \cdot)\rangle_a \exp\left(-(2\pi)^2 T\langle\xi, D(a)\xi\rangle\right).$$

Since $\exp[-(2\pi)^2 T\langle\xi, D(a)\xi\rangle]$ is the fundamental solution to the heat equation (15), from the definition of $\langle\cdot\rangle$, and after inverse Fourier transform we obtain (17). This completes the sketch of the calculation of the main term.

Acknowledgements

L. Erdős was partially supported by NSF grant DMS-0200235 and EU-IHP Network "Analysis and Quantum" HPRN-CT-2002-0027. H.-T. Yau was partially supported by NSF grant DMS-0307295 and MacArthur Fellowship.

References

1. M. Aizenman and S. Molchanov: Commun. Math. Phys. **157**, 245 (1993).
2. P. Anderson: Phys. Rev. **109**, 1492 (1958).
3. C. Boldrighrini, L. Bunimovich and Y. Sinai: J. Statis. Phys. **32** no. 3, 477 (1983).
4. J. Bourgain: *Random lattice Schrödinger operators with decaying potential: some higher dimensional phenomena* (Lecture Notes in Mathematics Vol. **1807** (Springer, Berlin Heidelberg New York 2003) pp 70–99.
5. L. Bunimovich, Y. Sinai: Commun. Math. Phys. **78**, 479 (1980/81).
6. T. Chen: *Localization Lengths and Boltzmann Limit for the Anderson Model at Small Disorders in Dimension 3*. Preprint (2003) http://xxx.lanl.gov/abs/math-ph/0305051.
7. D. Dürr, S. Goldstein, J. Lebowitz: Commun. Math. Phys. **78**, 507 (1980/81).
8. D. Dürr, S. Goldstein, J. Lebowitz: Commun. Math. Phys. **113**, 209 (1987).
9. L. Erdős: J. Statis. Phys. **107**, 1043 (2002).
10. L. Erdős and H.-T. Yau: Commun. Pure Appl. Math. **LIII**, 667 (2000).
11. L. Erdős, M. Salmhofer and H.-T. Yau: *Quantum diffusion of the random Schrödinger evolution in the scaling limit*. Preprints (2005) http://xxx.lanl.gov/abs/math-ph/0502025, 0512014, 0512015
12. J. Fröhlich and T. Spencer: Commun. Math. Phys. **88**, 151 (1983).
13. T.G. Ho, L.J. Landau and A.J. Wilkins: Rev. Math. Phys. **5**, 209 (1993).
14. G. Gallavotti: *Rigorous theory of the Boltzmann equation in the Lorentz model.* Nota interna n. 358, Physics Department, Universita di Roma, pp. 1–9 (1972) mp_arc@math.utexas.edu, # 93–304.
15. I.Ya. Goldsheid, S.A. Molchanov and L.A. Pastur: Funct. Anal. Appl. **11**, 1 (1997).
16. H. Kesten, G. Papanicolaou: Comm. Math. Phys. **78**, 19 (1980/81).
17. A. Klein: Math. Res. Lett. **1**, 399 (1994).
18. T. Komorowski, L. Ryzhik: *Diffusion in a weakly random Hamiltonian flow.* Preprint (2004).
19. O. Lanford: Astérisque **40**, 117 (1976).
20. P.A. Lee and T.V. Ramakrishnan: Rev. Mod. Phys. **57**, 287 (1985).
21. J. Lukkarinen and H. Spohn. *Kinetic limit for wave propagation in a random medium.* Preprint. http://xxx.lanl.gov/math-ph/0505075
22. I. Rodnianski and W. Schlag: Int. Math. Res. Not. **5**, 243 (2003).
23. E. Seiler, *Gauge Theories as a Problem of Constructive Quantum Field Theory and Statistical Mechanics.* Lecture Notes in Physics **159** (Springer, Berlin Heidelberg New York 1982).
24. W. Schlag, C. Shubin, T. Wolff: J. Anal. Math. **88**, 173 (2002).
25. H. Spohn: J. Statist. Phys. **17**, 385 (1977).
26. H. Spohn: Commun. Math. Phys. **60**, 277 (1978).
27. D. Vollhardt and P. Wölfle: Phys. Rev. B **22**, 4666 (1980).

Bose-Einstein Condensation
and Superradiance*

J.V. Pulé[1], A.F. Verbeure[2] and V.A. Zagrebnov[3]

[1] Department of Mathematical Physics, University College Dublin, Belfield,
Dublin 4, Ireland
Joe.Pule@ucd.ie
[2] Instituut voor Theoretische Fysika, Katholieke Universiteit Leuven,
Celestijnenlaan 200D, 3001 Leuven, Belgium
andre.verbeure@fys.kuleuven.ac.be
[3] Université de la Méditerranée and Centre de Physique Théorique,
Luminy-Case 907, 13288 Marseille, Cedex 09, France
zagrebnov@cpt.univ-mrs.fr

Abstract. We consider two models which exhibit equilibrium BEC superradiance.
They are related to two different types of superradiant scattering observed in recent
experiments. The first one corresponds to the amplification of matter-waves due to
Raman superradiant scattering from a cigar-shaped BE condensate, when the recoiled and the condensed atoms are in different internal states. The main mechanism
is stimulated Raman scattering in two-level atoms, which occurs in a superradiant
way. Our second model is related to the superradiant Rayleigh scattering from a
cigar-shaped BE condensate. This again leads to a matter-waves amplification but
now with the recoiled atoms in the same state as the atoms in the condensate.
Here the recoiling atoms are able to interfere with the condensate at rest to form
a matter-wave grating (interference *fringes*) which has been recently observed in
experiments.

1 Introduction

We report here our recent results on the equilibrium Bose-Einstein Condensation (BEC) superradiance motivated by discovery of the Dicke superradiance and BEC matter waves amplification [1]– [5]. In these experiments the
condensate is illuminated with a laser beam, the so called *dressing beam*.
The BEC atoms then scatter photons from this beam and receive the corresponding recoil momentum producing coherent *four-wave mixing* of light
and atoms [5]. The aim of our project is the construction of soluble statistical
mechanical models for these phenomena.

In the first paper [6], motivated by the principle of *four-wave mixing*
of light and atoms [5], we considered two models with a linear interaction
between Bose atoms and photons, one with a global gauge symmetry and
another one in which this symmetry is broken. We proved that there is

---

*Talk given by Valentin Zagrebnov.

J.V. Pulé et al.: *Bose-Einstein Condensation and Superradiance*, Lect. Notes Phys. **690**,
259–278 (2006)
www.springerlink.com

equilibrium superradiance and also that there is an enhancement of condensation compared with that occurring in the case of the free Bose gas.

In the second paper [7] we formalized the ideas described in [4,5] by constructing a thermodynamically stable model whose main ingredient is the *two-level* internal states of the Bose condensate atoms. We showed that our model is equivalent to a *bosonized* Dicke maser model. Besides determining its equilibrium states, we computed and analyzed the thermodynamic functions, again finding the existence of a cooperative effect between BEC and superradiance.

In our last paper [8] we study the effect of *momentum recoil* which was omitted in [6] and [7]. There we consider two models motivated by two different types of superradiant scattering observed in recent experiments carried out by the MIT group, see e.g. [1]– [3]. Our first model (*Model 1*) corresponds to the Raman superradiant scattering from a cigar-shaped BE condensate considered in [1]. This leads to the amplification of matter waves (recoiled atoms) in the situation when amplified and condensate atoms are in *different* internal states. The main mechanism is stimulated Raman scattering in two-level atoms, which occurs in a way similar to Dicke superradiance [7].

Our second model (*Model 2*) is related to the superradiant Rayleigh scattering from a cigar-shaped BE condensate [2,3]. This again leads to a matter-wave amplification but now with recoiled atoms in the *same* state as the condensate at rest. This is because the condensate is now illuminated by an off-resonant pump laser beam, so that for a long-pulse the atoms remain in their lower level states. In this case the (*non-Dicke*) superradiance is due to self-stimulated Bragg scattering [3].

From a theoretical point of view both models are interesting as they describe homogeneous systems in which there is *spontaneous breaking of translation invariance*. In the case of the Rayleigh superradiance this means that the phase transition corresponding to BEC is at the same time also a transition into a *matter-wave grating* i.e. a "frozen" spatial density wave structure, see Sect. 3. The fact that *recoiling* atoms are able to interfere with the condensate *at rest* to form a matter-wave grating (interference *fringes*) has been recently observed experimentally, see [3]– [5] and discussion in [9] and [10].

In the case of the Raman superradiance there is an important difference: the internal atomic states for condensed and recoiled bosons are *orthogonal*. Therefore these bosons are *different* and consequently cannot interfere to produce a matter-wave grating as in the first case. Thus the observed spatial modulation is not in the atomic density of interfering recoiled and condensed bosons, but in the *off-diagonal* coherence and photon condensate producing a one-dimensional (*corrugated*) optical lattice, see Sect. 3 and discussion in Sect. 4.

To make the definition of our models more exact consider a system of identical bosons of mass m enclosed in a cube $\Lambda \subset \mathbb{R}^\nu$ of volume $V = |\Lambda|$

centered at the origin. We impose periodic boundary conditions so that the momentum dual set is $\Lambda^* = \{2\pi p/V^{1/\nu}|p \in \mathbb{Z}^\nu\}$.

In *Model 1* the bosons have an internal structure which we describe by considering them as two-level atoms, the two levels being denoted by $\sigma = \pm$. For momentum k and level σ, $a^*_{k,\sigma}$ and $a_{k,\sigma}$ are the usual boson creation and annihilation operators with $[a_{k,\sigma}, a^*_{k',\sigma'}] = \delta_{k,k'}\delta_{\sigma,\sigma'}$. Let $\epsilon(k) = \|k\|^2/2m$ be the single particle kinetic energy and $N_{k,\sigma} = a^*_{k,\sigma}a_{k,\sigma}$ the operator for the number of particles with momentum k and level σ. Then the total kinetic energy is

$$T_{1,\Lambda} = \sum_{k \in \Lambda^*} \epsilon(k)(N_{k,+} + N_{k,-}) \tag{1}$$

and the total number operator is $N_{1,\Lambda} = \sum_{k \in \Lambda^*}(N_{k,+} + N_{k,-})$. We define the Hamiltonian $H_{1,\Lambda}$ for *Model 1* by

$$H_{1,\Lambda} = T_{1,\Lambda} + U_{1,\Lambda} \tag{2}$$

where

$$U_{1,\Lambda} = \Omega\, b^*_q b_q + \frac{g}{2\sqrt{V}}(a^*_{q+}a_{0-}b_q + a_{q+}a^*_{0-}b^*_q) + \frac{\lambda}{2V}N^2_{1,\Lambda}, \tag{3}$$

$g > 0$ and $\lambda > 0$. Here b_q, b^*_q are the creation and annihilation operators of the photons, which we take as a one-mode boson field with $[b_q, b^*_q] = 1$ and a frequency Ω. g is the coupling constant of the interaction of the bosons with the photon external field which, without loss of generality, we can take to be positive as we can always incorporate the sign of g into b. Finally the λ-term is added in (2) to obtain a thermodynamical stable system and to ensure the right thermodynamic behaviour.

In *Model 2* we consider the situation when the excited atoms have already irradiated photons, i.e. we deal only with de-excited atoms $\sigma = -$. In other words, we neglect the atom excitation and consider only *elastic* atom-photon scattering. This is close to the experimental situation [3]– [5], in which the atoms in the BE condensate are irradiated by *off-resonance* laser beam. Assuming that detuning between the optical fields and the atomic two-level resonance is much larger that the natural line of the atomic transition (superradiant Rayleigh regime [2, 3]) we get that the atoms always remain in their *lower* internal energy state. We can then ignore the internal structure of the atoms and let a^*_k and a_k be the usual boson creation and annihilation operators for momentum k with $[a_k, a^*_{k'}] = \delta_{k,k'}$, $N_k = a^*_k a_k$ the operator for the number of particles with momentum k,

$$T_{2,\Lambda} = \sum_{k \in \Lambda^*} \epsilon(k)N_k \tag{4}$$

the total kinetic energy, and $N_{2,\Lambda} = \sum_{k \in \Lambda^*} N_k$ the total number operator. We then define the Hamiltonian $H^{(2)}_\Lambda$ for *Model 2* by

$$H_{2,\Lambda} = T_{2,\Lambda} + U_{2,\Lambda} \tag{5}$$

where

$$U_{2,\Lambda} = \Omega\, b_q^* b_q + \frac{g}{2\sqrt{V}}(a_q^* a_0 b_q + a_q a_0^* b_q^*) + \frac{\lambda}{2V} N_{2,\Lambda}^2. \tag{6}$$

Our main results concerning the *Models 1* and *2* can be summarized as follows:

- We give a complete rigorous solution of the variational principle for the equilibrium state of *Model 1*, and we compute the pressure as a function of the temperature and the chemical potential. We show (see Sect. 2) that *Model 1* manifests the Raman superradiance and we prove for this model the occurrence of spontaneous breaking of translation invariance of the equilibrium state.
- The analysis of the *Model 2* for the Rayleigh superradiance is very similar and therefore we do not repeat it but simply state the results in Sect. 3.
- In Sects. 3 and 4 we relate the spontaneous breaking of translation invariance of the equilibrium state with spatial modulation of matter-waves (*matter-wave grating*) in both models.

We finish this introduction with the following comments concerning interpretation of the *Models 1* and *2*:

- In our models we do not use for effective photon-boson interaction the *four-wave mixing principle*, see [5, 6, 11]. The latter seems to be important for the geometry, when a linearly polarized pump laser beam is incident in a direction perpendicular to the long axis of a cigar-shaped BE condensate, inducing the "45°-recoil pattern" picture [1]– [3]. Instead as in [7], we consider a *minimal* photon-atom interaction only with superradiated photons, cf [12]. This corresponds to superradiance in a "one-dimensional" geometry, when a pump laser beam is collimated and aligned along the long axis of a cigar-shaped BE condensate, see [10, 13].
- In this geometry the superradiant photons and recoiled matter-waves propagate in the same direction as the incident pump laser beam. If one considers it as a classical "source" (see [5]), then we get a *minimal* photon-atom interaction [7] generalized to take into account the effects of recoil. Notice that the further approximation of the BEC operators by c-numbers leads to a bilinear photon-atom interaction studied in [5, 6].
- Here we study *equilibrium* BEC superradiance while the experimental situation (as is the case with Dicke superradiance [14]) is more accurately described by non-equilibrium statistical mechanics. However we believe that for the purpose of understanding the quantum coherence interaction between light and the BE condensate our analysis is as instructive and is in the same spirit as the rigorous study of the Dicke model in thermodynamic equilibrium, see e.g. [15]– [17].

– In spite of the simplicity of our exactly soluble *Models 1* and *2* they are able to demonstrate the main features of the BEC superradiance with recoil: the photon-boson condensate *enhancement* with formation of the *light corrugated optical lattice* and the *matter-wave grating*. The corresponding phase diagrams are very similar to those in [7]. However, though the type of behaviour is similar, this is now partially due to the *momentum recoil* and not entirely to the *internal* atomic level structure.

2 Solution of the Model 1

2.1 The Effective Hamiltonian

We start with the *stability* of Hamiltonian (2). Consider the term $U_{1,\Lambda}$ in (3). This gives

$$
U_{1,\Lambda} = \Omega \left(b_q^* + \frac{g}{2\Omega\sqrt{V}} a_{q+} a_{0-}^* \right) \left(b_q + \frac{g}{2\Omega\sqrt{V}} a_{q+}^* a_{0-} \right) - \frac{g^2}{4\Omega^2 V} N_0
$$

$$
-(N_{q+}+1) + \frac{\lambda}{2V} N_{1,\Lambda}^2 \geq \frac{\lambda}{2V} N_{1,\Lambda}^2 - \frac{g^2}{4\Omega^2 V} N_0 - (N_{q+}+1) . \quad (7)
$$

On the basis of the trivial inequality $4ab \leq (a+b)^2$, the last term in the lower bound in (7) is dominated by the first term if $\lambda > g^2/8\Omega$, that is if the stabilizing coupling λ is large with respect to the coupling constant g or if the external frequency is large enough. We therefore assume the *stability condition*: $\lambda > g^2/8\Omega$.

Since we want to study the equilibrium properties of the model (2) in the grand-canonical ensemble, we shall work with the Hamiltonian

$$
H_{1,\Lambda}(\mu) = H_{1,\Lambda} - \mu N_{1,\Lambda} \qquad (8)
$$

where μ is the chemical potential. Since $T_{1,\Lambda}$ and the interaction $U_{1,\Lambda}$ conserve the quasi-momentum, Hamiltonian (2) describes a *homogeneous* (translation invariant) system. To see this explicitly, notice that the external laser field possesses a natural quasi-local structure as the Fourier transform of the field operator $b(x)$:

$$
b_q = \frac{1}{\sqrt{V}} \int_\Lambda dx \, e^{iq \cdot x} b(x) . \qquad (9)
$$

If for $z \in \mathbb{R}^\nu$, we let τ_x be the translation automorphism $(\tau_z b)(x) = b(x+z)$, then $\tau_z b_q = e^{iq \cdot z} b_q$ and similarly $\tau_z a_{k,\sigma} = e^{ik \cdot z} a_{k,\sigma}$. Therefore, the Hamiltonian (2) is translation invariant.

Because the interaction (3) is not bilinear or quadratic in the creation and annihilation operators the system cannot be diagonalized by a standard symplectic or Bogoliubov transformation. One way to solve the *Model 1* is by applying the so-called *effective Hamiltonian* method, which is based on the fact that an equilibrium state is not determined by the Hamiltonian but by its *Liouvillian* [18]. The best route to prove the exactness of the effective Hamiltonian method in our case (see also [19]) is to use the characterization of the equilibrium state by means of the *correlation inequalities* [18,20]:

A state ω is an equilibrium state for $H_{1,\Lambda}(\mu)$ at inverse temperature β, if and only if for all local observables A, it satisfies

$$\lim_{V \to \infty} \beta \omega \left([A^*, [H_{1,\Lambda}(\mu), A]] \right) \geq \omega(A^*A) \ln \frac{\omega(A^*A)}{\omega(AA^*)} . \tag{10}$$

Clearly only the *Liouvillian* $[H_{1,\Lambda}(\mu), \cdot]$ of the Hamiltonian enters into these inequalities and therefore we can replace $H_{1,\Lambda}(\mu)$ by a simpler Hamiltonian, the *effective Hamiltonian*, which gives in the limiting state ω the same Liouvillian as $H_{1,\Lambda}(\mu)$ and then look for the equilibrium states corresponding to it. Now in our case for an extremal or mixing state ω we define the effective Hamiltonian $H_{1,\Lambda}^{\text{eff}}(\mu, \eta, \rho)$ such that for all local observables A and B

$$\lim_{V \to \infty} \omega \left(A, [H_{1,\Lambda}(\mu), B] \right) = \lim_{V \to \infty} \omega \left(A, [H_{1,\Lambda}^{\text{eff}}(\mu, \eta, \rho), B] \right) . \tag{11}$$

The significance of the *parameters* η and ρ will become clear below. One can then replace (10) by

$$\lim_{V \to \infty} \beta \omega \left([A^*, [H_{1,\Lambda}^{\text{eff}}(\mu, \eta, \rho), A]] \right) \geq \omega(A^*A) \ln \frac{\omega(A^*A)}{\omega(AA^*)} . \tag{12}$$

We can choose $H_{1,\Lambda}^{\text{eff}}(\mu, \eta, \rho)$ so that it can be diagonalized and thus (12) can be solved explicitly. For a given chemical potential μ, the inequalities (12) can have *more than one* solution. We determine the physical solution by minimizing the free energy density with respect to the set of states or equivalently by maximizing the grand canonical pressure on this set. Let

$$H_{1,\Lambda}^{\text{eff}}(\mu, \eta, \rho) = (\lambda\rho - \mu + \epsilon(q)) a_{q+}^* a_{q+} + (\lambda\rho - \mu) a_{0-}^* a_{0-} + \frac{g}{2} (\eta a_{q+}^* b_q$$

$$+ \bar{\eta} a_{q+} b_q^*) + \Omega \, b_q^* b_q + \frac{g\sqrt{V}}{2} \left(\zeta a_{0-} + \bar{\zeta} a_{0-}^* \right) + T'_{1,\Lambda} + (\lambda\rho - \mu) N'_{1,\Lambda} \tag{13}$$

where

$$T'_{1,\Lambda} = \sum_{k \in \Lambda^*, \, k \neq q} \epsilon(k) N_{k,+} + \sum_{k \in \Lambda^*, \, k \neq 0} \epsilon(k) N_{k,-} , \tag{14}$$

$$N'_{1,\Lambda} = \sum_{k \in \Lambda^*, \, k \neq q} N_{k,+} + \sum_{k \in \Lambda^*, \, k \neq 0} N_{k,-} , \tag{15}$$

η and ζ are complex numbers and ρ is a positive real number. Then one can easily check that (11) is satisfied if

$$\eta = \frac{\omega(a_{0-})}{\sqrt{V}}, \quad \zeta = \frac{\omega(a_{q+}^* b_q)}{V} \quad \text{and} \quad \rho = \frac{\omega(N_{1,\Lambda})}{V} , \tag{16}$$

where the state ω coincides with the equilibrium state $\langle \cdot \rangle_{H^{\text{eff}}_{1,\Lambda}(\mu,\eta,\rho)}$ defined by the effective Hamiltonian $H^{\text{eff}}_{1,\Lambda}(\mu,\eta,\rho)$. By virtue of (11) and (16) we then obtain the *self-consistency* equations

$$\eta = \frac{1}{\sqrt{V}}\langle a_{0-}\rangle_{H^{\text{eff}}_{1,\Lambda}(\mu,\eta,\rho)}, \quad \zeta = \frac{1}{V}\langle a^*_{q+}b_q\rangle_{H^{\text{eff}}_{1,\Lambda}(\mu,\eta,\rho)}, \quad \rho = \frac{1}{V}\langle N_{1,\Lambda}\rangle_{H^{\text{eff}}_{1,\Lambda}(\mu,\eta,\rho)}.$$
(17)

Note that since ζ is a function of η and ρ through (17), we do not need to label the effective Hamiltonian by ζ. The important simplification here is that $H^{\text{eff}}_{1,\Lambda}(\mu,\eta,\rho)$ can be diagonalized:

$$H^{\text{eff}}_{1,\Lambda}(\mu,\eta,\rho) = E_{+,\Lambda}(\mu,\eta,\rho)\alpha^*_1\alpha_1 + E_{-,\Lambda}(\mu,\eta,\rho)\alpha^*_2\alpha_2$$
(18)
$$+ (\lambda\rho - \mu)\alpha^*_3\alpha_3 + T'_{1,\Lambda} + (\lambda\rho - \mu)N'_{1,\Lambda} + \frac{g^2 V|\zeta|^2}{4(\mu - \lambda\rho)},$$

where

$$E_{+,\Lambda}(\mu,\eta,\rho) = \frac{1}{2}(\Omega - \mu + \lambda\rho + \epsilon(q)) + \frac{1}{2}\sqrt{(\Omega + \mu - \lambda\rho - \epsilon(q))^2 + g^2|\eta|^2},$$

$$E_{-,\Lambda}(\mu,\eta,\rho) = \frac{1}{2}(\Omega - \mu + \lambda\rho + \epsilon(q)) - \frac{1}{2}\sqrt{(\Omega + \mu - \lambda\rho - \epsilon(q))^2 + g^2|\eta|^2},$$
(19)

$$\alpha_1 = a_{q+}\cos\theta + b_q\sin\theta, \quad \alpha_2 = a_{q+}\sin\theta - b_q\cos\theta, \quad \alpha_3 = a_{0-} + \frac{g\sqrt{V}\zeta}{2(\lambda\rho - \mu)},$$
(20)

and

$$\tan 2\theta = -\frac{g|\eta|}{\Omega + \mu - \lambda\rho - \epsilon(q)}.$$
(21)

Note that the correlation inequalities (12) (see [20]) imply that

$$\lim_{V\to\infty} \omega\left(A^*, [H_{1,\Lambda}(\mu), A]\right) \geq 0$$
(22)

for all observables A. Applying (22) with $A = a^*_{0+}$, one gets $\lambda\rho - \mu \geq 0$. Similarly, one obtains the condition $\lambda\rho + \epsilon(q) - \mu \geq 0$ by applying (22) to $A = a^*_{q+}$. We also have that $E_{+,\Lambda}(\mu,\eta,\rho) \geq E_{-,\Lambda}(\mu,\eta,\rho)$ and $E_{-,\Lambda}(\mu,\eta,\rho) = 0$ when $|\eta|^2 = 4\Omega(\lambda\rho + \epsilon(q) - \mu)/g^2$ and then $E_{+,\Lambda}(\mu,\eta,\rho) = \Omega - \mu + \lambda\rho + \epsilon(q)$. Thus we have the constraint: $|\eta|^2 \leq 4\Omega(\lambda\rho + \epsilon(q) - \mu)/g^2$. We shall need this information to make sense of the thermodynamic functions below.

The (17) can be made explicit using (18):

$$\eta = \frac{g}{2(\mu - \lambda\rho)}\zeta,$$
(23)

$$\zeta = \frac{1}{2}\frac{g|\eta|}{V(E_+ - E_-)}\left\{\frac{1}{e^{\beta E_+} - 1} - \frac{1}{e^{\beta E_-} - 1}\right\}$$
(24)

and

$$\rho = |\eta|^2 + \frac{1}{V}\frac{1}{e^{-\beta(\mu-\lambda\rho)}-1} + \frac{1}{2V}\left\{\frac{1}{e^{\beta E_+}-1} + \frac{1}{e^{\beta E_-}-1}\right\}$$
$$-\frac{(\mu-\lambda\rho-\epsilon(q)+\Omega)}{2V(E_+-E_-)}\left\{\frac{1}{e^{\beta E_+}-1} - \frac{1}{e^{\beta E_-}-1}\right\}$$
$$+\frac{1}{V}\sum_{k\in\Lambda^*,\,k\neq q}\frac{1}{e^{\beta(\epsilon(k)-\mu+\lambda\rho)}-1} + \frac{1}{V}\sum_{k\in\Lambda^*,\,k\neq 0}\frac{1}{e^{\beta(\epsilon(k)-\mu+\lambda\rho)}-1}.$$

Combining (23) and (24) we obtain the consistency equation:

$$\eta = \frac{g^2|\eta|}{4(\mu-\lambda\rho)V(E_+-E_-)}\left\{\frac{1}{e^{\beta E_+}-1} - \frac{1}{e^{\beta E_-}-1}\right\}. \tag{25}$$

So, the equilibrium states are determined by the limiting form of the consistency equations (25) and (25).

Clearly if E_- does not tend to zero as $V \to \infty$ then the right-hand side of (25) tends to zero and $\eta = 0$. For $\eta \neq 0$ we must have $E_- \to 0$, that is $|\eta|^2 \to 4\Omega(\lambda\rho + \epsilon(q) - \mu)/g^2$. In fact for a finite limit, the large-volume asymptotic should to be

$$|\eta|^2 \approx \frac{4\Omega}{g^2}\left(\lambda\rho + \epsilon(q) - \mu - \frac{1}{\beta V\tau}\right), \tag{26}$$

where $\tau > 0$. This implies that

$$E_+ \to \Omega - \mu + \lambda\rho + \epsilon(q), \qquad E_- \approx \frac{\Omega}{\beta V\tau(\Omega - \mu + \lambda\rho + \epsilon(q))} \tag{27}$$

and (25) becomes in the limit:

$$\eta\left(1 - \frac{g^2\tau}{4(\lambda\rho-\mu)\Omega}\right) = 0. \tag{28}$$

The last equation has solutions:

$$\eta = 0, \quad \text{or} \quad \tau = \frac{4(\lambda\rho-\mu)\Omega}{g^2}. \tag{29}$$

For $\mu < 0$, let

$$\varepsilon_0(\mu) = \frac{1}{(2\pi)^3}\int_{\mathbb{R}^3} d^3k \frac{\epsilon(k) - \mu}{e^{\beta(\epsilon(k)-\mu)}-1}, \tag{30}$$

$$\rho_0(\mu) = \frac{1}{(2\pi)^3}\int_{\mathbb{R}^3} d^3k \frac{1}{e^{\beta(\epsilon(k)-\mu)}-1} \tag{31}$$

and

$$p_0(\mu) = -\frac{1}{(2\pi)^3}\int_{\mathbb{R}^3} d^3k \ln(1 - e^{-\beta(\epsilon(k)-\mu)}), \tag{32}$$

that is the grand-canonical energy density, the particle density and the pressure for the *free Bose-gas*. Let

$$s_0(\mu) = \beta(\varepsilon_0(\mu) + p_0(\mu)) , \qquad (33)$$

and note that $s_0(\mu)$ is an increasing function of μ. We shall denote the free Bose-gas *critical* density by $\rho_c := \rho_0(0)$. Recall that ρ_c is infinite for $\nu < 3$ and finite for $\nu \geq 3$.

Now we analyze in detail the solutions of (28) and compute their thermodynamic functions. We consider three cases:

Case 1. Suppose that in the thermodynamic limit one has: $\eta = 0$ and $\lambda\rho - \mu > 0$. By virtue of (24) in this case we have $\zeta = 0$, i.e. there is *no condensation*:

$$\lim_{V\to\infty} \frac{1}{V} \langle a_0^* a_0 - \rangle_{H_{1,A}^{\mathrm{eff}}(\mu,\eta,\rho)} = \lim_{V\to\infty} \frac{1}{V} \langle a_{q+}^* a_{q+} \rangle_{H_{1,A}^{\mathrm{eff}}(\mu,\eta,\rho)}$$

$$= \lim_{V\to\infty} \frac{1}{V} \langle b_q^* b_q \rangle_{H_{1,A}^{\mathrm{eff}}(\mu,\eta,\rho)} = 0 ,$$

and the photon and boson subsystems are *decoupled*. Now (25) takes the form

$$\rho = 2\rho_0(\mu - \lambda\rho) , \qquad (34)$$

the energy density is given by

$$\lim_{V\to\infty} \frac{1}{V} \langle H_{1,A}(\mu) \rangle_{H_{1,A}^{\mathrm{eff}}(\mu,\eta,\rho)} = 2\varepsilon_0(\mu - \lambda\rho) - \frac{1}{2}\lambda\rho^2$$

and the entropy density is equal to

$$s(\mu) = 2s_0(\mu - \lambda\rho) . \qquad (35)$$

Since the grand-canonical pressure is given by

$$p(\mu) = \frac{1}{\beta}s(\mu) - \lim_{V\to\infty} \frac{1}{V} \langle H_{1,A}(\mu) \rangle_{H_{1,A}^{\mathrm{eff}}(\mu,\eta,\rho)} , \qquad (36)$$

then

$$p(\mu) = 2p_0(\mu - \lambda\rho) + \frac{1}{2}\lambda\rho^2 . \qquad (37)$$

Case 2. Now suppose that in this case the thermodynamic limit gives: $\eta = 0$, $\lambda\rho - \mu = 0$. Then the solution of (25) has the asymptotic form:

$$\rho_V = \frac{\mu}{\lambda} + \frac{1}{V\lambda\tau'} + o\left(\frac{1}{V}\right) , \qquad (38)$$

with some $\tau' \geq 0$. So, (25) yields the identity

$$\rho = \tau' + 2\rho_c , \qquad (39)$$

which implies that $\rho \geq 2\rho_c$. Note that this case is possible only if $\nu \geq 3$. By explicit calculations one gets:

$$\lim_{V \to \infty} \frac{1}{V} \langle a_{0-}^* a_{0-} \rangle_{H_{1,\Lambda}^{\mathrm{eff}}(\mu,\eta,\rho)} = \tau' = \rho - 2\rho_c , \tag{40}$$

i.e., there is a possibility of the *mean-field* condensation of *zero-mode non-excited* bosons $\sigma = -$. Notice that the gauge invariance implies

$$\lim_{V \to \infty} \langle a_{0-}^{\#} \rangle_{H_{1,\Lambda}^{\mathrm{eff}}(\mu,0,\rho)} / \sqrt{V} = 0$$

Furthermore, we get also:

$$\lim_{V \to \infty} \frac{1}{V} \langle a_{q+}^* a_{q+} \rangle_{H_{1,\Lambda}^{\mathrm{eff}}(\mu,0,\rho)} = \lim_{V \to \infty} \frac{1}{V} \langle b_q^* b_q \rangle_{H_{1,\Lambda}^{\mathrm{eff}}(\mu,0,\rho)} = 0 , \tag{41}$$

i.e., there is *no condensation* in the $q \neq 0$ modes and the laser boson field. Hence again the contribution from the interaction term vanishes, i.e. the photon and boson subsystems are *decoupled*. In this case the energy density is given by:

$$\lim_{V \to \infty} \frac{1}{V} \langle H_{1,\Lambda}(\mu) \rangle_{H_{1,\Lambda}^{\mathrm{eff}}(\mu,\eta,\rho)} = 2\varepsilon_0(0) - \frac{\mu^2}{2\lambda} \tag{42}$$

and the entropy density has the form:

$$s(\mu) = 2s_0(0) = 2\beta(\varepsilon_0(0) + p_0(0)) . \tag{43}$$

Thus for the pressure one gets:

$$p(\mu) = 2p_0(0) + \frac{\mu^2}{2\lambda} . \tag{44}$$

Case 3. Suppose that $\eta \neq 0$. Then by diagonalization of (13) we obtain a *simultaneous* condensation of the excited/non-excited bosons and the laser photons in the q-mode:

$$\lim_{V \to \infty} \frac{1}{V} \langle a_{0-}^* a_{0-} \rangle_{H_{1,\Lambda}^{\mathrm{eff}}(\mu,\eta,\rho)} = |\eta|^2 = \frac{4\Omega(\lambda\rho + \epsilon(q) - \mu)}{g^2} , \tag{45}$$

$$\lim_{V \to \infty} \frac{1}{V} \langle a_{q+}^* a_{q+} \rangle_{H_{1,\Lambda}^{\mathrm{eff}}(\mu,\eta,\rho)} = \tau = \frac{4(\lambda\rho - \mu)\Omega}{g^2} , \tag{46}$$

$$\lim_{V \to \infty} \frac{1}{V} \langle b_q^* b_q \rangle_{H_{1,\Lambda}^{\mathrm{eff}}(\mu,\eta,\rho)} = \frac{4(\lambda\rho + \epsilon(q) - \mu)(\lambda\rho - \mu)}{g^2} . \tag{47}$$

Equation (25) becomes:

$$\rho = \frac{8\Omega}{g^2}(\lambda\rho - \mu + \epsilon(q)/2) + 2p_0(\mu - \lambda\rho) . \tag{48}$$

Using the diagonalization of (13) one computes also

$$\lim_{V\to\infty}\frac{1}{V}\langle a_{q+}^* b_q\rangle_{H_{1,\Lambda}^{\text{eff}}(\mu,\eta,\rho)}=\frac{4(\lambda\rho-\mu)\sqrt{\Omega(\lambda\rho+\epsilon(q)-\mu)}}{g^2}\,. \tag{49}$$

Then using (46) and (47), one obtains

$$\lim_{V\to\infty}\left|\frac{1}{V}\langle a_{q+}^* b_q\rangle_{H_{1,\Lambda}^{\text{eff}}(\mu,\eta,\rho)}\right|^2$$

$$=\lim_{V\to\infty}\frac{1}{V}\langle a_{q+}^* a_{q+}\rangle_{H_{1,\Lambda}^{\text{eff}}(\mu,\eta,\rho)}\lim_{V\to\infty}\frac{1}{V}\langle b_q^* b_q\rangle_{H_{1,\Lambda}^{\text{eff}}(\mu,\eta,\rho)}\,. \tag{50}$$

In this case the energy density is given by:

$$\lim_{V\to\infty}\frac{1}{V}\langle H_{1,\Lambda}(\mu)\rangle_{H_{1,\Lambda}^{\text{eff}}(\mu,\eta,\rho)}=\frac{4\Omega(\lambda\rho+\epsilon(q)-\mu)(\lambda\rho-\mu)}{g^2}+2\varepsilon_0(\mu-\lambda\rho)-\frac{1}{2}\lambda\rho^2\,.$$

The entropy density is again given by

$$s(\mu)=2s_0(\mu-\lambda\rho) \tag{51}$$

and the pressure becomes

$$p(\mu)=2p_0(\mu-\lambda\rho)+\frac{1}{2}\lambda\rho^2-\frac{4\Omega(\lambda\rho+\epsilon(q)-\mu)(\lambda\rho-\mu)}{g^2}\,. \tag{52}$$

Before proceeding to a detailed study of the pressure, we would like to comment an important conclusion which is implied by the fact that instead of a standard inequality we obtain the *equality* in (50).

Proposition 1. *The equality (50) implies that the limiting Gibbs state manifests a spontaneous space translation invariance breaking in direction of the vector q.*

The *proof* follows from a direct application of arguments proving Theorem III.3 in [21]. These arguments are developed to prove the spontaneous breaking of the gauge invariance, but by inspection one finds that they cover also the case of translation invariance. Instead of repeating them for our case, we prefer to describe our situation in an explicit manner, cf [21]:

For simplicity we take $q=(2\pi/\gamma)\mathbf{e}_1$, where $\mathbf{e}_1=(1,0,\ldots,0)\in\mathbb{R}^\nu$ and $\gamma>0$. Let $\omega(\cdot)=\lim_{V\to\infty}\langle\cdot\rangle_{H_{1,\Lambda}^{\text{eff}}(\mu,\eta,\rho)}$ denote our equilibrium state determined by the parameter η, the chemical potential μ and the density ρ, where in fact η and ρ are functions of μ through the self-consistency equations. Since the initial Hamiltonian (2) is translation invariant, so is the state ω. Note that (45), (46), (47) and (49) are expectation values in this state of *translation invariant* operators $a_{0-}^* a_{0-}$, $a_{q+}^* a_{q+}$, $b_q^* b_q$ and $a_{q+}^* b_q$. However the single operators $a_{q+}^\sharp$, and $b_q^\sharp$ are not translation invariant and their averages in the state ω are modulated with period γ in the $\mathbf{e}_1$ direction.

Let ω_γ denote an equilibrium state periodic in the $\mathbf{e}_1$ direction with period γ. Then (50) expresses that for ω_γ the analogue of the mixing property for ω takes the following form:

$$\lim_{V \to \infty} \frac{1}{V} \, \omega(a_{q+}^* b_q) = \lim_{V \to \infty} \frac{1}{V} \, \omega_\gamma(a_{q+}^* b_q) \tag{53}$$

$$= \lim_{V \to \infty} \frac{1}{\sqrt{V}} \, \omega_\gamma(a_{q+}^*) \lim_{V \to \infty} \frac{1}{\sqrt{V}} \, \omega_\gamma(b_q) \,,$$

implying that both factors in the right-hand side are non-zero. Similarly, we get, for example, that

$$\lim_{V \to \infty} \frac{1}{V} \, \omega(a_{0-} b_q) = \lim_{V \to \infty} \frac{1}{V} \, \omega_\gamma(a_{0-} b_q) \tag{54}$$

$$= \lim_{V \to \infty} \frac{1}{\sqrt{V}} \, \omega_\gamma(a_{0-}) \lim_{V \to \infty} \frac{1}{\sqrt{V}} \, \omega_\gamma(b_q) \,.$$

To get decoupling in (53) and (54) note that the quasi-local structure (9) implies that

$$\frac{1}{\sqrt{V}} b_q = \frac{1}{V} \int_\Lambda dx \, e^{iq \cdot x} b(x) \tag{55}$$

is an operator space-average, which in the limit is a c-number in the *periodic* state ω_γ. Therefore the emergence of *macroscopic occupation* of the laser q-mode (47) is accompanied by the creation of a one-dimensional *optical lattice* in the e_1 direction with period γ. We can then reconstruct the translation invariant state ω and thus recover (50), by averaging ω_γ:

$$\omega = \frac{1}{\gamma} \int_0^\gamma d\xi \, \omega_\gamma \circ \tau_{e_1 \xi} \,, \tag{56}$$

over elementary interval of the length γ.

2.2 The Pressure for Model 1

Having discussed the three Cases 1–3 we give the values of the chemical potential μ when they occur. This analysis involves a detailed study of the pressure.

Let $\kappa = 8\Omega\lambda/g^2 - 1$ and $\alpha = \epsilon(q)(\kappa + 1)/2$. From the condition for thermodynamic stability we know that $\kappa > 0$. Let x_0 be the unique value of $x \in [0, \infty)$ such that $2\lambda\rho_0'(-x) = \kappa$ and let $\mu_0 = 2\lambda\rho_0(-x_0) + \kappa x_0$. Note that $\mu_0 < 2\lambda\rho_c$.

- The case when $\mu_0 + \alpha \geq 2\lambda\rho_c$ is easy. In this situation Case 1 applies for $\mu \leq 2\lambda\rho_c$ and there exists $\mu_1 > \mu_0 + \alpha$ such that Case 2 applies for $2\lambda\rho_c < \mu < \mu_1$ and Case 3 for $\mu \geq \mu_1$.
- When $\mu_0 + \alpha < 2\lambda\rho_c$ the situation is more subtle. We show that there exists $\mu_1 > \mu_0 + \alpha$ such that Case 3 applies for $\mu \geq \mu_1$. However we are not able to decide on which side of $2\lambda\rho_c$, the point μ_1 lies. If $\mu_1 > 2\lambda\rho_c$ the situation is as in the previous subcase, while if $\mu_0 + \alpha < \mu_1 < 2\lambda\rho_c$ the *intermediate phase* where Case 2 obtains is *eliminated*. This the situation is similar to [7], where one has $\alpha = 0$.

– Note that for $\nu < 3$, Case 1 applies when $\mu < \mu_1$ and Case 3 when $\mu \geq \mu_1$.

Let $x = \lambda \rho - \mu$ and recall that $\kappa = 8\Omega\lambda/g^2 - 1$ and $\alpha = \epsilon(q)(\kappa+1)/2$. In terms of x and η the above classification takes the form:

Case 4. $\eta = 0$ and $x > 0$. Then (34) becomes

$$2\lambda\rho_0(-x) - x = \mu . \tag{57}$$

For $\mu \leq 2\lambda\rho_c$ this has a unique solution in x, denoted by $x_1(\mu)$, while for $\mu > 2\lambda\rho_c$ it has no solutions. Let

$$p_1(x,\mu) := 2p_0(-x) + \frac{(x+\mu)^2}{2\lambda} . \tag{58}$$

Then

$$p(\mu) = p_1(x_1(\mu),\mu) . \tag{59}$$

Case 5. $\eta = 0$ and $x = 0$. For $\mu > 2\lambda\rho_c$

$$p(\mu) = p_2(\mu) := 2p_0(0) + \frac{\mu^2}{2\lambda} . \tag{60}$$

Case 6. $\eta \neq 0$. Then (48) becomes

$$2\lambda\rho_0(-x) + \kappa x + \alpha = \mu . \tag{61}$$

Recall that x_0 is the unique value of $x \in [0,\infty)$ such that $2\lambda\rho_0'(-x) = \kappa$, $\mu_0 = 2\lambda\rho_0(-x_0) + \kappa x_0$ and that $\mu_0 < 2\lambda\rho_c$.

Then for $\mu < \mu_0 + \alpha$, (61) has *no* solutions. For $\mu_0 + \alpha \leq \mu \leq 2\lambda\rho_c + \alpha$ this equation has *two* solutions: $\tilde{x}_3(\mu)$ and $x_3(\mu)$, where $\tilde{x}_3(\mu) < x_3(\mu)$ if $\mu \neq \mu_0 + \alpha$, and $\tilde{x}_3(\mu_0 + \alpha) = x_3(\mu_0 + \alpha)$. Finally for $\mu > 2\lambda\rho_c + \alpha$ it has a *unique* solution $x_3(\mu)$. Let

$$p_3(x,\mu) := 2p_0(-x) + \frac{\{(x+\mu)^2 - (\kappa+1)x^2 - 2\alpha x\}}{2\lambda} . \tag{62}$$

Then

$$\frac{dp_3(\tilde{x}_3(\mu),\mu)}{d\mu} = \frac{\tilde{x}_3(\mu) + \mu}{\lambda} < \frac{x_3(\mu) + \mu}{\lambda} = \frac{dp_3(x_3(\mu),\mu)}{d\mu} \tag{63}$$

for $\mu \neq \mu_0 + \alpha$. Since $p_3(\tilde{x}_3(\mu_0 + \alpha), \mu_0 + \alpha) = p_3(x_3(\mu_0 + \alpha), \mu_0 + \alpha)$,

$$p_3(\tilde{x}_3(\mu),\mu) < p_3(x_3(\mu),\mu) \tag{64}$$

for $\mu_0 + \alpha < \mu \leq 2\lambda\rho_c + \alpha$. Therefore

$$p(\mu) = p_3(x_3(\mu),\mu) \tag{65}$$

for all $\mu \geq \mu_0 + \alpha$. Note that $\tilde{x}_3(2\lambda\rho_c + \alpha) = 0$ so that

$$p_3(\tilde{x}_3(2\lambda\rho_c + \alpha), 2\lambda\rho_c + \alpha) = p_1(x_1(2\lambda\rho_c), 2\lambda\rho_c) = p_2(2\lambda\rho_c) . \tag{66}$$

Therefore

$$p_3(x_3(\mu_0 + \alpha), \mu_0 + \alpha) = p_3(\tilde{x}_3(\mu_0 + \alpha), \mu_0 + \alpha) < 2p_0(0) + 2\lambda\rho_c^2 . \tag{67}$$

Also for large μ, $p_3(x_3(\mu), \mu) \approx (\mu^2/2\lambda)((\kappa + 1)/\kappa)$ while $p_2(\mu) \approx (\mu^2/2\lambda)$, so that $p_3(x_3(\mu), \mu) > p_2(\mu)$ eventually. We remark finally that the slope of $p_3(x_3(\mu), \mu)$ is greater than that of $p_2(\mu)$,

$$\frac{dp_3(x_3(\mu), \mu)}{d\mu} = \frac{x_3(\mu) + \mu}{\lambda} > \frac{\mu}{\lambda} = \frac{dp_2(\mu)}{d\mu} , \tag{68}$$

so that the corresponding curves intersect at most once.

The case $\alpha = 0$, i.e. $\epsilon(q = 0) = 0$, has been examined in [7]. For the case $\alpha > 0$ we have two *subcases*:

The subcase $\mu_0 + \alpha \geq 2\lambda\rho_c$ is easy. In this situation Case 1 applies for $\mu \leq 2\lambda\rho_c$. From (67) we see that

$$p_3(x_3(\mu_0 + \alpha), \mu_0 + \alpha) < 2p_0(0) + 2\lambda\rho_c^2 < p_2(\mu_0 + \alpha) \tag{69}$$

and therefore from the behaviour for large μ we can deduce that there exists $\mu_1 > \mu_0 + \alpha$ such that Case 2 applies for $2\lambda\rho_c < \mu < \mu_1$ and Case 3 for $\mu \geq \mu_1$.

The subcase $\mu_0 + \alpha < 2\lambda\rho_c$ is more complicated. We know that

$$p_3(\tilde{x}_3(2\lambda\rho_c), 2\lambda\rho_c) < p_3(\tilde{x}_3(2\lambda\rho_c + \alpha), 2\lambda\rho_c + \alpha) = p_1(x_1(2\lambda\rho_c), 2\lambda\rho_c) . \tag{70}$$

Therefore since the slope of $p_3(\tilde{x}_3(\mu), \mu)$ is greater than the slope of $p_1(x_1(\mu), \mu)$ for $\mu_0 + \alpha < \mu < 2\lambda\rho_c$:

$$\frac{dp_3(\tilde{x}_3(\mu), \mu)}{d\mu} = \frac{\tilde{x}_3(\mu) + \mu}{\lambda} > \frac{x_1(\mu) + \mu}{\lambda} = \frac{dp_1(x_1(\mu), \mu)}{d\mu} , \tag{71}$$

we can conclude that

$$p_3(x_3(\mu_0 + \alpha), \mu_0 + \alpha) = p_3(\tilde{x}_3(\mu_0 + \alpha), \mu_0 + \alpha) < p_1(x_1(\mu_0 + \alpha), \mu_0 + \alpha) . \tag{72}$$

We also know by the arguments above that there exists $\mu_1 > \mu_0 + \alpha$ such that Case 3 applies for for $\mu \geq \mu_1$. However we do know on which side of $2\lambda\rho_c$, the point μ_1 lies. If $\mu_1 > 2\lambda\rho_c$ the situation is as in the previous subcase while if $\mu_0 + \alpha < \mu_1 < 2\lambda\rho_c$ the intermediate phase where Case 2 obtains is eliminated.

3 Model 2 and Matter-Wave Grating

The analysis for this model is very similar to that of *Model 1*. Therefore, we only briefly summarize the main results. For *Model 2* the effective Hamiltonian is

$$H_{2,\Lambda}^{\mathrm{eff}}(\mu,\eta,\rho) = (\lambda\rho - \mu + \epsilon(q))a_q^* a_q + (\lambda\rho - \mu)a_0^* a_0 + \frac{g}{2}(\eta a_q^* b_q + \bar{\eta} a_q b_q^*)$$

$$+\Omega\, b_q^* b_q + \frac{g\sqrt{V}}{2}(\zeta a_0 + \bar{\zeta} a_0^*) + T_{2,\Lambda}' + (\lambda\rho - \mu)N_{2,\Lambda}' \quad (73)$$

where

$$T_{2,\Lambda}' = \sum_{k\in\Lambda^*,\, k\neq 0\, k\neq q} \epsilon(k)N_k\,, \quad (74)$$

$$N_{2,\Lambda}' = \sum_{k\in\Lambda^*,\, k\neq 0\, k\neq q} \epsilon(k)N_k\,. \quad (75)$$

The parameters η, ζ and ρ satisfy the *self-consistency* equations:

$$\eta = \frac{1}{\sqrt{V}}\langle a_0\rangle_{H_{2,\Lambda}^{\mathrm{eff}}(\mu,\eta,\rho)},\ \zeta = \frac{1}{V}\langle a_q^* b_q\rangle_{H_{2,\Lambda}^{\mathrm{eff}}(\mu,\eta,\rho)},\ \rho = \frac{1}{V}\langle N_{2,\Lambda}\rangle_{H_{2,\Lambda}^{\mathrm{eff}}(\mu,\eta,\rho)}\,.$$
$$(76)$$

Solving these equations we again have three cases:

Case 7. $\zeta = \eta = 0$ and $\lambda\rho - \mu > 0$. In this case there is *no condensation*:

$$\lim_{V\to\infty}\frac{1}{V}\langle a_0^* a_0\rangle_{H_{2,\Lambda}^{\mathrm{eff}}(\mu,\eta,\rho)} = \lim_{V\to\infty}\frac{1}{V}\langle a_q^* a_q\rangle_{H_{2,\Lambda}^{\mathrm{eff}}(\mu,\eta,\rho)}$$

$$= \lim_{V\to\infty}\frac{1}{V}\langle b_q^* b_q\rangle_{H_{2,\Lambda}^{\mathrm{eff}}(\mu,\eta,\rho)} = 0\,,$$

the density equation is

$$\rho = \rho_0(\mu - \lambda\rho) \quad (77)$$

and the pressure is

$$p(\mu) = p_0(\mu - \lambda\rho) + \frac{1}{2}\lambda\rho^2\,.$$

Case 8. $\eta = 0$, $\lambda\rho - \mu = 0$. Here $\rho \geq \rho_c$ and

$$\lim_{V\to\infty}\frac{1}{V}\langle a_0^* a_0\rangle_{H_{2,\Lambda}^{\mathrm{eff}}(\mu,\eta,\rho)} = \rho - \rho_c\,.$$

There is condensation in the $k = 0$ mode but there is *no condensation* in the $k = q$ mode and the photon laser field:

$$\lim_{V\to\infty}\frac{1}{V}\langle a_q^* a_q\rangle_{H_{2,\Lambda}^{\mathrm{eff}}(\mu,\eta,\rho)} = \lim_{V\to\infty}\frac{1}{V}\langle b_q^* b_q\rangle_{H_{2,\Lambda}^{\mathrm{eff}}(\mu,\eta,\rho)} = 0\,. \quad (78)$$

The pressure density is given by

$$p(\mu) = p_0(0) + \frac{\mu^2}{2\lambda}\,. \quad (79)$$

Case 9. $\eta \neq 0$. There is simultaneous condensation of the zero-mode and the q-mode bosons as well as the laser q-mode photons:

$$\lim_{V\to\infty}\frac{1}{V}\langle a_0^* a_0\rangle_{H_{2,\Lambda}^{\mathrm{eff}}(\mu,\eta,\rho)} = \frac{4\Omega(\lambda\rho + \epsilon(q) - \mu)}{g^2}\,, \quad (80)$$

$$\lim_{V \to \infty} \frac{1}{V} \langle a_q^* a_q \rangle_{H_{2,\Lambda}^{\text{eff}}(\mu,\eta,\rho)} = \frac{4(\lambda\rho - \mu)\Omega}{g^2} , \tag{81}$$

$$\lim_{V \to \infty} \frac{1}{V} \langle b_q^* b_q \rangle_{H_{2,\Lambda}^{\text{eff}}(\mu,\eta,\rho)} = \frac{4(\lambda\rho + \epsilon(q) - \mu)(\lambda\rho - \mu)}{g^2} , \tag{82}$$

$$\lim_{V \to \infty} \frac{1}{V} \langle a_q^* b_q \rangle_{H_{2,\Lambda}^{\text{eff}}(\mu,\eta,\rho)} = \frac{4(\lambda\rho - \mu)\sqrt{\Omega(\lambda\rho + \epsilon(q) - \mu)}}{g^2} \tag{83}$$

and

$$\lim_{V \to \infty} \left| \frac{1}{V} \langle a_q^* b_q \rangle_{H_{2,\Lambda}^{\text{eff}}(\mu,\eta,\rho)} \right|^2 \tag{84}$$

$$= \lim_{V \to \infty} \frac{1}{V} \langle a_q^* a_q \rangle_{H_{2,\Lambda}^{\text{eff}}(\mu,\eta,\rho)} \lim_{V \to \infty} \frac{1}{V} \langle b_q^* b_q \rangle_{H_{2,\Lambda}^{\text{eff}}(\mu,\eta,\rho)} .$$

The density is given by

$$\rho = \frac{8\Omega}{g^2}(\lambda\rho - \mu + \epsilon(q)/2) + \rho_0(\mu - \lambda\rho)$$

and pressure is

$$p(\mu) = p_0(\mu - \lambda\rho) + \frac{1}{2}\lambda\rho^2 - \frac{4\Omega(\lambda\rho + \epsilon(q) - \mu)(\lambda\rho - \mu)}{g^2} .$$

Note that relations between the values of μ and the three cases above are exactly the same as for *Model 1* apart from the fact that $2\rho_0$ is now replaced by ρ_0 and $2\rho_c$ by ρ_c. For that one has to compare the kinetic energy operators (1) and (4).

The recently observed phenomenon of *periodic spacial variation in the boson-density* is responsible for the light and matter-wave *amplification* in superradiant condensation, see [2–4, 13]. This so called *matter-wave grating* is produced by the interference of two different macroscopically occupied momentum states: the first corresponds to a macroscopic number of *recoiled* bosons and the second to *residual* BE condensate at rest.

To study the possibility of such interference in *Model 1* we recall that for system (2), with *two* species of boson atoms the local particle density operator has the form

$$\rho(x) := \frac{1}{V} \sum_{k \in \Lambda^*, \sigma = \pm} \rho_{k,\sigma} e^{ikx} , \tag{85}$$

where the Fourier transforms of the local particle densities for two species are

$$\rho_{k,\sigma} := \sum_{p \in \Lambda^*} a_{p+k,\sigma}^* a_{p,\sigma} . \tag{86}$$

If the limiting equilibrium states generated by the Hamiltonian (2) is translation invariant, the momentum conservation law yields:

$$\lim_{V\to\infty} \omega(a^*_{p+k,\sigma} a_{p,\sigma}) = \delta_{k,0} \lim_{V\to\infty} \omega(a^*_{p,\sigma} a_{p,\sigma}) . \tag{87}$$

So, in this state the equilibrium expectation of the local density

$$\lim_{V\to\infty} \omega(\rho(x)) = \tag{88}$$

$$\lim_{V\to\infty} \frac{1}{V} \sum_{k\in\Lambda^*,\sigma} e^{ikx} \sum_{p\in\Lambda^*} \omega(a^*_{p+k,\sigma} a_{p,\sigma}) = \lim_{V\to\infty} \omega(\rho(0))$$

is a *constant*. Since condensation may *break* the translation invariance in one direction (see Proposition 1), we get a corresponding non-homogeneity (*grating*) of the equilibrium total particle density in the extremal ω_γ state. This means that in the integral sum (88) over k, the $\pm q$-mode terms survive in the thermodynamic limit. By (53) and (54) we know that condensation occurs only in the *zero* mode for the $\sigma = -$ bosons and in the q-mode for the $\sigma = +$ bosons. and therefore we have the following relations:

$$\lim_{V\to\infty} \frac{1}{V}\omega_\gamma(a^*_{q,+} a_{0,+}) = \lim_{V\to\infty} \frac{1}{\sqrt{V}}\omega_\gamma(a^*_{q,+}) \lim_{V\to\infty} \frac{1}{\sqrt{V}}\omega_\gamma(a_{0,+}) , \tag{89}$$

$$\lim_{V\to\infty} \frac{1}{V}\omega_\gamma(a^*_{q,-} a_{0,-}) = \lim_{V\to\infty} \frac{1}{\sqrt{V}}\omega_\gamma(a^*_{q,-}) \lim_{V\to\infty} \frac{1}{\sqrt{V}}\omega_\gamma(a_{0,-}) . \tag{90}$$

To get decoupling in (89) and (90) one has to note that (as in (53), (54)) the operators $a_{0,\pm}/\sqrt{V}$, or $a^*_{q,\pm}/\sqrt{V}$, are space-averages, which in the limit are c-numbers in the periodic state ω_γ. Since there is no condensation of $\sigma = \pm$ bosons except in those two modes, the right-hand sides of both (89) and (90) are equal to zero. Noting that

$$\lim_{V\to\infty} \frac{1}{V}\omega_\gamma(a^*_{p+k,\sigma} a_{p,\sigma}) = 0 \tag{91}$$

for any other mode, one gets the *space homogeneity* of the equilibrium particle density in the extremal ω_γ state: $\lim_{V\to\infty} \omega_\gamma(\rho(x)) = const$. So, for *the Model 1* we have no particle density space variation even in the presence of the *light corrugated lattice of condensed photons*, cf Proposition 1.

Let us now look at the corresponding situation in *Model 2*. The important difference is that in this model the *same* boson atoms may condense in two states (see Case 3):

$$\lim_{V\to\infty} \frac{1}{\sqrt{V}}\omega_\gamma(a^*_q) \neq 0 \quad \text{and} \quad \lim_{V\to\infty} \frac{1}{\sqrt{V}}\omega_\gamma(a_0) \neq 0 . \tag{92}$$

Then (90) implies

$$\lim_{V\to\infty} \frac{1}{V}\omega_\gamma(a^*_q a_0) = \lim_{V\to\infty} \frac{1}{\sqrt{V}}\omega_\gamma(a^*_q) \lim_{V\to\infty} \frac{1}{\sqrt{V}}\omega_\gamma(a_0) \equiv \xi \neq 0 . \tag{93}$$

Therefore, the bosons of these two condensates may interfere. By virtue of (88) and (93) this gives the matter-wave grating formed by two macroscopically occupied momentum states:

$$\lim_{V\to\infty} \omega_\gamma(\rho(x)) = \lim_{V\to\infty} \frac{1}{V} \sum_{k\in\Lambda^*,\sigma} e^{ikx} \sum_{p\in\Lambda^*} \omega_\gamma(a^*_{p+k,\sigma}a_{p,\sigma}) \tag{94}$$

$$= (\xi e^{iqx} + \overline{\xi}e^{-iqx}) + \lim_{V\to\infty} \frac{1}{V} \sum_{k\neq\pm q} e^{ikx}\omega_\gamma(\rho^\circ_k) ,$$

where

$$\rho^\circ_k := \sum_{p\neq 0,q} a^*_{p+k}a_p . \tag{95}$$

Notice that by (93) and by (94) there is no matter-wave grating in the Case 2, when one of the condensates (for example the q-$condensate$) is empty, see (41).

4 Conclusion

It is clear that the absence of the matter-wave grating in *Model 1* and its presence in *Model 2* provides a physical distinction between Raman and Rayleigh superradiance. Note first that matter-wave amplification differs from light amplification in one important aspect: a matter-wave amplifier has to possess a *reservoir* of atoms. In *Models 1* and *2* this is the BE condensate. In both models the superradiant scattering transfers atoms from the condensate at rest to a recoil mode.

The *gain* mechanism for the *Raman amplifier* is superradiant Raman scattering in a two-level atoms, transferring bosons from the condensate into the recoil state [1]. The *Rayleigh amplifier* is in a sense even more effective. Since now the atoms in a recoil state interfere with the BE condensate at rest, the system exhibits a space *matter-wave grating* and the quantum-mechanical amplitude of transfer into the recoil state is proportional to the product $N_0(N_q + 1)$. Each time the momentum imparted by photon scattering is absorbed by the matter-wave grating by the coherent transfer of an atom from the condensate into the recoil mode. Thus, the variance of the grating grows, since the quantum amplitude for scattered atom to be transferred into a recoiled state is increasing [2–4, 13]. At the same time the dressing laser beam prepares from the BE condensate a gain medium able to amplify the light. The matter-wave grating diffracts the dressing beam into the path of the probe light resulting in the amplification of the latter [5].

In the case of equilibrium BEC superradiance the amplification of the light and the matter waves manifests itself in *Models 1* and *2* as a *mutual enhancement* of the BEC and the photons condensations, see Cases 3 in Sects. 2 and 3. Note that the corresponding formulæ for condensation densities for *Model 1* (45)–(47) and for *Model 2* (80), (81), (82) are *identical*. The same is true for the boson-photon correlations (*entanglements*) between recoiled bosons and photons, see (49), (83), as well as between photons and the BE condensate at rest:

$$\lim_{V \to \infty} \frac{1}{V} \langle a_0^* {}_- b_q \rangle_{H_{1,\Lambda}^{\mathrm{eff}}(\mu,\eta,\rho)} = \frac{1}{V} \langle a_0^* b_q \rangle_{H_{2,\Lambda}^{\mathrm{eff}}(\mu,\eta,\rho)} \tag{96}$$

$$= \frac{4\Omega(\lambda\rho + \epsilon(q) - \mu)\sqrt{\lambda\rho - \mu}}{g^2},$$

and for the *off-diagonal coherence* between recoiled atoms and the condensate at rest:

$$\lim_{V \to \infty} \frac{1}{V} \langle a_0 a_{q+}^* \rangle_{H_{1,\Lambda}^{\mathrm{eff}}(\mu,\eta,\rho)} = \frac{1}{V} \langle a_0 a_q^* \rangle_{H_{2,\Lambda}^{\mathrm{eff}}(\mu,\eta,\rho)} \tag{97}$$

$$= \frac{4\Omega\sqrt{(\lambda\rho + \epsilon(q) - \mu)(\lambda\rho - \mu)}}{g^2}.$$

As we have shown above, the *difference* between *Models 1* and *2* becomes visible only on the level of the wave-grating or spatial modulation of the *local* particle density (94).

Acknowledgement

We wish to thank Alain Joye and Joachim Asch for invitation to give this lecture on QMath9.

References

1. D. Schneble, G.K. Campbell, E.W. Streed et al. Phys. Rev. A **69**, 041601 (2004).
2. S. Inouye, A.P. Chikkatur, D.M. Stamper-Kurn et al. Science **285**, 571 (1999).
3. D. Schneble, Y. Toril, M. Boyd et al. Science **300**, 475 (2003).
4. W. Ketterle, S. Inouye: Phys. Rev. Lett. **89**, 4203 (2001).
5. W. Ketterle, S. Inouye: C.R. Acad. Sci. Paris série IV **2**, 339 (2001).
6. J.V. Pulé, A. Verbeure, V.A. Zagrebnov: J. Phys. A (Math. Gen.) **37**, L321 (2004).
7. J.V. Pulé, A.F. Verbeure, V.A. Zagrebnov: J. Stat. Phys. **118** (2005).
8. J.V. Pulé, A.F. Verbeure, V.A. Zagrebnov: J. Phys. A (Math. Gen.) (2005).
9. N. Piovella, R. Bonifacio, B.W.J. McNeil et al. Optics Commun. **187**, 165 (2001).
10. R. Bonifacio, F.S. Cataliotti, M. Cola et al. Optics Commun. **233**, 155 (2004).
11. M.G. Moore, P. Meystre: Phys. Rev. Lett. **86**, 4199 (2001).
12. M.G. Moore, P. Meystre: Phys. Rev. Lett. **83**, 5202 (1999).
13. M. Kozuma, Y. Suzuki, Y. Torii et al. Science **286**, 2309 (1999).
14. R.H. Dicke: Phys. Rev. **93**, 99 (1954).
15. K. Hepp, E.H. Lieb: Ann. Phys. **76**, 360 (1973).
16. J.G. Brankov, V.A. Zagrebnov, N.S. Tonchev: Theor. Math. Phys. **22**, 13 (1975).
17. M. Fannes, P.N.M. Sisson, A. Verbeure et al. Ann. Phys. **98**, 38 (1976).

278   J.V. Pulé et al.

18. O. Bratteli, D.W. Robinson: *Operator Algebras and Quantum Statistical Mechanics II*, 2nd edn (Springer, Berlin Heidelberg New York 1997).
19. M. Fannes, H. Spohn, A. Verbeure: J. Math. Phys. **21**, 355 (1980).
20. M. Fannes, A. Verbeure: Commun. Math. Phys. **57**, 165 (1977).
21. M. Fannes, J.V. Pulé, A. Verbeure: Helv.Phys.Acta **55**, 391 (1982).

Derivation of the Gross-Pitaevskii Hierarchy

Benjamin Schlein[*]

Department of Mathematics, Stanford University, Stanford, CA 94305, USA
schlein@math.stanford.edu

Abstract. We report on some recent results regarding the dynamical behavior of a trapped Bose-Einstein condensate, in the limit of a large number of particles. These results were obtained in [4], a joint work with L. Erdős and H.-T. Yau.

1 Introduction

In the last years, progress in the experimental techniques has made the study of dilute Bose gas near the ground state a hot topic in physics. For the first time, the existence of Bose-Einstein condensation for trapped gases at very low temperatures has been verified experimentally. The experiments were conducted observing the dynamics of Bose systems, trapped by strong magnetic field and cooled down at very low temperatures, when the confining traps are switched off. It seems therefore important to have a good theoretical description of the dynamics of the condensate. Already in 1961 Gross [7,8] and Pitaevskii [14] proposed to model the many body effects in a trapped dilute Bose gas by a nonlinear on-site self interaction of a complex order parameter (the condensate wave function u_t). They derived the Gross-Pitaevskii equation

$$i\partial_t u_t = -\Delta u_t + 8\pi a_0 |u_t|^2 u_t \tag{1}$$

for the evolution of u_t. Here a_0 is the scattering length of the pair interaction. A mathematically rigorous justification of this equation is still missing. The aim of this article is to report on recent partial results towards the derivation of (1) starting from the microscopic quantum dynamics in the limit of a large number of particles. Here we only expose the main ideas: for more details, and for all the proofs, we refer to [4].

Also in the mathematical analysis of dilute bosonic systems some important progress has recently been made. In [13], Lieb and Yngvason give a rigorous proof of a formula for the leading order contribution to the ground state energy of a dilute Bose gas (the correct upper bound for the energy was already obtained by Dyson in [2], for the case of hard spheres). This important result inspired a lot of subsequent works establishing different properties of the ground state of the Bose system. In [12], the authors give a proof of

---

[*]Supported by NSF Postdoctoral Fellowship

B. Schlein: *Derivation of the Gross-Pitaevskii Hierarchy*, Lect. Notes Phys. **690**, 279–293 (2006)
www.springerlink.com   © Springer-Verlag Berlin Heidelberg 2006

the asymptotic exactness of the Gross-Pitaevskii energy functional for the computation of the ground state energy of a trapped Bose gas. In [10], the complete condensation of the ground state of a trapped Bose gas is proven. For a review of recent results on dilute Bose systems we refer to [11]. All these works investigate the properties of the ground state of the system. Here, on the other hand, we are interested in the dynamical behavior.

Next, we want to describe our main result in some details. To this end, we need to introduce some notation. From now on we consider a system of N bosons trapped in a box $\Lambda \subset \mathbb{R}^3$ with volume one and we impose periodic boundary conditions. In order to describe the interaction among the bosons, we choose a positive, smooth, spherical symmetric potential $V(x)$ with compact support and with scattering length a_0.

Let us briefly recall the definition of the scattering length a_0 of the potential $V(x)$. To define a_0 we consider the radial symmetric solution $f(x)$ of the zero energy one-particle Schrödinger equation

$$\left(-\Delta + \frac{1}{2}V(x)\right) f(x) = 0 \,, \tag{2}$$

with the condition $f(x) \to 1$ for $|x| \to \infty$. Since the potential has compact support, we can define the scattering length a_0 associated to $V(x)$ by the equation $f(x) = 1 - a_0/|x|$ for x outside the support of $V(x)$ (this definition can be generalized by $a_0 = \lim_{r\to\infty} r(1 - f(r))$, if V has unbounded support but still decays sufficiently fast at infinity). Another equivalent characterization of the scattering length is given by the formula

$$\int \mathrm{d}x\, V(x)f(x) = 8\pi a_0 \,. \tag{3}$$

Physically, a_0 is a measure of the effective range of the potential $V(x)$.

The Hamiltonian of the N-boson system is then given by

$$H = -\sum_{j=1}^{N} \Delta_j + \sum_{i<j} V_a(x_i - x_j) \tag{4}$$

with $V_a(x) = (a_0/a)^2 V((a_0/a)x)$. By scaling, V_a has scattering length a. In the following we keep a_0 fixed (of order one) and we vary a with N, so that when N tends to infinity a approaches zero. In order for the Gross-Pitaevskii theory to be relevant we have to take a of order N^{-1} (see [12] for a discussion of other possible scalings). In the following we choose therefore $a = a_0/N$, and thus $V_a(x) = N^2 V(Nx)$. Note that, with this choice of a, the Hamiltonian (4) can be viewed as a special case of the mean-field Hamiltonian

$$H_{\mathrm{mf}} = -\sum_{j=1}^{N} \Delta_j + \frac{1}{N} \sum_{i<j} \beta^3 V(\beta(x_i - x_j)) \,. \tag{5}$$

The Gross-Pitaevskii scaling is recovered when $\beta = N$. We study the dynamics generated by (5) for other choices of β ($\beta = N^\alpha$, with $\alpha < 3/5$) in [3].

Since we have N particles in a box of volume one, the density is given by $\rho = N$. Hence, the total number of particles interacting at a given time with a fixed particle in the system is typically of order $\rho a^3 \simeq N^{-2} \ll 1$. This means that our system is actually a very dilute gas, scaled so that the total volume remains fixed to one.

The dynamics of the N-boson system is determined by the Schrödinger equation

$$i\partial_t \psi_{N,t} = H\psi_{N,t} \tag{6}$$

for the wave function $\psi_{N,t} \in L^2(\mathbb{R}^{3N}, d\mathbf{x})$. Instead of describing the quantum mechanical system through its wave-function we can describe it by the corresponding density matrix $\gamma_{N,t} = |\psi_{N,t}\rangle\langle\psi_{N,t}|$ which is the orthogonal projection onto $\psi_{N,t}$. We choose the normalization so that $\mathrm{Tr}\,\gamma_{N,t} = 1$. The Schrödinger equation (6) takes the form

$$i\partial_t \gamma_{N,t} = [H, \gamma_{N,t}] . \tag{7}$$

For large N this equation becomes very difficult to solve, even numerically. Therefore, it is desirable to have an easier description of the dynamics of the system in the limit $N \to \infty$, assuming we are only interested in its macroscopic behavior, resulting from averaging over the N particles. In order to investigate the macroscopic dynamics, we introduce the marginal distributions associated to the density matrix $\gamma_{N,t}$. The k-particle marginal distribution $\gamma_{N,t}^{(k)}$ is defined by taking the partial trace over the last $N - k$ variables. That is, the kernel of $\gamma_{N,t}^{(k)}$ is given by

$$\gamma_{N,t}^{(k)}(\mathbf{x}_k; \mathbf{x}_k') = \int d\mathbf{x}_{N-k}\, \gamma_{N,t}(\mathbf{x}_k, \mathbf{x}_{N,k}; \mathbf{x}_k', \mathbf{x}_{N-k})$$

where $\gamma_{N,t}(\mathbf{x}; \mathbf{x}')$ denotes the kernel of the density matrix $\gamma_{N,t}$. Here and in the following we use the notation $\mathbf{x} = (x_1, \ldots, x_N)$, $\mathbf{x}_k = (x_1, \ldots, x_k)$, $\mathbf{x}_{N-k} = (x_{k+1}, \ldots, x_N)$, and analogously for the primed variables. By definition, the k-particle marginal distributions satisfy the normalization condition

$$\mathrm{Tr}\,\gamma_{N,t}^{(k)} = 1 \qquad \text{for all} \quad k = 1, \ldots, N .$$

In contrast to the density matrix $\gamma_{N,t}$, one can expect that, for fixed k, the marginal distribution $\gamma_{N,t}^{(k)}$ has a well defined limit $\gamma_{\infty,t}^{(k)}$ for $N \to \infty$ (with respect to some suitable weak topology), whose dynamics can be investigated. In particular, the Gross-Pitaevskii equation (1) is expected to describe the time evolution of the limit $\gamma_{\infty,t}^{(1)}$ of the one-particle marginal distribution, provided $\gamma_{\infty,t}^{(1)} = |u_t\rangle\langle u_t|$ is a pure state. Equation (1) can be generalized, for $\gamma_{\infty,t}^{(1)}$ describing a mixed state, to

$$i\partial_t\gamma_{\infty,t}^{(1)}(x;x') = (-\Delta + \Delta')\gamma_{\infty,t}^{(1)}(x;x')$$
$$+ 8\pi a_0 \left(\gamma_{\infty,t}^{(1)}(x;x) - \gamma_{\infty,t}^{(1)}(x';x')\right)\gamma_{\infty,t}^{(1)}(x;x') \,, \tag{8}$$

which we again denote as the Gross-Pitaevskii equation.

To understand the origin of (8), we start from the dynamics of the marginals $\gamma_{N,t}^{(k)}$, for finite N. From the Schrödinger equation (7), we can easily derive a hierarchy of N equations, commonly called the BBGKY hierarchy, describing the evolution of the distributions $\gamma_{N,t}^{(k)}$, for $k = 1,\ldots,N$:

$$i\partial_t\gamma_{N,t}^{(k)}(\mathbf{x}_k;\mathbf{x}_k') = \sum_{j=1}^{k}(-\Delta_{x_j} + \Delta_{x_j'})\gamma_{N,t}^{(k)}(\mathbf{x}_k;\mathbf{x}_k')$$

$$+ \sum_{j\neq\ell}^{k}(V_a(x_j - x_\ell) - V_a(x_j' - x_\ell'))\gamma_{N,t}^{(k)}(\mathbf{x}_k;\mathbf{x}_k')$$

$$+ (N-k)\sum_{j=1}^{k}\int dx_{k+1}(V_a(x_j - x_{k+1}) - V_a(x_j' - x_{k+1})) \tag{9}$$

$$\times \gamma_{N,t}^{(k+1)}(\mathbf{x}_k,x_{k+1};\mathbf{x}_k',x_{k+1}) \,.$$

Here we use the convention that $\gamma_{N,t}^{(k)} = 0$, for $k > N$. Hence, the one-particle marginal density $\gamma_{N,t}^{(1)}$ satisfies

$$i\partial_t\gamma_{N,t}^{(1)}(x_1;x_1') = (-\Delta_{x_1} + \Delta_{x_1'})\gamma_{N,t}^{(1)}(x_1;x_1')$$
$$+ (N-1)\int dx_2 \,(V_a(x_1 - x_2) - V_a(x_1' - x_2))\,\gamma_{N,t}^{(2)}(x_1,x_2;x_1',x_2) \,. \tag{10}$$

In order to get a closed equation for $\gamma_{N,t}^{(1)}$ we need to assume some relation between $\gamma_{N,t}^{(1)}$ and $\gamma_{N,t}^{(2)}$. The most natural assumption consists in taking the two particle marginal to be the product of two identical copies of the one particle marginal. Although this kind of factorization cannot be true for finite N, it may hold in the limit $N \to \infty$. We suppose therefore that $\gamma_{\infty,t}^{(k)}$, for $k = 1,2$, is a limit point of $\gamma_{N,t}^{(k)}$, with respect to some weak topology, with the factorization property

$$\gamma_{\infty,t}^{(2)}(x_1,x_2;x_1',x_2') = \gamma_{\infty,t}^{(1)}(x_1;x_1')\gamma_{\infty,t}^{(1)}(x_2;x_2') \,.$$

Under this assumption we could naively guess that, in the limit $N \to \infty$, (10) takes the form

$$i\partial_t\gamma_{\infty,t}^{(1)}(x_1;x_1') = (-\Delta_{x_1} + \Delta_{x_1'})\gamma_{\infty,t}^{(1)}(x_1;x_1')$$
$$+ (Q_t(x_1) - Q_t(x_1'))\gamma_{\infty,t}^{(1)}(x_1;x_1') \tag{11}$$

with

$$Q_t(x_1) = \lim_{N \to \infty} N \int dx_2 \, V_a(x_1 - x_2) \gamma_{\infty,t}^{(1)}(x_2; x_2)$$

$$= \lim_{N \to \infty} \int dx_2 \, N^3 V(N(x_1 - x_2)) \gamma_{\infty,t}^{(1)}(x_2; x_2)$$

$$= b_0 \gamma_{\infty,t}^{(1)}(x_1; x_1)$$

where we defined $b_0 = \int dx \, V(x)$. Using the last equation, (11) can be rewritten as

$$i\partial_t \gamma_{\infty,t}^{(1)}(x_1; x_1') = (-\Delta_{x_1} + \Delta_{x_1'}) \gamma_{\infty,t}^{(1)}(x_1; x_1')$$
$$+ b_0 \left(\gamma_{\infty,t}^{(1)}(x_1; x_1) - \gamma_{\infty,t}^{(1)}(x_1'; x_1') \right) \gamma_{\infty,t}^{(1)}(x_1; x_1') \tag{12}$$

which is exactly the Gross-Pitaevskii equation (8), but with the wrong coupling constant in front of the non-linear term (b_0 instead of $8\pi a_0$). The fact that we get the wrong coupling constant suggests that something was not completely correct with the naif argument leading from (10) to (12). Reconsidering the argument, the origin of the error is quite clear: when passing to the limit $N \to \infty$ we first replaced $\gamma_{N,t}^{(2)}$ with $\gamma_{\infty,t}^{(2)}$ and only after this replacement we took the limit $N \to \infty$ in the potential. This procedure gives the wrong result because the marginal distribution $\gamma_{N,t}^{(2)}$ has a short scale structure living on the scale $1/N$, which is the same length scale characterizing the potential $V_a(x)$. The short scale structure of $\gamma_{N,t}^{(2)}$ (which describes the correlations among the particles) disappears when the weak limit is taken, so that $\gamma_{\infty,t}^{(2)}$ lives on a length scale of order one. Therefore, in (12) we get the wrong coupling constant because we erroneously disregarded the effect of the correlations present in $\gamma_{N,t}^{(2)}$. It is hence clear that in order to derive the Gross-Pitaevskii equation (8) with the correct coupling constant $8\pi a_0$, we need to take into account the short scale structure of $\gamma_{N,t}^{(2)}$. To this end we begin by studying the ground state of the system.

A good approximation for the ground state wave function of the N boson system is given by

$$W(\mathbf{x}) = \prod_{i<j}^{N} f(N(x_i - x_j))$$

where $f(x)$ is defined by (2) (then $f(Nx)$ solves the same equation (2) with V replaced by V_a). Since we assumed the potential to be compactly supported (let R denote the radius of its support), we have $f(x) = 1 - a_0/|x|$, for $|x| > R$, and thus $f(Nx) = 1 - a_0/N|x| = 1 - a/|x|$, for $|x| > Ra$. A similar ansatz for the ground state wave function was already used by Dyson in [2] to prove his upper bound on the ground state energy. In order to describe states of the condensate, it seems appropriate to consider wave functions of the form

$$\psi_N(\mathbf{x}) = W(\mathbf{x})\phi_N(\mathbf{x})$$

where $\phi_N(\mathbf{x})$ varies over distances of order one, and is approximately factorized, that is $\phi_N(\mathbf{x}) \simeq \prod_{j=1}^{N} \phi(x_j)$. Assuming for the moment that this form is preserved under the time-evolution we have

$$\gamma_{N,t}^{(2)}(x_1, x_2; x_1', x_2') \simeq f(N(x_1 - x_2))f(N(x_1' - x_2'))\gamma_{N,t}^{(1)}(x_1; x_1')\gamma_{N,t}^{(1)}(x_2; x_2') \ .$$

Thus, for finite N, $\gamma_{N,t}^{(2)}$ is not exactly factorized and has a short scale structure given by the function $f(Nx)$. When we consider the limit $N \to \infty$ of the second term on the right hand side of (10) we obtain

$$\lim_{N \to \infty} N \int dx_2\, V_a(x_1 - x_2)\gamma_{N,t}^{(2)}(x_1, x_2; x_1', x_2)$$

$$= \lim_{N \to \infty} N^3 \int dx_2\, V(N(x_1 - x_2))f(N(x_1 - x_2))\gamma_{\infty,t}^{(1)}(x_1; x_1')\gamma_{\infty,t}^{(1)}(x_2; x_2)$$

$$= 8\pi a_0 \gamma_{\infty,t}^{(1)}(x_1; x_1')\gamma_{\infty,t}^{(1)}(x_1; x_1)$$

$$(13)$$

where we used (3) and the fact that $\gamma_{N,t}^{(1)}$ lives on a scale of order one (and thus we can replace it by $\gamma_{\infty,t}^{(1)}$ without worrying about the correlations). This leads to the Gross-Pitaevskii equation for $\gamma_{\infty,t}^{(1)}$,

$$i\partial_t \gamma_{\infty,t}^{(1)}(x_1; x_1') = \left(-\Delta_{x_1} + \Delta_{x_1'}\right)\gamma_{\infty,t}^{(1)}(x_1; x_1')$$
$$+ 8\pi a_0 \left(\gamma_{\infty,t}^{(1)}(x_1; x_1) - \gamma_{\infty,t}^{(1)}(x_1'; x_1')\right)\gamma_{\infty,t}^{(1)}(x_1; x_1')$$

which has the correct coupling constant in front of the non-linear term.

Note that the factorization

$$\gamma_{\infty,t}^{(2)}(x_1, x_2; x_1', x_2') = \gamma_{\infty,t}^{(1)}(x_1; x_1')\gamma_{\infty,t}^{(1)}(x_2; x_2')$$

still holds true, because the short scale structure of $\gamma_{N,t}^{(2)}$ vanishes when the weak limit $N \to \infty$ is taken. The short scale structure only shows up in the Gross-Pitaevskii equation due to the singularity of the potential.

In order to make this heuristic argument for the derivation of the Gross-Pitaevskii equation rigorous, we are faced with two major steps.

(i) In the first step we have to prove that the k-particle marginal density in the limit $N \to \infty$ really has the short scale structure we discussed above. That is we have to prove that, for large N,

$$\gamma_{N,t}^{(k+1)}(\mathbf{x}_{k+1}; \mathbf{x}_{k+1}') \simeq \left(\prod_{i<j}^{k+1} f(N(x_i - x_j))f(N(x_i' - x_j'))\right)$$
$$\times \gamma_{\infty,t}^{(k+1)}(\mathbf{x}_{k+1}; \mathbf{x}_{k+1}') \quad (14)$$

where $\gamma_{\infty,t}^{(k+1)}$ is the limit of $\gamma_{N,t}^{(k+1)}$ with respect to some suitable weak topology (in the heuristic argument above we considered the case $k = 1$, here k is an arbitrary fixed integer $k \geq 1$). Equation (14) would then imply that, as $N \to \infty$, the last term on the r.h.s. of the BBGKY hierarchy (9) converges to

$$\lim_{N \to \infty} N \int dx_{k+1} V_a(x_j - x_{k+1}) \gamma_{N,t}^{(k+1)}(\mathbf{x}_{k+1}; \mathbf{x}'_{k+1})$$
$$= 8\pi a_0 \gamma_{\infty,t}^{(k+1)}(\mathbf{x}_k, x_j; \mathbf{x}'_k, x_j).$$

Therefore, if we could also prove that the second term on the r.h.s. of (9) vanishes in the limit $N \to \infty$ (as expected, because formally of the order N^{-1}), then it would follow that the family $\gamma_{\infty,t}^{(k)}$ satisfies the Gross-Pitaevskii hierarchy

$$i\partial_t \gamma_{\infty,t}^{(k)}(\mathbf{x}_k; \mathbf{x}'_k) = \sum_{j=1}^{k} \left(-\Delta_j + \Delta'_j \right) \gamma_{\infty,t}^{(k)}(\mathbf{x}_k; \mathbf{x}'_k)$$

$$+ 8\pi a_0 \sum_{j=1}^{k} \int dx_{k+1} \left(\delta(x_{k+1} - x_j) - \delta(x_{k+1} - x'_j) \right)$$

$$\times \gamma_{\infty,t}^{(k+1)}(\mathbf{x}_k, x_{k+1}; \mathbf{x}'_k, x_{k+1}) \tag{15}$$

for all $k \geq 1$. We already know that this infinite hierarchy of equation has a solution. In fact the factorized family of densities $\gamma_{\infty,t}^{(k)}(\mathbf{x}_k; \mathbf{x}'_k) = \prod_{j=1}^{k} \gamma_{\infty,t}^{(1)}(x_j; x'_j)$ is a solution of (15) if and only if $\gamma_{\infty,t}^{(1)}$ solves the Gross-Pitaevskii equation (8).

(ii) Secondly, we need to prove that the densities $\gamma_{\infty,t}^{(k)}$ factorize, that is, that, for all $k \geq 1$,

$$\gamma_{\infty,t}^{(k)}(\mathbf{x}_k; \mathbf{x}'_k) = \prod_{j=1}^{k} \gamma_{\infty,t}^{(1)}(x_j; x'_j) . \tag{16}$$

Then, from (15) and (16), it would follow that $\gamma_{\infty,t}^{(1)}$ is a solution of the Gross-Pitaevskii equation (8). Note that, since we already know that (15) has a factorized solution, in order to prove (16) it is enough to prove the uniqueness of the solution of the infinite hierarchy (15).

Unfortunately, due to the singularity of the δ-function, we are still unable to prove that (15) has a unique solution and thus we cannot prove part (ii) (the best result in this direction is the proof of the uniqueness for the hierarchy with a Coulomb singularity, see [5]). On the other hand we can complete part i) of our program, that is, we can prove that any limit point $\{\gamma_{\infty,t}^{(k)}\}_{k \geq 1}$ of the family $\{\gamma_{N,t}^{(k)}\}_{k=1}^{N}$ (with respect to an appropriate weak topology), satisfies the infinite hierarchy (15), provided we replace the original Hamiltonian

H with a slightly modified version $\widetilde{H}$, where we artificially modify the interaction when a large number of particles come into a region with diameter
much smaller than the typical inter-particle distance. Since H agrees with
$\widetilde{H}$, apart in the very rare event (rare with respect to the expected typical
distribution of the particles) that many particles come very close together,
we don't expect this modification to change the macroscopic dynamics of the
system: but unfortunately we cannot control this effect rigorously.

Note that the Gross-Pitaevskii equation (1) is a nonlinear Hartree equation

$$i\partial_t u_t = -\Delta u_t + (V * |u_t|^2)u_t \tag{17}$$

in the special case $V(x) = 8\pi a_0 \delta(x)$. In the literature there are several works
devoted to the derivation of (17) from the N-body Schrödinger equation. The
first results were obtained by Hepp in [9], for a smooth potential $V(x)$, and
by Spohn in [15], for bounded $V(x)$. Later, Ginibre and Velo extended these
results to singular potentials in [6]: their result is limited to coherent initial
states, for which the number of particles cannot be fixed. In [5], Erdős and
Yau derived (17) for the Coulomb potential $V(x) = \pm 1/|x|$. More recently,
Adami, Bardos, Golse and Teta obtained partial results for the potential
$V(x) = \delta(x)$, which leads to the Gross-Pitaevskii equation, in the case of
one-dimensional systems; see [1].

2 The Main Result

In this section we explain how we need to modify the Hamiltonian and then
we state our main theorem. In order to derive (15) it is very important to
find a good approximation for the wave function of the ground state of the N
boson system. We need an approximation which reproduces the correct short
scale structure and, at the same time, does not become too singular (so that
error terms can be controlled). Our first guess

$$W(\mathbf{x}) = \prod_{i<j} f_a(x_i - x_j) = \prod_{i<j} f(N(x_i - x_j)) \tag{18}$$

is unfortunately not good enough. First of all we need to cutoff the correla-
tions at large distances (we want $f_a(x) = 1$ for $|x| \gg a$). To this end we fix
a length scale $\ell_1 \gg a$, and we consider the Neumann problem on the ball
$\{x : |x| \leq \ell_1\}$ (we will choose $\ell_1 = N^{-2/3+\kappa}$ for a small $\kappa > 0$). We are
interested in the solution of the ground state problem

$$(-\Delta + 1/2V_a(x))(1 - w(x)) = e_{\ell_1}(1 - w(x))$$

on $\{x : |x| \leq \ell_1\}$, with the normalization condition $w(x) = 0$ for $|x| = \ell_1$. Here e_{ℓ_1} is the lowest possible eigenvalue. It is easy to check that, up
to contributions of lower order, $e_{\ell_1} \simeq 3a/\ell_1^3$. We can extend $w(x)$ to be
identically zero, for $|x| \geq \ell_1$. Then

$$(-\Delta + 1/2V_a(x))(1 - w(x)) = q(x)(1 - w(x)), \quad \text{with}$$

$$q(x) \simeq \frac{3a}{\ell_1^3}\chi(|x| \le \ell_1) . \tag{19}$$

For $a \ll |x| \ll \ell_1$, the function $1 - w(x)$ still looks very much like $1 - a/|x|$, but now it equals one, for $|x| \ge \ell_1$. Replacing $f_a(x_i - x_j)$ by $1 - w(x_i - x_j)$ in (18) is still not sufficient for our purposes. The problem is that the wave function $\prod_{i<j}(1 - w(x_i - x_j))$ becomes very singular when a large number of particles come very close together. We introduce another cutoff to avoid this problem. We fix a new length scale $\ell \gg \ell_1 \gg a$, such that $\ell \ll N^{-1/3}$ (that is ℓ is still much smaller than the typical inter-particle distance: we will choose $\ell = N^{-2/5-\kappa}$ for a small $\kappa > 0$). Then, for fixed indices i and j, and for an arbitrary fixed number $K \ge 1$, we cutoff the correlation between particles i and j (that is we replace $1 - w(x_i - x_j)$ by one) whenever at least K other particles come inside a ball of radius ℓ around i and j. In order to keep our exposition as clear as possible we choose $K = 1$, that is we cutoff correlations if at least three particles come very close together. But there is nothing special about $K = 1$: what we really need to avoid are correlations among a macroscopic number of particles, all very close together. To implement our cutoff we introduce, for fixed indices i, j, a function $F_{ij}(\mathbf{x})$ with the property that

$$F_{ij}(\mathbf{x}) \cong 1 \quad \text{if} \quad \begin{cases} |x_i - x_m| \gg \ell \\ |x_j - x_m| \gg \ell \end{cases} \quad \text{for all } m \ne i, j$$

$$F_{ij}(\mathbf{x}) \cong 0 \quad \text{otherwise.}$$

Instead of using the wave function $\prod_{i<j}(1 - w(x_i - x_j))$ we will approximate the ground state of the N boson system by

$$W(\mathbf{x}) = \prod_{i<j}(1 - w(x_i - x_j)F_{ij}(\mathbf{x})) , \tag{20}$$

(the exact definition of $W(\mathbf{x})$ is a little bit more complicated; see [4], Sect. 2.3). The introduction of the cutoffs F_{ij} in the wave function $W(\mathbf{x})$ forces us to modify the Hamiltonian H. To understand how H has to be modified, we compute its action on $W(\mathbf{x})$. We have, using (19),

$$W(\mathbf{x})^{-1}(HW)(\mathbf{x}) = \sum_{i,j} q(x_i - x_j)$$

$$+ \sum_{i,j}((1/2)V_a(x_i - x_j) - q(x_i - x_j))(1 - F_{ij}(\mathbf{x}))$$

$$+ \text{lower order contributions.}$$

The "lower order contributions" are terms containing derivatives of F_{ij}: they need some control, but they are not very dangerous for our analysis. On the other hand, the second term on the r.h.s. of the last equation, whose presence is due to the introduction of the cutoffs F_{ij}, still contains the potential V_a and unfortunately we cannot control it with our techniques. Therefore, we artificially remove it, defining a new Hamiltonian $\widetilde{H}$, by

$$\widetilde{H} = H - \sum_{i,j} \left((1/2) V_a(x_i - x_j) - q(x_i - x_j) \right) (1 - F_{ij}(\mathbf{x})) \ .$$

Note that the new Hamiltonian $\widetilde{H}$ equals the physical Hamiltonian H unless three or more particles come at distances less than $\ell \ll N^{-1/3}$. This is a rare event, and thus we don't expect the modification of the Hamiltonian H to change in a macroscopic relevant way the dynamics of the system.

Before stating our main theorem, we still have to specify the topology we use in taking the limit $N \to \infty$ of the marginal distributions $\gamma_{N,t}^{(k)}$. It is easy to check that, for every $k \geq 1$, $\gamma_{N,t}^{(k)}(\mathbf{x}_k; \mathbf{x}_k') \in L^2(\Lambda^k \times \Lambda^k)$. This motivates the following definition. For $\Gamma = \{\gamma^{(k)}\}_{k \geq 1} \in \bigoplus_{k \geq 1} L^2(\Lambda^k \times \Lambda^k)$, and for a fixed $\nu > 1$, we define the two norms

$$\|\Gamma\|_- := \sum_{k \geq 0} \nu^{-k} \|\gamma^{(k)}\|_2 \quad \text{and} \quad \|\Gamma\|_+ := \sup_{k \geq 1} \nu^k \|\gamma^{(k)}\|_2 \tag{21}$$

where $\|.\|_2$ denotes the L^2-norm on $\Lambda^k \times \Lambda^k$. We have to introduce the parameter $\nu > 1$ to make sure that, for $\Gamma_{N,t} = \{\gamma_{N,t}^{(k)}\}_{k=1}^N$, the norm $\|\Gamma_{N,t}\|_-$ is finite (choosing ν large enough, we find $\|\Gamma_{N,t}\|_- \leq 1$, uniformly in N and t). We also define the Banach spaces

$$\mathcal{H}_- := \{ \Gamma = \{\gamma^{(k)}\}_{k \geq 0} \in \bigoplus_{k \geq 1} L^2(\Lambda^k \times \Lambda^k) : \|\Gamma\|_- < \infty \}$$

and

$$\mathcal{H}_+ := \{ \Gamma = \{\gamma^{(k)}\}_{k \geq 0} \in \bigoplus_{k \geq 1} L^2(\Lambda^k \times \Lambda^k) : \lim_{k \to \infty} \nu^k \|\gamma^{(k)}\|_2 = 0 \} \ .$$

Then we have $(\mathcal{H}_-, \|.\|_-) = (\mathcal{H}_+, \|.\|_+)^*$. This induces a weak* topology on $\mathcal{H}_-$, with respect to which the unit ball $\mathcal{B}_-$ of $\mathcal{H}_-$ is compact (Banach-Alaouglu Theorem). Since the space $\mathcal{H}_+$ is separable, the weak* topology on the unit ball $\mathcal{B}_-$ is metrizable: we can find a metric ρ on $\mathcal{H}_-$, such that a sequence $\Gamma_n \in \mathcal{B}_-$ converges with respect to the weak* topology if and only if it converges with respect to the metric ρ. For a fixed time T, we will also consider the space $C([0,T], \mathcal{B}_-)$ of functions of $t \in [0,T]$, with values in the unit ball $\mathcal{B}_- \subset \mathcal{H}_-$, which are continuous with respect to the metric ρ (or equivalently with respect to the weak* topology of $\mathcal{H}_-$). We equip $C([0,T], \mathcal{B}_-)$ with the metric

$$\widetilde{\rho}(\Gamma_1(t), \Gamma_2(t)) = \sup_{t \in [0,T]} \rho(\Gamma_1(t), \widetilde{\Gamma}_2(t)) \ .$$

In the following we will consider the families $\Gamma_{N,t} = \{\gamma_{N,t}^{(k)}\}_{k=1}^{N}$ as elements of $C([0,T], \mathcal{B}_-)$, and we will study their convergence and their limit points with respect to the metric $\tilde{\rho}$. We are now ready to state our main theorem.

Theorem 1. *Assume* $a = a_0/N$, $\ell_1 = N^{-2/3+\kappa}$, $\ell = N^{-2/5-\kappa}$, *for some sufficiently small* $\kappa > 0$. *Assume*

$$(\psi_{N,0}, \tilde{H}^2 \psi_{N,0}) \leq C N^2,$$

where $(.,.)$ *denotes the inner product on* $L^2(\mathbb{R}^{3N}, \mathrm{d}\mathbf{x})$. *Let* $\psi_{N,t}$, *for* $t \in [0,T]$, *be the solution of the Schrödinger equation*

$$i\partial_t \psi_{N,t} = \tilde{H}\psi_{N,t} \tag{22}$$

with initial data $\psi_{N,0}$. *Then, if* $\alpha = (\|V\|_1 + \|V\|_\infty)$ *is small enough (of order one) and* $\nu > 1$ *is large enough (recall that* ν *enters the definition of the norms (21)), we have:*

(i) $\Gamma_{N,t} = \{\gamma_{N,t}^{(k)}\}_{k=1}^{N}$ *has at least one (non-trivial) limit point* $\Gamma_{\infty,t} = \{\gamma_{\infty,t}^{(k)}\}_{k \geq 1} \in C([0,T], \mathcal{B}_-)$ *with respect to the metric* $\tilde{\rho}$.

(ii) *For any limit point* $\Gamma_{\infty,t} = \{\gamma_{\infty,t}^{(k)}\}_{k \geq 1}$ *and for all* $k \geq 1$, *there exists a constant* C *such that*

$$Tr(1 - \Delta_i)(1 - \Delta_j)\gamma_{\infty,t}^{(k)} \leq C \tag{23}$$

for all $i \neq j$, $t \in [0,T]$.

(iii) *Any limit point* $\Gamma_{\infty,t}$ *satisfies the infinite Gross-Pitaevskii hierarchy (15) when tested against a regular function* $J^{(k)}(\mathbf{x}_k; \mathbf{x}_k')$:

$$\langle J^{(k)}, \gamma_{\infty,t}^{(k)} \rangle = \langle J^{(k)}, \gamma_{\infty,0}^{(k)} \rangle - i \sum_{j=1}^{k} \int_0^t \mathrm{d}s \langle J^{(k)}, (-\Delta_j + \Delta_j')\gamma_{\infty,s}^{(k)} \rangle$$

$$- 8i\pi a_0 \sum_{j=1}^{k} \int_0^t \mathrm{d}s \int \mathrm{d}\mathbf{x}_k \mathrm{d}\mathbf{x}_k' \, J^{(k)}(\mathbf{x}_k; \mathbf{x}_k') \int \mathrm{d}x_{k+1} \tag{24}$$

$$\times \left(\delta(x_j - x_{k+1}) - \delta(x_j' - x_{k+1})\right) \gamma_{\infty,s}^{(k+1)}(\mathbf{x}_k, x_{k+1}; \mathbf{x}_k', x_{k+1}) \ .$$

Here we use the notation $\langle J^{(k)}, \gamma^{(k)} \rangle = \int \mathrm{d}\mathbf{x}_k \mathrm{d}\mathbf{x}_k' \, \overline{J^{(k)}(\mathbf{x}_k; \mathbf{x}_k')} \, \gamma^{(k)}(\mathbf{x}_k; \mathbf{x}_k')$.

Remarks.

(i) The main assumption of the theorem is the requirement that the expectation of $\tilde{H}^2$ at $t = 0$ is of order N^2. One can prove that this condition is satisfied for $\psi_{N,0}(\mathbf{x}) = W(\mathbf{x})\phi_N(\mathbf{x})$ and ϕ_N sufficiently smooth (see [4], Lemma D1). Physically, this assumption guarantees that the initial wave function $\psi_{N,0}(\mathbf{x})$ has the short-scale structure characteristic of $W(\mathbf{x})$ and, hence, that it describes, locally, a condensate.

(ii) It is a priori not clear that the action of the delta-functions in the Gross-Pitaevskii hierarchy (24) is well defined. This fact follows by the bound (23), which makes sure that $\gamma_{\infty,t}^{(k)}$ is sufficiently smooth.

(iii) We also need to assume that $\alpha = (\|V\|_\infty + \|V\|_1)$ is small enough (but still of order one). This technical assumption is needed in the proof of the energy estimate, Proposition 1.

3 Sketch of the Proof

In this section we explain some of the main ideas used in the proof of Theorem 1. Let $\psi_{N,t}$ be the solution of the Schrödinger equation (22) (with the modified Hamiltonian $\tilde{H}$). We can decompose $\psi_{N,t}$ as

$$\psi_{N,t}(\mathbf{x}) = W(\mathbf{x})\phi_{N,t}(\mathbf{x}) \, ,$$

where $W(\mathbf{x})$ is the approximation for the ground state wave function defined in (20). This decomposition is always possible because $W(\mathbf{x})$ is strictly positive.

The main tool in the proof of Theorem 1 is an estimate for the L^2-norm of the second derivatives of $\phi_{N,t}$. This bound follows from the following *energy estimate*.

Proposition 1. *Assume* $a = a_0/N$, $\ell_1 = N^{-2/3+\kappa}$ *and* $\ell = N^{-2/5-\kappa}$ *for* $\kappa > 0$ *small enough, and suppose* $\alpha = (\|V\|_1 + \|V\|_\infty)$ *is sufficiently small. Then there exists a constant* $C > 0$ *such that*

$$\int d\mathbf{x}\,|(\tilde{H}W\phi)(\mathbf{x})|^2 \geq (C - o(1)) \sum_{i,j=1}^N \int d\mathbf{x}\,W^2(\mathbf{x})|\nabla_i\nabla_j\phi(\mathbf{x})|^2$$

$$- o(1)\left(N\sum_{i=1}^N \int d\mathbf{x}\,W^2(\mathbf{x})|\nabla_i\phi(\mathbf{x})|^2 + N^2\int d\mathbf{x}\,W^2(\mathbf{x})|\phi(\mathbf{x})|^2\right) \, ,$$

where $o(1) \to 0$ *as* $N \to \infty$.

Remark. The proof of this proposition is the main technical difficulty in our analysis. It is in order to prove this proposition that we need to introduce the cutoffs F_{ij} in the approximate ground state wave function $W(\mathbf{x})$, and that we need to modify the Hamiltonian.

Using the assumption that, at $t = 0$, $(\psi_{N,0}, \tilde{H}^2\psi_{N,0}) \leq CN^2$, the conservation of the energy, and the symmetry with respect to permutations, we immediately get the following corollary.

Corollary 1. *Suppose the assumptions of Proposition 1 are satisfied. Suppose moreover that the initial data* $\psi_{N,0}$ *is symmetric with respect to permutations and* $(\psi_{N,0}, \tilde{H}^2\psi_{N,0}) \leq CN^2$. *Then there exists a constant* C *such that*

$$\int W^2(\mathbf{x}) |\nabla_i \nabla_j \phi_{N,t}(\mathbf{x})|^2 \le C \tag{25}$$

for all $i \ne j$, t and all N large enough.

Remark. The bound (25) is not an estimate for the derivatives of the whole wave function $\psi_{N,t}$. The inequality

$$\int d\mathbf{x} \, |\nabla_i \nabla_j \psi(\mathbf{x})|^2 < C \tag{26}$$

is wrong, if ψ satisfies $(\psi, \tilde{H}^2 \psi) \le CN^2$. In fact, in order for $(\psi, \tilde{H}^2 \psi)$ to be of order N^2, the wave function $\psi(\mathbf{x})$ needs to have the short scale structure characterizing $W(\mathbf{x})$. This makes (26) impossible to hold true uniformly in N. Only after the singular part $W(\mathbf{x})$ has been factorized out, we can prove bounds like (25) for the derivatives of the remainder. One of the consequences of our energy estimate, and one of the possible interpretation of our result, is that the separation between the singular part of the wave function (living on the scale $1/N$) and its regular part is preserved by the time evolution.

Next we show how the important bound (25) can be used to prove Theorem 1. According to the decomposition $\psi_{N,t}(\mathbf{x}) = W(\mathbf{x}) \phi_{N,t}(\mathbf{x})$, we define, for $k = 1, \ldots, N$, the densities $U_{N,t}^{(k)}(\mathbf{x}_k; \mathbf{x}_k')$, for $k = 1, \ldots, N$, to be, roughly, the k-particle marginal density corresponding to the wave function $\phi_{N,t}$ (the exact definition is a little bit more involved, see [4], Sect. 4). The estimate (25) for the second derivatives of $\phi_{N,t}$ translates into a bound for the densities $U_{N,t}^{(k)}$:

$$\text{Tr} \, (1 - \Delta_i)(1 - \Delta_j) U_{N,t}^{(k)} \le C \tag{27}$$

for all $i, j \le N$ with $i \ne j$, for all t and for all N large enough.

Moreover, we can show that, for $\nu > 1$ large enough (recall that the parameter ν enters the definition of the norms (21)), the families $U_{N,t} = \{U_{N,t}^{(k)}\}_{k=1}^N$ define an equicontinuous sequence in the space $C([0,T], \mathcal{B}_-)$ (this follows from a careful analysis of the BBGKY hierarchy associated to the Schrödinger equation (22); see [4], Sects 9.1 and 9.2 for more details). Applying standard results (Arzela-Ascoli Theorem), it follows that the sequence $U_{N,t}$ has at least one limit point, denoted $U_{\infty,t} = \{U_{\infty,t}^{(k)}\}_{k \ge 1}$, in the space $C([0,T], \mathcal{B}_-)$. The bound (27) can then be passed to the limit $N \to \infty$, and we obtain

$$\text{Tr} \, (1 - \Delta_i)(1 - \Delta_j) U_{\infty,t}^{(k)} \le C$$

for all $i \ne j$ and $t \in [0, T]$.

Next we go back to the family $\Gamma_{N,t} = \{\gamma_{N,t}^{(k)}\}_{k=1}^N$. Clearly, the densities $\gamma_{N,t}^{(k)}$ do not satisfy the estimate (27). In fact, $\gamma_{N,t}^{(k)}$ still contains the short scale structure of $W(\mathbf{x})$ (which, on the contrary, has been factorized out from $U_{N,t}^{(k)}$), and thus cannot have the smoothness required by (27).

It is nevertheless clear that the short scale structure of $\Gamma_{N,t}$ disappears when we consider the limit $N \to \infty$ (in the weak sense specified by Theorem 1). In fact, one can prove the convergence of an appropriate subsequence of $\Gamma_{N,t}$ to the limit point $U_{\infty,t}$ of $U_{N,t}$. In other words one can show that limit points of $\Gamma_{N,t}$, denoted by $\Gamma_{\infty,t}$, coincide with the limit points of $U_{N,t}$. Therefore, even though $\Gamma_{N,t}$, for finite N, does not satisfies the bound (27), its limit points $\Gamma_{\infty,t} = \{\gamma_{\infty,t}^{(k)}\}_{k\geq 1}$ do. For every $k \geq 1$ we have

$$\mathrm{Tr}\,(1 - \Delta_i)(1 - \Delta_j)\gamma_{\infty,t}^{(k)} \leq C \qquad (28)$$

for all $i \neq j$ and $t \in [0, T]$. This proves part i) and ii) of Theorem 1 (the non-triviality of the limit follows by showing that $\mathrm{Tr}\,\gamma_{\infty,t}^{(1)} = 1$). The bound (28) can then be used to prove part iii) of Theorem 1, that is to prove that the family $\Gamma_{\infty,t}$ satisfies the infinite Gross-Pitaevskii hierarchy (24). In fact, having control over the derivatives of $\gamma_{\infty,t}^{(k)}$ allows us to prove the convergence of the potential to a delta-function (that is, it allows us to make (13) rigorous). To this end we use the following lemma (see [4], Sect. 8).

Lemma 1. *Suppose* $\delta_\beta(x) = \beta^{-3}h(x/\beta)$, *for some regular function* h, *with* $\int h(x) = 1$. *Then, for any* $1 \leq j \leq k$, *and for any regular function* $J(\mathbf{x}_k; \mathbf{x}'_k)$, *we have*

$$\left| \int \mathrm{d}\mathbf{x}_k \mathrm{d}\mathbf{x}'_k \mathrm{d}x_{k+1}\, J(\mathbf{x}_k; \mathbf{x}'_k)(\delta_\beta(x_j - x_{k+1}) - \delta(x_j - x_{k+1})) \right.$$
$$\left. \times\, \gamma^{(k+1)}(\mathbf{x}_k, x_{k+1}; \mathbf{x}'_k, x_{k+1}) \right|$$
$$\leq C_J \sqrt{\beta}\, \mathrm{Tr}(1 - \Delta_j)(1 - \Delta_{k+1})\gamma^{(k+1)}\,.$$

Part (iii) of Theorem 1 can then be proven combining this lemma with the estimates (27) and (28) (see [4], Sect. 9.4, for more details).

References

1. R. Adami, C. Bardos, F. Golse and A. Teta: *Towards a rigorous derivation of the cubic nonlinear Schrödinger equation in dimension one.* Asymptot. Anal. (2) **40** (2004), 93–108.

2. F.J. Dyson: *Ground-state energy of a hard-sphere gas.* Phys. Rev. (1) **106** (1957), 20–26.

3. A. Elgart, L. Erdős, B. Schlein, and H.-T. Yau: *Gross-Pitaevskii equation as the mean field limit of weakly coupled bosons.* Preprint math-ph/0410038. To appear in Arch. Rat. Mech. Anal.

4. L. Erdős, B. Schlein, and H.-T. Yau: *Derivation of the Gross–Pitaevskii hierarchy for the dynamics of a Bose-Einstein condensate.* Preprint math-ph/0410005.

5. L. Erdős and H.-T. Yau: *Derivation of the nonlinear Schrödinger equation from a many body Coulomb system.* Adv. Theor. Math. Phys. (6) **5** (2001), 1169–1205.

6. J. Ginibre and G. Velo: *The classical field limit of scattering theory for non-relativistic many-boson systems. I and II.* Commun. Math. Phys. **66** (1979), 37–76 and **68** (1979), 45–68.

7. E.P. Gross: *Structure of a quantized vortex in boson systems.* Nuovo Cimento **20** (1961), 454–466.

8. E.P. Gross: *Hydrodynamics of a superfluid condensate.* J. Math. Phys. **4** (1963), 195–207.

9. K. Hepp: *The classical limit for quantum mechanical correlation functions.* Commun. Math. Phys. **35** (1974), 265–277.

10. E.H. Lieb and R. Seiringer: *Proof of Bose-Einstein Condensation for Dilute Trapped Gases.* Phys. Rev. Lett. **88** (2002), 170409-1-4.

11. E.H. Lieb, R. Seiringer, J.P. Solovej, and J. Yngvason: *The Quantum-Mechanical Many-Body Problem: Bose Gas.* Preprint math-ph/0405004.

12. E.H. Lieb, R. Seiringer, J. Yngvason: *Bosons in a Trap: A Rigorous Derivation of the Gross-Pitaevskii Energy Functional.* Phys. Rev A **61** (2000), 043602.

13. E.H. Lieb and J. Yngvason: *Ground State Energy of the low density Bose Gas.* Phys. Rev. Lett. **80** (1998), 2504–2507.

14. L.P. Pitaevskii: *Vortex lines in an imperfect Bose gas.* Sov. Phys. JETP **13** (1961), 451–454.

15. H. Spohn: *Kinetic Equations from Hamiltonian Dynamics.* Rev. Mod. Phys. **52** no. 3 (1980), 569–615.

Towards a Microscopic Derivation of the Phonon Boltzmann Equation

Herbert Spohn

Zentrum Mathematik and Physik Department, TU München, 85747 Garching, Boltzmannstr. 3, Germany
spohn@ma.tum.de

1 Introduction

The thermal conductivity of insulating (dielectric) crystals is computed almost exclusively on the basis of the phonon Boltzmann equation. We refer to [1] for a discussion more complete than possible in this contribution. On the microscopic level the starting point is the Born-Oppenheimer approximation (see [2] for a modern version), which provides an effective Hamiltonian for the slow motion of the nuclei. Since their deviation from the equilibrium position is small, one is led to a wave equation with a *weak* nonlinearity. As already emphasized by R. Peierls in his seminal work [3], physically it is of importance to retain the structure resulting from the atomic lattice, which forces the discrete wave equation.

On the other hand, continuum wave equations with weak nonlinearity appear in the description of the waves in the upper ocean and in many other fields. This topic is referred to as weak turbulence. Again the theoretical treatment of such equations is based mostly on the phonon Boltzmann eqation, see e.g. [4]. In these applications one considers scales which are much larger than the atomistic scale, hence quantum effects are negligible. For dielectric crystals, on the other side, quantum effects are of importance at low temperatures. We refer to [1] and discuss here only the classical discrete wave equation with a small nonlinearity.

If one considers crystals with a single nucleus per unit cell, then the displacement field is a 3-vector field over the crystal lattice Γ. The nonlinearity results from the weakly non-quadratic interaction potentials between the nuclei. As we will see, the microscopic mechanism responsible for the validity of the Boltzmann equation can be understood already in case the displacement field is declared to be scalar, the nonlinearity to be due to an on-site potential, and the lattice $\Gamma = \mathbb{Z}^3$. This is the model I will discuss in my notes.

As the title indicates there is no complete proof available for the validity of the phonon Boltzmann equation. The plan is to explain the kinetic scaling and to restate our conjecture in terms of the asymptotics of certain Feynman diagrams.

H. Spohn: *Towards a Microscopic Derivation of the Phonon Boltzmann Equation*, Lect. Notes Phys. **690**, 295–304 (2006)
www.springerlink.com   © Springer-Verlag Berlin Heidelberg 2006

2 Microscopic Model

We consider the simple cubic crystal $\mathbb{Z}^3$. The displacement field is denoted by

$$q_x \in \mathbb{R}, \quad x \in \mathbb{Z}^3 , \tag{1}$$

with the canonically conjugate momenta

$$p_x \in \mathbb{R}, \quad x \in \mathbb{Z}^3 . \tag{2}$$

We use units in which the mass of the nuclei is $m = 1$. The particles interact harmonically and are subject to an on-site potential, which is divided into a quadratic part and a non-quadratic correction. Thus the Hamiltonian of the system reads

$$H = \frac{1}{2} \sum_{x \in \mathbb{Z}^3} \left(p_x^2 + \omega_0^2 q_x^2 \right) + \frac{1}{2} \sum_{x,y \in \mathbb{Z}^3} \alpha(x-y) q_x q_y + \sum_{x \in \mathbb{Z}^3} V(q_x) = H_0 + \sum_{x \in \mathbb{Z}^3} V(q_x) . \tag{3}$$

The coupling constants have the properties

$$\alpha(x) = \alpha(-x) , \tag{4}$$

$$|\alpha(x)| \leq \alpha_0 e^{-\gamma|x|} \tag{5}$$

for suitable $\alpha_0, \gamma > 0$, and

$$\sum_{x \in \mathbb{Z}^3} \alpha(x) = 0 , \tag{6}$$

because of the invariance of the interaction between the nuclei under the translation $q_x \rightsquigarrow q_x + a$.

For the anharmonic on-site potential we set

$$V(u) = \sqrt{\varepsilon} \frac{1}{3} \lambda u^3 + \varepsilon (\lambda^2/18\omega_0^2) u^4 , \ u \in \mathbb{R} . \tag{7}$$

ε is the dimensionless scale parameter, eventually $\varepsilon \to 0$. The quartic piece is added so to make sure that $H \geq 0$. In the limit $\varepsilon \to 0$ its contribution will vanish and for simplicity of notation we will omit it from the outset. Then the equations of motion are

$$\frac{d}{dt} q_x(t) = p_x(t) ,$$

$$\frac{d}{dt} p_x(t) = - \sum_{y \in \mathbb{Z}^3} \alpha(y-x) q_y(t) - \omega_0^2 q_x(t) - \sqrt{\varepsilon} \lambda q_x(t)^2 , \quad x \in \mathbb{Z}^3 . \tag{8}$$

We will consider only finite energy solutions. In particular, it is assumed that $|p_x| \to 0$, $|q_x| \to 0$ sufficiently fast as $|x| \to \infty$. In fact, later on there will be the need to impose random initial data, which again are assumed to be supported on finite energy configurations. In the kinetic limit the average energy will diverge as ε^{-3}.

It is convenient to work in Fourier space. For $f : \mathbb{Z}^3 \to \mathbb{R}$ we define

$$\widehat{f}(k) = \sum_{x \in \mathbb{Z}^3} e^{-i2\pi k \cdot x} f_x , \tag{9}$$

$k \in \mathbb{T}^3 = [-\frac{1}{2}, \frac{1}{2}]^3$, with inverse

$$f_x = \int_{\mathbb{T}^3} dk e^{i2\pi k \cdot x} \widehat{f}(k) , \tag{10}$$

dk the 3-dimensional Lebesgue measure. The dispersion relation for the harmonic part H_0 is

$$\omega(k) = \left(\omega_0^2 + \widehat{\alpha}(k)\right)^{1/2} \geq \omega_0 > 0 , \tag{11}$$

since $\widehat{\alpha}(k) > 0$ for $k \neq 0$ because of the mechanical stability of the harmonic lattice with vanishing on-site potential.

In Fourier space the equations of motion read

$$\frac{\partial}{\partial t}\widehat{q}(k, t) = \widehat{p}(k, t) ,$$

$$\frac{\partial}{\partial t}\widehat{p}(k, t) = -\omega(k)^2 \widehat{q}(k, t)$$
$$- \sqrt{\varepsilon}\lambda \int_{\mathbb{T}^6} dk_1 dk_2 \delta(k - k_1 - k_2)\widehat{q}(k_1, t)\widehat{q}(k_2, t) \tag{12}$$

with $k \in \mathbb{T}^3$. Here δ is the δ-function on the unit torus, to say, $\delta(k')$ carries a point mass whenever $k' \in \mathbb{Z}^3$.

It will be convenient to concatenate q_x and p_x into a single complex-valued field. We set

$$a(k) = \frac{1}{\sqrt{2}}\left(\sqrt{\omega}\widehat{q}(k) + i\frac{1}{\sqrt{\omega}}\widehat{p}(k)\right) \tag{13}$$

with the inverse

$$\widehat{q}(k) = \frac{1}{\sqrt{2}}\frac{1}{\sqrt{\omega}}\left(a(k) + a(-k)^*\right) , \quad \widehat{p}(k) = \frac{1}{\sqrt{2}}i\sqrt{\omega}\left(-a(k) + a(-k)^*\right) . \tag{14}$$

To have a concise notation, we introduce

$$a(k, +) = a(k)^* , \quad a(k, -) = a(k) . \tag{15}$$

Then the a-field evolves as

$$\frac{\partial}{\partial t}a(k, \sigma, t) = i\sigma\omega(k)a(k, \sigma, t) + i\sigma\sqrt{\varepsilon}\lambda \sum_{\sigma_1, \sigma_2 = \pm 1} \int_{\mathbb{T}^6} dk_1 dk_2$$

$$(8\omega(k)\omega(k_1)\omega(k_2))^{-1/2}\delta(-\sigma k + \sigma_1 k_1 + \sigma_2 k_2)a(k_1, \sigma_1, t)a(k_2, \sigma_2, t) . \tag{16}$$

3 Kinetic Limit and Boltzmann Equation

The kinetic limit deals with a special class of initial probability measures. Their displacement field has a support of linear size ε^{-1} and average energy of order ε^{-3}. More specifically, these probability measures have the property of being locally Gaussian and almost stationary under the dynamics. Because of the assumed slow variation in space the covariance of such probability measures changes only slowly, i.e. on the scale ε^{-1}, in time.

Let us assume then that the initial data for (16) are random and specified by a Gaussian probability measure on phase space. It is assumed to have mean

$$\langle a(k,\sigma)\rangle_\varepsilon^{\mathrm{G}} = 0 \,, \tag{17}$$

and for the covariance we set

$$\langle a(k,\sigma)a(k',\sigma)\rangle_\varepsilon^{\mathrm{G}} = 0 \,, \tag{18}$$

$$W^\varepsilon(y,k) = \varepsilon^3 \int_{(\mathbb{T}/\varepsilon)^3} d\eta e^{i2\pi y\cdot\eta}\langle a(k-\varepsilon\eta/2,+)a(k+\varepsilon\eta/2,-)\rangle_\varepsilon^{\mathrm{G}} \,, \tag{19}$$

$y \in (\varepsilon\mathbb{Z})^3$, which defines the *Wigner function* rescaled to the lattice $(\varepsilon\mathbb{Z})^3$. Local stationarity is ensured by the condition

$$\lim_{\varepsilon\to 0} W^\varepsilon(\lfloor r\rfloor_\varepsilon, k) = W^0(r,k) \,, \tag{20}$$

where $\lfloor r\rfloor_\varepsilon$ denotes integer part modulo ε. Note that W^ε is normalized as

$$\sum_{y\in(\varepsilon\mathbb{Z})^3} \int_{\mathbb{T}^3} dk W^\varepsilon(y,k) = \int_{\mathbb{T}^3} dk\langle a(k,+)a(k,-)\rangle_\varepsilon^{\mathrm{G}} \,. \tag{21}$$

The condition that the limit in (20) exists thus implies that the average phonon number increases as ε^{-3}, equivalently the average total energy increases as

$$\left\langle \int_{\mathbb{T}^3} d^3 k\omega(k)a(k,+)a(k,-) \right\rangle_\varepsilon^{\mathrm{G}} = \langle H_0\rangle_\varepsilon^{\mathrm{G}} = \mathcal{O}(\varepsilon^{-3}) \,. \tag{22}$$

Let $\langle\cdot\rangle_t$ be the time-evolved measure at time t. Its rescaled Wigner function is

$$W^\varepsilon(y,k,t) = \varepsilon^3 \int_{(\mathbb{T}/\varepsilon)^3} d\eta e^{i2\pi y\cdot\eta}\langle a(k-\varepsilon\eta/2,+)a(k+\varepsilon\eta/2,-)\rangle_{t/\varepsilon} \,. \tag{23}$$

Kinetic theory claims that

$$\lim_{\varepsilon\to 0} W^\varepsilon(\lfloor r\rfloor_\varepsilon, k, t) = W(r,k,t) \,, \tag{24}$$

where $W(r,k,t)$ is the solution of the phonon Boltzmann equation

$$\frac{\partial}{\partial t}W(r,k,t) + \frac{1}{2\pi}\nabla\omega(k)\cdot\nabla_r W(r,k,t)$$

$$= \frac{\pi}{2}\lambda^2 \sum_{\sigma_1,\sigma_2=\pm1} \int_{\mathbb{T}^6} dk_1 dk_2 (\omega(k)\omega(k_1)\omega(k_2))^{-1}\delta(\omega(k)+\sigma_1\omega(k_1)+\sigma_2\omega(k_2))$$

$$\delta(k+\sigma_1 k_1+\sigma_2 k_2)\big(W(r,k_1,t)W(r,k_2,t)$$

$$+\sigma_1 W(r,k,t)W(r,k_2,t) + \sigma_2 W(r,k,t)W(r,k_1,t)\big) \quad (25)$$

to be solved with the initial condition $W(r,k,0) = W^0(r,k)$.

The free streaming part is an immediate consequence of the evolution of W as generated by H_0. The strength of the cubic nonlinearity was assumed to be of order $\sqrt{\varepsilon}$, which results in an effect of order 1 on the kinetic time scale. The specific form of the collision operator will be explained in the following section. It can be brought into a more familiar form by performing the sum over σ_1, σ_2. Then the collision operator has two terms. The first one describes the merging of two phonons with wave number k and k_1 into a phonon with wave number $k_2 = k + k_1$, while the second term describes the splitting of a phonon with wave number k into two phonons with wave numbers k_1 and k_2, $k = k_1 + k_2$. In such a collision process energy is conserved and wave number is conserved modulo an integer vector.

In (25) the summand with $\sigma_1 = 1 = \sigma_2$ vanishes trivially. However it could be the case that the condition for energy conservation,

$$\omega(k) + \omega(k') - \omega(k + k') = 0, \quad (26)$$

has also no solution. If so, the collision operator vanishes. In fact, for nearest neighbor coupling only, $\alpha(0) = 6$, $\alpha(e) = -1$ for $|e| = 1$, $\alpha(x) = 0$ otherwise, it can be shown that (26) has no solution whenever $\omega_0 > 0$. To have a non-zero collision term we have to require

$$\int dk \int dk' \delta(\omega(k) + \omega(k') - \omega(k + k')) > 0, \quad (27)$$

which is an implicit condition on the couplings $\alpha(x)$. A general condition to ensure (27) is not known. A simple example where (27) can be checked by hand is

$$\omega(k) = \omega_0 + \sum_{\alpha=1}^{3}(1 - \cos(2\pi k^\alpha)), \quad k = (k^1, k^2, k^3). \quad (28)$$

It corresponds to suitable nearest and next nearest neighbor couplings.

There is a second more technical condition which requires that

$$\sup_k \int dk' \delta(\omega(k) + \omega(k') - \omega(k + k')) = c_0 < \infty. \quad (29)$$

It holds for the dispersion relation (28). This uniform bound allows for a simple proof that the Boltzmann equation has a unique solution for short times provided $W^0(r,k)$ is bounded.

4 Feynman Diagrams

Denoting by $\langle \cdot \rangle_t$ the average with respect to the measure at time t (in microscopic units), the starting point of the time-dependent perturbation series is the identity

$$
\left\langle \prod_{j=1}^{n} a(k_j, \sigma_j) \right\rangle_t = \exp\left[it \left(\sum_{j=1}^{n} \sigma_j \omega(k_j) \right) \right] \left\langle \prod_{j=1}^{n} a(k_j, \sigma_j) \right\rangle^{G}
$$

$$
+ i\sqrt{\varepsilon} \int_0^t ds \exp\left[i(t-s) \left(\sum_{j=1}^{n} \sigma_j \omega(k_j) \right) \right]
$$

$$
\left(\sum_{\ell=1}^{n} \sum_{\sigma', \sigma'' = \pm 1} \sigma_\ell \int_{\mathbb{T}^6} dk' dk'' \phi(k_\ell, k', k'') \delta(-\sigma_\ell k_\ell + \sigma' k' + \sigma'' k'') \right.
$$

$$
\left\langle \left(\prod_{\substack{j=1 \\ j \neq \ell}}^{n} a(k_j, \sigma_j) \right) a(k', \sigma') a(k'', \sigma'') \right\rangle_s \right) . \tag{30}
$$

Here

$$
\phi(k, k', k'') = \lambda(8\omega(k)\omega(k')\omega(k''))^{-1/2} \tag{31}
$$

One starts with $n = 2$ and $\sigma_1 = 1$, $\sigma_2 = -1$. Then on the right hand side of (30) there is the product of three a's. One resubstitutes (30) with $n = 3$, etc. Thereby one generates an infinite series, in which only the average over the initial Gaussian measure $\langle \cdot \rangle^{G}$ appears.

To keep the presentation transparent, let me assume that $\langle \cdot \rangle^{G}$ is a translation invariant Gaussian measure with

$$
\langle a(k, \pm) \rangle^{G} = 0, \quad \langle a(k, \sigma) a(k', \sigma) \rangle^{G} = 0 ,
$$

$$
\langle a(k, +) a(k', -) \rangle^{G} = \delta(k - k') W(k) . \tag{32}
$$

Then the measure at time t is again translation invariant. Kinetic scaling now merely amounts to considering the long times t/ε. The Wigner function at that time is then represented by the infinite series

$$
\langle a(q, -) a(p, +) \rangle_{t/\varepsilon} = \delta(q - p) \left(W(q) + \sum_{n=1}^{\infty} W_n^\varepsilon(q, t) \right) . \tag{33}
$$

The infinite sum is only formal. Taking naively the absolute value at iteration $2n$ one finds that

$$
|W_n^\varepsilon(q, t)| \le \varepsilon^n (t/\varepsilon)^{2n} ((2n)!)^{-1} (2n)! c^{2n} ((2n+2)!/2^{n+1}(n+1)!) . \tag{34}
$$

Here $\varepsilon^n = (\sqrt{\varepsilon})^{2n}$, $(t/\varepsilon)^{2n}/(2n)!$ comes from the time integration, $(2n)!$ from the sum over ℓ in (30), c^{2n} from the k-integrations and the initial $W(k)$, and the factor $(2n + 2)!/2^{n+1}(n + 1)!$ from the Gaussian pairings in the initial measure. Thus even at fixed ε there are too many terms in the sum.

Since no better estimate is available at present, we concentrate on the structure of a single summand $W_n^\varepsilon(q, t)$. $\delta(q - p)W_n^\varepsilon(q, t)$ is a sum of integrals. The summation comes from

– the sum over σ', σ'' in (30)
– the sum over ℓ in (30)
– the sum over all pairings resulting from the average with respect to the initial Gaussian measure $\langle \cdot \rangle^G$.

Since each single integral has a rather complicated structure, it is convenient to visualize them as *Feynman diagrams*.

A Feynman diagram is a graph with labels. Let us first explain the graph. The graph consists of two binary trees. It is convenient to draw them on a "backbone" consisting of $2n + 2$ equidistant horizontal level lines which are labelled from 0 (bottom) to $2n + 1$ (top). The two roots of the tree are two vertical bonds from line $2n + 1$ to level line $2n$. At level m there is *exactly one* branch point with two branches in either tree. Thus there are exactly $2n$ branch points. At level 0 there are then $2n + 2$ branches. They are connected according to the pairing rule, see figure below.

In the Feynman graph each bond is oriented with arrows pointing either up ($\sigma = +1$) or down ($\sigma = -1$). The left root is down while the right root is up. If there is no branching the orientation is inherited from the upper level. At a pairing the orientation must be maintained. Thus at level 0 a branch with an up arrow can be paired only with a branch with a down arrow, see (32). Every internal line in the graph must terminate at either end by a branch point. Every such internal line admits precisely two orientations.

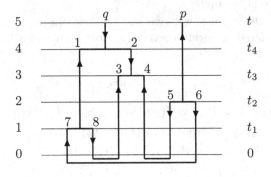

Next we insert the labels. The level lines 0 to $2n+1$ are labelled by times $0 < t_1 \ldots < t_{2n} < t$. The left root carries the label q while the right root carries the label p. Each internal line is labelled with a wave number k.

To each Feynman diagram one associates an integral through the following steps.

(i) The time integration is over the simplex $0 \leq t_1 \ldots \leq t_{2n} \leq t$ as $dt_1 \ldots dt_{2n}$.

(ii) The wave number integration is over all internal lines as $\int dk_1 \ldots \int dk_\kappa$, where $\kappa = 3n - 1$ is the number of internal lines.

(iii) One sums over all orientations of the internal lines.

The integrand is a product of three factors.

(iv) There is a product over all branch points. At each branchpoint there is a root, say wave vector k_1 and orientation σ_1, and there are two branches, say wave vectors k_2, k_3 and orientations σ_2, σ_3. Then each branch point carries the weight

$$\delta(-\sigma_1 k_1 + \sigma_2 k_2 + \sigma_3 k_3)\sigma_1 \phi(k_1, k_2, k_3) . \tag{35}$$

If one regards the wave vector k as a current with orientation σ, then (35) expresses Kirchhoff's rule for conservation of the current.

(v) By construction each bond carries a time difference $t_{m+1} - t_m$, a wave vector k, and an orientation σ. Then to this bond one associates the phase factor

$$\exp[i(t_{m+1} - t_m)\sigma\omega(k)/\varepsilon] . \tag{36}$$

The second factor is the product of such phase factors over all bonds.

(vi) The third factor of the integrand is given by

$$\prod_{j=1}^{n+1} W(k_j) , \tag{37}$$

where $k_1, \ldots, k_{n+1}$ are the wave numbers of the bonds between level 0 and level 1.

(vii) Finally there is the prefactor $(-1)^n \varepsilon^{-n}$.

To illustrate these rules we give an example for $n = 2$, see figure above. The associated integral is given by, more transparently keeping the δ-functions from the pairings,

$$\varepsilon^{-2} \int_{0 \leq t_1 \leq \ldots \leq t_4 \leq t} dt_1 \ldots dt_4 \int_{\mathbb{T}^{24}} dk_1 \ldots dk_8$$

$$\delta(q + k_1 - k_2)\delta(k_2 + k_3 + k_4)\delta(-p - k_5 - k_6)\delta(-k_1 + k_7 - k_8)$$

$$\phi(q, k_1, k_2)\phi(k_2, k_3, k_4)\phi(p, k_5, k_6)\phi(k_1, k_7, k_8)$$

$$\delta(k_7 - k_6)W(k_7)\delta(k_8 - k_3)W(k_8)\delta(k_4 - k_5)W(k_4)$$

$$\exp \big[\{i(t - t_4)(-\omega_q + \omega_p) + i(t_4 - t_3)(\omega_1 - \omega_2 + \omega_p)$$

$$+i(t_3 - t_2)(\omega_1 + \omega_3 + \omega_4 + \omega_p) + i(t_2 - t_1)(\omega_1 + \omega_3 + \omega_4 - \omega_5 - \omega_6)$$

$$+it_1(\omega_7 - \omega_8 + \omega_3 + \omega_4 - \omega_5 - \omega_6)\}/\varepsilon\big] \tag{38}$$

with $\omega_q = \omega(q)$, $\omega_p = \omega(p)$, $\omega_j = \omega(k_j)$.

$\delta(q - p)W_n^\varepsilon(q, t)$ is the sum over all Feynman diagrams with $2n + 2$ levels and thus is a sum of oscillatory integrals. In the limit $\varepsilon \to 0$ only a few leading terms survive while all remainders vanish. E.g., the Feynman diagram above is subleading. In fact, the conjecture of kinetic theory can be stated rather concisely:

Kinetic Conjecture: *In a leading Feynman diagram the Kirchhoff rule never forces an identification of the form $\delta(k_j)$ with some wave vector k_j. In addition, the sum of the $2(n - m + 1)$ phases from the bonds between level lines $2m$ and $2m+1$ vanishes for every choice of internal wave numbers. This cancellation must hold for $m = 0, \ldots, n$.*

Since we assumed the initial data to be spatially homogeneous, the phonon Boltzmann equation (25) simplifies to

$$\frac{\partial}{\partial t}W(k, t)$$
$$= 4\pi\lambda^2 \sum_{\sigma_1,\sigma_2=\pm 1} \int_{\mathbb{T}^6} dk_1 dk_2 \phi(k, k_1, k_2)^2 \delta(\omega(k) + \sigma_1\omega(k_1) + \sigma_2\omega(k_2))$$
$$\delta(k + \sigma_1 k_1 + \sigma_2 k_2)\big(W(k_1, t)W(k_2, t) + 2\sigma_2 W(k, t)W(k_1, t)\big), \quad (39)$$

where we used the symmetry with respect to (k_1, σ_1) and (k_2, σ_2). To (39) we associate the Boltzmann hierarchy

$$\frac{\partial}{\partial t}f_n = C_{n,n+1}f_{n+1}, \; n = 1, 2, \ldots , \quad (40)$$

acting on the symmetric functions $f_n(k_1, \ldots, k_n)$ with

$$C_{n,n+1}f_{n+1}(k_1, \ldots, k_n) = 4\pi\lambda^2 \sum_{\ell=1}^{n} \sum_{\sigma',\sigma''=\pm 1} \int_{\mathbb{T}^6} dk' dk'' \phi(k_\ell, k', k'')^2$$
$$\delta(\omega(k_\ell) + \sigma'\omega(k') + \sigma''\omega(k''))\delta(k_\ell + \sigma'k' + \sigma''k'')$$
$$[f_{n+1}(k_1, \ldots, k', k_{\ell+1}, \ldots, k'') + 2\sigma'' f_{n+1}(k_1, \ldots, k_n, k')] . \quad (41)$$

Under the condition (29) and provided $\|W\|_\infty < \infty$, the hierarchy (40) has a unique solution for short times. In case

$$f_n(k_1, \ldots, k_n, 0) = \prod_{j=1}^{n} W(k_j) , \quad (42)$$

the factorization is maintained in time and each factor agrees with the solution of the Boltzmann equation (39). From (40) one easily constructs the perturbative solution to (39) with the result

$$W(k,t) = W(k) + \sum_{n=1}^{\infty} \frac{1}{n!} t^n (\mathcal{C}_{1,2} \ldots \mathcal{C}_{n,n+1} W^{\otimes n+1})(k)$$

$$= W(k) + \sum_{n=1}^{\infty} W_n(k,t) . \tag{43}$$

The series in (43) converges for t sufficiently small.

For $n = 1, 2$ the oscillating integrals can be handled by direct inspection with the expected results $\lim_{\varepsilon \to 0} W_1^{\varepsilon}(k,t) = W_1(k,t)$, $\lim_{\varepsilon \to 0} W_2^{\varepsilon}(k,t) = W_2(k,t)$. If the leading terms are as claimed in the Kinetic Conjecture, then they agree with the series (43). The complete argument is a somewhat tricky counting of diagrams, which would lead us too far astray. Thus the most immediate project is to establish that all subleading diagrams vanish in the limit $\varepsilon \to 0$. This would be a step further when compared to the investigations [5], [6].

Of course a complete proof must deal with the uniform convergence of the series in (33).

Acknowledgements

I thank Jani Lukkarinen for instructive discussions and Gianluca Panati for a careful reading.

References

1. H. Spohn, The Phonon Boltzmann Equation, Properties and Link to Weakly Anharmonic Lattice Dynamics, preprint.
2. S. Teufel, Adiabatic Perturbation Theory in Quantum Dynamics, Lecture Notes in Mathematics 1821, Springer-Verlag, Berlin, Heidelberg, New York 2003.
3. R.E. Peierls, Zur kinetischen Theorie der Wärmeleitung in Kristallen, Annalen Physik **3**, 1055–1101 (1929).
4. V.E. Zakharov, V.S. L'vov, and G. Falkovich, Kolmogorov Spectra of Turbulence: I Wave Turbulence. Springer, Berlin 1992.
5. D. Benedetto, F. Castella, R. Esposito, and M. Pulvirenti, Some considerations on the derivation of the nonlinear quantum Boltzmann equation, J. Stat. Phys. **116**, 381–410 (2004).
6. L. Erdös, M. Salmhofer, and H.T. Yau, On the quantum Boltzmann equation, J. Stat. Phys. **116**, 367–380 (2004).

Part IV

Disordered Systems and Random Operators

This part consists in five contributions devoted to spectral or dynamical properties of certain random operators.

The texts by Elgart and Combes-Germinet-Hislop address properties of two-dimensional magnetic Schrödinger operators in presence of electric potentials defined separately in two half planes. The quantity of interest is the edge conductance at the interface between the half planes, in this typical Quantum Hall geometry, for various deterministic and random potentials. Recent progresses about the mathematical definition of the edge conductance, its quantization properties in presence of disorder and its equality with the bulk conductance in a strong dynamical localization regime are reported in these contributions.

Examples of quasi-periodic Schrödinger operators exhibiting purely singular continuous spectrum for "generic" sets of parameters are known. In the paper by Lenz and Stollmann it is shown that purely singular continuous spectrum is true for solid state models of Schrödinger operators with ions located on "generic" Delone crystallographic sets. This situation is also called geometric disorder.

A quantity of importance for one-dimensional random Schrödinger operators is the Lyapunov exponent, or inverse localization length. Materials characterized by a small density of strong impurities can be described by means of a random potential whose distribution deviates slightly from a Dirac peak. Schultz-Baldes analyzes the low density limit of the Lyapunov exponent for energies in the interior of the spectrum of the free Laplacian, and thereby confirms physical non-rigorous observations about specific properties of the Lyapunov exponent at energies corresponding to rational quasi-momenta.

The contribution of Soshnikov deals with the asymptotic spectral properties of certain ensembles of random matrices as the size of these matrices goes to infinity. Recent results about the Poisson statistics of the largest eigenvalues of different classes of random matrices are explained.

On the Quantization of Hall Currents
in Presence of Disorder

Jean-Michel Combes[1], François Germinet[2], and Peter D. Hislop[3]

[1] Département de Mathématiques, Université de Toulon-Var, BP 132, 83957 La
Garde Cédex, France
`combes@cpt.univ-mrs.fr`
[2] Département de Mathématiques, Université de Cergy-Pontoise, Site de
Saint-Martin, 2 avenue Adolphe Chauvin, 95302 Cergy-Pontoise Cédex, France
`germinet@math.u-cergy.fr`
[3] Department of Mathematics, University of Kentucky, Lexington, KY
40506-0027, USA
`hislop@ms.uky.edu`

Abstract. We review recent results of two of the authors concerning the quantization of Hall currents, in particular a general quantization formula for the difference of edge Hall conductances in semi-infinite samples with and without a confining wall. We then study the case where the Fermi energy is located in a region of localized states and discuss new regularizations. We also sketch the proof of localization for 2D-models with constant magnetic field with random potential located in a half-plane in two different situations: (1) with a zero potential in the other half plane and for energies away from the Landau levels and (2) with a confining potential in the other half plane and on an interval of energies that covers an arbitrary number of Landau levels.

1 The Edge Conductance
and General Invariance Principles

Quickly after the discovery of the integer quantum Hall effect (IQHE) by von Klitzing et. al. [33], then Halperin [28] put the accent on the crucial role of quantum currents flowing at the edges of the (finite) sample. Such edge currents, carried by edge states, should be quantized, and the quantization should agree with the one of the transverse (Hall) conductance. While edge currents have been widely studied in the physics literature since the early eighties, e.g. [1, 11, 20, 35, 40] (see also [34, 37] and references therein), it is only recently that a mathematical understanding of the existence of such edge currents has been obtained [9, 10, 13, 17–19, 21]. The study of the quantization of the edge Hall conductance at a mathematical level is even more recent [3, 15, 16, 31, 32, 38].

We consider the simplest model for quantum devices exhibiting the IQHE. This consists of an electron confined to the 2-dimensional plane considered as the union of two complementary semi-infinite regions supporting potentials V_1 and V_2, respectively, and under the influence of a constant magnetic field

J.-M. Combes et al.: *On the Quantization of Hall Currents in Presence of Disorder*, Lect. Notes Phys. **690**, 307–323 (2006)
`www.springerlink.com`

B orthogonal to the sample. In the absence of potentials V_1 and V_2, the free electron is described by the free Landau Hamiltonian $H_L = p_x^2 + (p_y - Bx)^2$. The spectrum of H_L consists of the well-known Landau levels $B_N = (2N - 1)B$, $N \geq 1$, with the convention $B_0 = -\infty$. To introduce the half-plane potentials, we let 1_- and 1_+ be the characteristic functions of, respectively, $\{x \leq 0\}$ and $\{x > 0\}$. Then, if V_1, V_2 are two potentials bounded from below and in the Kato class [12], the Hamiltonian of the system is given, in suitable units and Landau gauge, by

$$H(V_1, V_2) := H_L + V_1 1_- + V_2 1_+ , \tag{1}$$

as a self-adjoint operator acting on $L^2(\mathbb{R}^2, dxdy)$, where $H_L = H(0,0)$ in this notation. For technical reasons it is convenient to assume that V_1, respectively V_2, does not grow faster than polynomially as $x \to -\infty$, respectively, as $x \to +\infty$.

We shall say that V_1 is a left *confining potential* with respect to the interval $I = [a, b] \subset \mathbb{R}$ if, in addition to the previous conditions, the following holds: There exists $R > 0$, s.t.

$$\forall x \leq -R, \ \forall y \in \mathbb{R}, \ V_1(x, y) > b . \tag{2}$$

The "hard wall" case, i.e. $V_1 = +\infty$ and $H = H_L + V_2$ acting on $L^2(\mathbb{R}^+ \times \mathbb{R}, dxdy)$ with Dirichlet boundary conditions, can be considered as well.

As typical examples for $H(V_1, V_2)$ one may think of the right potential V_2 as an impurity potential and the left potential V_1 as either a wall, confining the electron to the right half-plane and generating an edge current near $x = 0$, or as the zero potential. In this latter case, the issue is to determine whether or not V_2 is strong enough to create edge currents by itself. We will discuss this in Sect. 3.1. Another example is the strip geometry, where both V_1 and V_2 are confining potentials outside of a strip $x \in [-R, R]$, where the electron is localized.

We define a "switch" function as a smooth real valued *increasing* function equal to 1 (resp. 0) at the right (resp. left) of some bounded interval. Following [3, 15, 31, 38], we define the (Hall) edge conductance as follows.

Definition 1. *Let* $\mathcal{X} \in C^\infty(\mathbb{R}^2)$ *be a x-translation invariant switch function with* $\text{supp}\,\mathcal{X}' \subset \mathbb{R} \times [-\frac{1}{4}, \frac{1}{4}]$, *and let* $-g \in C^\infty(\mathbb{R})$ *be switch a function with* $\text{supp}\,g' \subset I = [a, b]$ *a compact interval. The* edge conductance *of* $H(V_1, V_2)$ *in the interval* I *is defined as*

$$\sigma_e(g, V_1, V_2) = -\text{tr}(g'(H(V_1, V_2))i[H_L, \mathcal{X}]) \tag{3}$$

whenever the trace is finite (we shall also use the notation $\sigma_e(g, H) = \sigma_e(g, V_1, V_2)$ *if* $H = H(V_1, V_2)$*).*

Note that in the situations of interest $\sigma_e(g, V_1, V_2)$ will turn out to be independent of the particular shape of the switch function $\mathcal{X}$ and also of the switch function g, provided $\text{supp}\,g'$ does not contain any Landau level.

We turn to the description of the results of [3].

Let us assume that I lies in between two successive Landau levels, say the Nth and the $(N+1)^{\text{th}}$. While clearly $\sigma_e(g,0,0) = 0$, for any g as above since $g'(H_L) = 0$, a straightforward computation shows that $\sigma_e(g,V_1,0) = N$, provided $V_1(x,y) = V_1(x)$ is such that $\lim_{x\to-\infty} V_1(x) > b$ (see, for example, [3, Proposition 1]). The first result tells us that the edge conductance is stable under a perturbation by a potential W located in a strip $[L_1, L_2] \times \mathbb{R}$ of finite width.

Theorem 1. ([3, Theorem 1]) *Let $H = H(V_1, V_2)$ be as in (1), and let W be a bounded potential supported in a strip $[L_1, L_2] \times \mathbb{R}$, with $-\infty < L_1 < L_2 < +\infty$. Then the operator $(g'(H+W) - g'(H))i[H_L, \mathcal{X}]$ is trace class, and*

$$\operatorname{tr}((g'(H+W) - g'(H))i[H_L, \mathcal{X}]) = 0 . \tag{4}$$

As a consequence:
(i) $\sigma_e(g, H_L + W) = 0$.
(ii) Assume V_1 is a y-invariant potential, i.e. $V_1(x,y) = V_1(x)$, that is left confining with respect to $I \supset \operatorname{supp} g'$. If $I \subset]B_N, B_{N+1}[$, for some $N \geq 0$, then

$$\sigma_e(g, H_L + V_1 + W) = N . \tag{5}$$

We note that Theorem 1 extends perturbations W that decay polynomially fast in the x-direction. In particular, it allows for more general confining potentials than y-invariant ones. But, it is easy to see that Theorem 1 does not hold for all perturbations in the x direction. For example, if supp $g' \subset]B_N, B_{N+1}[$, then $\sigma_e(g,0,0) = 0$, so for if $W_\ell(x,y) = \nu_0 \mathbf{1}_{[0,\ell]}(x)$, with $\nu_0 > B_{N+1}$ and $0 < \ell < \infty$, then $\sigma_e(g,0,W_\ell) = 0$. On the other hand, a simple calculation shows (e.g. [3]) that for $W_\infty = \nu_0 \mathbf{1}_{[0,\infty[}(x)$, we have $\sigma_e(g,0,W_\infty) = -N$. However, one has the following invariance principle, which is a consequence of a more general sum rule given in [3, Theorem 2].

Theorem 2. ([3, Corollary 3]) *Let g be s.t. supp $g' \subset]B_N, B_{N+1}[$, for some $N \geq 0$. Let V_1 be a y-invariant left confining potential with respect to $\operatorname{supp} g'$. Then the operator $\{g'(H(V_1, V_2)) - g'(H(0, V_2))\}i[H_L, \mathcal{X}]$ is trace class and*

$$- \operatorname{tr}(\{g'(H(V_1, V_2)) - g'(H(0, V_2))\}i[H_L, \mathcal{X}]) = N . \tag{6}$$

In particular, if either $\sigma_e(g, V_1, V_2)$ or $\sigma_e(g, 0, V_2)$ is finite, then both are finite, and

$$\sigma_e(g, V_1, V_2) - \sigma_e(g, 0, V_2) = N . \tag{7}$$

An immediate, but important, consequence of Theorem 2, if $\|V_2\|_\infty < B$, then $\sigma_e(g, V_1, V_2) = N$, whenever $I \subset]B_N + \|V_2\|_\infty, B_{N+1} - \|V_2\|_\infty[$, recovering a recent result of Kellendonk and Schulz-Baldes [32].

In general, neither term in (6) is separately trace class but a meaning can be given to each term through an appropriate regularization. Various regularizations of the edge conductance were discussed in [3], and two others are presented in Sect. 2. It is proved in [3] that the regularized edge conductance $\sigma_e^{reg}(g, V_0, V)$ satisfies a sum rule similar to (7):

$$\sigma_e^{reg}(g, V_0, V) = N + \sigma_e^{reg}(g, 0, V) . \tag{8}$$

With reference to this, note that $\sigma_e^{reg}(g, 0, V) \neq 0$ would imply the existence of current carrying states solely due to the impurity potential. Since (8) would yield $\sigma_e(g, V_1, V_2) \neq N$, we see that such "edge currents without edges" are responsible for the deviation of the Hall conductance from its ideal value N. Typically this is expected to happen in a regime of strong disorder (with respect to the magnetic strength B). As an example of this phenomenon, a model studied by S. Nakamura and J. Bellissard [36] is revisited in [3] and it is shown that in this case $\sigma_e(g, V_0, V_2) = 0$ and thus $\sigma_e(g, 0, V_2) = -N$. In Sect. 3.2, we present another example for which localization in the strong disorder regime implies that $\sigma_e^{reg}(0, V)$ is quantized so that there are edge currents without edges. As a counterpart, in the weak disorder regime (i.e. weak impurities in the region $x_1 \geq 0$ and no electric potential in the left half-plane), one expects that no current will flow near the region $x_1 = 0$. After a regularizing procedure, we argue in Sect. 3.1 that this is exactly what happens with the model studied in [5, 25, 39] if I lies in a region of the spectrum where localization has been shown.

Notation: Throughout this note $\mathbf{1}_X = \mathbf{1}_{(x,y)}$ will denote the characteristic function of a unit cube centered $X = (x, y) \in \mathbb{Z}^2$. If A is a subset of $\mathbb{R}^2$, then $\mathbf{1}_A$ will denote the characteristic function of this set. We recall that $\mathbf{1}_-$ and $\mathbf{1}_+$ stand, respectively, for $\mathbf{1}_{x \leq 0}$ and $\mathbf{1}_{x > 0}$.

2 Regularizing the Edge Conductance in Presence of Impurities

2.1 Generalities

Let $V_2 = V$ be a potential located in the region $x \geq 0$. If the operator $H(0, V)$ has a spectral gap and if the interval I falls into this gap, then the edge conductance is quantized as mentioned above. However such a situation is not physically relevant, since the quantization of the Hall conductance can only be related to the quantum Hall effect in presence of impurities that create the well-known "plateaus" [2, 34]. If I falls into a region of localized states of $H(0, V)$, then the conductances may not be well-defined, and a regularization is needed. In this section, we briefly recall the regularization procedure described in [3], and we then propose new candidates.

Assume supp $g' \subset I \subset]B_N, B_{N+1}[$. Let $(J_R)_{R>0}$ be a family of operators s.t.

C1. $\|J_R\| = 1$ and $\lim_{R \to \infty} J_R \psi = \psi$, for all $\psi \in E_{H(0,V)}(I)L^2(\mathbb{R}^2)$.

C2. J_R regularizes $H(0,V)$ in the sense that $g'(H(0,V))i[H_L,\mathcal{X}]J_R$ is trace class for all $R > 0$, and $\lim_{R\to\infty} \mathrm{tr}(g'(H(0,V))i[H_L,\mathcal{X}]J_R)$ exists and is finite.

Then if $V_1 = V_0$ is a y-invariant left confining potential with respect to I, it follows from Theorem 2 that

$$\lim_{R\to\infty} -\mathrm{tr}\left(\{g'(H(V_0,V)) - g'(H(0,V))\}i[H_L,\mathcal{X}]J_R\right) = N\ .$$

In other terms, if **C1** and **C2** hold, then J_R also regularizes $H(V_0,V)$. Defining the regularized edge conductance by

$$\sigma_e^{\mathrm{reg}}(g,V_1,V_2) := -\lim_{R\to\infty} \mathrm{tr}(g'(H(V_1,V_2))i[H_L,\mathcal{X}]J_R)\ , \qquad (9)$$

whenever the limit exists, we get the analog of Theorem 2:

$$\sigma_e^{\mathrm{reg}}(g,V_0,V) = N + \sigma_e^{\mathrm{reg}}(g,0,V)\ . \qquad (10)$$

In particular, if we can show that $\sigma_e^{\mathrm{reg}}(g,0,V) = 0$, for instance, under some localization property, then the edge quantization for $H(V_0,V)$ follows:

$$\sigma_e^{\mathrm{reg}}(g,V_0,V) = -\lim_{R\to\infty} \mathrm{tr}(g'(H(V_0,V))i[H_L,\mathcal{X}]J_R) = N\ . \qquad (11)$$

Let us now consider

$$H_\omega = H(0,V_{\omega,+}) = H_L + V_{\omega,+},\quad V_{\omega,+} = \sum_{i\in\mathbb{Z}^{+*}\times\mathbb{Z}} \omega_i u(x-i)\ , \qquad (12)$$

a random Schrödinger operator modeling impurities located on the positive half-plane (the $(\omega_i)_i$ are i.i.d. (independent, identically distributed) random variables, and u is a smooth bump function). If H_ω has pure point spectrum in I for $\mathbb{P}$-a.e. ω, then denoting by $(\varphi_{\omega,n})_{n\geq 1}$ a basis of orthonormal eigenfunctions of H_ω with energies $E_{\omega,n} \in \mathrm{supp}\, g' \subset I$, one has, whenever the regularization holds,

$$\sigma_e^{\mathrm{reg}}(g,0,V_{\omega,+}) = -\lim_{R\to\infty} \sum_n g'(E_{\omega,n})\langle\varphi_{\omega,n}, i[H_\omega,\mathcal{X}]J_R\varphi_{\omega,n}\rangle\ . \qquad (13)$$

If $J_R = \mathbf{1}_{x\leq R}$, the limit (13) actually exists [3, Proposition 2], but it is very likely that it will not be zero, even under strong localization properties of the eigenfunctions such as (SULE) (see [14]) or (SUDEC) (see Definition 2 below and [26]). We refer to [16] for a concrete example. This can be understood as follows: Because the cut-off J_R (even a smooth version of it) cuts classical orbits living near $x = R$, it will create spurious contributions to the total current, and the latter will no longer be zero. The quantum counter part of this picture is that although the expectation of $i[H_\omega,\mathcal{X}]$ in an eigenstate of H_ω is zero by the Virial Theorem, this is no longer true if this commutator is multiplied by J_R. Of course, the sum in (13) is zero if J_R commutes with H_ω, as in [3, Theorem 3]. In the next two sections we investigate new regularizations that commute with H_ω only asymptotically (as $R \to \infty$).

2.2 A Time Averaged Regularization
for a Dynamically Localized System

We assume that the operator $H = H(0, V)$ exhibits dynamical localization in an open interval $I \in]B_N, B_{N+1}[$. This means that for any $p \geq 0$, there exists a nonnegative constant $C_p < \infty$ such that for any Borel function f on I, with $|f| \leq 1$, and for any $X_1, X_2 \in \mathbb{R}^2$,

$$\sup_{t \in \mathbb{R}} \| \mathbf{1}_{X_1} f(H) e^{-itH} \mathbf{1}_{X_2} \|_2 \leq C_p \min(1, |X_1 - X_2|)^{-p}) . \tag{14}$$

We used the Hilbert-Schmidt norm. For random Schrödinger operators H_ω, this assumption is one of the standard conclusions of multiscale analysis [23, 24]. We show in Sect. 3.1 that as long as $I \in]B_N, B_{N+1}[$ such an analysis applies to the Hamiltonian $H_\omega = H(0, V_\omega)$ as in (12).

Note that it follows from (14) that if $X \in \mathbb{R}^2$ and A is a subset of $\mathbb{R}^2$ (A may contain X) then, for any $p > 0$, there exists a (new) constant $0 \leq C_p < \infty$, such that

$$\sup_{t \in \mathbb{R}} \| \mathbf{1}_A f(H) e^{-itH} \mathbf{1}_X \|_2 \leq C_p \min(1, \text{dist}(\{X\}, A)^{-p}) . \tag{15}$$

For $R < \infty$, $\eta > 0$ and $\gamma > 0$, we set, with $H = H(0, V)$ and $X = (x, y)$,

$$J_R = \eta \int_0^\infty E_H(I) e^{itH} \mathbf{1}_{x \leq R} e^{-itH} E_H(I) e^{-\eta t} dt, \text{ with } R = \eta^{-\gamma} . \tag{16}$$

Theorem 3. *Let J_R as in (16), with $\gamma \in]0, 1[$. Assume that $H(0, V)$ exhibits dynamical localization (i.e. (14)) in $I \subset]B_N, B_{N+1}[$ for some $N \geq 0$. Then J_R regularizes $H(0, V)$, and thus also $H(V_0, V)$, in the sense that $C1$ and $C2$ hold. Moreover the edge conductances take the quantized values: $\sigma_e^{\text{reg}}(g, 0, V) = 0$ and $\sigma_e^{\text{reg}}(g, V_0, V) = N$.*

Remark 1. In [16], a similar regularization is considered, where $\gamma = 1$ and H is the *bulk* Hamiltonian $H(V, V)$. We also note that if R and η are independent variables, then one recovers the regularization [3, (7.13)], see [3, Remark 13].

Proof. Since $\|J_R\| \leq 1$, to get $C1$ it is enough to check $\lim_{R \to \infty} J_R E_H$ $(I) \psi E_H(I) \psi$ for compactly supported states, and it is thus enough to note that by (15),

$$\| (1 - J_R) E_H(I) \mathbf{1}_0 \| \leq \eta \int_0^\infty \sup_t \| \mathbf{1}_{x > R} e^{-itH} E_H(I) \mathbf{1}_0 \| e^{-\eta t} dt \leq C_p R^{-p} . \tag{17}$$

We turn to $C2$. As in [3], we write $i[H, \mathcal{X}] = i[H, \mathcal{X}] \mathbf{1}_{|y| \leq \frac{1}{2}}$, with $\mathbf{1}_{|y| \leq \frac{1}{2}} = \sum_{x_2 \in \mathbb{Z}} \mathbf{1}_{(x_2, 0)}$. Note that, by hypothesis on I, $g'(H) = g'(H) - g'(H_L)$, so that terms that are far in the left half plane will give small contributions. To see this, we develop

$$\left\| (g'(H) - g'(H_L))i[H_L, \mathcal{X}]E_H(I)e^{-itH}\mathbf{1}_{x \leq -R} \right\|_1 \tag{18}$$

using the Helffer-Sjöstrand formula [29, 30] and the resolvent identity with $R_L(z) = (H_L - z)^{-1}$ and $R(z) = (H - z)^{-1}$. It is thus enough to control terms of the form, with $\mathrm{Im}\, z \neq 0$, $x_1, y_1, x_2 \in \mathbb{Z}$, $X_1 = (x_1, y_1)$,

$$\left\| R(z)V\mathbf{1}_{x \geq 0}R_L(z)i[H_L, \mathcal{X}]\mathbf{1}_{|y| \leq \frac{1}{2}}E_H(I)e^{-itH}\mathbf{1}_{x \leq -R} \right\|_1 \leq \sum_{x_1 \geq 0, y_1, x_2} \tag{19}$$

$$\times \left\| R(z)V\mathbf{1}_{X_1} \right\|_2 \left\| \mathbf{1}_{X_1}R_L(z)i[H_L, \mathcal{X}]\mathbf{1}_{(x_2, 0)} \right\| \left\| \mathbf{1}_{(x_2, 0)}E_H(I)e^{-itH}\mathbf{1}_{x \leq -R} \right\|_2$$

But, as a well-known fact, $R(z)\mathbf{1}_{X_1}$ is Hilbert-Schmidt in dimension 2, e.g. [25, Lemma A.4], and its Hilbert-Schmidt norm is bounded by $C[\mathrm{dist}(z, \sigma(H))]^{-1} \leq C\Im z^{-1}$, uniformly in X_1. Then, for some $\kappa \geq 1$,

$$(19) \leq \frac{C}{(\Im z)^\kappa} \sum_{x_1 \geq 0, y_1, x_2 \leq -\frac{R}{2}} \left\| \mathbf{1}_{X_1}R_L(z)i[H_L, \mathcal{X}]\mathbf{1}_{(x_2, 0)} \right\| \tag{20}$$

$$+ \frac{C}{(\Im z)^\kappa} \sum_{x_2 \geq -\frac{R}{2}} \left\| \mathbf{1}_{(x_2, 0)}E_H(I)e^{-itH}\mathbf{1}_{x \leq -R} \right\|_2 \tag{21}$$

$$\leq \frac{C}{(\Im z)^\kappa}R^{-p} \,, \tag{22}$$

where the latter follows from (15) and from the fact that $\| \mathbf{1}_{X_1}R_L(z)$ $i[H_L, \mathcal{X}]\mathbf{1}_{(x_2, 0)}\|$ decays faster than any polynomial in $|X_1 - X_2|$, as can be seen by a Combes-Thomas estimate together with standard computations (e.g. [3, Lemma 3]). As a consequence the trace norm (18) is finite and goes to zero as $R \to \infty$, uniformly in η.

The next step is to control contributions coming from terms living far from the support of $\mathcal{X}'$, i.e. terms s.t. $|y| \geq R^\nu$ with $\nu \in]0, 1]$. Set $S(R, \nu) = \{(x, y) \in \mathbb{R}^2, |x| \leq R, |y| \geq R^\nu\}$. Then, using (15),

$$\left\| g'(H)i[H_L, \mathcal{X}]E_H(I)e^{-itH}\mathbf{1}_{S(R, \nu)} \right\|_1 \tag{23}$$

$$\leq \sum_{x_2 \in \mathbb{Z}} \left\| g'(H)i[H_L, \mathcal{X}]\mathbf{1}_{(x_2, 0)} \right\|_2 \left\| \mathbf{1}_{(x_2, 0)}E_H(I)e^{-itH}\mathbf{1}_{S(R, \nu)} \right\|_2 \tag{24}$$

$$\leq C \sum_{x_2 \in \mathbb{Z}} \min(1, \mathrm{dist}(\{(x_2, 0)\}, S(R, \nu))^{-p}) \leq C(2R)(R^\nu)^{-p} \,. \tag{25}$$

Contributions from (23) are thus negligible as $R \to \infty$, uniformly in η. Letting $K_{R,\nu}$ denote the compact set

$$K_{R,\nu} = \{(x, y) \in \mathbb{R}^2, |x| \leq R, |y| \leq R^\nu\} \,,$$

we are left so far with the evaluation of

$$\eta \int_0^\infty e^{-\eta t}dt\, g'(H)i[H_L, \mathcal{X}]E_H(I)e^{itH}\mathbf{1}_{K_{R,\nu}}e^{-itH}E_H(I) \,, \tag{26}$$

which is clearly now a trace class operator (the integral is absolutely convergent in trace norm). In other terms $g'(H)i[H_L, \mathcal{X}]J_R$ is thus trace class. It remains to show that its trace goes to zero as R goes to infinity. But on the account of (18) and (23), it remains to show that the trace of (26) goes to zero. By cyclicity,

$$
\begin{aligned}
\text{tr}(26) &= -\eta \int_0^\infty e^{-\eta t} \text{tr}\left\{ g'(H)\mathcal{X}e^{-itH}i[H_L, E_H(I)\mathbf{1}_{K_{R,\nu}}E_H(I)]e^{itH} \right\} dt \\
&= \eta \int_0^\infty e^{-\eta t}\frac{d}{dt}\text{tr}\left\{ g'(H)\mathcal{X}e^{-itH}E_H(I)\mathbf{1}_{K_{R,\nu}}E_H(I)e^{itH} \right\} dt \\
&= \eta\text{tr}\left\{ g'(H)\mathcal{X}E_H(I)\mathbf{1}_{K_{R,\nu}}E_H(I) \right\} \\
&\quad -\eta^2 \int_0^\infty e^{-\eta t}\text{tr}\left\{ g'(H)\mathcal{X}e^{-itH}E_H(I)\mathbf{1}_{K_{R,\nu}}E_H(I)e^{itH}E_H(I) \right\} dt.
\end{aligned}
$$

Thus $|\text{tr}(26)| \leq C\eta|K_{R,\nu}| = C\eta R^{1+\nu}$. Since $R = \eta^{-\gamma}$, the trace goes to zero if $\gamma < \frac{1}{1+\nu}$.

2.3 Regularization Under a Stronger form of Dynamical Localization

In this section, we consider

$$
J_R = E_{H(0,V)}(I)\mathbf{1}_{x \leq R}E_{H(0,V)}(I) . \tag{27}
$$

Note that the regularization (16) studied in Sect. (2.2) is the time average of (27). The effect of the time averaging is to provide a control on the cross terms arising in (27) if one expands $E_{H(0,V)}(I)$ over a basis of eigenfunctions. In [3, (7.13)], cross terms were suppressed from the very definition of J_R. By showing that J_R, given in (27), regularizes $H(0, V_\omega)$ under, basically, the same assumption as in [3, Theorem 3], we strengthen [3]'s result.

Let T be the multiplication operator by $T(X) = \langle X \rangle^\nu$, $\nu > \frac{d}{2} = 1$, with $\langle X \rangle = (1 + |X|^2)^{\frac{1}{2}}$, for $X \in \mathbb{R}^2$. It is well known for Schrödinger operators that $\text{tr}(T^{-1}E_{H(0,V)}(I)T^{-1}) < \infty$, if I is compact (e.g. [25]).

Definition 2 (SUDEC). *Assume H has pure point spectrum in I with eigenvalues E_n and corresponding normalized eigenfunctions φ_n, listed with multiplicities. We say that H has* Summable Uniform Decay of Eigenfunction Correlations *(SUDEC) in I, if there exist $\zeta \in {]0,1[}$ and a finite constant $c_0 > 0$ such that for any $E_n \in I$ and $X_1, X_2 \in \mathbb{Z}^2$,*

$$
\|\mathbf{1}_{X_1}\varphi_n\|\|\mathbf{1}_{X_2}\varphi_n\| \leq c_0\alpha_n\|T\mathbf{1}_{X_1}\|^2\|T\mathbf{1}_{X_2}\|^2 e^{-|X_1-X_2|^\zeta} , \tag{28}
$$

where $\alpha_n = \|T^{-1}\varphi_n\|^2$.

Note that,

$$
\sum_n \alpha_n = \text{tr}(T^{-1}E_{H(0,V)}(I)T^{-1}) < \infty . \tag{29}
$$

Remark 2. Property (28) (or a modified version of it) was called (WULE) in [3] and was introduced in [22]. The more accurate acronym (SUDEC) comes from [26] and Property (SUDEC) is used in [27] as a very natural signature of localization in order to get the quantization of the bulk conductance.

Theorem 4. *Assume that $H(0,V)$ has (SUDEC) in $I \subset]B_N, B_{N+1}[$ for some $N \geq 0$. Then J_R, given in (27), regularizes $H(0,V)$, and thus also $H(V_0,V)$, in the sense that **C1** and **C2** hold. Moreover the edge conductances take the quantized values: $\sigma_e^{\mathrm{reg}}(g,0,V) = 0$ and $\sigma_e^{\mathrm{reg}}(g,V_0,V) = N$.*

Proof. That the operator $g'(H(0,V))i[H(0,V),\mathcal{X}]J_R$ is trace class follows from the comparison $g'(H(0,V)) = g'(H(0,V)) - g'(H_L)$. In order to control the region $x \leq 0$, and the immediate estimate, let P_n be the eigenprojector on the eigenfunction φ_n, and write

$$\|\mathbf{1}_X E_{H(0,V)}(I)\mathbf{1}_Y\|_2 \leq \sum_n \|\mathbf{1}_X P_n \mathbf{1}_Y\|_2 = \sum_n \|\mathbf{1}_X \varphi_n\| \|\mathbf{1}_Y \varphi_n\| \quad (30)$$

$$\leq c_0 \left(\sum_n a_n\right) \|T\mathbf{1}_X\|^2 \|T\mathbf{1}_Y\|^2 e^{-|X-Y|^\varsigma}, \quad (31)$$

where we used the assumption (28) (and recall (29)). We proceed and set $\Lambda_{2,R} = \mathbf{1}_{x \leq R}$. We are looking at

$$\sigma_E^{(reg)}(g,R) = \mathrm{tr}(g'(H(0,V))i[H(0,V),\mathcal{X}]J_R). \quad (32)$$

The operator being trace class, we expand the trace in the basis of eigenfunctions of $H(0,V)$ in the interval I and get

$$|\sigma_E^{(reg)}(g,R)| = \left|\sum_{n \neq m} g'(E_n)(E_n - E_m)\langle \varphi_n, \mathcal{X}\varphi_m\rangle\langle \varphi_m, \Lambda_{2,R}\varphi_n\rangle\right| \quad (33)$$

$$\leq C(g,I) \sum_{n \neq m} |\langle \varphi_n, \mathcal{X}\varphi_m\rangle| |\langle \varphi_m, \Lambda_{2,R}\varphi_n\rangle|. \quad (34)$$

It remains to show that the double sum in (34) is convergent (i.e. the trace is absolutely convergent). If it is so, then we can interchange the limit in R and the double sum to get zero due to the orthogonality of the eigenfunctions. In full generality, dynamical localization is not enough to show that (34) is absolutely convergent. This is the case if $H = H(0,0) = H_L$ and I contains a Landau level. The sum (34) diverges, even though H_L exhibits dynamical localization. But if one has (SULE) or (SUDEC), then the sum converges absolutely. We have, writing Λ_2 instead of $\Lambda_{2,R}$:

(34)   (35)

$$\leq \sum_{n \neq m} |\langle \varphi_n, \mathcal{X}\varphi_m \rangle| \, |\langle \varphi_m, \Lambda_2 \varphi_n \rangle|$$ (36)

$$= \sum_{n \neq m} |\langle \varphi_n, \mathcal{X}\varphi_m \rangle|^{\frac{1}{2}} |\langle \varphi_n, (1-\mathcal{X})\varphi_m \rangle|^{\frac{1}{2}} |\langle \varphi_m, \Lambda_2 \varphi_n \rangle|^{\frac{1}{2}} |\langle \varphi_m, (1-\Lambda_2)\varphi_n \rangle|^{\frac{1}{2}}$$

$$\leq \sum_{n \neq m} \Big(\|\sqrt{\mathcal{X}}\varphi_n\| \|\sqrt{\mathcal{X}}\varphi_m\| \|\sqrt{1-\mathcal{X}}\varphi_n\| \|\sqrt{1-\mathcal{X}}\varphi_m\|$$ (37)

$$\times \|\sqrt{\Lambda_2}\varphi_n\| \|\sqrt{\Lambda_2}\varphi_m\| \|\sqrt{1-\Lambda_2}\varphi_n\| \|\sqrt{1-\Lambda_2}\varphi_m\| \Big)^{\frac{1}{2}}$$ (38)

$$\leq \sum_{n} \Big(\|\sqrt{\mathcal{X}}\varphi_n\| \|\sqrt{1-\mathcal{X}}\varphi_n\| \|\sqrt{\Lambda_2}\varphi_n\| \|\sqrt{1-\Lambda_2}\varphi_n\| \Big)^{\frac{1}{2}}$$ (39)

$$\times \sum_{m} \Big(\|\sqrt{\mathcal{X}}\varphi_m\| \|\sqrt{1-\mathcal{X}}\varphi_m\| \|\sqrt{\Lambda_2}\varphi_m\| \|\sqrt{1-\Lambda_2}\varphi_m\| \Big)^{\frac{1}{2}}$$ (40)

It remains to show that (28) implies

$$\sum_{n} \Big(\|\sqrt{\mathcal{X}}\varphi_n\| \|\sqrt{1-\mathcal{X}}\varphi_n\| \|\sqrt{\Lambda_2}\varphi_n\| \|\sqrt{1-\Lambda_2}\varphi_n\| \Big)^{\frac{1}{2}} < \infty \,.$$ (41)

We consider division of $\mathbb{R}^2$ into four quadrants given by the supports of the various localization functions: $I = \operatorname{supp} \mathcal{X}\Lambda_2$, $II = \operatorname{supp} (1-\mathcal{X})\Lambda_2$, $III = \operatorname{supp} (1-\mathcal{X})(1-\Lambda_2)$, and $IV = \operatorname{supp} \mathcal{X}(1-\Lambda_2)$. We first note that summing (28) over two opposite quadrants $(\mathbf{I})(\mathbf{III})$ yields a constant:

$$\|\sqrt{\mathcal{X}}\sqrt{\Lambda_2}\varphi_n\| \|\sqrt{1-\mathcal{X}}\sqrt{1-\Lambda_2}\varphi_n\| \leq c\alpha_n \,,$$ (42)

and summing over the opposite quadrants $(\mathbf{II})(\mathbf{IV})$ yields,

$$\|\sqrt{\mathcal{X}}\sqrt{1-\Lambda_2}\varphi_n\| \|\sqrt{1-\mathcal{X}}\sqrt{\Lambda_2}\varphi_n\| \leq c\alpha_n$$ (43)

We write a term in (41) as

$$\|\sqrt{\mathcal{X}}\varphi_n\| \|\sqrt{1-\mathcal{X}}\varphi_n\| \|\sqrt{\Lambda_2}\varphi_n\| \|\sqrt{1-\Lambda_2}\varphi_n\|$$ (44)

$$\leq (\|\sqrt{\mathcal{X}}\sqrt{\Lambda_2}\varphi_n\| + \|\sqrt{\mathcal{X}}\sqrt{1-\Lambda_2}\varphi_n\|)$$

$$(\|\sqrt{1-\mathcal{X}}\sqrt{\Lambda_2}\varphi_n\| + \|\sqrt{1-\mathcal{X}}\sqrt{1-\Lambda_2}\varphi_n\|)$$

$$(\|\sqrt{\mathcal{X}}\sqrt{\Lambda_2}\varphi_n\| + \|\sqrt{1-\mathcal{X}}\sqrt{\Lambda_2}\varphi_n\|)$$

$$(\|\sqrt{\mathcal{X}}\sqrt{1-\Lambda_2}\varphi_n\| + \|\sqrt{1-\mathcal{X}}\sqrt{1-\Lambda_2}\varphi_n\|)$$

$$= (\mathbf{I}+\mathbf{IV})(\mathbf{II}+\mathbf{III})(\mathbf{I}+\mathbf{II})(\mathbf{III}+\mathbf{IV}) \,.$$ (45)

This decomposition yields 16 terms, each of them having at least one product of the form (42) or (43), i.e. with opposite terms: $(\mathbf{I})(\mathbf{III})$ and $(\mathbf{II})(\mathbf{IV})$. If we now bound the other factors by one, we get $\sum_n \sqrt{\alpha_n}$ in (41), while our assumption only ensures that $\sum_n \alpha_n < \infty$. To get the missing factor $\sqrt{\alpha_n}$ we have to be a bit more careful. First, obviously, terms of the form $(\mathbf{I})^2(\mathbf{III})^2$, $(\mathbf{II})^2(\mathbf{IV})^2$ and $(\mathbf{I})(\mathbf{II})(\mathbf{III})(\mathbf{IV})$ will directly yield the desired α_n. It remains to study terms of the form $(\mathbf{I})^2(\mathbf{II})(\mathbf{III})$, $(\mathbf{I})^2(\mathbf{II})(\mathbf{IV})$, and $(\mathbf{I})^2(\mathbf{II})(\mathbf{IV})$, and the 9 remaining terms beginning with $(\mathbf{II})^2$, $(\mathbf{III})^2$, and $(\mathbf{IV})^2$. Let us treat the first case, the other two terms being similar. Note that $(\mathbf{I})^2 = \sum_{x_1 \leq R, y_1 \leq 0} \|\mathbf{1}_{X_1} \varphi_n\|^2$, with $X_1 = (x_1, y_1)$. Then going back to (28), with obvious notations, we have

$$(\mathbf{I})^2(\mathbf{II})(\mathbf{III}) \leq \sum_{X_1, X_2, X_3} \|\mathbf{1}_{X_1} \varphi_n\|^2 \|\mathbf{1}_{X_2} \varphi_n\| \|\mathbf{1}_{X_3} \varphi_n\| \tag{46}$$

$$\leq \sum_{X_1, X_2, X_3} (\|\mathbf{1}_{X_1} \varphi_n\| \|\mathbf{1}_{X_2} \varphi_n\|)(\|\mathbf{1}_{X_1} \varphi_n\| \|\mathbf{1}_{X_3} \varphi_n\|)$$

$$\leq (c_0 \alpha_n)^2 \sum_{X_1, X_2, X_3} \langle X_1 \rangle^{4\nu} \langle X_2 \rangle^{2\nu} \langle X_3 \rangle^{2\nu}$$

$$\times e^{-\frac{1}{4}(|y_1|^\varsigma - |y_2|^\varsigma - |x_1 - x_2|^\varsigma)} e^{-\frac{1}{4}(|x_1|^\varsigma - |x_3|^\varsigma - |y_1|^\varsigma - |y_3|^\varsigma)}$$

$$\leq C(R)\alpha_n^2 . \tag{47}$$

3 Localization for the Landau Operator with a Half-Plane Random Potential

We describe some results concerning the localization properties of the Hamiltonians $H(V_1, V_2)$ of interest to the IQHE. First, we sketch the proof of localization for $H(0, V_2)$, with V_2 random, in the large B regime, a result mentioned in Sect. 2.2 and announced in [3, Remark 12]. We then sketch the proof of localization for $H(V_1, V_2)$, where V_1 is a left confining potential, and V_2 is a random Anderson-type potential in the large disorder regime and with a covering condition on the single-site potentials. In the large disorder regime, this provides an example of edge currents without edges. Other results for such special models of interest to edge conductance and the IQHE are discussed in [4].

3.1 A Large Magnetic Field Regime

The aim of this section is to justify Remark 12 in [3] where localization for $H(0, V)$ away from the Landau levels is claimed. We let $X = (x, y) \in \mathbb{R}^2$, and consider, for $\lambda > 0$, $B > 0$ given, the Hamiltonian

$$H_\omega = H(0, V_\omega), \text{ with } V_\omega = \lambda \sum_{i \in \mathbb{Z}^+ \times \mathbb{Z}} \omega_i u(X - i) . \tag{48}$$

The assumptions on the random variables ω_i, $i \in \mathbb{Z}^+ \times \mathbb{Z}$, and on the single site potential u are the one considered in [5,25]. Namely, the ω_i's are i.i.d.

random variables with a common law $\mu(dt) = g(t)dt$, where g is an even bounded function with support in $[-M, M]$, $M > 0$, with, in addition, the condition $\mu([0, t]) \geq c \min(t, M)^\varsigma$, for some $\varsigma > 0^4$. In order to apply the percolation estimate as in [5] we require that

$$\text{supp } u \in B(0, 1/\sqrt{2}) \,. \tag{49}$$

With no loss we assume that $\|u\|_\infty = 1$, so that the spectrum of H_ω satisfies

$$\sigma(H_\omega) \subset \bigcup_{n \geq 1} [B_n - M, B_n + M] \,.$$

We note that thanks to the ergodicity of H_ω with respect to integer translations in the y-direction, the spectrum equals a deterministic set for almost all $\omega = (\omega_i)_{i \in \mathbb{Z}^+ \times \mathbb{Z}}$. For convenience we shall extend the ω_i's to the left half plane by setting $\omega_i = 0$ if $i \in \mathbb{Z}^- \times \mathbb{Z}$.

The only difference between the present model and that of [5, 25] is that the random potential in the left half-plane is replaced by a zero potential. This absence of a potential creates a classically forbidden region in the spectral sense for energies between Landau levels. This situation is different from a classically forbidden region created by a wall. The intuition is that looking at a given distance of a Landau level (in the energy axis), the absence of potential should help for localization. One may think of [5, 25, 39]'s result as a weak disorder result. The disorder is kept fixed and localization is obtained for large B. In this spirit putting $\omega_i = 0$ should be even better, for one creates fewer states at a given distance from the Landau level. One might think that the interface at $x = 0$ between the random potential in the right half-plane and the absence of potential in the left one would create some current along the interface. It could be so for energies very close to the Landau level where the above reasoning breaks down.

To get localization, one has to investigate how the Wegner estimate, the multiscale analysis (MSA), and the starting estimates of the MSA are affected by the new geometry of the random potential. In particular, since we broke translation invariance in the x direction, we have to check things for all boxes, regardless of the position with respect to the interface $x = 0$.

The Wegner estimate: It is immediately seen that the proof of the Wegner estimate given in [5] is still valid with this geometry. Indeed, if a box $\Lambda_L(x, y)$ is such that $x < L/2$, then $\Lambda_L(x, y)$ overlaps the left half-plane (it may even be contained in it). Then, in [5, (3.8)] the sum is restricted to sites $i = (i_1, i_2)$ where $\omega_i \neq 0$ (i.e. $i_1 > 0$). The rest of the proof is unchanged, and as a result the volume factor one gets at the end is $|\Lambda_L(x, y) \cap (\mathbb{R}^+ \times \mathbb{R})|$ rather than $|\Lambda_L(x, y)|$. In particular one gets zero if $\Lambda_L(x, y) \subset \mathbb{R}^- \times \mathbb{R}$, as expected. So (W) and (NE) of [24] hold.

---

[4] This last hypothesis is not necessary to prove localization at a given fixed distance, independent of B, from the Landau levels

The multiscale analysis: The deterministic part of the MSA (properties (SLI) and (EDI) in [24]) is not sensitive to changes of the random variables. Independence of far separated boxes (property (IAD) in [24]) is still true. In fact, what happens in the probabilistic estimates that appear in the MSA is that we shall estimate probabilities of bad events related to boxes which have an overlap with the left half-plane as if they where contained in the right half-plane, and thus by a bigger (thus worse) probability. In particular, if a box is totally included in the left half-plane, the probability of having a singular box is zero, and we shall estimate it by a polynomially (or sub-exponentially) small factor in the size of the box.

The starting estimate: We follow the argument given in [5]. Let us focus on energies $E \in]B_n, B_n + M]$, the other case $E \in [B_n - M, B_n[$ being similar. We thus set $E = B_n + 2a$, $a > 0$. We say that a site $i \in \mathbb{Z}^2$ is *occupied* if $\omega_i \in [-M, a]$, in other words, $\text{dist}(E, B_n + \omega_i) \geq a$ (recall $\|u\|_\infty = 1$). Note that by hypothesis on ω_i, for any $a > 0$,

$$\mathbb{P}(\omega_i \in [-M, a]) \geq \frac{1}{2} + ca^\varsigma .$$

In particular, the probability is $\mathbb{P}(\omega_i \in [-M, a]) = 1$, if $i \in \mathbb{Z}^- \times \mathbb{Z}$. We are thus above the critical bond percolation threshold $p_c = \frac{1}{2}$ (in dimension 2) for all $i \in \mathbb{Z}^2$. Consequently, bonds percolate, and [5, Proposition 4.1] follows. The rest of the proof leading to the initial length scale estimate [5, Proposition 5.1] is the same.

At this stage Theorem 4.1 in [25] applies, and one has Anderson localization, (SULE), and strong Hilbert-Schmidt dynamical localization as described in [24], as well as (SUDEC) (following the proof of [22]; see also [26]).

We note that the above arguments are not restricted to the particular half-plane geometry of the random potential we discussed here. Any random potential of the form $V_\omega = \sum_{i \in \mathcal{J}} \omega_i u(X - i)$, where $\mathcal{J} \subset \mathbb{Z}^2$ has an infinite cardinal would yield the same localization result.

3.2 A Large Disorder Regime

We next consider the random Landau Hamiltonian defined in (48) with a left constant confining potential $V_0(x, y) = V_0 \mathbf{1}_-$ (see (2)) so that $H(V_0, V_\omega) = H_L + V_0 \mathbf{1}_- + \lambda V_\omega \mathbf{1}_+$, for large values of the disorder parameter λ. The random potential in the right half-plane V_ω, as in (48), has i.i.d. random variables $\omega_i' s$ with a common *positively* supported distribution, say on $[0, 1]$. We also impose the condition that the single site potential $u \in \mathcal{C}_c^\infty(\mathbb{R})$ satisfies the following covering condition: If $\Lambda \subset \mathbb{R}^+ \times \mathbb{R}$,

$$\sum_{i \in \Lambda} u(X - i) \geq C_0 \mathbf{1}_\Lambda . \tag{50}$$

We show that if the disorder is large enough, then at low energy, no edge current will exist along the interface $x = 0$ in the sense that the regularized edge conductance $\sigma_e^{reg}(g, V_0, \lambda V_\omega)$ of $H(V_0, V_\omega)$ will be zero. As consequence of (10), however, the regularized edge conductance of $H(0, V_\omega)$ will be quantized to a non zero value, i.e. $\sigma_e^{reg}(g, 0, \lambda V_\omega) = -N$. In other terms the random potential λV_ω is strong enough to create "edge currents without edges" (as in [17]). Such a situation is similar to the model studied by S. Nakamura and J. Bellissard [36], and revisited in [3] from the "edge" point of view.

The strategy to prove localization for $H(V_0, V_\omega)$ is the same as the one exposed in Sect. 3.1, i.e. use a modified multiscale analysis taking into account the new geometry of the problem. Here the potential in the left half-plane is no longer zero but a constant $V_0 > b$, if $I = [0, b]$ is the interval where we would like to prove localization. As in Sect. 3.1, the modifications of the Wegner estimate, of the starting estimates of the the multiscale analysis (MSA), and of the MSA itself, have to be checked separately. While the comments made in Sect. 3.1 concerning the MSA are still valid, the new geometry requires new specific arguments for the Wegner estimate and the starting estimate.

The Wegner estimate: Its proof can no longer be borrowed from [5] as in Sect. 3.1, and one has to explicitly take into account the effect of the confining potential $V_0 1_-$. We shall modify the argument given in [8] as follows. The only case we have to discuss is the one of a box $\Lambda_L(X)$, with $|x| < \frac{L}{2}$ so that it overlaps both types of potentials. We set

$$\tilde{V}_L = \sum_{i \in \Lambda_L(X) \cap \mathbb{Z}^2} \omega_i u(X - i) .$$

Let H_L denote the restriction of $H(V_0, V_\omega)$ to the box $\Lambda_L(X)$ with self-adjoint boundary conditions (e.g. [27]). By Chebychev's inequality, the proof of the Wegner estimate is reduced to an upper bound on the expectation of the trace of the spectral projector $E_{H_L}(I)$ for the interval I. Following [7], we write

$$\text{tr}E_{H_L}(I) = \text{tr}1_{\Lambda_L(X)}E_{H_L}(I) \tag{51}$$

$$\leq \frac{1}{V_0}\text{tr}V_0 1_{\Lambda_L(X)} 1_- E_{H_L}(I) + \frac{\lambda}{C_0}\text{tr}\tilde{V}_L E_{H_L}(I) \tag{52}$$

$$\leq \frac{1}{V_0}\text{tr}1_{-,H_L}E_{H_L}(I) + \frac{\lambda}{C_0}\text{tr}\tilde{V}_L E_{H_L}(I) \tag{53}$$

$$\leq \frac{b}{V_0}\text{tr}E_{H_L}(I) + \frac{\lambda}{C_0}\text{tr}\tilde{V}_L E_{H_L}(I) , \tag{54}$$

so that, with $V_0 > b$ by assumption,

$$\text{tr}E_{H_L}(I) \leq \frac{\lambda}{C_0} \left(1 - \frac{b}{V_0}\right)^{-1} \text{tr}\tilde{V}_L E_{H_L}(I) . \tag{55}$$

At this point, the proof follows the usual strategy, as in [6–8].

The starting estimate: The initial estimate follows from the analysis, at large disorder, given in [25, Section 3]. Since in the left half-plane, the potential is already very high ($V_0 > b$), it is enough to estimate the probability that all the random variables ω_i in the right part of the box is higher than say $b/2$. Doing this creates a gap in the spectrum of the finite volume operator H_L. This spectral gap, occurring with good probability, can be used to obtain the exponential decay of the (finite volume) resolvent thanks to a Combes-Thomas argument.

References

1. Avron, J.E., Seiler, R.: Quantization of the Hall conductance for general, multiparticle Schrödinger operators, Phys. Rev. Lett. **54** 259–262 (1985)
2. Bellissard, J., van Elst, A., Schulz-Baldes, H.: The non commutative geometry of the quantum Hall effect. J. Math. Phys. **35**, 5373–5451 (1994)
3. Combes, J.M., Germinet, F.: Edge and Impurity Effects on Quantization of Hall Currents, Commun. Math. Phys. **256**, 159–180 (2005)
4. Combes, J.M., Germinet, F., Hislop, P.D: in preparation.
5. Combes, J.M., Hislop, P.D.: Landau Hamiltonians with random potentials: localization and the density of states Commun. Math. Phys. **177**, 603–629 (1996)
6. Combes, J.M., Hislop, P.D., Klopp, F. Nakamura, S.: The Wegner estimate and the integrated density of states for some random operators. Spectral and inverse spectral theory (Goa, 2000), Proc. Indian Acad. Sci. Math. Sci. **112**, 31–53 (2002)
7. Combes, J.M., Hislop, P.D., Klopp, F.: Hölder continuity of the integrated density of states for some random operators at all energies. IMRN **4**, 179–209 (2003)
8. Combes, J.M., Hislop, P.D., Nakamura, S.: The L^p-theory of the spectral shift function, the Wegner estimate, and the integrated density of states for some random operators Commun. Math. Phys. **218**, 113–130 (2001)
9. Combes, J.-M., Hislop, P.D., Soccorsi, E.: Edge states for quantum Hall Hamiltonians. Mathematical results in quantum mechanics (Taxco, 2001), 69–81, Contemp. Math., 307, Amer. Math. Soc., Providence, RI, 2002
10. Combes, J.-M., Hislop, P.D., Soccorsi, E.: in preparation
11. Cresti, A., Fardrioni, R., Grosso, G., Parravicini, G.P.: Current distribution and conductance quantization in the integer quantum Hall regime, J. Phys. Condens. Matter **15**, L377–L383 (2003)
12. Cycon, H.L., Froese, R.G., Kirsch, W., Simon, B.: *Schrödinger operators*. Heidelberg: Springer-Verlag, 1987
13. De Bièvre, S., Pulé, J.: Propagating Edge States for a Magnetic Hamiltonian. Math. Phys. Elec. J. **vol. 5**, paper 3
14. Del Rio, R., Jitomirskaya, S., Last, Y., Simon, B.: Operators with singular continuous spectrum. IV. Hausdorff dimensions, rank one perturbations, and localization. J. Anal. Math. **69**, 153–200 (1996)
15. Elbau, P., Graf., G.M.: Equality of Bulk and Edge Hall Conductance Revisited. Commun. Math. Phys. **229**, 415–432 (2002)
16. Elgart, A., Graf, G.M., Schenker, J.: Equality of the bulk and edge Hall conductances in a mobility gap. Comm. Math. Phys. **259**, 185–221 (2005)
17. Exner, P., Joye, A., Kovarik, H.: Edge currents in the absence of edges, Physics Letters A, **264**, 124–130 (1999)

18. Exner, P., Joye, A., Kovarik, H.: Magnetic transport in a straight parabolic channel, J. of Physics A: Math. Gen., **34**, 9733–9752 (2001)

19. Ferrari, C., Macris, N.: Intermixture of extended edge and localized bulk levels in macroscopic Hall systems. J. Phys. A: Math. Gen. **35**, 6339–6358 (2002)

20. Fleischmann, R., Geisel, T,. Ketzmerick, R., Phys. Rev. Lett. **68**, 1367 (1992).

21. Fröhlich, J., Graf, G.M., Walcher, J.: On the extended nature of edge states of quantum Hall Hamiltonians. Ann. H. Poincaré **1**, 405–444 (2000)

22. Germinet, F.: Dynamical localization II with an Application to the Almost Mathieu Operator. J. Stat. Phys. **95**, 273–286 (1999)

23. Germinet, F., De Bièvre, S.: Dynamical localization for discrete and continuous random Schrödinger operators. Commun. Math. Phys. **194**, 323–341 (1998)

24. Germinet, F., Klein, A.: Bootstrap Multiscale Analysis and Localization in Random Media. Commun. Math. Phys. **222**, 415–448 (2001)

25. Germinet, F, Klein, A.: Explicit finite volume criteria for localization in continuous random media and applications. Geom. Funct. Anal. **13**, 1201–1238 (2003)

26. Germinet, F, Klein, A.: New characterizations of the region of dynamical localization for random Schrödinger operators, to appear in J. Stat. Phys.

27. Germinet, F, Klein, A., Schenker, J.: Dynamical delocalization in random Landau Hamiltonians, to appear in Annals of Math.

28. Halperin, B.I.: Quantized Hall conductance, current carrying edge states and the existence of extended states in a two-dimensional disordered potential. Phys. Rev. B **25**, 2185–2190 (1982)

29. Helffer, B., Sjöstrand, J.: Équation de Schrödinger avec champ magnétique et équation de Harper, in *Schrödinger operators*, H. Holden and A. Jensen eds., LNP **345**, 118–197 (1989)

30. Hunziker W., Sigal, I.M.: Time-dependent scattering theory for N-body quantum systems. Rev. Math. Phys. **12**, 1033–1084 (2000)

31. Kellendonk, J., Richter, T., Schulz-Baldes, H.: Edge Current channels and Chern numbers in the integer quantum Hall effect. Rev. Math. Phys. **14**, 87–119 (2002)

32. Kellendonk, T., Schulz-Baldes, H.: Quantization of Edge Currents for continuous magnetic operators. J. Funct. Anal. **209**, 388–413 (2004)

33. von Klitzing, K, Dorda, G, Pepper, N.: New method for high-accuracy determination of the fine structure constant based on quantized Hall resistance, Phys. Rev. Lett **45**, 494 (1980).

34. von Klitzing, 25 years of QHE, in Séminaire Henri Poincaré (2004)

35. Mac Donald, A.H., Streda, P.: Quantized Hall effect and edge currents. Phys. Rev. B **29**, 1616–1619 (1984)

36. Nakamura, S., Bellissard, J.: Low Energy Bands do not Contribute to Quantum Hall Effect. Commun. Math. Phys. **131**, 283–305 (1990)

37. Prange, Girvin, *The Quantum Hall Effect*, Graduate texts in contemporary Physics, Springer-Verlag, N.Y. 1987

38. Schulz-Baldes, H., Kellendonk, J., Richter, T.: Simultaneous quantization of edge and bulk Hall conductivity. J. Phys. A **33**, L27–L32 (2000)

39. Wang, W.-M.: Microlocalization, percolation, and Anderson localization for the magnetic Schrödinger operator with a random potential. J. Funct. Anal. **146**, 1–26 (1997)
40. Zozoulenko, I.V., Maao, F.A., Hauge, E.H., Phys. Rev. B**53**, 7975 (1996).

Equality of the Bulk and Edge Hall Conductances in $2D$

A. Elgart

Department of Mathematics, Stanford University, Stanford, CA 94305-2125
elgart@math.stanford.edu

1 Introduction and Main Result

Von Klitzing [15] observed that a two dimensional electron gas at very low temperatures and strong magnetic field displays a quantization of the Hall conductance, that is the conductance measured in the direction transversal to the applied current. Specifically, the conductance plotted as a function of the magnetic field shows extremely flat plateaux at integer multiples of e^2/h (e is a charge of electron and h is Planck's constant). Two pictures were introduced for a description of the Quantum Hall Effect: "Edge currents picture" and "Bulk currents picture". The edge current picture suggests that the Hall current flows in the narrow regions along the sample boundaries (we will denote the corresponding conductance by σ_E), so that the Hall voltage drops entirely in these regions. On the other hand, the description in terms of bulk currents suggests that the Hall voltage drops gradually across the sample (and let σ_B denote the Hall conductance associated with this regime). It was proposed by Halperin [13] that in reality one should expect an intermix of these two pictures, and that $\sigma_E = \sigma_B$. In [14], $\sigma_{E,B}$ were linked for a Harper's model with rational flux. In more general setup, the equality of the edge and the bulk conductances was recently rigorously established [10,16,18], provided that there is a spectral gap Δ at Fermi energy of the single-particle (bulk) Hamiltonian H_B.

The aim of this article is to report on a recent joint result with G. M. Graf and J. H. Schenker [11], and to expose some of the ideas involved. There we proved this equality in the more general setting in which H_B exhibits Anderson localization in Δ – more precisely, strong dynamical localization (see (2) below). The result applies to Schrödinger operators which are random, but does not depend on that property. We therefore formulate the result for deterministic operators. A full account of this work with detailed proofs of all the results stated below is provided in [11].

The bulk system is described by the lattice $\mathbb{Z}^2 \ni x = (x_1, x_2)$ with Hamiltonian $H_B = H_B^*$ on $\ell^2(\mathbb{Z}^2)$. We assume its matrix elements $H_B(x, x')$, $x, x' \in \mathbb{Z}^2$, to be of short range, namely

$$\sup_{x \in \mathbb{Z}^2} \sum_{x' \in \mathbb{Z}^2} |H_B(x, x')| \left(e^{\mu|x-x'|} - 1 \right) =: C_1 < \infty \tag{1}$$

A. Elgart: *Equality of the Bulk and Edge Hall Conductances in 2D*, Lect. Notes Phys. **690**, 325–332 (2006)
www.springerlink.com

for some $\mu > 0$, where $|x| = |x_1| + |x_2|$. Our hypothesis on the bounded open interval $\Delta \subset \mathbb{R}$ is that for some $\nu \geq 0$

$$\sup_{g \in B_1(\Delta)} \sum_{x,x' \in \mathbb{Z}^2} |g(H_B)(x,x')| (1 + |x|)^{-\nu} e^{\mu|x-x'|} =: C_2 < \infty \quad (2)$$

where $B_1(\Delta)$ denotes the set of Borel measurable functions g which are constant in $\{\lambda | \lambda < \Delta\}$ and in $\{\lambda | \lambda > \Delta\}$ with $|g(x)| \leq 1$ for every x.

In particular C_2 is a bound when g is of the form $g_t(\lambda) = e^{-it\lambda} E_\Delta(\lambda)$ and the supremum is over $t \in \mathbb{R}$, which is a statement of dynamical localization. By the RAGE theorem this implies that the spectrum of H_B is pure point in Δ. We denote the corresponding eigen-projections by $E_{\{\lambda\}}(H_B)$ for $\lambda \in \mathcal{E}_\Delta$, the set of eigenvalues $\lambda \in \Delta$. We assume that no eigenvalue in $\mathcal{E}_\Delta$ is infinitely degenerate,

$$\dim E_{\{\lambda\}}(H_B) < \infty, \quad \lambda \in \mathcal{E}_\Delta. \quad (3)$$

The validity of these assumptions is discussed below.

The zero temperature bulk Hall conductance at Fermi energy λ is defined by the Kubo-Středa formula [5]

$$\sigma_B(\lambda) = -i \operatorname{tr} P_\lambda \left[[P_\lambda, \Lambda_1], [P_\lambda, \Lambda_2] \right], \quad (4)$$

where $P_\lambda = E_{(-\infty,\lambda)}(H_B)$ and $\Lambda_i(x)$ is the characteristic function of

$$\{x = (x_1, x_2) \in \mathbb{Z}^2 \mid x_i < 0\}.$$

Under the above assumptions $\sigma_B(\lambda)$ is well-defined for $\lambda \in \Delta$, and shows a plateau (see [11] for further details). We remark that (3) is essential for a plateau: for the Landau Hamiltonian (though defined on the continuum rather than on the lattice) (1, 2) hold if properly interpreted, but (3) fails in an interval containing a Landau level, where indeed $\sigma_B(\lambda)$ jumps.

The edge system is described as a half-plane $\mathbb{Z} \times \mathbb{Z}_a$, where $\mathbb{Z}_a = \{n \in \mathbb{Z} \mid n \geq -a\}$, with the height $-a$ of the edge eventually tending to $-\infty$. The Hamiltonian $H_a = H_a^*$ on $\ell^2(\mathbb{Z} \times \mathbb{Z}_a)$ is obtained by restriction of H_B under some largely arbitrary boundary condition. More precisely, we assume that

$$E_a = J_a H_a - H_B J_a : \ell^2(\mathbb{Z} \times \mathbb{Z}_a) \to \ell^2(\mathbb{Z}^2) \quad (5)$$

satisfies

$$\sup_{x \in \mathbb{Z}^2} \sum_{x' \in \mathbb{Z} \times \mathbb{Z}_a} |E_a(x,x')| e^{\mu(|x_2+a| + |x_1-x_1'|)} \leq C_3 < \infty, \quad (6)$$

where $J_a : \ell^2(\mathbb{Z} \times \mathbb{Z}_a) \to \ell^2(\mathbb{Z}^2)$ denotes extension by 0. For instance with Dirichlet boundary conditions, $H_a = J_a^* H_B J_a$, we have

$$E_a = (J_a J_a^* - 1) H_B J_a,$$

i.e.,

$$E_a(x, x') = \begin{cases} -H_B(x, x'), & x_2 < -a, \\ 0, & x_2 \geq -a, \end{cases}$$

whence (6) follows from (1). We remark that (1) is inherited by H_a with a constant C_1 that is uniform in a, but not so for (2) as a rule.

The definition of the edge Hall conductance requires some preparation. The current operator across the line $x_1 = 0$ is $-\mathrm{i}[H_a, \Lambda_1]$. Matters are simpler if we temporarily assume that Δ is a gap for H_B, i.e., if $\sigma(H_B) \cap \Delta = \emptyset$, in which case one may set [18]

$$\sigma_E := -\mathrm{i}\,\mathrm{tr}\,\rho'(H_a)\,[H_a, \Lambda_1], \tag{7}$$

where $\rho \in C^\infty(\mathbb{R})$ satisfies

$$\rho(\lambda) = \begin{cases} 1, & \lambda < \Delta, \\ 0, & \lambda > \Delta. \end{cases} \tag{8}$$

The operator in (7) is trace class essentially because $\mathrm{i}[H, \Lambda_1]$ is relevant only on the states near $x_1 = 0$, and $\rho'(H_a)$ only near the edge $x_2 = -a$, so that the intersection of the two strips is compact (see Sect. 2 for more details). In the situation (2) considered in this paper the operator appearing in (7) is not trace class, since the bulk operator may have spectrum in Δ, which can cause the above stated property to fail for $\rho'(H_a)$. In search of a proper definition of σ_E, we consider only the current flowing across the line $x_1 = 0$ within a finite window $-a \leq x_2 < 0$ next to the edge. This amounts to modifying the current operator to be

$$-\frac{\mathrm{i}}{2}\left(\Lambda_2\,[H_a, \Lambda_1] + [H_a, \Lambda_1]\,\Lambda_2\right) = -\frac{\mathrm{i}}{2}\left\{[H_a, \Lambda_1], \Lambda_2\right\}, \tag{9}$$

with which one may be tempted to use

$$\lim_{a \to \infty} -\frac{\mathrm{i}}{2}\,\mathrm{tr}\,\rho'(H_a)\left\{[H_a, \Lambda_1], \Lambda_2\right\} \tag{10}$$

as a definition for σ_E. Though we show that this limit exists, it is not the physically correct choice. In particular, the edge conductance should be non-zero on average for the Harper Hamiltonian with an i.i.d. random potential [11].

The basic fact that the net current of a bound eigenstate ψ_λ of H_B is zero,

$$-\mathrm{i}\,(\psi_\lambda, [H_B, \Lambda_1]\,\psi_\lambda) = 0, \tag{11}$$

can be preserved by the regularization provided the spatial cutoff Λ_2 is time averaged. In fact, let

$$A_{T,a}(X) = \frac{1}{T}\int_0^T e^{\mathrm{i}H_a t}\,X\,e^{-\mathrm{i}H_a t}\,\mathrm{d}t \tag{12}$$

be the time average over $[0, T]$ of a (bounded) operator X with respect to the Heisenberg evolution generated by H_a, with a finite or $a = B$. If a limit $\Lambda_2^\infty = \lim_{T \to \infty} A_{T,B}(\Lambda_2)$ were to exist, it would commute with H_B so that

$$-\frac{i}{2} \left(\psi_\lambda, \{ [H_B, \Lambda_1], \Lambda_2^\infty \} \psi_\lambda \right) = 0 .$$

This motivates our definition,

$$\sigma_E^1 := \lim_{T \to \infty} \lim_{a \to \infty} -\frac{i}{2} \operatorname{tr} \rho'(H_a) \{ [H_a, \Lambda_1], A_{T,a}(\Lambda_2) \} . \tag{13}$$

We have the following result [11].

Theorem 1. *Under the assumptions (1, 2, 3, 6, 8) the limit in (13) exists, and*

$$\sigma_E^1 = \sigma_B .$$

In particular (13) depend neither on the choice of ρ nor on that of E_a.

Remark 1. (i) The hypotheses (1, 2) hold almost surely for ergodic Schrödinger operators whose Green's function $G(x, x'; z) = (H_B - z)^{-1}(x, x')$ satisfies a moment condition [3] of the form

$$\sup_{E \in \Delta} \limsup_{\eta \to 0} \mathbb{E} \left(|G(x, x'; E + i\eta)|^s \right) \leq C e^{-\mu |x - x'|} \tag{14}$$

for some $s < 1$. It implies the dynamical localization bound

$$\mathbb{E} \left(\sup_{g \in B_1(\Delta)} |g(H_B)(x, x')| \right) \leq C e^{-\mu |x - x'|} , \tag{15}$$

although (2) has also been obtained by different means, e.g., [12]. The implication (14) $\Rightarrow$ (15) was proved in [1] (see also [2, 4, 9]). The bound (15) may be better known for $\operatorname{supp} g \subset \Delta$, but is true as stated since it also holds [2, 6] for the projections $g(H_B) = P_\lambda$, $P_\lambda^\perp = 1 - P_\lambda$, ($\lambda \in \Delta$).

(ii) Condition (3), in fact simple spectrum, follows from the arguments in [19], at least for operators with nearest neighbor hopping, $H_B(x, y) = 0$ if $|x - y| > 1$.

(iii) When $\sigma(H_B) \cap \Delta = \emptyset$, the operator appearing in (7) is known to be trace class. In this case, the conductance σ_E^1 coincides with σ_E defined in (7). This statement follows from Theorem 1 and the known equality $\sigma_E = \sigma_B$ [10, 18], but can also be proven directly.

(iv) A recent paper of Combes and Germinet [7] contains results which are topically related to but substantially different from the ones presented here. The detailed comparison can be found in [11].

In the second part of the paper we will outline the simple proof of equality $\sigma_B = \sigma_E$, with σ_E defined in (7), when Δ is a spectral gap for H_B. Even though this result was established before by a number of authors, the new proof highlights some of the ideas used in our generalization of this result, namely in [11].

2 Proof of $\sigma_B = \sigma_E$

2.1 Some Preliminaries

We are going to use the Helffer-Sjöstrand representation for function $\rho(H_a)$:

$$\rho(H_a) = \frac{1}{2\pi} \int dm(z) \partial_{\bar{z}} \rho(z) R_a(z) \tag{16a}$$

$$\rho'(H_a) = -\frac{1}{2\pi} \int dm(z) \partial_{\bar{z}} \rho(z) R_a^2(z) , \tag{16b}$$

with $R_a(z) = (H_a - z)^{-1}$. The integral is over $z = x + iy \in \mathbb{C}$ with measure $dm(z) = dxdy$, $\partial_{\bar{z}} = \partial_x + i\partial_y$, and $\rho(z)$ is a quasi-analytic extension of $\rho(x)$ which for given n can be chosen so that

$$\int dm(z) \, |\partial_{\bar{z}} \rho(z)| \, |y|^{-p-1} \leq C \sum_{k=0}^{n+2} \left\| \rho^{(k)} \right\|_{k-p-1} \tag{17}$$

for $p = 1, ..., n$, provided the appearing norms $\|f\|_k = \int dx (1+x^2)^{\frac{k}{2}} |f(x)|$ are finite. This is the case for ρ with $\rho' \in C_0^\infty(\mathbb{R})$. For $p = 1$ this shows that (16b) is norm convergent. The integral (16a), which would correspond to the case $p = 0$, is nevertheless a strongly convergent improper integral, see e.g., (A.12) of [10]. In particular, we obtain

$$[\rho(H_a), \Lambda_1] = -\frac{1}{2\pi} \int dm(z) \partial_{\bar{z}} \rho(z) R_a(z) [H_a, \Lambda_1] R_a(z) \tag{18a}$$

$$\rho'(H_a) [H_a, \Lambda_1] = -\frac{1}{2\pi} \int dm(z) \partial_{\bar{z}} \rho(z) R_a^2(z) [H_a, \Lambda_1] . \tag{18b}$$

A further preliminary is the Combes-Thomas bound [8]

$$\left\| e^{\delta \ell(x)} R_a(z) e^{-\delta \ell(x)} \right\| \leq \frac{C}{|\mathrm{Im}z|} , \tag{19}$$

where δ can be chosen as

$$\delta^{-1} = C \left(1 + |\mathrm{Im}z|^{-1} \right) \tag{20}$$

for some (large) $C > 0$ and $\ell(x)$ is any Lipschitz function on $\mathbb{Z}^2$ with

$$|\ell(x) - \ell(y)| \leq |x - y| . \tag{21}$$

We will also use

$$\mathrm{tr}\, AB = \mathrm{tr}\, BA \quad \text{if } AB , BA \in \mathfrak{J}_1 . \tag{22}$$

2.2 Convergence and Trace Class Properties

Let $R_B(z) := (H_B - z)^{-1}$. Note that

$$\mathcal{J}_a R_a(z) \mathcal{J}_a^* - R_B(z) = -\left(\mathcal{J}_a R_a(z) E_a^* + 1 - \mathcal{J}_a \mathcal{J}_a^*\right) R_B(z) \xrightarrow[a \to \infty]{s} 0 \quad (23)$$

because $E_a^* \xrightarrow[a \to \infty]{s} 0$ by (6) and because $1 - \mathcal{J}_a \mathcal{J}_a^* \xrightarrow[a \to \infty]{s} 0$ is the projection onto states supported in $\{x_2 < -a\}$. This implies [17, Theorem VIII.20]

$$\operatorname*{s-lim}_{a \to \infty} \mathcal{J}_a f(H_a) \mathcal{J}_a^* = f(H_B) \quad (24)$$

for any bounded continuous function f. Moreover, using Combes-Thomas bound (19) and $\mathcal{J}_a^* \mathcal{J}_a = 1$, together with Holmgren's bound one can infer that[1]

Lemma 1. *For any a, a'*

$$[H_a, \Lambda_1] \Lambda_2 \in \mathfrak{I}_1 , \quad \rho'(H_a) [H_a, \Lambda_1] \in \mathfrak{I}_1 , \quad [\rho(H_B), \Lambda_1] [\rho(H_B), \Lambda_1] \in \mathfrak{I}_1 ; \quad (25)$$

$$\operatorname{tr} \rho'(H_a) [H_a, \Lambda_1] = \operatorname{tr} \rho'(H_{a'}) [H_{a'}, \Lambda_1] = \lim_{a \to \infty} \operatorname{tr} \rho'(H_a) [H_a, \Lambda_1] \Lambda_2 . \quad (26)$$

2.3 Edge – Bulk Interpolation

We can now use (18a), (18b) and cyclicity of the trace to rewrite the expression appearing in the last equation as

$$\operatorname{tr} \rho'(H_a) [H_a, \Lambda_1] \Lambda_2 = -\frac{1}{2\pi} \int \mathrm{d}m(z) \partial_{\bar{z}} \rho(z) \operatorname{tr} R_a(z) [H_a, \Lambda_1] \Lambda_2 R_a(z)$$

$$= \operatorname{tr} [\rho(H_a), \Lambda_1] \Lambda_2 - \frac{1}{2\pi} \int \mathrm{d}m(z) \partial_{\bar{z}} \rho(z) \operatorname{tr} R(z)$$

$$[H_a, \Lambda_1] R_a(z) [H_a, \Lambda_2] R_a(z) . \quad (27)$$

The first contribution, computed in the position basis, is seen to vanish. As for the second contribution, one can use (23) and Combes-Thomas estimate to obtain

$$\lim_{a \to \infty} \operatorname{tr} R_a(z) [H_a, \Lambda_1] R_a(z) [H_a, \Lambda_2] R_a(z)$$

$$= \operatorname{tr} R_B(z) [H_B, \Lambda_1] R_B(z) [H_B, \Lambda_2] R_B(z) , \quad (28)$$

for $\operatorname{Im} z \neq 0$. We deduce that

$$\sigma_E = -i \operatorname{tr} \rho'(H_a) [H_a, \Lambda_1]$$

$$= \frac{i}{2\pi} \int \mathrm{d}m(z) \partial_{\bar{z}} \rho(z) \operatorname{tr} R_B(z) [H_B, \Lambda_1] R_B(z) [H_B, \Lambda_2] R_B(z) . \quad (29)$$

---

[1] See [11] for details.

2.4 $\sigma_E = \sigma_B$

To finish the proof we need to reduce the rhs of (29) into the form of (4). To this end, we write

$$\frac{i}{2\pi} \int dm(z) \partial_{\bar{z}} \rho(z) \, \mathrm{tr} \, R_B(z) \left[H_B, \Lambda_1 \right] R_B(z) \left[H_B, \Lambda_2 \right] R_B(z)$$

$$= \frac{i}{2\pi} \int dm(z) \partial_{\bar{z}} \rho(z) \, \mathrm{tr} \, P_\lambda R_B(z) \left[H_B, \Lambda_1 \right] R_B(z) \left[H_B, \Lambda_2 \right] R_B(z) P_\lambda$$

$$+ \frac{i}{2\pi} \int dm(z) \partial_{\bar{z}} \rho(z) \, \mathrm{tr} \, \bar{P}_\lambda R_B(z) \left[H_B, \Lambda_1 \right] R_B(z) \left[H_B, \Lambda_2 \right] R_B(z) \bar{P}_\lambda \, , \quad (30)$$

with $\bar{P}_\lambda := 1 - P_\lambda$. The first term is equal to

$$\frac{i}{2\pi} \int dm(z) \partial_{\bar{z}} \rho(z) \, \mathrm{tr} \, P_\lambda R_B(z) \left[H_B, \Lambda_1 \right] R_B(z) \left[H_B, \Lambda_2 \right] R_B(z) P_\lambda$$

$$= \frac{i}{2\pi} \int dm(z) \partial_{\bar{z}} \rho(z) \, \mathrm{tr} \, (P_\lambda \Lambda_1 R_B(z) \left[H_B, \Lambda_2 \right] R_B(z) P_\lambda$$

$$+ \, P_\lambda R_B(z) \Lambda_1 \left[H_B, \Lambda_2 \right] R_B(z) P_\lambda)$$

$$= -\frac{i}{2\pi} \int dm(z) \partial_{\bar{z}} \rho(z) \, \mathrm{tr} \, P_\lambda \Lambda_1 \left[R_B(z), \Lambda_2 \right] P_\lambda$$

$$= -i \, \mathrm{tr} \, P_\lambda \Lambda_1 \left[\rho(H_B), \Lambda_2 \right] P_\lambda$$

$$= i \, \mathrm{tr} \, P_\lambda \Lambda_1 \left[\bar{P}_\lambda, \Lambda_2 \right] P_\lambda = i \, \mathrm{tr} \, P_\lambda \Lambda_1 \bar{P}_\lambda \Lambda_2 P_\lambda \, , \quad (31)$$

where the term on the third line vanishes by integration by parts since $P_\lambda R_B(z) = R_B(z) P_\lambda$ is analytic on the support of $1 - \rho(z)$. Similar consideration shows that

$$\frac{i}{2\pi} \int dm(z) \partial_{\bar{z}} \rho(z) \, \mathrm{tr} \, \bar{P}_\lambda R_B(z) \left[H_B, \Lambda_1 \right] R_B(z) \left[H_B, \Lambda_2 \right] R_B(z) \bar{P}_\lambda$$

$$= -i \, \mathrm{tr} \, \bar{P}_\lambda \Lambda_1 \left[P_\lambda, \Lambda_2 \right] \bar{P}_\lambda = -i \, \mathrm{tr} \, \bar{P}_\lambda \Lambda_1 P_\lambda \Lambda_2 \bar{P}_\lambda \, . \quad (32)$$

Since $P_\lambda \Lambda_{1,2} \bar{P}_\lambda = P_\lambda \left[\rho(H_B), \Lambda_{1,2} \right] \bar{P}_\lambda$, one can deduce from (25) that both contribution that appear in (31) and (32) are trace class, and one can use cyclicity of the trace and $P_\lambda^2 = P_\lambda$ to conclude that

$$\frac{i}{2\pi} \int dm(z) \partial_{\bar{z}} \rho(z) \, \mathrm{tr} \, R_B(z) \left[H_B, \Lambda_1 \right] R_B(z) \left[H_B, \Lambda_2 \right] R_B(z)$$

$$= i \, \mathrm{tr} \, P_\lambda \Lambda_1 \bar{P}_\lambda \Lambda_2 P_\lambda - i \, \mathrm{tr} \, \bar{P}_\lambda \Lambda_1 P_\lambda \Lambda_2 \bar{P}_\lambda$$

$$= i \, \mathrm{tr} \, P_\lambda \Lambda_1 \bar{P}_\lambda \Lambda_2 P_\lambda - i \, \mathrm{tr} \, P_\lambda \Lambda_2 \bar{P}_\lambda \Lambda_2 P_\lambda = \sigma_B(\lambda) \, . \quad (33)$$

332   A. Elgart

References

1. M. Aizenman: Rev. Math. Phys. **6**, 1163 (1994).
2. M. Aizenman and G. M. Graf: J. Phys. A **31**, 6783 (1998).
3. M. Aizenman and S. Molchanov: Commun. Math. Phys. **157**, 245 (1993).
4. M. Aizenman, J. H. Schenker, R. M. Friedrich, and D. Hundertmark: Commun. Math. Phys. **224**, 219 (2001).
5. J.E. Avron, R. Seiler, and B. Simon: Comm. Math. Phys. **159**, 399 (1994).
6. J. Bellissard, A. van Elst, and H. Schulz-Baldes: J. Math. Phys. **35**, 5373 (1994).
7. J.-M. Combes and F. Germinet: Comm. Math. Phys., to appear.
8. J.-M. Combes and L. Thomas: Comm. Math. Phys. **34**, 251 (1973).
9. R. del Rio, S. Jitomirskaya, Y. Last, and B. Simon: J. Anal. Math. **69**, 153 (1996).
10. P. Elbau and G.M. Graf: Comm. Math. Phys. **229**, 415 (2002).
11. A. Elgart, G.M. Graf and J.H. Schenker: Comm. Math. Phys., to appear.
12. F. Germinet and S. De Bièvre: Comm. Math. Phys. **194**, 323 (1998).
13. B.I. Halperin: Phys. Rev. B **25**, 2185 (1982).
14. Y. Hatsugai: Phys. Rev. L. **71**, 3697 (1993).
15. K.v. Klitzing, G. Dorda, and M. Pepper: Phys. Rev. Lett. **45**, 494 (1980).
16. N. Macris: preprint, 2003.
17. M. Reed and B. Simon: *Methods of modern mathematical physics. I*, 2nd edn (Academic Press Inc., New York 1980).
18. H. Schulz-Baldes, J. Kellendonk, and T. Richter: J. Phys. A **33**, L27 (2000).
19. B. Simon: Rev. Math. Phys. **6**, 1183 (1994).

Generic Subsets in Spaces of Measures and Singular Continuous Spectrum

Daniel Lenz and Peter Stollmann

Fakultät für Mathematik, Technische Universität, 09107 Chemnitz, Germany
p.stollmann@mathematik.tu-chemnitz.de
d.lenz@@mathematik.tu-chemnitz.de

Abstract. We discuss recent results of ours showing that geometric disorder leads to some purely singularly continuous spectrum generically. This is based on a slight extension of Simons Wonderland theorem. Our approach to this theorem relies on the study of generic subsets of certain spaces of measures. In this article, we elaborate on this purely measure theoretic basis of our approach.

Key words: Random Schrödinger operators, Delone sets, spaces of measures, 81Q10,35J10,82B44, 28A33,28C15

1 Introduction

In this article we review our work [8] and elaborate on certain measure theoretic parts of it. The starting point of [8] is the study of operators

$$H_\omega := -\Delta + \sum_{x \in \omega} v(\cdot - x)$$

on $L^2(\mathbb{R}^d)$, for suitable functions v and certain uniformly discrete subsets ω of $\mathbb{R}^d$ called Delone sets. Such operators arise in the study of disordered solids. More precisely, they can be thought to model geometric disorder. It is shown that these operators have a purely singularly continuous spectral component generically. Here, generic refers to a topology on the set of all Delone sets (see below for details).

The abstract operator theoretic tool behind our reasoning is a slight strengthening of a result due to B. Simon called the "Wonderland theorem" [11]. Our method of proving this is different from Simons. It consists of two steps. We first prove that certain subsets of spaces of measures are generic. This generalizes the corresponding results of Simon [11] and Zamfirescu [13] for $\mathbb{R}$ to rather general measure spaces. In the second step, the Wonderland theorem follows by considering spectral measures as continuous maps from the space of selfadjoint operators to measures on the real line.

Given our version of the Wonderland theorem, our result on generic singularly continuous spectrum follows from geometric considerations. There, we approximate arbitrary ω by essentially periodic ones in various ways.

D. Lenz and P. Stollmann: *Generic Subsets in Spaces of Measures and Singular Continuous Spectrum*, Lect. Notes Phys. **690**, 333–341 (2006)
www.springerlink.com

In the present article, we particularly focus on the measure theoretic side of things. This is discussed in the next two sections. In particular, Sect. 3 elaborates on the last part of Sect. 2 of [8] and presents a generalization of Corollary 2.8 given there.

The subsequent discussion of the Wonderland theorem in Sect. 4 and the application to geometric disorder in Sect. 5 then follows [8] quite closely.

2 Generic Subsets in Spaces of Measures

We will be concerned with subsets of the set of positive, regular Borel measures $\mathcal{M}_+(S)$ on some locally compact, σ-compact, separable metric space S. The closed ball around $x \in S$ with radius r is denoted by $B_r(x)$. The space $\mathcal{M}_+(S)$ is endowed with the weak topology from $C_c(S)$, also called the vague topology. We refer the reader to [3] for standard results concerning the space of measures. We will use in particular that the vague topology is metrizable such that $\mathcal{M}_+(S)$ becomes a complete metric space. Thus, the Baire category theorem becomes applicable. For the application to spectral theory, S is just an open subset U of the real line.

We call a measure $\mu \in \mathcal{M}_+(S)$ *continuous* if its *atomic* or *pure point* part vanishes, i.e. if $\mu(\{x\}) = 0$ for every $x \in S$. (We prefer the former terminology in the abstract framework and the latter for measures on the real line.) The set of all continuous measures on S is denoted by $\mathcal{M}_c(S)$. A measure μ is called a point measure if there exists a countable set Y in X with $\mu(X \setminus Y) = 0$. The set of all point measures on S is denoted by $\mathcal{M}_p(S)$. This set is dense in $\mathcal{M}_+(S)$. Two measures are said to be *mutually singular*, $\mu \perp \nu$, if there exists a set $C \subset S$ such that $\mu(C) = 0 = \nu(S \setminus C)$. The set of all measures on S which are singular with respect to a measure λ is denoted by $\mathcal{M}_{\lambda,sing}(S)$.

Our main generic result on spaces of measures reads as follows.

Theorem 1. *Let S be a locally compact, σ-compact, separable complete metric space. Then, the following holds:*

(1) The set $\mathcal{M}_c(S)$ is a G_δ-set in $\mathcal{M}_+(S)$.
(2) For any $\lambda \in \mathcal{M}_+(S)$, the set $\mathcal{M}_{\lambda,sing}(S)$ is a G_δ-set in $\mathcal{M}_+(S)$.
(3) For any closed $F \subset S$ the set $\{\mu \in \mathcal{M}_+(S) | F \subset \text{supp}(\mu)\}$ is a G_δ-set in $\mathcal{M}_+(S)$.

3 Singular Continuity of Measures

As a first application of Theorem 1, we obtain a result on genericity of singular continuous spectrum. This generalizes the corresponding results of Simon [11] and Zamfirescu [13, 14] for measures on the real line. The first result of this section elaborates on Corollary 2.8 of [8] and generalizes the result mentioned there. The second result seems to be new.

Theorem 2. *Let S be be locally compact, σ-compact, separable complete metric space. Let λ be a continuous measure on S with $\operatorname{supp} \lambda = S$. Then,*

$$\{\mu \in \mathcal{M}_+(S) \mid \mu \text{ continuous and } \mu \perp \lambda\}$$

is a dense G_δ-set in $\mathcal{M}_+(S)$.

Proof. The set in question is the intersection of $\mathcal{M}_c(S)$ and $\mathcal{M}_{\lambda,sing}(S)$. By Theorem 1, $\mathcal{M}_c(S)$ and $\mathcal{M}_{\lambda,sing}(S)$ are G_δ-sets. Thus, by the Baire category theorem, it remains to show that $\mathcal{M}_c(S)$ and $\mathcal{M}_{\lambda,sing}(S)$ are dense. Denseness of these two sets follows from continuity of λ: By continuity of λ, $\mathcal{M}_{\lambda,sing}(S)$ contains all point measures and is therefore dense. Also, by continuity of λ, $\mathcal{M}_c(S)$ is dense (see e.g. (ii) of the next theorem). $\qquad \blacksquare$

The previous theorem assumes the existence of a suitable measure λ. Such a λ does not always exist. Instead the following is valid.

Theorem 3. *Let S be a locally compact σ-compact separable complete metric space. Then, the following assertions are equivalent:*

(i)   There exists a continuous measure λ on S with $\operatorname{supp} \lambda = S$.
(ii) The set $\mathcal{M}_c(S)$ is dense in $\mathcal{M}_+(S)$.
(iii) S has no isolated points.

Proof. (i)$\Longrightarrow$ (ii): Let $x \in S$ be arbitrary. Then, for any $n \in \mathbb{N}$, the measure μ_n with

$$\mu_n := \lambda(B_{1/n}(x))^{-1} \lambda|_{B_{1/n}(x)}$$

is continuous. Moreover, the sequence (μ_n) converges towards the unit point mass at x. Thus, every point measure can be approximated by continuous measures. As the point measures are dense, so are the continuous measures.

(ii)$\Longrightarrow$ (iii): This is clear.

(iii)$\Longrightarrow$ (i): We start with the following intermediate result:

Claim. For each closed ball B in S with positive radius, there exists a continuous probability measure μ_B whose support is contained in B.

Proof of claim. Denote the metric on S by d. Let $x \in S$ and $r > 0$ with $B = B_r(x)$ be given. As S is locally compact, we can assume without loss of generality that B is compact.

Define $\delta_0 := r/4$ and consider $B_{\delta_0}(x)$. By (iii), $B_{\delta_0}(x)$ contains two different points $x_1^{(1)}$ and $x_2^{(1)}$. Set $\delta_1 := 1/4\, d(x_1^1, x_2^{(1)})$ and consider $B_{\delta_1}(x_2^{(1)})$ and $B_{\delta_1}(x_2^{(1)})$. These balls are disjoint and by (ii) each contains two different points, i.e. there exist $x_1^{(2)}$, $x_2^{(2)}$ in $B_{\delta_1}(x_1^{(1)})$ and $x_3^{(2)}$, $x_4^{(2)}$ in $B_{\delta_1}(x_2^{(1)})$.

Set $\delta_2 := 1/4 \min\{d(x_i^{(2)}, x_j^{(2)}) : i, j = 1, \ldots, 4\}$. Then, the balls $B_{\delta_2}(x_j^{(2)})$, $j = 1, \ldots, 4$, are disjoint. Proceeding inductively, we can construct for each $n \in \mathbb{N}$ a $\delta_n > 0$ and a set

$$X_n := \{x_1^{(n)}, \ldots, x_{2^n}^{(n)}\}$$

with 2^n elements such that

$$\delta_{n+1} \leq \frac{\delta_n}{2}, \ X_n \subset B,$$

and for each $y \in B$, and $n \geq k$, the ball $B_{\delta_k}(y)$ contains at most 2^{n-k} points of X_n. Now, consider, for each $n \in \mathbb{N}$, the measure

$$\mu_n := \frac{1}{2^n} \sum_{j=1}^{2^n} \delta_{x_j^{(n)}}.$$

Then each μ_n is a probability measure supported in B and

$$\mu_n(B_{\delta_k}(y)) \leq \frac{1}{2^n} \times 2^{n-k} = 2^{-k} \tag{1}$$

for every $k \leq n$ and $y \in S$. By the Theorem of Banach/Alaoglu, the sequence (μ_n) has a converging subsequence. Thus, without loss of generality, we can assume that the sequence itself converges to a measure μ. As each μ_n is a probability measure supported in B and B is compact, μ is a probability measure supported in B as well. Moreover, by (1), μ is continuous. This finishes the proof of the claim.

Let D be a countable dense subset of S. For each $x \in D$ and $n \in \mathbb{N}$, we can find a continuous probability measure $\mu_{n,x} := \mu_{B_{\frac{1}{n}}(x)}$ supported in $B_{\frac{1}{n}}(x)$ according to the claim. Now, choose numbers $c_{n,x} > 0$, $x \in D$ and $n \in \mathbb{N}$, with

$$\sum_{n \in \mathbb{N}, x \in D} c_{n,x} < \infty.$$

Then,

$$\lambda := \sum_{n \in \mathbb{N}, x \in D} c_{n,x} \mu_{n,x}$$

is a continuous measure whose support contains all $x \in D$. As D is dense, the support of λ equals S and (i) follows.

4 Selfadjoint Operators and the Wonderland Theorem

In this section, we discuss a consequence of Theorem 2 on generic appearance of a purely singular continuous component in the spectrum for certain Hamiltonians.

This provides a slight strengthening of Simon's "Wonderland Theorem" from [11]. The main point of our discussion, however, is not this strengthening but rather the new proof we provide.

In order to formulate our result, let us introduce the following notation:

For a fixed separable Hilbert space $\mathcal{H}$ consider the space $\mathcal{S} = \mathcal{S}(\mathcal{H})$ of self-adjoint operators in $\mathcal{H}$. For $\xi \in \mathcal{H}$ and $A \in \mathcal{S}$ let the spectral measure ρ_ξ^A be defined by

$$\rho_\xi^A(\varphi) := \langle \xi, \varphi(A)\xi \rangle$$

for each continuous φ on $\mathbb{R}$ with compact support.

We endow $\mathcal{S}$ with the *strong resolvent topology* τ_{srs}, the weakest topology for which all the mappings

$$\mathcal{S} \to \mathbb{C}, A \mapsto (A+i)^{-1}\xi \quad (\xi \in \mathcal{H})$$

are continuous. Therefore, a sequence (A_n) converges to A w.r.t. τ_{srs} if and only if

$$(A_n + i)^{-1}\xi \to (A+i)^{-1}\xi$$

for all $\xi \in \mathcal{H}$. Thus, for each $\xi \in \mathcal{H}$, the mapping

$$\rho_\xi : \mathcal{S} \longrightarrow \mathcal{M}(\mathbb{R}), \quad A \mapsto \rho_\xi^A,$$

is continuous.

The spectral subspaces of $A \in \mathcal{S}$ are defined by

$$\mathcal{H}_{ac}(A) = \{\xi \in \mathcal{H} | \ \rho_\xi^A \text{ is absolutely continuous}\}$$
$$\mathcal{H}_{sc}(A) = \{\xi \in \mathcal{H} | \ \rho_\xi^A \text{ is singular continuous}\},$$
$$\mathcal{H}_c(A) = \{\xi \in \mathcal{H} | \ \rho_\xi^A \text{ is continuous}\}$$
$$\mathcal{H}_{pp}(A) = \mathcal{H}_c(A)^\perp, \mathcal{H}_s(A) = \mathcal{H}_{ac}(A)^\perp .$$

These subspaces are closed and invariant under A. $\mathcal{H}_{pp}(A)$ is the closed linear hull of the eigenvectors of A. Recall that the spectra $\sigma_*(A)$ are just the spectra of A restricted to $\mathcal{H}_*(A)$.

Theorem 4. *Let (X, ρ) be a complete metric space and $H : (X, \rho) \to (\mathcal{S}, \tau_{srs})$ a continuous mapping. Assume that, for an open set $U \subset \mathbb{R}$,*

(1) the set $\{x \in X | \ \sigma_{pp}(H(x)) \cap U = \emptyset\}$ is dense in X,
(2) the set $\{x \in X | \ \sigma_{ac}(H(x)) \cap U = \emptyset\}$ is dense in X,
(3) the set $\{x \in X | \ U \subset \sigma(H(x))\}$ is dense in X.

Then, the set

$$\{x \in X | \ U \subset \sigma(H(x)), \sigma_{ac}(H(x)) \cap U = \emptyset, \sigma_{pp}(H(x)) \cap U = \emptyset\}$$

is a dense G_δ-set in X.

A *proof* can be sketched as follows (see [8] for details): By assumption H is continuous. Furthermore, the restriction $r_U : \mathcal{M}(\mathbb{R}) \longrightarrow \mathcal{M}(U)$, $\mu \mapsto \mu|_U$, can easily be seen to be continuous as well. Finally, as discussed above, for each $\xi \in \mathcal{H}$, the map ρ_ξ is continuous. Thus, the composition

$$\mu_\xi := r_U \circ \rho_\xi \circ H : X \longrightarrow \mathcal{M}(U), \quad x \mapsto \rho_\xi^{H(x)}|_U,$$

is a continuous map. Thus, the inverse image of a G_δ-set in $\mathcal{M}(U)$ under μ_ξ is a G_δ-set in X. Thus, by Theorem 1, the sets $\{x \in S : \mu_\xi(x) \text{ is continuous}\}$, $\{x \in S : \mu_\xi(x) \text{ is singular w.r.t. Lebesgue measure}\}$ and $\{x \in S : \operatorname{supp}\mu_\xi(x)$ contains $U\}$ are all G_δ-sets. Moreover, by assumption they are dense. Thus, their intersection is a dense G_δ-set by the Baire category theorem. One more intersection over a countable dense subset of $\xi \in \mathcal{H}$ yields the desired result. This finishes the proof of the theorem.

We say that the spectrum of $A \in \mathcal{S}$ is pure point (purely absolutely continuous, purely singularly continuous) on $U \subset \mathbb{R}$ if the restrictions $\rho_\xi^A|U$ have the corresponding properties. Of course, if A has pure point (purely absolutely continuous) spectrum on U it does not have any absolutely continuous (pure point) spectrum on U. As in [11], the theorem has then the following immediate but remarkable corollary.

Corollary 1. *Let (X, ρ) be a complete metric space and $H : (X, \rho) \to (\mathcal{S}, \tau_{srs})$ a continuous mapping. Assume that, for an open set $U \subset \mathbb{R}$,*

(1) the set $\{x \in X \mid H(x) \text{ has pure point spectrum in } U\}$ is dense in X,
(2) the set $\{x \in X \mid H(x) \text{ has purely absolutely continuous spectrum in } U\}$ is dense in X,
(3) the set $\{x \in X \mid U \subset \sigma(H(x))\}$ is dense in X.

Then, the set

$$\{x \in X \mid U \subset \sigma(H(x)), \sigma_{ac}(H(x)) \cap U = \emptyset, \sigma_{pp}(H(x)) \cap U = \emptyset\}$$

is a dense G_δ-set in X.

5 Operators Associated to Delone Sets

In this section we discuss an application of Theorem 4 to geometric disorder.

We start by recalling the necessary notation. A key notion is the notion of *Delone set*, named after B.N. Delone (Delaunay), [6]. The Euclidean norm on $\mathbb{R}^d$ is denoted by $\|\cdot\|$. We replace B by U to denote open balls.

Definition 1. *A set $\omega \subset \mathbb{R}^d$ is called an (r, R)-set if*

- $\forall x, y \in \omega, x \neq y : \|x - y\| > r$,
- $\forall p \in \mathbb{R}^d \, \exists x \in \omega : \|x - p\| \leq R$.

By $\mathbb{D}_{r,R}(\mathbb{R}^d) = \mathbb{D}_{r,R}$ *we denote the set of all* (r, R)-*sets. We say that* $\omega \subset \mathbb{R}^d$ *is a* Delone set, *if it is an* (r, R)-*set for some* $0 < r \leq R$ *so that* $\mathbb{D}(\mathbb{R}^d) = \mathbb{D} = \bigcup_{0 < r \leq R} \mathbb{D}_{r,R}(\mathbb{R}^d)$ *is the set of all Delone sets.*

Delone sets turn out to be quite useful in the description of quasicrystals and more general aperiodic solids; see also [4], where the relation to discrete operators is discussed. In fact, if we regard an infinitely extended solid whose ions are assumed to be fixed, then the positions are naturally distributed according to the points of a Delone set. Fixing an effective potential v for all the ions this leads us to consider the Hamiltonian

$$H(\omega) := -\Delta + \sum_{x \in \omega} v(\cdot - x) \text{ in } \mathbb{R}^d ,$$

where $\omega \in \mathbb{D}$. Let us assume, for simplicity that v is bounded, measurable and compactly supported.

In order to apply our analysis above, we need to introduce a suitable topology on $\mathbb{D}$. This can be done in several ways, cf. [4,7]. The emerging topology is called the *natural topology*. It defines a compact, complete metrizable topology on the set of all closed subsets of $\mathbb{R}^d$ for which $\mathbb{D}_{r,R}(\mathbb{R}^d)$ is a compact, complete space. We refrain from giving the exact definition of this topology here and refer to the cited literature. Instead we note the following lemma, which describes convergence w.r.t the natural topology.

Lemma 1. *A sequence* (ω_n) *of Delone sets converges to* $\omega \in \mathbb{D}$ *in the natural topology if and only if there exists for any* $l > 0$ *an* $L > l$ *such that the* $\omega_n \cap U_L(0)$ *converge to* $\omega \cap U_L(0)$ *with respect to the Hausdorff distance as* $n \to \infty$.

Given the lemma, it is not hard to see that the map

$$H : \mathbb{D}_{r,R}(\mathbb{R}^d) \longrightarrow \mathcal{S}(L^2((\mathbb{R}^d)), \quad \omega \mapsto H(\omega),$$

is continuous.

Finally, we recall that a Delone set γ on $\mathbb{R}^d$ is called crystallographic if the set of its periods

$$Per(\gamma) := \{t \in \mathbb{R}^d : t + \gamma = \gamma\}$$

is a lattice of full rank in $\mathbb{R}^d$. Now our result on generic singularly continuous spectrum can be stated as follows.

Theorem 5. *Let* $r, R > 0$ *with* $2r \leq R$ *and* v *be given such that there exist crystallographic* $\gamma, \tilde{\gamma} \in \mathbb{D}_{r,R}$ *with* $\sigma(H(\gamma)) \neq \sigma(H(\tilde{\gamma}))$. *Then*

$$U := (\sigma(H(\gamma))^{\circ} \setminus \sigma(H(\tilde{\gamma}))) \cup (\sigma(H(\tilde{\gamma}))^{\circ} \setminus \sigma(H(\gamma)))$$

is nonempty and there exists a dense G_{δ}-*set* $\Omega_{sc} \subset \mathbb{D}_{r,R}$ *such that for every* $\omega \in \Omega_{sc}$ *the spectrum of* $H(\omega)$ *contains* U *and is purely singular continuous in* U.

A *proof* can be sketched as follows (see [8] for details):

We let $U_1 := \sigma(H(\gamma))^\circ \setminus \sigma(H(\tilde{\gamma}))$ and $U_2 := \sigma(H(\tilde{\gamma}))^\circ \setminus \sigma(H(\gamma))$. Since $\gamma, \tilde{\gamma}$ are crystallographic, the corresponding operators are periodic and their spectra are consequently purely absolutely continuous and consist of a union of closed intervals with only finitely many gaps in every compact subset of the reals. Hence, under the assumption of the theorem U_1 or U_2 is nonempty. Thus, U is nonempty.

We now consider the case that U_1 is nonempty. We will verify conditions (1)-(3) from Theorem 4.

Ad (1): Fix $\omega \in \mathbb{D}_{r,R}$. For $n \in \mathbb{N}$ consider $\nu_n := \omega \cap Q(n)$. We can then periodically extend ν_n, i.e. we find crystallographic ω_n in $\mathbb{D}_{r,R}$ with $\omega_n \cap Q(n) = \nu_n$. For given $L > 0$ we get that $\omega_n \cap U_L(0) = \omega \cap U_L(0)$ if n is large enough. Therefore, by Lemma 1, we find that $\omega_n \to \omega$ with respect to the natural topology. On the other hand, $\sigma_{pp}(H(\omega_n)) = \emptyset$ since the potential of $H(\omega_n)$ is periodic. Consequently,

$$\{\omega \in \mathbb{D}_{r,R} | \sigma_{pp}(H(\omega)) \cap U_1 = \emptyset\}$$

is dense in $\mathbb{D}_{r,R}$.

Ad (2): We have to show denseness of ω for which $\sigma_{ac}(H(\omega)) \cap U_1 = \emptyset$. Thus, fix $\omega \in \mathbb{D}_{r,R}$. Then, we can construct ω_n which agree with ω around zero and with $\tilde{\gamma}$ away from zero. More precisely, for $n \in \mathbb{N}$ large enough, find $\omega_n \in \mathbb{D}_{r,R}$ such that

$$\omega_n \cap U_n(0) = \omega \cap U_n(0) \text{ and } \omega_n \cap U_{2n}(0)^c = \tilde{\gamma} \cap U_{2n}(0)^c.$$

In virtue of the last property, $H(\omega_n)$ and $H(\tilde{\gamma})$ only differ by a compactly supported, bounded potential, so that $\sigma_{ac}(H(\omega_n)) = \sigma_{ac}(H(\tilde{\gamma})) \subset U_1^c$. Again, $\omega_n \to \omega$ yields condition (2) of Theorem 4.

Ad (3): This can be checked with a similar argument as (2), this time with $\tilde{\gamma}$ instead of γ. More precisely, fix $\omega \in \mathbb{D}_{r,R}$. For $n \in \mathbb{N}$ large enough, we find $\omega_n \in \mathbb{D}_{r,R}$ such that

$$\omega_n \cap U_n(0) = \omega \cap U_n(0) \text{ and } \omega_n \cap U_{2n}(0)^c = \gamma \cap U_{2n}(0)^c.$$

In virtue of the last property, $H(\omega_n)$ and $H(\gamma)$ only differ by a compactly supported, bounded potential, so that $\sigma_{ac}(H(\omega_n)) = \sigma_{ac}(H(\gamma)) \supset U_1$. By $\omega_n \to \omega$, we obtain (3) of Theorem 4.

As a consequence of these considerations, Theorem 4 gives that

$$\{\omega \in \mathbb{D}_{r,R} | \sigma_{pp}(H(\omega)) \cap U_1 = \emptyset, \sigma_{ac}(H(\omega)) \cap U_1 = \emptyset, U_1 \subset \sigma(H(\omega))\}$$

is a dense G_δ-set if U_1 is not empty. An analogous argument shows the same statement with U_2 instead of U_1. This proves the assertion if only one of the U_i, $i = 1, 2$, is not empty. Otherwise, the assertion follows after intersecting the two dense G_δ's. This finishes the proof of the theorem.

Acknowledgment

The research presented above has been partly supported by the DFG. Useful comments of J. Voigt are gratefully acknowledged. We would also like to take this opportunity to thank the organizers of QMath9 for the stimulating and relaxed atmosphere.

References

1. M. Baake, R.V. Moody (eds), Directions in mathematical quasicrystals, CRM Monogr. Ser., 13, Amer. Math. Soc., Provicence, RI, 2000.
2. S. Banach, Über die Bairesche Kategorie gewisser Funktionenmengen. *Studia Math.* **3** (1931), 174–179.
3. H. Bauer, *Maß- und Integrationstheorie*, De Gruyter, Berlin, 1990.
4. J. Bellissard, D.J.L. Hermann, and M. Zarrouati, Hulls of Aperiodic Solids and Gap Labeling Theorem, In: [1], pp. 207–258.
5. D. Buschmann, G. Stolz, Two-parameter spectral averaging and localization for non-monotonic random Schrödinger operators, *Trans. Amer. Math. Soc.* **353** (2001), 635–653.
6. B. Delaunay [B.N. Delone], Sur la sphère vide, *Izvestia Akad Nauk SSSR Otdel. Mat. Sov. Nauk.* 7, 1934, pp. 793-800.
7. D. Lenz and P. Stollmann, Delone dynamical systems and associated random operators, Proceedings, Constanta (Romania), 2001, eds. J.-M. Combes et al., Theta Foundation, `eprint: arXiv:math-ph/0202042`.
8. D. Lenz and P. Stollmann, Generic sets in spaces of measures and generic singular continuous spectrum for Delone Hamiltonians, `eprint: arXiv:math-ph/0410021`, to appear in *Duke Math. J.*.
9. S. Mazurkiewicz, Sur les fonctions non dérivables, *Studia Math.* **3** (1931), 92–94.
10. R.V. Moody (ed.), The mathematics of long-range aperiodic order. NATO Advanced Science Institutes Series C: Mathematical and Physical Sciences, 489. Kluwer Academic Publishers Group, Dordrecht, 1997.
11. B. Simon, Operators with singular continuous spectrum: I. General operators. *Annals of Math.* **141**, 131–145 (1995).
12. J. Weidmann, *Lineare Operatoren in Hilberträumen. Teil I: Grundlagen.* B.G. Teubner, Stuttgart, 2000.
13. T. Zamfirescu, Most monotone functions are singular, *Amer. Math. Monthly* **88** (1981), 456–458.
14. T. Zamfirescu, Typical monotone continuous functions, *Arch. Math.* **42** (1984), 151–156.

Low Density Expansion
for Lyapunov Exponents

Hermann Schulz-Baldes

Mathematisches Institut, Universität Erlangen-Nürnberg
schuba@mi.uni-erlangen.de

1 Introduction

A perturbative formula for the Lyapunov exponent of a one-dimensional random medium for weakly coupled disorder was first given by Thouless [12] and then proven rigorously by Pastur and Figotin [9]. Anomalies in the perturbation theory at the band center were discovered by Kappus and Wegner [7] and further discussed by various other authors [2,3,11]. The Lyapunov exponent is then identified with the inverse localization length of the system. This short note concerns the behavior of the Lyapunov exponent for a low density of impurities, each of which may, however, be large. The presented method is as [6,10,11] a further application of diagonalizing the transfer matrices without perturbation (here the low density of impurities) and then rigorously controlling the error terms by means of oscillatory sums of rotating modified Prüfer phases. Some of the oscillatory sums remain large if the rotation phases (here the quasi-momenta) are rational. This leads to supplementary contributions of the Kappus-Wegner type.

The calucalations are carried through explicitly for the one-dimensional Anderson model, but the method transposes also to more complicated models with a periodic background as well as low-density disorder with correlations similar to the random polymer model [6]. Extension to a quasi-one-dimensional situation as in [11] should be possible, but is even more cumbersome on a calculatory level. It is also straightforward to calculate and control higher order terms in the density.

As one motivation for this study (apart from a mathematical one), let us indicate that a low density of strong impurities seems to describe materials like carbon nanotubes more adequately than a small coupling limit of the Anderson model. Indeed, these materials have perfect cristaline structure over distances of microns which leads to a ballisistic transport over such a distance [5]. The existing few defects are, on the other hand, quite large. Coherent transport within a one-particle framework should then be studied by a model similar to the one considered here. However, it is possible that the impurties rather play the role of quantum dots so that Coulomb blockade is the determining effect for the transport properties [8] rather than the coherent transport studied here.

H. Schulz-Baldes: *Low Density Expansion for Lyapunov Exponents*, Lect. Notes Phys. **690**, 343–350 (2006)
www.springerlink.com

2 Model and Preliminaries

The standard one-dimensional Anderson Hamiltonian is given by

$$(H_\omega\psi)_n = -\psi_{n+1} - \psi_{n-1} + v_n\psi_n, \qquad \psi \in \ell^2(\mathbb{Z}) .$$

Here $\omega = (v_n)_{n\in\mathbb{Z}}$ is a sequence of independent and identically distributed real random variables. The model is determined by their probability distribution $\mathbf{p}$ depending on a given density $\rho \in [0,1]$:

$$\mathbf{p} = (1-\rho)\,\delta_0 + \rho\,\tilde{\mathbf{p}} , \tag{1}$$

where $\tilde{\mathbf{p}}$ is a fixed compactly supported probability measure on $\mathbb{R}$. This measure may simply be a Dirac peak if there is only one type of impurity, but different from δ_0. Set $\Sigma = \mathrm{supp}(\mathbf{p})$ and $\tilde{\Sigma} = \mathrm{supp}(\tilde{\mathbf{p}})$. The expectation w.r.t. the $\mathbf{p}$'s will be denoted by $\mathbf{E}$, that w.r.t. the $\tilde{\mathbf{p}}$'s by $\tilde{\mathbf{E}}$, while $\mathbf{E}_v$ and $\tilde{\mathbf{E}}_v$ is the expectation w.r.t. $\mathbf{p}$ and $\tilde{\mathbf{p}}$ respectively over one random variable $v \in \Sigma$ only.

In order to define the Lyapunov exponent, one rewrites the Schrödinger equation $H_\omega\psi = E\psi$ using transfer matrices

$$\begin{pmatrix} \psi_{n+1} \\ \psi_n \end{pmatrix} = T_n^E \begin{pmatrix} \psi_n \\ \psi_{n-1} \end{pmatrix} . \qquad T_n^E = \begin{pmatrix} v_n - E & -1 \\ 1 & 0 \end{pmatrix} .$$

We also write T_v^E for T_n^E if $v_n = v$. Then the Lyapunov exponent at energy $E \in \mathbb{R}$ associated to products of random matrices chosen independently according to $\mathbf{p}$ from the family $(T_v^E)_{v\in\Sigma}$ of $\mathrm{SL}(2,\mathbb{R})$-matrices is given by

$$\gamma(\rho, E) = \lim_{N\to\infty} \frac{1}{N}\, \mathbf{E} \log \left(\left\| \prod_{n=1}^N T_n^E \right\| \right) . \tag{2}$$

The aim is to study the asymptotics of $\gamma(\rho, E)$ in small ρ for $|E| < 2$.

In order to state our results, let us introduce, for $E = -2\cos(k)$ with $k \in (0,\pi)$, the basis change $M \in \mathrm{SL}(2,\mathbb{R})$ and the rotation matrix R_k by the quasi-momentum k:

$$M = \frac{1}{\sqrt{\sin(k)}} \begin{pmatrix} \sin(k) & 0 \\ -\cos(k) & 1 \end{pmatrix} , \qquad R_k = \begin{pmatrix} \cos(k) & -\sin(k) \\ \sin(k) & \cos(k) \end{pmatrix} .$$

It is then a matter of computation to verify

$$MT_v^E M^{-1} = R_k(1 + P_v^E) , \qquad P_v^E = -\frac{v}{\sin(k)} \begin{pmatrix} 0 & 0 \\ 1 & 0 \end{pmatrix} .$$

Next we introduce another auxiliary family of random matrices. Set $\hat{\Sigma} = [-\frac{\pi}{2}, \frac{\pi}{2}) \times \tilde{\Sigma}$ and, for $(\psi, v) \in \hat{\Sigma}$:

$$\hat{T}^E_{\psi,v} = R_\psi M T^E_v M^{-1} .$$

The following probability measures on $\hat{\Sigma}$ will play a role in the sequel: $\hat{\mathbf{p}}_\infty = \frac{d\psi}{\pi} \otimes \tilde{\mathbf{p}}$ and $\hat{\mathbf{p}}_q = \left(\frac{1}{q} \sum_{p=1}^q \delta_{\frac{\pi}{2}(\frac{p}{q}-\frac{q+1}{2q})} \right) \otimes \tilde{\mathbf{p}}$ for $q \in \mathbb{N}$. The Lyapunov exponents associated to these families of random matrices are denoted by $\hat{\gamma}_\infty(E)$ and $\hat{\gamma}_q(E)$ respectively. It is elementary to check that the subgroups generated by matrices corresponding to the supports of $\hat{\mathbf{p}}_\infty$ and $\hat{\mathbf{p}}_q$ are non-compact and strongly irreducible, which implies [1] that the corresponding Lyapunov exponents are strictly positive.

The matrices T^E_v and $\hat{T}^E_{\psi,v}$ induce actions $\mathcal{S}_{E,v}$ and $\hat{\mathcal{S}}_{E,\psi,v}$ on $\mathbb{R}$ via

$$e_{\mathcal{S}_{E,v}(\theta)} = \frac{M T^E_v M^{-1} e_\theta}{\| M T^E_v M^{-1} e_\theta \|} , \qquad e_{\hat{\mathcal{S}}_{E,\psi,v}(\theta)} = \frac{\hat{T}^E_{\psi,v} e_\theta}{\| \hat{T}^E_{\psi,v} e_\theta \|} . \qquad (3)$$

where the freedom of phase is fixed by $\mathcal{S}_{E,0}(\theta) = \theta + k$ and $\hat{\mathcal{S}}_{E,\psi,0}(\theta) = \theta + k + \psi$ as well as the continuity in v. Invariant measures ν^E, $\hat{\nu}^E_\infty$ and $\hat{\nu}^E_q$ for these actions and the probability measures $\mathbf{p}$, $\hat{\mathbf{p}}_\infty$ and $\hat{\mathbf{p}}_q$ are defined by

$$\int_0^\pi d\nu^E(\theta) f(\theta) = \int_0^\pi d\nu^E(\theta) \, \mathbf{E}_v \, f(\mathcal{S}_{E,v}(\theta) \mathrm{mod} \pi) , \qquad f \in C(\mathbb{R}/\pi\mathbb{Z}) ,$$

and similar formulas for $\hat{\nu}^E_\infty$ and $\hat{\nu}^E_q$. Due to a theorem of Furstenberg [1], the invariant measures exist and are unique whenever the associated Lyapunov exponent is positive. Let us note that one can easily verify that the invariant measure $\hat{\nu}^E_\infty$ is simply given by the Lebesgue measure $\frac{d\theta}{\pi}$. Furthermore $\hat{\nu}^E_\infty$ and $\hat{\nu}^E_q$ do not depend on ρ (but ν^E does).

Next let us write out a more explicit formula for the new Lyapunov exponent $\hat{\gamma}_\infty(E)$. First of all, according to Furstenberg's formula [1,6],

$$\hat{\gamma}_\infty(E) = \hat{\mathbf{E}}_{\psi,v} \int_0^\pi d\hat{\nu}^E_\infty(\theta) \, \log(\| \hat{T}^E_{\psi,v} e_\theta \|) , \qquad \text{where } e_\theta = \begin{pmatrix} \cos(\theta) \\ \sin(\theta) \end{pmatrix} .$$

As $\hat{\nu}^E_\infty$ is the Lebesgue measure, rotations are orthogonal and the integrand is π-periodic, one gets

$$\hat{\gamma}_\infty(E) = \frac{1}{2} \tilde{\mathbf{E}}_v \int_0^{2\pi} \frac{d\theta}{2\pi} \, \log\left(\langle e_\theta | (M T^E_v M^{-1})^* (M T^E_v M^{-1}) | e_\theta \rangle \right) . \qquad (4)$$

Now $(M T^E_v M^{-1})^* (M T^E_v M^{-1}) = |\mathbf{1} + P^E_v|^2$ is a positive matrix with eigenvalues $\lambda_v \geq 1$ and $1/\lambda_v$ where $\lambda_v = 1 + \frac{a}{2} + \sqrt{a + \frac{a^2}{4}}$ with $a = \frac{v^2}{\sin^2(k)}$. As it can be diagonalized by an orthogonal transformation leaving the Lebesgue measure invariant, we deduce that

$$\hat{\gamma}_\infty(E) = \frac{1}{2} \tilde{\mathbf{E}}_v \int_0^{2\pi} \frac{d\theta}{2\pi} \, \log\left(\lambda_v \cos^2(\theta) + \frac{1}{\lambda_v} \sin^2(\theta) \right)$$

$$= \frac{1}{2} \int d\tilde{\mathbf{p}}(v) \, \log\left(\frac{1+\lambda_v^2}{2\lambda_v} \right) .$$

This formula shows immediately that $\hat{\gamma}_\infty(E) > 0$ unless $\tilde{p} = \delta_0$ (in which case $\lambda_v = \lambda_0 = 1$).

3 Result on the Lyapunov Exponent

Theorem. *Let* $E = -2\cos(k)$ *and* $k \in (0, \pi)$ *with either* $\frac{k}{\pi}$ *rational or* k *satisfying the weak diophantine condition*

$$\left| 1 - e^{2\imath mk} \right| \geq c e^{-\xi'|m|}, \qquad \forall\, m \in \mathbb{Z}, \tag{5}$$

for some $c > 0$ *and* $\xi' > 0$. *Then*

$$\gamma(\rho, E) = \begin{cases} \rho\,\hat{\gamma}_\infty(E) + \mathcal{O}(\rho^2) & k \text{ satisfies } (5), \\ \rho\,\hat{\gamma}_q(E) + \mathcal{O}(\rho^2) & k = \pi\,\frac{p}{q}, \end{cases}$$

where p *and* q *are relatively prime. Furthermore, for* ξ *depending only on* $\tilde{p}$,

$$\left| \hat{\gamma}_q(E) - \hat{\gamma}_\infty(E) \right| \leq c\, e^{-\xi|q|}.$$

The result can be interpreted as follows: if the density of the impurities is small, then the incoming (Prüfer) phase at the impurity is uniformly distrubuted for a sufficiently irrational rotation angle (*i.e.* quasi-momentum) because the sole invariant measure of an irrational rotation is the Lebesgue measure. For a rational rotation, the mixing is to lowest order in ρ perfect over the orbits of the rational rotation, which leads to the definition of the family $(\hat{T}^E_{p,\sigma})_{(p,\sigma)\in\hat{\Sigma}_q}$ and its distribution $\hat{p}_q$. As indicated above, the proof that this is the correct image is another simple application of modified Prüfer phases and an oscillatory sum argument.

Let us note that $\hat{\gamma}_q(E) \neq \hat{\gamma}_\infty(E)$; more detailed formulas for the difference are given below. As a result, one can expect a numerical curve of the energy dependence of the Lyapunov exponent at a given fixed low density to have spikes at energies corresponding to rational quasimomenta with small denominators. Moreover, the invariant measures ν^E and $\hat{\nu}^E_q$ are *not* close to the Lebesgue measure, but have higher harmonics as is typical at Kappus-Wegner anomalies. Furthermore, let us add that at the band center $E = 0$ the identity $\gamma(\rho, 0) = \rho\,\hat{\gamma}_0(0)$ holds with no higher order correction terms and where $\hat{\gamma}_0(0)$ is the center of band Lyapunov exponent of the usual Anderson model with distribution $\tilde{p}$.

Finally, let us compare the above result with that obtained for a weak-coupling limit of the Anderson model [6, 9]: the Lyapunov exponent grows quadratically in the coupling constant of the disordered potential, while it grows linearly in the density. The reason is easily understood if one thinks of the change of the coupling constant also rather as a change of the probability distribution on the space of matrices. At zero coupling, the measure is supported on one *critical* point (or more generally, on a commuting subset), and the weight in its neighborhood grows as the coupling constant grows. In (1) the weight may grow far from the critical point, and this leads to the faster increase of the Lyapunov exponent.

4 Proof

For fixed energy E, configuration $(v_n)_{n \in \mathbb{N}}$ and $(\psi_n)_{n \in \mathbb{N}}$, as well as an initial condition θ_0, let us define iteratively the seqences

$$\theta_n = \mathcal{S}_{E,v_n}(\theta_{n-1}), \qquad \hat{\theta}_n = \hat{\mathcal{S}}_{E,\psi_n,v_n}(\hat{\theta}_{n-1}) = \mathcal{S}_{E,v_n}(\hat{\theta}_{n-1}) + \psi_n . \quad (6)$$

When considered modulo π, these are also called the modified Prüfer phases. They can be efficiently used in order to calculate the Lyapunov exponent as well as the density of states. For the Lyapunov exponent, let us first note that one can make a basis change in (2) at the price of boundary terms vanishing at the limit, and furthermore, that according to [1, A.III.3.4] it is possible to introduce an arbitrary initial vector, so that

$$\gamma(\rho, E) = \lim_{N \to \infty} \frac{1}{N} \mathbf{E} \log \left(\left\| \left(\prod_{n=1}^N M T_n^E M^{-1} \right) e_\theta \right\| \right) . \quad (7)$$

Now using the modified Prüfer phases with initial condition $\theta_0 = \theta$, this can be developed into a telescopic sum:

$$\gamma(\rho, E) = \lim_{N \to \infty} \frac{1}{N} \mathbf{E} \sum_{n=1}^N \log \left(\left\| M T_n^E M^{-1} e_{\theta_{n-1}} \right\| \right)$$

$$= \rho \lim_{N \to \infty} \frac{1}{N} \mathbf{E} \sum_{n=1}^N \tilde{\mathbf{E}}_v \log \left(\left\| M T_v^E M^{-1} e_{\theta_{n-1}} \right\| \right) ,$$

where in the second step we have evaluated the partial expectation over the last random variable v_n by using the fact that for $v_n = 0$ the contribution vanishes. Next let us note that the function $e^{i\theta} \mapsto \tilde{\mathbf{E}}_v \log(\|M T_v^E M^{-1} e_\theta\|)$ has an analytic extension to $\mathbb{C} \backslash \{0\}$, contains only even frequencies so that its Fourier series

$$\tilde{\mathbf{E}}_v \log \left(\left\| M T_v^E M^{-1} e_\theta \right\| \right) = \sum_{m \in \mathbb{Z}} a_m \, e^{2im\theta} ,$$

has coefficients satisfying for any $\xi > 0$ a Cauchy estimate of the form

$$|a_m| \leq c_\xi \, e^{-\xi |m|} . \quad (8)$$

Comparing with (4), we deduce

$$a_0 = \hat{\gamma}_\infty(E) .$$

Introducing now the oscillatory sums

$$I_m(N) = \mathbf{E} \frac{1}{N} \sum_{n=1}^{N} e^{2\imath m\theta_n} , \qquad \hat{I}_m(N) = \hat{\mathbf{E}} \frac{1}{N} \sum_{n=1}^{N} e^{2\imath m\hat{\theta}_n} ,$$

the Lyapunov exponent now reads

$$\gamma(\rho, E) = \rho \sum_{m \in \mathbb{Z}} a_m \lim_{N \to \infty} I_m(N) . \tag{9}$$

Hence we need to calculate $I_m(N)$ perturbatively in ρ. Clearly $I_0(N) = 1$. Furthermore, integrating over the initial condition w.r.t. the invariant measure gives for all $N \in \mathbb{N}$

$$\int d\nu^E(\theta) \, I_m(N) = \int d\nu^E(\theta) \, e^{2\imath m\theta} .$$

Hence calculating $I_m(N)$ perturbatively also gives the harmonics of ν^E perturbatively (similar statements hold for $\hat{I}_m(N)$, of course). Going back in history once, one gets

$$I_m(N) = \frac{1}{N} \mathbf{E} \sum_{n=1}^{N} \left((1 - \rho) e^{2\imath mk} e^{2\imath m\theta_{n-1}} + \rho \, \tilde{\mathbf{E}}_v \big(e^{2\imath m \mathcal{S}_{E,v}(\theta_{n-1})} \big) \right)$$

$$= (1 - \rho) e^{2\imath mk} I_m(N) + \mathcal{O}(\rho, N^{-1}) .$$

For k satisfying (5), one deduces

$$|I_m(N)| \leq \frac{1}{|1 - (1 - \rho) e^{2\imath mk}|} \mathcal{O}(\rho, N^{-1}) \leq c \, e^{\xi'|m|} \, \mathcal{O}(\rho, N^{-1}) .$$

Replacing this and (8) with $\xi > \xi'$ into (9) concludes the proof in this case because only the term $m = 0$ gives a contribution to order ρ.

If now $k = \pi\frac{p}{q}$, the same argument implies

$$I_{nq+r}(N) = \mathcal{O}(\rho, N^{-1}) , \qquad \forall \, n \in \mathbb{Z} , \; r = 1, \dots, q-1 . \tag{10}$$

Setting

$$\tilde{\mathbf{E}}_v \big(e^{2\imath m \mathcal{S}_{E,v}(\theta)} \big) = \sum_{l \in \mathbb{Z}} b_l^{(m)} \, e^{2\imath(m+l)\theta} ,$$

and

$$\hat{\mathbf{E}}_{\psi,v} \big(e^{2\imath m \hat{\mathcal{S}}_{E,\psi,v}(\theta)} \big) = \sum_{l \in \mathbb{Z}} \hat{b}_l^{(m)} \, e^{2\imath(m+l)\theta} ,$$

one deduces for the remaining cases

$$I_{nq}(N) = (1 - \rho)\, I_{nq}(N) + \mathcal{O}(N^{-1}) + \rho \sum_{l \in \mathbb{Z}} b_l^{(nq)} \left(I_{nq+l}(N) + \mathcal{O}(N^{-1}) \right) \ .$$

Due to (10), this gives the following equations

$$I_{nq}(N) = \sum_{r \in \mathbb{Z}} b_{rq}^{(nq)} I_{(n+r)q}(N) + \mathcal{O}((\rho N)^{-1}, \rho) \ .$$

They determine the invariant measure ν^E to lowest order in ρ. This shows, in particular, that ν^E is already to lowest order not given by the Lebesgue measure. We will not solve these equations, but rather show that the oscillatory sums $\hat{I}_{nq}(N)$ satisfy the same equations, and hence, up to higher order corrections, the measure $\hat{\nu}_q^E$ can be used instead of ν^E in order to calculate the Lyapunov exponent. Indeed, it follows directly from (6) and the definition of $\hat{\mathbf{p}}_q$ that

$$\hat{b}_l^{(m)} = \delta_{m \bmod q = 0}\, b_l^{(m)} \ .$$

In particular, $\hat{I}_m(N) = 0$ if $m \bmod q \neq 0$. Thus we deduce

$$\hat{I}_{nq}(N) = \sum_{r \in \mathbb{Z}} b_{rq}^{(nq)}\, \hat{I}_{(n+r)q}(N) + \mathcal{O}(N^{-1}) \ .$$

Comparing the equations for $I_{nq}(N)$ and $\hat{I}_{nq}(N)$ (which have a unique solution becaue the invariant measures are unique by Furstenberg's theorem), it follows that

$$\hat{I}_{nq}(N) = I_{nq}(N) + \mathcal{O}(\rho, (\rho N)^{-1}) \ .$$

Replacing this into (9), one deduces

$$\gamma(\rho, E) = \rho \sum_{m \in \mathbb{Z}} a_m \int d\hat{\nu}_q^E(\theta)\, e^{2\imath m\theta} + \mathcal{O}(\rho^2)$$

$$= \rho \int d\hat{\nu}_q^E(\theta)\, \tilde{\mathbf{E}}_v \log\left(\left\| M T_v^E M^{-1} e_\theta \right\| \right) + \mathcal{O}(\rho^2) \ .$$

Now due to the orthogonality of rotations one may replace $M T_v^E M^{-1}$ by $\hat{T}_{\psi,v}^E$, and then the r.h.s. contains exactly the Furstenberg formula for $\hat{\gamma}_q(E)$ as claimed. The estimate comparing $\hat{\gamma}_q(E)$ and $\hat{\gamma}_\infty(E)$ follows directly from the Cauchy estimate (8).

5 Result on the Density of States

Another ergodic quantity of interest is the integrated density of states, defined by

$$\mathcal{N}(\rho, E) = \lim_{N \to \infty} \frac{1}{N}\, \mathbf{E}\, \#\left\{ \text{neg. eigenvalues of } H_\omega - E \text{ on } \ell^2(\{1, \ldots, N\}) \right\} \ .$$

Recall, in particular, that $\mathcal{N}(0, E) = k$ if $E = -2\cos(k)$. Defining the mean phase shift of the impurities by

$$\tilde{\varphi}(\theta) = \tilde{\mathbf{E}}_v(\mathcal{S}_{E,v}(\theta) - \theta) ,$$

the low density expansion of the density of states reads as follows:

$$\mathcal{N}(\rho, E) = \begin{cases} (1-\rho)\,k + \rho \int \frac{d\theta}{2\pi}\, \tilde{\varphi}(\theta) + \mathcal{O}(\rho^2) & k \text{ satisfies (5)} , \\[2mm] (1-\rho)\,k + \rho \int d\hat{v}_q^E(\theta)\, \tilde{\varphi}(\theta) + \mathcal{O}(\rho^2) & k = \pi\frac{p}{q} \end{cases}$$

with p and q relatively prime. The proof of this is completely analogous to the above when the rotation number calculation (e.g. [6] for a proof) is applied:

$$\mathcal{N}(\rho, E) = \lim_{N\to\infty} \frac{1}{N}\, \mathbf{E} \sum_{n=1}^{N} \left(\mathcal{S}_{E,v}(\theta_{n-1}) - \theta_{n-1}\right)$$

$$= (1-\rho)\,k + \rho \lim_{N\to\infty} \frac{1}{N}\, \mathbf{E} \sum_{n=1}^{N} \tilde{\varphi}(\theta_{n-1}) .$$

References

1. P. Bougerol, J. Lacroix, *Products of Random Matrices with Applications to Schrödinger Operators*, (Birkhäuser, Boston, 1985).
2. A. Bovier, A. Klein, *Weak disorder expansion of the invariant measure for the one-dimensional Anderson model*, J. Stat. Phys. **51**, 501–517 (1988).
3. M. Campanino, A. Klein, *Anomalies in the one-dimensional Anderson model at weak disorder*, Commun. Math. Phys. **130**, 441–456 (1990).
4. B. Derrida, E.J. Gardner, *Lyapunov exponent of the one dimensional Anderson model: weak disorder expansion*, J. Physique **45**, 1283 (1984).
5. S. Frank, P. Poncharal, Z.L. Wang, W. DeHeer, *Carbon Nanotube Quantum Resistors*, Science **280**, 1744–1746 (1998).
6. S. Jitomirskaya, H. Schulz-Baldes, G. Stolz, *Delocalization in random polymer chains*, Commun. Math. Phys. **233**, 27–48 (2003).
7. M. Kappus, F. Wegner, *Anomaly in the band centre of the one-dimensional Anderson model*, Z. Phys. B **45**, 15–21 (1981).
8. P.L. McEuen, M. Bockrath, D.H. Cobden, Y.-G. Yoon, S. Louie, *Disorder, pseudospins, and backscattering in carbon nanotubes*, Phys. Rev. Lett. **83**, 5098 (1999).
9. L. Pastur, A. Figotin, *Spectra of Random and Almost-Periodic Operators*, (Springer, Berlin, 1992).
10. R. Schrader, H. Schulz-Baldes, A. Sedrakyan, *Perturbative test of single parameter scaling for 1D random media*, Ann. H. Poincaré, **5**, 1159–1180 (2004).
11. H. Schulz-Baldes, *Perturbation theroy for Lyapunov exponents of an Anderson model on a strip*, GAFA **14**, 1089–1117 (2004).
12. D.J. Thouless, in *Ill-Condensed Matter*, Les Houches Summer School, 1978, edited by R. Balian, R. Maynard, G. Toulouse (North-Holland, New York, 1979).

Poisson Statistics for the Largest Eigenvalues in Random Matrix Ensembles

Alexander Soshnikov

University of California at Davis, Department of Mathematics, Davis, CA 95616, USA
soshniko@math.ucdavis.edu

1 Introduction

The two archetypal ensembles of random matrices are Wigner real symmetric (Hermitian) random matrices and Wishart sample covariance real (complex) random matrices. In this paper we study the statistical properties of the largest eigenvalues of such matrices in the case when the second moments of matrix entries are infinite. In the first two subsections we consider Wigner ensemble of random matrices and its generalization – band random matrices.

1.1 Wigner Random Matrices

A real symmetric Wigner random matrix is defined as a square symmetric $n \times n$ matrix with i.i.d. entries up from the diagonal

$$A = (a_{jk}), a_{jk} = a_{kj}, 1 \leq j \leq k \leq n, \{a_{jk}\}_{j<k} - \text{i.i.d. real random variables .} \tag{1}$$

The diagonal entries $\{a_{ii}\}$, $1 \leq i \leq n$, are usually assumed to be i.i.d. random variables, independent from the off-diagonal entries. A Hermitian Wigner random matrix is defined in a similar way, namely as a square $n \times n$ Hermitian matrix with i.i.d. entries up from the diagonal

$$A = (a_{jk}), a_{jk} = \overline{a_{kj}}, 1 \leq j \leq k \leq n, \{a_{jk}\}_{j<k} - \text{i.i.d. complex random variables .} \tag{2}$$

As in the real symmetric case, it is usually assumed that the diagonal entries $\{a_{ii}\}$, $1 \leq i \leq n$, are i.i.d. (real) random variables independent from the off-diagonal entries.

Ensembles (1) and (2) were introduced in mathematical physics by Eugene Wigner in the 1950s ([44–46]). Wigner viewed these ensembles as a mathematical model to study the statistics of the excited energy levels of heavy nuclei.

The famous Wigner's semicircle law can be formulated as follows. Let the matrix entries in (1) or (2) be centered random variables with the tail of distribution decaying sufficiently fast, so that all moments exist. Denote by $\lambda_1 \geq \lambda_2 \geq \ldots \geq \lambda_n$ the eigenvalues of a random matrix $n^{-1/2}A$. Then the

A. Soshnikov: *Poisson Statistics for the Largest Eigenvalues in Random Matrix Ensembles*, Lect. Notes Phys. **690**, 351–364 (2006)
www.springerlink.com

empirical distribution function of the eigenvalues converges, as $n \to \infty$, to a non-random limit

$$\frac{1}{n}\#(\lambda_i \leq x, \ 1 \leq i \leq n) \to F(x) = \int_{-\infty}^{x} f(t)dt \ , \tag{3}$$

where the density of the semicircle law is given by $f(t) = \frac{1}{\pi\sigma^2}\sqrt{2\sigma^2 - x^2}$, for $t \in [-\sqrt{2}\sigma, \sqrt{2}\sigma]$, and σ^2 is the second moment of matrix entries.

This result was subsequently strengthened by many mathematicians (see e.g. [1, 12, 32]). In its general form (due to Pastur and Girko), the theorem holds if the matrix entries of A satisfy the Lindeberg-Feller condition: $\frac{1}{n^2}\sum_{1\leq i \leq j \leq n}\int_{|x|>\tau\sqrt{n}}x^2 dF_{ij}(x) \to 0$, where $F_{ij}(x)$ is the distribution function of $a_{ij}^{(n)}$.

From the analytical point of view, the simplest examples of Wigner random matrices are given by the so-called Gaussian Orthogonal and Unitary Ensembles (GOE and GUE for short). The GUE is defined as the ensemble of $n \times n$ Hermitian matrices with the Gaussian entries $\Re a_{jk} \sim N(0, 1/2)$, $\Im a_{jk} \sim N(0, 1/2)$, $1 \leq j < k \leq n$; $a_{ii} \sim N(0, 1)$, $1 \leq i \leq n$ (see ([27], Chap. 6). The joint distribution of the matrix entries has the form

$$P(dA) = const_n \exp\left(-\frac{1}{2}Tr(A^2)\right) dA \ , \tag{4}$$

where $dA = \prod_{j\leq k} d\Re a_{ij} d\Im a_{jk} \prod_{i=1}^{n} da_{ii}$ is the Lebesgue measure on the space of n-dimensional Hermitian matrices. The joint distribution of the eigenvalues is given by its density

$$p_n(x_1, \ldots, x_n) = Z_n^{-1} \prod_{1\leq i < j \leq n} (x_i - x_j)^2 \exp\left(-\frac{1}{2}\sum_i x_i^2\right) . \tag{5}$$

The normalization constants in (4) and (5) are known. What is more, one can calculate explicitly the k-point correlation functions (see [27], Chap. 6). This allows one to study the local distribution of the eigenvalues, both in the bulk of ths spectrum and at its edges in great detail. In particular, a celebrated result of Tracy and Widom (see [40]) states that

$$\lim_{n\to\infty} \ Pr\left(\lambda_{max} \leq 2\sqrt{n} + \frac{s}{n^{1/6}}\right) = F_2(s) = \exp\left(-\int_s^{+\infty}(x-s)q^2(x)dx\right) , \tag{6}$$

where $q(x)$ is the solution of the Painléve II differential equation

$$q''(x) = xq(x) + 2q^3(x)$$

with the asymptotics at infinity $q(x) \sim Ai(x)$ at $x = +\infty$.

The limiting k-point correlation function at the edge of spectrum is given by the formula

$$\rho_k(x_1, \ldots, x_k) = \det\left(K(x_i, x_j)\right)_{1 \le i, j \le k}, \tag{7}$$

where

$$K(x, y) = K_{Airy}(x, y) = \frac{Ai(x)Ai'(y) - Ai'(x)Ai(y)}{x - y} \tag{8}$$

is a so-called Airy kernel. We refer the reader to [40] and [13] for the details. We recall that the k-point correlation function is defined in such a way that for any disjoint subintervals of the real line $I_1, I_2, \ldots, I_k$, one has

$$E \prod_{i=1}^{k} \#(I_i) = \int_{I_1} \cdots \int_{I_k} \rho_k(x_1, \ldots, x_k) dx_1 \ldots dx_k,$$

where $\#(I)$ denotes the number of the eigenvalues in I. A probabilistic interpretation of the above formula is that $\rho_k(x_1, \ldots, x_k)dx_1 \ldots dx_k$ is the probability to find an eigenvalue in each of the k infinitesimal intervals $[x_i, x_i + dx_i], i = 1, \ldots, k$.

The Gaussian Orthogonal Ensemble (GOE) is defined as the ensemble of $n \times n$ Wigner real symmetric random matrices with the Gaussian entries. More precisely, we assume that $a_{ij}, 1 \le i \le j \le n$, are independent Gaussian $N(0, 1 + \delta_{ij})$ random variables (see e.g. [27], Chap. 7). The joint distribution of the matrix entries has the form

$$P(dA) = c_n \exp\left(-\frac{1}{4}Tr(A^2)\right) dA, \tag{9}$$

where $dA = \prod_{i \le j} da_{ij}$ is the Lebesgue measure on the space of n-dimensional real symmetric matrices. The distribution (9) induces the joint distribution of the eigenvalues of the GOE matrix, given by its density

$$p_n(x_1, \ldots, x_n) = Z_n^{-1} \prod_{1 \le i < j \le n} |x_i - x_j| \exp\left(-\frac{1}{4}\sum_i x_i^2\right). \tag{10}$$

The limiting distribution of the (normalized) largest eigenvalue of a GOE matrix was calculated by Tracy and Widom in ([41]).

$$\lim_{n \to \infty} \Pr\left(\lambda_{max} \le 2\sqrt{n} + \frac{s}{n^{1/6}}\right) = F_1(s)$$

$$= \exp\left(-\frac{1}{2}\int_s^{+\infty} q(x) + (x - s)q^2(x)dx\right), \tag{11}$$

The Tracy-Widom distribution (11) was obtained by studying the asymptotic properties of the k-point correlation functions at the edge of the spectrum. The k-point correlation function in the GOE ensemble has the pfaffian form. In the limit $n \to \infty$ the k-point correlation function at the edge of the spectrum is given by the following formula

$$\rho_k(x_1, \ldots, x_k) = \left(\det \left(K(x_i, x_j) \right)_{1 \leq i,j \leq k} \right)^{1/2}, \tag{12}$$

where $K(x, y)$ is a 2×2 matrix kernel such that

$$K_{11}(x, y) = K_{22}(y, x) = K_{Airy}(x, y) + \frac{1}{2} Ai(x) \int_{-\infty}^{y} Ai(t) dt, \tag{13}$$

$$K_{12}(x, y) = -\frac{1}{2} Ai(x) Ai(y) - \frac{\partial}{\partial y} K_{Airy}(x, y), \tag{14}$$

$$K_{21}(x, y) = \int_0^{+\infty} \left(\int_{x+u}^{+\infty} Ai(v) dv \right) Ai(x + u) du - \epsilon(x - y)$$

$$+ \frac{1}{2} \int_y^x Ai(u) du + \frac{1}{2} \int_x^{+\infty} Ai(u) du \int_{-\infty}^{y} Ai(v) dv, \tag{15}$$

where $\epsilon(z) = \frac{1}{2} sign(z)$.

1.2 Band Random Matrices

A band random matrix is a generalization of a Wigner random matrix ensemble (1), (2). A real symmetric (aperiodic) band random matrix is defined as a square symmetric $n \times n$ matrix $A = (a_{jk})$ such that $a_{ij} = 0$ unless $|i - j| \leq d_n$, and

$$\{a_{jk}, \ j \leq k; \ |j - k| \leq d_n\} - \text{ i.i.d. real random variables}. \tag{16}$$

A Hermitian band random matrix is defined in a similar way, namely as a square $n \times n$ Hermitian matrix $A = (a_{jk})$, such that $a_{ij} = 0$ unless $|i - j| \leq d_n$, and

$$\{a_{jk}, \ j \leq k; \ |j - k| \leq d_n\} - \text{ i.i.d. complex random variables}. \tag{17}$$

If $d_n = n - 1$, we obtain the Wigner ensemble of random matrices. A matrix is called a periodic band matrix if $|i - j|$ is replaced above by $|i - j|_1 = \min(|i - j|, n - |i - j|)$. Band random matrices have been studied in the last fifteen years (see for example [6, 7, 17, 30]). In the periodic case, the limiting distribution of the eigenvalues of $d_n^{-1} A$ is given by the semi-circle law, provided matrix entries have a finite second moment. In the aperiodic case, the limiting distribution of the eigenvalues is different from the semi-circle law, unless $d_n/n \to 0$ (see e.g. [30]). One of the most interesting problems involving band random matrices is the localization/delocalization properties of the eigenvalues. It is conjectured in physical literature, that the eigenvalues of band random matrices are localized if $d_n = O(n^{1/2})$. As far as we know, there are no rigorous results yet in this direction.

1.3 Sample Covariance Random Matrices

Sample covariance random matrices have been studied in mathematical statistics for the last seventy-five years. We refer to [31], [43] and [19] for the applications of spectral properties of Wishart random matrices in multivariate statistical analysis.

Let A be a large $m \times n$ real rectangular random matrix with independent identically distributed entries. In applications, one is often interested in the statistical behavior of the singular values of A in the limit $m \to \infty$, $n \to \infty$. This is equivalent to studying the eigenvalues of a positive-definite matrix $M = A^t A$ in the limit of large dimensions. Without loss of generality, one can assume that $m \geq n$ (since the spectrum of AA^t differs from the spectrum of $A^t A$ only by a zero eigenvalue of multiplicity $m - n$.

The analogue of the Wigner semicircle law was proved by Marchenko and Pastur ([24]). Let $m \to \infty$, $n \to \infty$ in such a way that $m/n \to \gamma \geq 1$. Assume $E|a_{ij}|^{2+\epsilon} < +\infty$, where $\epsilon > 0$ is an arbitrary positive number. Then the empirical distribution function of the eigenvalues of $\frac{1}{m} A^t A$ converges to a non-random limit, known as the Marchenko-Pastur distribution

$$\frac{1}{n}\#(\lambda_i \leq x, \ i = 1, \ldots, n) \to G_\gamma(x) = \int_{-\infty}^x g_\gamma(t)dt , \qquad (18)$$

where the spectral density $g(t)$ is supported on the interval $[a, b]$, $a = \sigma^2(1 - \gamma^{-1/2})^2$, $b = \sigma^2(1 + \gamma^{-1/2})^2$, $\sigma^2 = Ea_{11}^2$, and $g(t) = \frac{1}{2\pi t \gamma \sigma^2}\sqrt{(b-t)(t-a)}$, $t \in [a, b]$.

The case $a_{ij} \sim N(0, 1)$ $1 \leq i, j \leq n$, is known in the literature as the Wishart (Laguerre) ensemble of real sample covariance matrices. The joint distribution of the eigenvalues of M is defined by its density. Similarly to the Gaussian ensembles of real symmetric and Hermitian matrices discussed in Subsect. 1.1, many important statistical quantities in the Wishart ensemble can be calculated explicitly. For example, the joint probability density of the eigenvalues is given by the formula

$$p_n(x_1, \ldots, x_n) = Z_{n,m}^{-1} \prod_{1 \leq i < j \leq n} |x_i - x_j| \prod_{i=1}^n x_i^{m-n-1} \exp(-x_i/2) . \qquad (19)$$

It was shown by Johnstone ([21]), that the largest eigenvalue of a Wishart random matrix converges, after a proper rescaling, to the Tracy-Widom distribution F_1. Namely, let $m \to \infty$, $n \to \infty$, $m/n \to \gamma$ and $\mu_{m,n} = (n^{1/2} + m^{1/2})^2$, $\sigma_{m,n} = (n^{1/2} + m^{1/2})(n^{-1/2} + m^{-1/2})^{1/3}$. Then

$$\Pr\left(\lambda_{max}(A^t A) \leq \mu_{m,n} + s\sigma_{m,n}\right) \to F_1(s) \qquad (20)$$

One can also show (see [37]), that the rescaled k-point correlation function at the edge of the spectrum converge in the limit to (12).

Finally, we want to remark, that there is a long-standing interest in nuclear physics in the spectral properties of the complex sample covariance matrices A^*A, where the entries of a reactangular matrix A are independent identically distributed complex random variables (see e.g. [2,5,16,42,46]). We refer the reader to [37] and the references therein for additional information.

1.4 Universality in Random Matrices

The universality conjecture in Random Matrix Theory states, loosely speaking, that the local statistical properties of a few eigenvalues in the bulk or at the edge of the spectrum are independent of the distribution of individual matrix entries in the limit of large dimension. The only thing that should matter is, whether the matrix is real symmetric, Hermitian or self-dual quaternion Hermitian.

For Wigner random matrices, the conjecture was rigorously proven at the edge of the spectrum, both for real symmetric and Hermitian case in [36], provided that all moments of matrix entries exist and do not grow faster than the moments of a Gaussian distribution, and the odd moments vanish. In particular, it was shown that the largest eigenvalue, after proper rescaling, converges in distribution to the Tracy-Widom law. In the bulk of the spectrum, the conjecture was proven by Johansson ([20]) for Wigner Hermitian matrices, provided the marginal distribution of a matrix entry has a Gaussian component. We refer to [11] and references therein for the universality results in the unitary ensembles of random matrices.

The situation for sample covariance random matrices is quite similar (see papers by Soshnikov [37] and Ben Arous and Péché [3]).

The natural question is how general such results are? What happens if matrix entries have only a finite number of moments? In this article we consider the extreme case when the entries of A do not have a finite second moment. In the next section, we discuss spectral properties of Wigner random matrices and, more generally, band random matrices when marginal distribution of matrix entries has heavy tails. As was shown in [38], the statistics of the largest eigenvalues of such matrices are given by a Poisson inhomogeneous random point process. In Sect. 3 we discuss a similar result (although in a weaker form) for the largest eigenvalues of sample covariance random matrices with Cauchy entries. Section 4 is devoted to conclusions.

2 Wigner and Band Random Matrices with Heavy Tails of Marginal Distributions

In this section we consider ensembles of Wigner real symmetric and Hermitian matrices (1) and (2), and band real symmetric and Hermitian random matrices (16), (2) with the additional condition on the tail of the marginal distribution

$$G(x) = \Pr\left(|a_{jk}| > x\right) = \frac{h(x)}{x^\alpha}, \qquad (21)$$

where $0 < \alpha < 2$ and $h(x)$ is a slowly varying function at infinity in a sense of Karamata ([22, 34]). In other words, $h(x)$ is a positive function for all $x > 0$, such that $\lim_{x \to \infty} \frac{h(tx)}{h(x)} = 1$ for all $t > 0$. The condition (21) means that the distribution of $|a_{ij}|$ belongs to the domain of the attraction of a stable distribution with the index α (see e.g. [18], Theorem 2.6.1).

Without loss of generality, we restrict our attention to the real symmetric case. The results in the Hermitian case are practically the same. Wigner random matrices (1), (2) with the heavy tails (21), in the special case when limit $\lim_{x \to \infty} h(x) > 0$ exists, were considered on a physical level of rigor by Cizeau and Bouchaud in [8]. They argued, that the typical eigenvalues of A are of the order of $n^{1/\alpha}$. Cizeau and Bouchaud also suggested a formula for the limiting spectral density of the empirical distribution function of the eigenvalues of $n^{-1/\alpha}A$. Unlike the Wigner semicircle and Marchenko-Pastur laws, the conjectured limiting spectral density is supported on the whole real line. It was given as

$$f(x) = L_{\alpha/2}^{C(x), \beta(x)}(x), \qquad (22)$$

where $L_\alpha^{C,\beta}$ is a density of a centered Lévy stable distribution defined through its Fourier transform $\hat{L}(k)$:

$$L_\alpha^{C,\beta} = \frac{1}{2\pi} \int dk \hat{L}(k) e^{ikx}, \qquad (23)$$

$$\ln\hat{L}(k) = -C|k|^\alpha \left(1 + i\beta sgn(k) \tan(\pi\alpha/2)\right), \qquad (24)$$

and functions $C(x), \beta(x)$ satisfy a system of integral equations

$$C(x) = \int_{-\infty}^{+\infty} |y|^{\frac{\alpha}{2} - 2} L_{\alpha/2}^{C(y), \beta(y)} \left(x - \frac{1}{y}\right) dy, \qquad (25)$$

$$\beta(x) = \int_{x}^{+\infty} L_{\alpha/2}^{C(y), \beta(y)} \left(x - \frac{1}{y}\right) dy. \qquad (26)$$

We would like to draw the reader's attention to the fact that the density in (22) is not a density of a Lévy stable distribution, since $C(x), \beta(x)$ are functions of x. Cizeau and Bouchaud argued, that the density $f(x)$ should decay as $\frac{1}{x^{1+\alpha}}$ at infinity, thus suggesting that the largest eigenvalues of A (in the case $h(x) = const$) should be of order $n^{\frac{2}{\alpha}}$, and not $n^{\frac{1}{\alpha}}$, which is the order of typical eigenvalues.

Even though originally proven in [38] in the Wigner case, the theorem written below holds in the general case of band random real symmetric (or Hermitian) random matrices (16), (17).

Let N_n be the number of independent (i.e. $i \leq j$), non-zero matrix entries a_{ij} in A. In other words, let $N_n = \#(1 \leq i \leq j \leq n, \ |i - j| \leq d_n)$ in the aperiodic band case, and $N_n = \#(1 \leq i \leq j \leq n, \ |i - j|_1 \leq d_n)$ in the periodic band case. It is not difficult to see, that $N_n = \frac{n(n+1)}{2}$ in the Wigner case, $N_n = n(d_n + 1)$ in the periodic band case, and $N_n = n \times (d_n + 1) - \frac{d_n(d_n+1)}{2}$ in the aperiodic band case. Let us define a normalization constant b_n in such a way that

$$\lim_{n \to \infty} N_n G(b_n x) = \frac{1}{x^\alpha}, \tag{27}$$

for all positive $x > 0$, where the tail distribution G has been defined in (21). Normalization b_n naturally appears (see [23] and Remark 1 below), when one studies the extremal values of a sequence of N_n independent identically distributed random variables (21). In particular, one can choose

$$b_n = \inf\{t : G(t - 0) \geq \frac{1}{N_n} \geq G(t + 0)\}. \tag{28}$$

It follows from (27) and (28), that $N_n^{\alpha-\delta} \ll b_n \ll N_n^{\alpha+\delta}$ for arbitrary small positive δ, and $\frac{N_n h(b_n)}{b_n^\alpha} \to 1$ as $n \to \infty$.

Theorem 1 claims that the largest eigenvalues of A have Poisson statistics in the limit $n \to \infty$.

Theorem 1. *Let A be a band real symmetric (16) or Hermitian (17) random matrix with a heavy tail of the distribution of matrix entries (21). Then the random point configuration composed of the positive eigenvalues of $b_n^{-1}A$ converges in distribution on the cylinder sets to the inhomogeneous Poisson random point process on $(0, +\infty)$ with the intensity $\rho(x) = \frac{\alpha}{x^{1+\alpha}}$.*

In other words, let $0 < x_1 < y_1 < x_2 < y_2 < \ldots x_k < y_k \leq +\infty$, and $I_j = (x_j, y_j)$, $j = 1, \ldots k$, be disjoint intervals on the positive half-line. Then the counting random variables $\#(I_j) = \#(1 \leq i \leq n : \lambda_i \in I_j)$, $j = 1, \ldots, k$, are independent in the limit $n \to \infty$, and have a joint Poisson distribution with the parameters $\mu_j = \int_{I_j} \rho(x) dx$, i.e.

$$\lim_{n \to \infty} \text{Pr} \left(\#(I_j) = s_j, \ j = 1, \ldots, k \right) = \prod_{j=1}^{k} \frac{\mu_j^{s_j}}{s_j!} e^{-\mu_j}. \tag{29}$$

For the additional information on Poisson random point processes we refer the reader to [10].

Corollary 1. *Let λ_k be the k-th largest eigenvalue of $b_n^{-1}A$, then*

$$\lim_{n \to \infty} \text{Pr} \left(\lambda_k \leq x \right) = \exp(-x^{-\alpha}) \sum_{l=0}^{k-1} \frac{x^{-l\alpha}}{l!}. \tag{30}$$

In particular, $\lim_{n \to \infty} \text{Pr} \left(\lambda_1 \leq x \right) = \exp(-x^{-\alpha})$.

Remark 1. The equivalent formulation of the theorem is the following. Let k be a finite positive integer. Then the joint distribution of the first k largest eigenvalues of $b_n^{-1} A$ is asymptotically (in the limit $n \to \infty$) the same as the joint distribution of the first k order statistics of $\{b_n^{-1}|a_{ij}|, \ 1 \le i \le j \le n\}$. It is a classical result, that extremal values of the sequence of independent identically distributed random variables with heavy tails distributions (21) have Poisson statistics (see e.g. [23], Theorem 2.3.1).

Theorem 1 was proven in [38] in the Wigner (i.e. full matrix) case (1), (2). The proof in the general (band matrix) case is essentially the same. However, it should be noted, that the original proof of Theorem 1 in [38] contained a little mistake, which could be easily corrected. The corrections are due in two places.

First of all, the correct formulation of the part c) of Lemma 4 from [38] (p. 87) should state, that for any positive constant $\delta > 0$, with probability going to 1 there is no no row $1 \le i \le n$, that contains at least two entries greater in absolute value than $b_n^{\frac{3}{4}+\delta}$. In other words, the exponent $\frac{1}{2} + \delta$ in $b_n^{\frac{1}{2}+\delta}$ in part c) of Lemma 4 must be replaced by $\frac{3}{4} + \delta$. After this correction, the statement is true. Indeed, the probability that there is a row with at least two entries greater than $b_n^{\frac{3}{4}+\delta}$ can be estimated from above by $n^3 \left(G(b_n^{\frac{3}{4}+\delta}) \right)^2$. It follows from (21), (27) and (28), that this probability goes to zero.

Also, the formula (28) in Lemma 5 (p. 88) should read

$$\Pr \left\{ \exists i, \ 1 \le i \le n : \ \max_{1 \le j \le n} |a_{ij}| > b_n^{\frac{3}{4}+\frac{\alpha}{8}}, \right.$$

$$\left(\sum_{1 \le j \le n} |a_{ij}| \right) - \max_{1 \le j \le n} |a_{ij}| > b_n^{\frac{3}{4}+\frac{\alpha}{8}} \right\} \to 0 \tag{31}$$

as $n \to \infty$. In other words, the exponent $\frac{1}{2} + \frac{\alpha}{4}$ in $b_n^{\frac{1}{2}+\frac{\alpha}{4}}$ must be replaced by $\frac{3}{4} + \frac{\alpha}{8}$. The key step of the proof of Lemma 5 was to show, that for any fixed row i and arbitrary small positive δ, the probability $\Pr \left(\sum_{j:|a_{ij}| \le b_n^{\frac{1}{2}+\delta}} |a_{ij}| \ge b_n^{\frac{1}{2}+2\delta} \right)$ can be estimated from above by $\exp(-n^\epsilon)$, where $\epsilon = \epsilon(\delta, \alpha) > 0$. We then concluded, that with probability going to 1, there is no row i such that $\sum_{j:|a_{ij}| \le b_n^{\frac{1}{2}+\delta}} |a_{ij}| \ge b_n^{\frac{1}{2}+2\delta}$. To establish (31), it is enough to prove that for any fixed row i

$$\Pr \left(\sum_{j:b_n^{1/2+\delta} \le |a_{ij}| \le b_n^{\frac{3}{4}+\delta}} |a_{ij}| \ge b_n^{\frac{3}{4}+2\delta} \right) < \exp(-n^\epsilon), \tag{32}$$

for sufficiently small positive ϵ. The proof is very similar to the argument presented in Lemma 5, and is left to the reader.

3 Real Sample Covariance Matrices with Cauchy Entries

Let A be a rectangular $m \times n$ matrix with independent identically distributed entries with the marginal probability distribution of matrix entries satisfying (21). Based on the results in the last section, one can expect that the largest eigenvalues have Poisson statistics as well. At this point, we have been able to prove it only in a weak form, and only when matrix entries have Cauchy distribution.

We recall, that the probability density of the Cauchy distribution is given by the formula $f(x) = \frac{1}{\pi(1+x^2)}$. Cauchy distribution is very important in probability theory (see e.g. [15]). In particular, Cauchy distribition is a $(1,1,0)$ stable distribution, i.e. the scale parameter is 1, the index of the distribution $\alpha = 1$ and the symmetry parameter is zero (see [18], [23]).

The following theorem was proven by Fyodorov and Soshnikov in [39]

Theorem 2. *Let A be a random rectangular $m \times n$ matrix ($m \geq n$) with i.i.d. Cauchy entries and let z be a complex number with a positive real part. Then, as $n \to \infty$ we have*

$$\lim_{n\to\infty} E\left(\det\left(1 + \frac{z}{m^2 n^2} A^t A\right)\right)^{-1/2} = \exp\left(-\frac{2}{\pi}\sqrt{z}\right) = \mathbf{E}\prod_{i=1}^{\infty}(1 + zx_i)^{-1/2},$$

(33)

where we consider the branch of $\sqrt{z}$ on $D = \{z : \Re z > 0\}$ such that $\sqrt{1} = 1$, E denotes the mathematical expectation with respect to the random matrix ensemble defined above, $\mathbf{E}$ denotes the mathematical expectation with respect to the inhomogeneous Poisson random point process on the positive half-axis with the intensity $\frac{1}{\pi x^{3/2}}$, and the convergence is uniform inside D (i.e. it is uniform on compact subsets of D). For a real positive $z = t^2$, $t \in R^1$, one can estimate the rate of convergence, namely

$$\lim_{n\to\infty} E\left(\det\left(1 + \frac{t^2}{m^2 n^2} A^t A\right)\right)^{-1/2} = \exp\left(-\frac{2}{\pi}|t|\left(1 + o(n^{-1/2+\epsilon})\right)\right),$$

(34)

where ϵ is an arbitrary small positive number and the convergence is uniform on the compact subsets of $[0, +\infty)$.

The result of Theorem 2 allows a generalization to the case of a sparse random matrix with Cauchy entries. Let, as before, $\{a_{jk}\}$, $1 \leq j \leq m$, $1 \leq k \leq n$, be i.i.d. Cauchy random variables, and $B = (b_{jk})$ be a $m \times n$ non-random rectangular $0 - 1$ matrix such that the number of non-zero entries in each column is fixed and equals to d_n. Let d_n grow polynomially, i.e. $b_n \geq n^\alpha$, for some $0 < \alpha \leq 1$. Also assume that $\ln m$ grows much slower than than any power of n.

Define a $m \times n$ rectangular matrix A with the entries $A_{jk} = b_{jk} a_{jk}$, $1 \leq j \leq m$, $1 \leq k \leq n$. Let $\lambda_1 \geq \lambda_2 \ldots \geq \lambda_n$ denote the eigenvalues of $A^t \times A$. The appropriate rescaling for the largest eigenvalues in this case is $\tilde{\lambda}_i = \frac{\lambda_i}{m^2 d_n^2}$, $i = 1, \ldots, n$.

Theorem 3. *Let A be a sparse random rectangular $m \times n$ matrix $(m \geq n)$ defined as above, and let z be a complex number with a positive real part. Then, as $n \to \infty$ we have*

$$\lim_{n \to \infty} E \left(\det \left(1 + \frac{z}{m^2 d_n^2} A^t A \right) \right)^{-1/2} = \lim_{n \to \infty} E \prod_{i=1}^{n} (1 + z \tilde{\lambda}_i)^{-1/2} \quad (35)$$

$$= \exp \left(-\frac{2}{\pi} \sqrt{z} \right) = \mathbf{E} \prod_{i=1}^{\infty} (1 + z x_i)^{-1/2} , \quad (36)$$

where, as in Theorem 1.1, we consider the branch of $\sqrt{z}$ on $D = \{z : \Re z > 0\}$ such that $\sqrt{1} = 1$. $\mathbf{E}$ denotes the mathematical expectation with respect to the inhomogeneous Poisson random point process on the positive half-axis with the intensity $\frac{1}{\pi x^{3/2}}$, and the convergence is uniform inside D (i.e. it is uniform on the compact subsets of D). For a real positive $z = t^2$, $t \in R^1$, one can get an estimate on the rate of convergence, namely

$$E \left(\det \left(1 + \frac{t^2}{m^2 d_n^2} A^t A \right) \right)^{-1/2} = \exp \left(-\frac{2}{\pi} t (1 + o(d_n^{-1/2 + \epsilon})) \right) , \quad (37)$$

where ϵ is an arbitrary small positive number and the convergence is uniform on the compact subsets of $[0, +\infty)$.

The proof relies on the following property of the Gaussian integrals:

$$(\det(B))^{-1/2} = \left(\frac{1}{\pi} \right)^N \int x \exp(-x B x^t) d^N , \quad (38)$$

where B is an N-dimensional matrix with a positive definite Hermitian part (i.e. all eigenvalues of $B + B^*$ are positive), $x = (x_1, \ldots, x_N) \in R^N$, and $x B x^t = \sum_{ij} b_{ij} x_i x_j$.

Let $B = B(t) = \begin{pmatrix} Id & tiA \\ tiA^t & Id \end{pmatrix}$. Then, one can write

$$\left(\det(1 + t^2 A^t A) \right)^{-1/2} = \left(\det \begin{pmatrix} 1 & tiA \\ tiA^t & 1 \end{pmatrix} \right)^{-1/2} = (\det(B))^{-1/2} , \quad (39)$$

and apply (38) to the r.h.s. of (39). Assuming that the entries of A are independent, one can significantly simplify the expression, using the fact that entries of A appear linearly in $B(t)$ (see Proposition 1 of [39]). In the Cauchy case, one can simplify the calculations even further, and prove that $\lim_{n \to \infty} E \left(\det(1 + \frac{z}{m^2 d_n^2} A^t A) \right)^{-1/2}$ exists and equals $\exp \left(-\frac{2}{\pi} \sqrt{z} \right)$.

On the other side, for Poisson random point processes the mathematical expectations of the type $\mathbf{E} \prod_{i=1}^{\infty} (1 + f(x_i))$ can be calculated explicitly

$$\mathbf{E}\prod_{i=1}^{\infty}(1+f(x_i)) = 1 + \sum_{k=1}^{\infty}\mathbf{E}\sum_{1\le i_1<i_2<...<i_k}\prod_{j=1}^{k}f(x_{i_j}) \tag{40}$$

$$= \sum_{k=0}^{\infty}\frac{1}{k!}\int_{(0,+\infty)^k}\prod_{j=1}^{k}f(x_j)\rho_k(x_1,\dots,x_k)dx_1\cdots dx_k \tag{41}$$

$$= \sum_{k=0}^{\infty}\frac{1}{k!}\left(\int_{(0,+\infty)}f(x)\rho(x)dx\right)^k = \exp\left(\int_{(0,+\infty)}f(x)\rho(x)dx\right). \tag{42}$$

In the equations above, ρ_k denotes the k-point correlation function, and ρ denotes the one-point correlation function (also known as intensity). It is a characteristic property of a Poisson random point process that the k-point correlation function factorizes as a product of one-point correlation functions, i.e. $\rho_k(x_1,\dots,x_k) = \prod_{i=1}^{k}\rho(x_i)$. In the context of Theorems 2 and 3, test function f has the form $f(x) = (1+zx)^{-1/2} - 1$. When the intensity ρ equals $\frac{1}{\pi x^{3/2}}$, one obtains

$$\int_{(0,+\infty)}f(x)\rho(x)dx = \int_{(0,+\infty)}((1+zx)^{-1/2}-1)\frac{1}{\pi x^{3/2}}dx = -\frac{2}{\pi}\sqrt{z},$$

which finishes the proof.

The fact, that the intensity $\rho(x) = \frac{1}{\pi x^{3/2}}$ of the Poisson random point process diverges at zero and is summable at $+\infty$, means, that the the vast majority of the eigenvalues of the normalized matrix converge to zero in the limit.

Remark 2. It should be pointed out, that the results of Theorem 2 and 3 do not imply that the statistics of the largest eigenvalues of a normalized sample covariance matrix with Gaussian entries are Poisson in the limit of $n \to \infty$. Indeed, to prove the Poisson statistics in the limit one has to show that

$$\lim_{n\to\infty}\mathbf{E}\prod_{i=1}^{n}(1+f(\tilde{\lambda}_i)) = \mathbf{E}\prod_{i=1}^{+\infty}(1+f(x_i)) \tag{43}$$

for a sufficiently large class of the test functions f, e.g. for step functions with compact support. As we already pointed out, the results of Theorems 2 and 3 claim that (43) is valid for $f(x) = (1+zx)^{-1/2} - 1$ for all z such that $\Re z > 0$.

4 Conclusion

It is known in the theory of random Schrödinger operators, that the statistics of the eigenvalues is Poisson in the localization regime (see e.g. [28,29]). It seems, that the same mechanism is responsible for the Poisson statistics for the largest eigenvalues in the random matrix models described above. The interesting next problem is to find a phase transition between the Tracy-Widom regime (when all moments of matrix entries exist) and the Poisson regime (when second moment does not exist).

It is also worth to point out, that there is a vast literature on the Poisson statistics of the energy levels of quantum sysytems in the case of the regular underlying dynamics (see e.g. [4, 9, 25, 26, 33, 35]).

References

1. L. Arnold: J. Math. Anal. Appl. **20**, 262 (1967).
2. C.W.J. Beenakker: Rev. Mod. Phys., **69**, 731, (1997).
3. G. Ben Arous, S. Péché: Commun. Pure Appl. Math., to appear, (2005).
4. M.V. Berry, M. Tabor: Proc. R. Soc. London Ser. A **356**, 375 (1977).
5. B.V. Bronk: J. Math. Phys., **6**, (1965).
6. A. Casati, L. Molinari, and F. Izrailev: Phys Rev. Lett. **64**, 1851 (1990).
7. A. Casati and V.L. Girko: Rand. Oper. Stoch. Equations, **1**, 15 (1991).
8. P. Cizeau, J.P. Bouchaud: Phys Rev E, **50**, 1810 (1994).
9. Z. Cheng, J.L. Lebowitz and P. Major: Prob. Theo. Rel. Fields, **100**, 253 (1994).
10. D.J. Daley, D. Vere-Jones: *An Introduction to the Theory of Point Processes*, vol.I, 2nd edn, (Springer, Berlin Heidelberg New York 2003).
11. P. Deift **Orthogonal Polynomials and Random Matrices: A Riemann-Hilbert Approach**, Courant Lecture Notes in Mathematics, Vol. 3, New York, 1999.
12. Z. Füredi and J. Komlós: Combinatorica, **1**, 233 (1981).
13. P. Forrester: Nucl. Phys. B, **402**, 709 (1994).
14. Y.V. Fyodorov and G. Akemann: JETP Lett. **77**, 438 (2003).
15. W. Feller: *An Introduction to Probability Theory and Its Applications*, Vol. II. 2nd edn. (John Wiley and Sons, Inc., New York, London, Sydney 1971).
16. Y.V. Fyodorov and H.-J.Sommers: J.Phys.A:Math.Gen. **36**, 3303 (2003).
17. A. Guionnet: Ann. Inst. H. Poincare Probab. Statist. **38**, 341 (2002).
18. I.A. Ibragimov, Yu.V. Linnik, *Independent and Stationary Sequences of Random Variables*, translation from the Russian edited by J.F.C. Kingman, (Wolters-Noordhoff Publishing, Groningen, 1971).
19. A.T. James: Ann. Math. Stat., **35**, (1964).
20. K. Johansson: Commun. Math. Phys., **215**, 683, (2001).
21. I.M. Johnstone: Ann. Stat., **29**, 297 (2001).
22. J. Karamata: Mathematica (Cluj), **4**, 38, (1930).
23. M.R. Leadbetter, G.Lindgren and H. Rootzén: *Extremes and Related Properties of Random Sequences and Processes*, (Springer-Verlag, New York 1983).
24. V.A. Marchenko, L.A. Pastur: Math. USSR-Sb. **1**, 457, (1967).
25. J. Marklof: Annals of Mathematics, **158**, 419, (2003).
26. J. Marklof: The Berry-Tabor conjecture. In: *Proceedings of the 3rd European Congress of Mathematics, Barcelona 2000*, (Progress in Mathematics 202 (2001)), pp 421–427.
27. M.L. Mehta: *Random Matrices*, (Academic Press, New York 1991).
28. N. Minami: Commun. Math. Phys., **177**, 709, (1996).

29. S.A. Molchanov: Commun. Math. Phys. **78**, 429, (1981).
30. S.A. Molchanov, L.A. Pastur and A.M. Khorunzhy: Theor. Math. Phys. **90**, 108 (1992).
31. R.J. Muirhead, *Aspects of Multivariate Statistical Theory*, (Wiley, New York 1982).
32. L.A. Pastur: Teor. Mat. Fiz., **10**, 102, (1972).
33. P. Sarnak: Values at integers of binary quadratic forms. In *Harmonic Analysis and Number Theory* (Montreal, PQ, 1996), CMS Conf. Proc. **21**, (Amer. Math. Soc., Providence, RI, 1997), pp 181–203.
34. E. Seneta: *Regularly Varying Functions*, Lecture Notes in Mathematics, **508** (eds. A.Dold and B.Eckmann), (Springer, New York, 1976).
35. Ya. Sinai: Adv. Sov. Math., AMS Publ., **3**, 199, (1991).
36. A. Soshnikov: Commun. Math. Phys., **207**, 697, (1999).
37. A. Soshnikov: J. Stat. Phys., **108**, 1033, (2002).
38. A. Soshnikov: Elec. Commun. Probab., **9**, 82, (2004).
39. A. Soshnikov, Y.Fyodorov: to appear in J. Math Phys. (2005), arXiv preprint math.PR/0403425.
40. C.A. Tracy, H. Widom: Commun. Math. Phys., **159**, 151, (1994).
41. C.A. Tracy, H. Widom: Commun. Math. Phys., **177**, 724, (1996).
42. J.J. Verbaarschot and T. Wettig : Annu.Rev.Nucl.Part.Sci, **50**, 343 (2000).
43. S.S. Wilks: *Mathematical Statistics*, (Princeton University Press, Princeton 1943).
44. E. Wigner: Ann. of Math., **62**, 548, (1955).
45. E. Wigner: Ann. of Math., **67**, 325, (1958).
46. E. Wigner: SIAM Rev., **9**, 1, (1967).

Part V

Semiclassical Analysis and Quantum Chaos

An important controversy of Quantum Chaos is about whether or not there exists sequences of eigenfunctions in $\hbar$ which concentrate on subsets of the energy shell which are invariant by the flow of the corresponding classical mechanical system. De Bièvre gives an overview on the state of knowledge about this question, with a focus on Quantum maps.

One dimensional quasi-periodic systems have an extremely rich spectral structure (see the contribution of Avila-Jitomirskaya in the spectral theory section). Fedotov, Klopp announce new results on a slowly varying two-frequency model. They explain the nature of the spectrum and level repulsion in terms of the phenomenon of tunneling in phase space.

Helffer presents a precise analysis of the low lying eigenvalues of the semi-classical Witten Laplacian using semiclassical quasimode constructions. This contributes to the application of quantum physics methods to a problem of geometric nature : the proof of the Morse inequalities.

The number of eigenstates of a closed system can, in certain limits, be approximated by the volume of the phase space available for the corresponding classical system. Nonnenmacher reports on progress towards proving such a Weyl type law for systems which are open and have a chaotic classical counterpart, thus relating the number of long living resonances to the fractal dimension of the set of classically trapped trajectories.

Mantoiu and Purice give an introduction to their results on quantization of classical observable in the presence of a non constant magnetic field. This theory is gauge invariant. Their algebraic methods lead to results on the location of the essential spectrum and to information on propagation properties of such operators.

The presence of constant magnetic field alters significantly the scattering properties of quantum systems. Raikov surveys recent results concerning in particular the asymptotic behavior of the spectral shift function measuring the scattering determinant for strong magnetic fields, near the Landau levels of the planar problem and at high energies.

Zelditch reports on recent mathematical progress in quantum gravity: the problem of counting the vacuum states allowed for our universe according to string/M theory is mathematically a counting problem of critical points of holomorphic sections of certain line bundles; so the realm of studying statistics of zeros of random holomorphic sections is joined. As for now the results are asymptotic. The problem of counting solutions of the black-hole attractor equation is discussed.

Recent Results on Quantum Map Eigenstates

S. De Bièvre

Université des Sciences et Technologies de Lille, UFR de Mathématiques et
Laboratoire Painlevé, 59655 Villeneuve d'Ascq
Stephan.De-Bievre@math.univ-lille1.fr

1 Introduction

One of the central problems in quantum chaos is to obtain a good understanding of the semi-classical behaviour of the eigenfunctions of quantum systems that have a chaotic Hamiltonian system as their classical limit. The pivotal result in this context, and the only general one to date, is the Schnirelman Theorem. Loosely speaking, it states that, if the underlying classical dynamics is ergodic on the appropriate energy surface, then "most" eigenfunctions of the quantum system equidistribute on this energy surface. The challenge is to go beyond this theorem. I will report here on the progress that has been made in this direction in recent years for the special case of quantum maps on the torus.

Although the present text is meant to be reasonably self-contained, some familiarity with quantum chaos, for example at the level of [11], is assumed. I will also allow myself a rather loose discursive style, and present some speculations and open problems, referring the interested reader to the cited literature for the hard facts. I have limited the references almost exclusively to those concerning the quantum maps under study here. For a recent review of the analogous questions in the context of the geodesic flow and the Laplace-Beltrami operator on compact Riemannian manifolds, one may consult [26].

The very statement of the Schnirelman theorem immediately invites several questions. Given some class of models, one may first of all wonder whether the equidistribution property does indeed only hold for "most" sequences of eigenfunctions or if, on the contrary, it holds for all. If you think the second alternative is true, you are a believer in what has been baptized "unique quantum ergodicity". It means you think the proof of the Schnirelman can be improved, and the restriction to "most" eigenfunctions is due to a shortcoming in the proof. Alternatively, you may believe that the theorem can not in general be improved, and that exceptional eigensequences exist, that have a limit that does not equidistribute. In that case, you should try to provide examples. For billiards, for example, there does exist some numerical and theoretical evidence for "scars". This means for eigenfunction sequences that concentrate – at least partially – on periodic orbits of the dynamics, an observation that goes back to [19]. The precise sense in which this concentration

S. De Bièvre: *Recent Results on Quantum Map Eigenstates*, Lect. Notes Phys. **690**, 367–381
(2006)
www.springerlink.com

takes place and if it is compatible or not with unique quantum ergodicity is not known, however.

I will explain in these pages what the answer is that has emerged in the case of a simple class of quantum maps, the so-called Continuous Automorphisms of the Torus, or CAT maps. I will also present some recent results on their perturbations.

2 Perturbed CAT Maps: Classical Dynamics

Let me recall what CAT maps are. Consider the two-torus $\mathbb{T}^2 = \mathbb{R}^2/\mathbb{Z}^2$ and think of it as a toy classical phase space, which means that you equip the C^∞ functions on $\mathbb{T}^2$ with the usual Poisson bracket

$$\{f, g\}(x) = \partial_q f(x)\partial_p g(x) - \partial_p f(x)\partial_q g(x), \quad \text{where } x = (q, p) \in \mathbb{T}^2$$

or equivalently, that you think of it as a symplectic space with symplectic form $dq \wedge dp$. This implies in particular that you can associate to every smooth function g a Hamiltonian flow Φ_t^g, $(t \in \mathbb{R})$ which is defined by $\Phi_t^g(q, p) = (q(t), p(t))$, where $(q(t), p(t))$ is the solution of the Hamiltonian equations of motion $\dot{q}(t) = \partial_p g(x(t))$, $\dot{p}(t) = -\partial_q g(x(t))$ with initial condition $q(0) = q, p(0) = p$. Note that the Φ_t^g are symplectic maps (or canonical transformations), meaning they preserve the Poisson brackets:

$$\{f_1 \circ \Phi_t^g, f_2 \circ \Phi_t^g\} = \{f_1, f_2\} \circ \Phi_t^g, \quad \forall f_1, f_2 \in C^\infty(\mathbb{T}^2).$$

Now I will actually be interested in *discrete time* dynamical systems on $\mathbb{T}^2$. For that purpose, consider a matrix $A \in \mathrm{SL}(2, \mathbb{Z})$. That's a two by two matrix with integer entries and determinant one. Clearly A defines a linear map on $\mathbb{R}^2$ that passes naturally to the quotient $\mathbb{T}^2 = \mathbb{R}^2/\mathbb{Z}^2$ because it has integer entries. Because its determinant is one, it is easily checked that it is symplectic in the above sense:

$$\{f_1 \circ A^t, f_2 \circ A^t\} = \{f_1, f_2\} \circ A^t, \quad \forall f_1, f_2 \in C^\infty(\mathbb{T}^2),$$

where this time t is an integer rather than a real number. So A provides us with a dynamical system that is obtained by folding a linear map onto a torus and is therefore necessarily quite simple. When $|\mathrm{Tr}A| > 2$, such maps nevertheless have surprisingly rich properties. Indeed, in that case A has two real eigenvectors $v_\pm$ and two real eigenvalues $\exp \pm \gamma_0$, with the so-called Lyapounov exponent $\gamma_0 > 0$. This implies the dynamical system is exponentially unstable. Roughly speaking, if $x, x' \in \mathbb{T}^2$ are a small distance ϵ apart, then generically $A^t x$ and $A^t x'$ will be a distance $\epsilon \exp \gamma_0 t$ apart: the dynamics displays *sensitive dependence on initial conditions*, a well known source of chaotic behaviour. To put it graphically, if you know that initially the system point is in some set of linear size ϵ, all you can say at a time of order $\frac{1}{\gamma_0}|\ln\epsilon|$ later is that the system is in a set of size 1, which is not saying much more than that it is *somewhere* in the torus, and that is not saying very much at all. In other words, after a time of order $\frac{1}{\gamma_0}|\ln\epsilon|$ you lost all information on the whereabouts of the system. This is a very short time indeed, as you can see by imagining $\epsilon = \epsilon_0 10^{-k}$: it then grows only linearly in k. This instability has the consequence that the dynamics is exponentially mixing. This means that $\forall f_1, f_2 \in C^\infty(\mathbb{T}^2)$,

$$| \int_{\mathbb{T}^2} (f_1 \circ A^t)(x) f_2(x) dx - \int_{\mathbb{T}^2} f_1(x) dx \int_{\mathbb{T}^2} f_2(x) dx | \leq C_{A,f_1} \parallel \nabla f_2 \parallel_1 e^{-\gamma_0 t} .$$

Replacing f_1 and f_2 by characteristic functions of subsets B_1 and B_2 of the torus, this can be interpreted as follows: at large times, the probability that a trajectory starting in some set B_2 ends up in B_1 converges to the size of B_1, independently of where B_1 and B_2 are located on the torus. This is again a way of saying that all information on the whereabouts of the system is rapidly lost over time. Exponential mixing plays an important role in the following analysis. It implies furthermore that the dynamics is ergodic, meaning that for almost all initial conditions $x_0 \in \mathbb{T}^2$,

$$\frac{1}{T} \sum_{t=1}^{T} f(A^t x_0) \to \int_{\mathbb{T}^2} f(x) dx .$$

Note that this can not possibly be true for all initial conditions since there exists a dense set of periodic points for the dynamics: every point on the torus that has two rational coordinates is periodic. All these periodic orbits are unstable, of course. The simplest one is the fixed point $x = 0$.

These discrete dynamical systems are arguably the simplest fully chaotic Hamiltonian systems one can find: in particular, the ergodic properties I mentioned are readily proved by some simple Fourier analysis. This is in sharp contrast to what happens for other systems. Proving the Sinai or Bunimovich billiards are ergodic (let alone mixing) is very hard work and the geodesic flow on a negative curvature Riemannian manifold is also not a particularly simple object to study.

This is presumably why Hannay and Berry [18] came up with the idea of constructing a quantum mechanical equivalent to these simple systems, that I will describe in the next section.

Before turning to that task, let me point out that all the above properties of the dynamics are preserved if it is perturbed. Let g be a fixed smooth function on $\mathbb{T}^2$ and consider

$$\Phi_\epsilon = \Phi_\epsilon^g \circ A .$$

For small ϵ, this is still hyperbolic and exponentially mixing, with exponent γ_ϵ, $\lim_{\epsilon \to 0} \gamma_\epsilon = \gamma_0$ [3].

3 Quantum Maps

What are the discrete quantum dynamical systems that correspond to the discrete classical dynamical systems described in the previous section? I will give a much simplified description and refer for details to [5] [11]. To avoid inessential technicalities, I will consider only the case where A is of the form ($m \in \mathbb{N}_*$)

$$A = \begin{pmatrix} 2m & 1 \\ 4m^2 - 1 & 2m \end{pmatrix}.$$

Introducing for each $a = (a_1, a_2) \in \mathbb{R}^2$ the phase space translation operator

$$U(a)\psi(y) = e^{-\frac{i}{2\hbar}a_1 a_2} e^{\frac{i}{\hbar}a_2 y} \psi(y - a_1) = e^{-\frac{i}{\hbar}(a_1 P - a_2 Q)} \psi(y),$$

the quantum Hilbert space of the system is defined by

$$\mathcal{H}_\hbar = \{\psi \in \mathcal{S}'(\mathbb{R}) \mid U(1,0)\psi = \psi = U(0,1)\psi\}.$$

This space is N dimensional when $2\pi\hbar N = 1$ and zero-dimensional otherwise. Then any $\psi \in \mathcal{H}_\hbar$ can be written as follows:

$$\psi(y) = \sum_{\ell \in \mathbb{Z}} c_\ell \delta\left(y - \frac{\ell}{N}\right); \quad c_{\ell+N} = c_\ell.$$

If A were not of the above form, the definition of $\mathcal{H}_\hbar$ involves some extra phases that clutter up the page, which is the only reason why I avoid treating the general case here. That $\mathcal{H}_\hbar$ is finite dimensional is not surprising: since the classical phase space has finite area, and since the uncertainty principle tells us every quantum state "takes up" an area $2\pi\hbar$, it is normal the Hilbert space is finite dimensional. This is reminiscent also of what happens when treating spin, where the classical phase space is a sphere, and the quantum Hilbert spaces are also finite dimensional.

The relation $2\pi\hbar N = 1$ will always be assumed to hold from now on. So when both $\hbar$ and N show up in the same formula, they are always related in that manner. Consequently, the semi-classical limit corresponds to taking N to infinity. In this spirit, I will occasionally find it useful to write $\mathcal{H}_N$ rather than $\mathcal{H}_\hbar$.

The standard Weyl quantization is readily adapted to the present situation. For $f \in C^\infty(\mathbb{T}^2)$, $x = (q, p) \in \mathbb{T}^2$, write

$$f(x) = \sum_{n \in \mathbb{Z}^2} f_n e^{-i2\pi(n_1 p - n_2 q)}$$

and define

$$\mathrm{Op}^{\mathrm{W}} f = \hat{f} = \sum_{n \in \mathbb{Z}^2} f_n e^{-i2\pi(n_1 P - n_2 Q)} = \sum_{n \in \mathbb{Z}^2} f_n U\left(\frac{n}{N}\right) : \mathcal{H}_\hbar \to \mathcal{H}_\hbar.$$

The quantum dynamics is now defined as follows. For $A \in \mathrm{SL}(2,\mathbb{Z})$, $|\operatorname{Tr}A| > 2$, construct

$$M(A)\psi(y) = \left(\frac{i}{2\pi\hbar}\right)^{1/2} \int_{\mathbb{R}} e^{\frac{i}{2\hbar}(2my^2 - yy' + 2my'^2)} \psi(y')dy' .$$

Then, for all $t \in \mathbb{Z}$,

$$M(A)\mathcal{H}_\hbar = \mathcal{H}_\hbar \quad \text{and} \quad M(A)^{-t}\, \mathrm{Op}^{\mathrm{W}} f\, M(A)^t - \mathrm{Op}^{\mathrm{W}}(f \circ A^t) = 0 .$$

Now, for $\epsilon > 0$ define the unitary operator

$$U_\epsilon = e^{-\frac{i}{\hbar}\epsilon \mathrm{Op}^{\mathrm{W}} g} M(A) : \mathcal{H}_\hbar \to \mathcal{H}_\hbar .$$

This is the quantum map we wish to study. It is naturally related to the discrete Hamiltonian dynamics on $\mathbb{T}^2$ obtained by iterating $\Phi_\epsilon = \Phi_\epsilon^g \circ A$. It acts on the N dimensional spaces $\mathcal{H}_\hbar$ and we are interested in the behaviour of its eigenfunctions and eigenvalues in the $N \to \infty$ limit:

$$U_\epsilon \psi_j^{(N)} = e^{i\theta_j^{(N)}} \psi_j^{(N)}, \quad j = 1 \dots N .$$

4 What is Known?

Let me start by recalling a precise statement of the Schnirelman theorem in the present context:

Theorem 1. [5] Let $\epsilon \geq 0$ and small. Then, for "almost all" sequences $\psi_{j_N}^{(N)} \in \mathcal{H}_\hbar$, so that $U_\epsilon \psi_{j_N}^{(N)} = e^{i\theta_{j_N}^{(N)}} \psi_{j_N}^{(N)}$,

$$\langle \psi_{j_N}^{(N)} \mid \mathrm{Op}^{\mathrm{W}} f \mid \psi_{j_N}^{(N)} \rangle \overset{N \to +\infty}{\to} \int_{\mathbb{T}^2} f(x)dx, \quad \forall f \in C^\infty(\mathbb{T}^2) . \tag{1}$$

Note that here, for each N, a fixed basis of eigenfunctions has been chosen, and j_N is then an index in the set $\{1, \dots N\}$. The "almost all" statement refers to the fact that one has to choose the j_N from a subset $G_N \subset \{1, \dots, N\}$ such that $\#G_N/N \to 1$. As I already mentioned, this result is 'robust': it holds quite generally whenever the classical system is ergodic. It is proven in each different situation by adapting arguments that are used to prove it for other systems, such as Laplace-Beltrami operators on negatively curved manifolds. This is why it holds also for $\epsilon \neq 0$ and on higher dimensional tori. With (sometimes considerable) additional work, the proof can be even be adapted for maps that are not continuous, such as the Baker map [12, 14]. A version of the theorem for certain systems with a mixed phase space also exists [25].

As pointed out in the introduction, this theorem invites some immediate and obvious questions:

Question 1. Do there exist exceptional sequences of eigenfunctions meaning ones that do not converge semi-classically to Lebesgue measure (this means (1) does not hold).

Question 2. Can you characterize all possible semi-classical measures? Here any measure ν on the torus for which there exists some sequence of eigenfunctions ψ_{N_ℓ} satisfying

$$\langle \psi_{N_\ell} | Op^W f | \psi_{N_\ell} \rangle \stackrel{N_\ell \to +\infty}{\longrightarrow} \int_{\mathbb{T}^2} f(x) d\nu, \quad \forall f \in C^\infty(\mathbb{T}^2) \tag{2}$$

is called a semi-classical measure. Note that in this definition the sequence is not necessarily picked from a pre-assigned basis as is the case in the Schnirelman theorem.

There are of course still other questions that readily come to mind, related to the speed of convergence in the Schnirelman theorem, as well as to the fluctuations of the matrix elements that I shall not discuss here. Answers to the above two questions are not available in general systems. Only in some restricted classes of models have partial answers been obtained. For the (Hecke) eigenfunctions of the Laplace-Beltrami operator of a (class) of constant negative curvature surfaces the answer to this first question has been proven to be "NO!" [24]. On the other hand, for quantized hyperbolic toral automorphisms, the answer has been proven to be "YES!":

Theorem 2. *[15] Let $\epsilon = 0$. Let $0 \le \alpha \le \frac{1}{2}$, then there exists $N_k \to \infty$ and eigenfunctions $\psi_{N_k} \in \mathcal{H}_{N_k}$ so that*

$$\langle \psi_{N_k} | Op^W f | \psi_{N_k} \rangle \stackrel{N_k \to +\infty}{\longrightarrow} \alpha f(0) + (1-\alpha) \int_{\mathbb{T}^2} f(x) dx, \quad \forall f \in C^\infty(\mathbb{T}^2). \tag{3}$$

This theorem holds for any choice of A with $|\mathrm{Tr} A| > 2$, not just for the particular ones discussed here. The sequence N_k depends on A. The construction (sketched below) yields such a sequence of eigenfunctions for any sequence of eigenangles $\theta_{j N_k}^{(N_k)}$ of the quantum maps.

Note that this theorem is only stated for the quantized cat maps themselves, not for their perturbations. I strongly believe the result does not survive perturbation, a claim I will try to corroborate below.

Let me first sketch some of the ingredients that go into the proof of the result. I will indicate how the eigenfunctions with the above property are constructed in the case $\alpha = 1/2$. First of all, it is well known that the quantum maps $M(A)$ have a period: for each N, there exists an integer T_N so that $M(A)^{T_N} = e^{i\varphi_N} \mathbb{1}_{\mathcal{H}_N}$. This implies that the eigenvalues $\exp i\theta_j^{(N)}$ have the property:

$$\theta_j^{(N)} \in \left\{ \frac{2\pi}{T_N} \ell + \frac{\varphi_N}{T_N} | \ell \in \mathbb{Z} \right\} .$$

Consequently, the N eigenangles $\theta_j^{(N)}$ can take on at most T_N distinct equally spaced values. Let me note in passing that it can be shown that the eigenvalues equidistribute around the circle in the classical limit [6]: this is the analog of Weyl's law in the present context. Now it is easy to see that

$$P_k = \frac{1}{T_N} \sum_{j=-T_N/2+1}^{T_N/2} e^{-\frac{i}{T_N}(2\pi jk + \varphi_N j)} M(A)^j$$

is a projector. I am supposing T_N is even; if it is odd, you sum from $-(T_N - 1)/2$ to $(T_N - 1)/2$. *If this projector is non-zero*, it projects onto the eigenspace corresponding to the eigenvalue $e^{\frac{i}{T_N}(2\pi k + \varphi_N)}$. In particular, if you can find some non-zero vector ϕ_N for which $P_k \phi_N \neq 0$, it certainly is an eigenvector. This suggests it is easy to construct at least some eigenvectors. Now, we are hunting for eigenvectors that are well localized around the origin in phase space. The idea is then to take for ϕ_N a coherent state centered on the origin. Recall that in the ordinary quantum mechanics of a system with one degree of freedom, those are of the form

$$\eta_z(y) = \left(\frac{\mathrm{Im}\, z}{\pi \hbar}\right)^{1/4} \exp i\frac{z}{2\hbar} y^2 \,,$$

where z is a complex number in the upper half plane. By periodicizing this Gaussian in both the y variable and the dual (Fourier) variable, one obtains a family $|0, z\rangle$ of states in $\mathcal{H}_N$ that have two crucial properties for our purposes. First, they are localized at the origin in the sense that, as $N \to \infty$,

$$|\langle 0, z| Op^{\mathrm{W}} f |0, z\rangle - f(0)| \leq C_f \frac{1 + |z|^2}{\mathrm{Im}\, z N} \,. \tag{4}$$

Next, they behave nicely under the dynamics:

$$M(A)|0, z\rangle = e^{i\sigma_N(A)}|0, A \cdot z\rangle \,. \tag{5}$$

Here $\sigma_N(A)$ is a phase that we will not play any role in what follows and

$$A \cdot z = \frac{2mz + (4m^2 - 1)}{z + 2m} \,.$$

Let us now compute the matrix element of an observable $Op^{\mathrm{W}} f$ in an evolved coherent state: $\langle 0, A^t \cdot z| Op^{\mathrm{W}} f |0, A^t \cdot z\rangle$. Since $\mathrm{Im}(A^t \cdot z)$ behaves like $\exp -2\gamma_0 |t|$ as $|t| \to \infty$ (the real part being uniformly bounded in time), the upper bound in (4) then guarantees that, for times t satisfying $|t| < \frac{1}{2}\tau_N$, where

$$\tau_N = \frac{1}{\gamma_0} \ln N \,,$$

this still converges to $f(0)$:

$$\langle 0, z| M(A)^{-t} Op^{\mathrm{W}} f M(A)^t |0, z\rangle \stackrel{N \to +\infty}{\to} f(0) \,. \tag{6}$$

But what happens for longer times? It was proven in [8–10] that, for times $|t|$ between $\frac{1}{2}\tau_N$ and $\frac{3}{2}\tau_N$, one has instead

$$\langle 0, z | M(A)^{-t} \mathrm{Op}^{\mathrm{W}} f M(A)^t | 0, z \rangle \to \int_{\mathbb{T}^2} f(x) dx . \tag{7}$$

This is a consequence of the familiar "spreading of the wavepacket", together with the exponential instability of the dynamics. One can indeed think of a coherent state as being concentrated in a region of linear size $\sqrt{\hbar}$ in both the position and momentum variables. Now in the semiclassical regime, evolving the coherent state with the exponentially unstable dynamics A stretches its support along the unstable direction v_+ to a size $\sqrt{\hbar} \exp \gamma_0 t$: this is no longer microscopic if t is of order $\frac{1}{2\gamma_0} \ln N$ or more so that the estimate (6) breaks down. Since the unstable manifold through 0, which is just the line $\mathbb{R} v_+$ through the origin, wraps itself ergodically around the torus, estimate (7) follows. To understand the reason for its breakdown beyond $\frac{3}{2}\tau_N$ we need to return to the quantum periods T_N.

The quantum periods T_N are very irregular functions of N, that grow roughly linearly in N [20], but that can, for some N, be very much smaller than that. It is shown in [8] that there exists a sequence N_k for which this period behaves like $2\tau_N$. But this means that, for those values of N

$$\langle 0, z | M(A)^{-(2\tau_N - s)} \mathrm{Op}^{\mathrm{W}} f M(A)^{2\tau_N - s} | 0, z \rangle = \langle 0, z | M(A)^s \mathrm{Op}^{\mathrm{W}} f M(A)^{-s} | 0, z \rangle$$

and consequently, for times s so that $|s| < \frac{1}{2}\tau_N$, and $N \to \infty$,

$$\langle 0, z | M(A)^{-(2\tau_N - s)} \mathrm{Op}^{\mathrm{W}} f M(A)^{2\tau_N - s} | 0, z \rangle \to f(0) .$$

This explains the breakdown of (7) for times beyond $\frac{3}{2}\tau_N$.

This breakdown can also be understood intuitively [8]. The exponential instability of the classical dynamics explained in Sect. 2 implies that *in the naïvest semi-classical picture of the quantum dynamics* the support of the Wigner function of the evolved coherent states is stretched out over a length $L_t = \sqrt{\hbar} \exp \gamma_0 t$ along the unstable direction. Now, this support is wound around the torus, and so successive windings are a distance $D_t = \hbar^{-1/2} \exp -\gamma_0 t$ apart. But on the torus, for any pair of conjugate variables Q, P, the uncertainty principle says $\Delta Q \Delta P \geq \hbar$, and, since we are on the torus $\Delta Q, \Delta P \leq 1$, so that both ΔQ and ΔP are *greater* than $\hbar$! This means, that on the torus, the Wigner function can not resolve details finer than on a scale $\hbar$. Hence one may expect that the above simple semi-classical picture of the evolution will break down when D_t is of order $\hbar$, and this is precisely when t is of order $\frac{3}{2}\tau_N$. At this time, interference effects start to take over. In the case of CAT maps, this leads eventually to the very "unclassical" recurrence at the quantum period, in which the original wave packet is reconstituted completely for $t = T_N$, and which may be as short as $2\tau_N$.

Let me now write

$$\psi_N = \sqrt{T_N} P_k |0, z\rangle .$$

These are our candidate eigenfunctions that will be proven to satisfy (3). The extra factor $\sqrt{T_N}$ is a normalization factor the origin of which will become clear in a second. It is enough to look at (4) and (7) to understand

(i) why this ought to work when $N = N_k$ and
(ii) what the remaining problems are.

First of all, let's be optimistic and suppose that in computing the matrix element $\langle \psi_N, Op^W f \psi_N \rangle$ we have to deal only with the diagonal matrix elements. Let us write

$$\langle \psi_{N_k} | Op^W f | \psi_{N_k} \rangle = \frac{1}{T_{N_k}} \sum_{t=-\frac{T_{N_k}}{2}+1}^{\frac{T_{N_k}}{2}} \langle 0, A^t \cdot z | Op^W f | 0, A^t \cdot z \rangle + \text{OffD Terms} .$$

(8)

Now taking $N \to \infty$ and using (6) and (7), the sum of the diagonal terms above converges to the desired $\frac{1}{2} f(0) + \frac{1}{2} \int_{\mathbb{T}^2} f(x) dx$. This then proves Theorem 2 provided one can control the off diagonal terms, which requires more work.

One further idea that goes into this, and that is easily understood intuitively, is the following. For positive times t_+ larger than $\frac{1}{2}\tau_N$, the evolved coherent states stretches along the unstable manifold of $x = 0$ in the sense that its Wigner or Husimi function has a support that is concentrated along $\mathbb{R}v_+$ over a length $L_{t_+} = \sqrt{\hbar} \exp \gamma_o t_+$. Similarly, in the past, for negative times $-t_-$ so that $t_- \geq \frac{1}{2}\tau_N$, the evolved coherent state stretches along the stable manifold $\mathbb{R}v_-$ over a length L_{t_-}. Now these two intervals intersect, for each such time in only a finite (but growing!) set of homoclinic points. Since a finite number of points on a line should carry no weight (if all goes well), this suggests that the evolved states $M(A)^{t_+} |0, z\rangle$ and $M(A)^{-t_-} |0, z\rangle$ are almost orthogonal. This is indeed what is proven in [15], for times $t_- + t_+ < 2\tau_N$.

It should by now be clear why I claim the construction of strongly scarred eigenstates is particular to quantum CAT maps. Without the existence of quantum periods (which is already exceptional), and without the rather special arithmetic properties guaranteeing the existence of very short quantum periods, the construction would generate only quasimodes, not eigenvectors. The construction of strongly scarred eigenstates even breaks down on higher dimensional tori, since then the maps with short quantum periods do not exist [7]. All of this indicates that these strongly scarred eigenstates on isolated unstable periodic orbits are a particular feature of the unperturbed two-dimensional cat maps.

Having answered Question 1 positively, we can now turn to Question 2: what measures on $\mathbb{T}^2$ can be obtained as limits of the type $\langle \psi_N | Op^W f | \psi_N \rangle$, for *some* sequence of eigenfunctions ψ_N. It is easy to see that such measures have to be invariant under the dynamics generated by A. Elaborating a little on the above ideas, one can show that, if μ is an A-invariant probability measure on $\mathbb{T}^2$ that is singular with respect to Lebesgue measure any probability measure on $\mathbb{T}^2$ of the form

$$\alpha\mu + (1 - \alpha)\mathrm{dx} \tag{9}$$

is a limit measure provided $\alpha \leq 1/2$. This follows from a diagonalization trick and the well known fact that all invariant probability measures are weak limits of pure point measures [16]. This is optimal in the following sense:

Theorem 3. *[9, 16] Let $\epsilon = 0$. Let μ be an A-invariant pure point measure. Let $\alpha > \frac{1}{2}$. Then there does not exist $N_k \to \infty$ and eigenfunctions $\psi_{N_k} \in \mathcal{H}_{N_k}$ so that*

$$\langle \psi_{N_k}, \mathrm{Op}^{\mathrm{W}} f \psi_{N_k} \rangle \overset{N_k \to +\infty}{\to} \alpha\mu(f) + (1 - \alpha) \int_{\mathbb{T}^2} f(x)dx, \quad \forall f \in C^\infty(\mathbb{T}^2) \,.$$

This result was obtained in [9] with a suboptimal bound $\alpha > 2/3$, which was then optimized in [16]. Still, this does not completely answer Question 2. In particular, it is not known what happens to this result if μ is purely singular continuous or, more specifically, if a purely singular measure can be a semi-classical limit of the above type. This theorem does continue to hold on higher dimensional tori, but it is not clear if it is then optimal, for the reasons explained above.

The sequence N_k in the above construction is very sparse. It behaves roughly like $\exp(\gamma_0 k/2)$. It is legitimate to ask what happens if you stay away from this sequence: are exceptional sequences of eigenfunctions then still possible? First, one could consider sequences of N along which T_N behaves like σT_N, with $\sigma > 2$. In that cases it is tempting to guess that any measure of the form (9) is now a semi-classical measure provided $\alpha \leq \sigma^{-1}$. The idea is that, in the eigenfunction construction sketched above, the fraction of terms that concentrates at the origin is σ^{-1}. Note however that our estimates do not imply such a result immediately, since we did not control the dynamics for times longer than $2\tau_N$. Still following the same line of thought, whenever the sequence of N is chosen so that $T_N/\ln N \to \infty$, one expects that the system should be uniquely quantum ergodic along that sequence. Here is a result that corroborates this conjecture:

Theorem 4. *[22] If $A \in SL(2, \mathbb{Z})$ is hyperbolic and $a_{11}a_{12} \equiv 0 \equiv a_{21}a_{22}$ mod 2, then there exists a density one sequence of integers $(\tilde{N}_\ell)_{\ell \in \mathbb{N}}$ along which*

$$\langle \psi_{\tilde{N}_\ell} | \mathrm{Op}^{\mathrm{W}} f | \psi_{\tilde{N}_\ell} \rangle \overset{\tilde{N}_\ell \to +\infty}{\to} \int_{\mathbb{T}^2} f(x)dx, \quad \forall f \in C^\infty(\mathbb{T}^2)$$

for any sequence of eigenfunctions $\psi_{\tilde{N}_\ell}$.

This is a generalization of the result of [13], where a result of this type is proven along a subsequence of primes. Clearly, the sequence $\tilde{N}_\ell$ showing up here is disjoint from the sequence N_k in Theorem 2. In fact $T_{\tilde{N}_\ell}$ is larger than $\sqrt{\tilde{N}_\ell}$. This result is of course a considerable strengthening of the Schnirelman theorem for the sequence of N concerned, because it says equidistribution holds for *all* sequences of eigenfunctions, not necessarily picked from a pre-assigned basis, and without any "density one" restriction, as opposed to what happens in the statement of the Schnirelman theorem.

To end this discussion, I would like to point out one further result that is fun to compare to Theorem 2:

Theorem 5. *[23] If $A \in SL(2, \mathbb{Z})$ is hyperbolic and $A \equiv \mathbb{I}_2$ mod 4, then there exists for each N a basis $\{\psi_1, \psi_2, \dots \psi_N\}$ of eigenfunctions of $M(A)$ so that*

$$\langle \psi_{j_N} | \mathrm{Op}^{\mathrm{W}} f | \psi_{j_N} \rangle \overset{N \to +\infty}{\rightrightarrows} \int_{\mathbb{T}^2} f(x) dx, \quad \forall f \in C^\infty(\mathbb{T}^2) \tag{10}$$

holds for any sequence ψ_{j_N}.

This again is an obvious strengthening of the Schnirelman theorem for the particular class of A considered. Note that the result holds for all N, without exceptions, so that a superficial reading of this result makes it look as if it is in contradiction with Theorem 2. The point is that the basis for which the result holds is explicitly described in the paper. Now, the eigenvalues of $M(A)$ may be degenerate so that it is possible that exceptional sequences of eigenfunctions not belonging to the above basis have a different semi-classical limit. This is precisely what happens with Theorem 2. Actually, since the quantum Hilbert spaces are N dimensional, the fact that the period is of order $\ln N_k$ for the special values of N implies that at least some of the eigenvalues must be highly degenerate in this case, with degeneracies of order $N_k/\ln N_k$. Although our construction of the exceptional eigenfunctions does not exploit this degeneracy explicitly, it clearly does so implicitly. Note finally that the sequence $\tilde{N}_\ell$ of Theorem 4 is such that the quantum periods $T_{\tilde{N}_\ell}$ are longer than $\sqrt{\tilde{N}_\ell}$. In fact, the authors of that paper show that, whenever the periods are at least this long along a sequence of integers, quantum unique ergodicity holds along that sequence. This implies that degeneracies as large as $\sqrt{\tilde{N}_\ell}$ do not suffice to produce strong scarring.

5 Perturbed Cat Maps

CAT maps are rather special: their linearity makes them easy to treat classically and endows their quantum counterparts with a number of arithmetic properties that can be exploited to prove a variety of results going beyond the Schnirelman theorem, as in [13, 22, 23] and to a lesser degree in [8, 9, 15, 16]. Since such results are not available in more complicated systems, the CAT maps provide a theoretical laboratory in which one can investigate with relative ease a certain number of questions in quantum chaos. But a real under-standing of the issues of quantum chaos requires proofs that rely exclusively on semi-classical analysis and on the ergodic properties of the underlying classical dynamics of the systems considered. Indeed, only such proofs have a fighting chance to give insight into more complicated models, such as billiards or geodesic flows on negatively curved manifolds, and ultimately more realistic models with mixed phase space. Before attacking these systems, a good starting point is to look at the perturbed CAT maps described in Sect. 2. They are non-linear and preserve none of the simple arithmetic properties of the CAT maps themselves, once quantized. Their classical mechanics is sufficiently complicated that detailed information on their classical mixing rate has become available only rather recently [3]. On the quantum side, and contrary to what happens for quantized CAT maps, their eigenvalue statistics has been shown numerically to be generic [21].

Beyond the Schnirelman theorem, the only theorem in the previous section the proof of which has a chance to generalize to the case of perturbed maps is Theorem 3. The simplest version one could hope to prove would be:

If $\psi_N \in \mathcal{H}_N$ is a sequence of eigenfunctions of U_ϵ, then there exist a function f so that

$$\lim_{N \to \infty} \langle \psi_N, \mathrm{Op}^{\mathrm{W}} f \psi_N \rangle \neq f(0).$$

In other words, this statement says that for a perturbed cat map, the delta measure at the origin is not a semi-classical measure of the system.

A stragegy of proof of such a result is suggested by the main ingredient of the proof of Theorem 3 in [10]. The following result is proven there:

Theorem 6. *Let $\epsilon = 0$. Let $a_0 \in \mathbb{T}^2$. If $\varphi_N \in \mathcal{H}_\hbar$ is some sequence (not necessarily eigenfunctions!) with the property that, for all $f \in C^\infty(\mathbb{T}^2)$*

$$\langle \varphi_N | \mathrm{Op}^{\mathrm{W}} f | \varphi_N \rangle \overset{N \to +\infty}{\to} f(a_0)$$

then there exists a sequence of times $t_N \to \infty$ so that

$$\langle \varphi_N | U_0^{-t_N} \mathrm{Op}^{\mathrm{W}} f \, U_0^{t_N} | \varphi_N \rangle \overset{N \to +\infty}{\to} \int_{\mathbb{T}^2} f(x) dx .$$

If you can prove this result for $\epsilon \neq 0$, then that would imply immediately that a sequence of eigenfunctions can not concentrate on $x = 0$, since for eigenfunctions

$$\langle \psi_N | \mathrm{Op}^{\mathrm{W}} f | \psi_N \rangle = \langle \psi_N | M(A)^{-t} \mathrm{Op}^{\mathrm{W}} f M(A)^t | \psi_N \rangle .$$

The time scales t_N involved in the theorem are logarithmic in $\hbar$, more precisely, they are of the form $(1 + \delta)\frac{1}{\gamma_0} \ln N$. So if you want to prove something like this for $\epsilon \neq 0$, you need to be able to control $U_\epsilon^{t_N}$ at such time scales. Recall that for the Schnirelman theorem, such long time control is not needed since there, you take $\hbar \to 0$ first, and only after that exploit ergodicity by taking $t \to \infty$.

In [4], we made some progress on this question. To state the result, we need a definition:

Definition 1. *A sequence $\varphi_N \in \mathcal{H}_\hbar$ is said to localize at $a_0 \in \mathbb{T}^2$ if*

$$\langle \varphi_N | \mathrm{Op}^{\mathrm{W}} f | \varphi_N \rangle \overset{N \to +\infty}{\longrightarrow} f(a_0), \qquad \forall f \in C^\infty(\mathbb{T}^2).$$

This implies there exists $r_\hbar \to 0$ so that

$$\int_{|x-a| \geq r_\hbar} |\langle \varphi_N, x | z \rangle|^2 \frac{dx}{2\pi\hbar} \overset{N \to +\infty}{\longrightarrow} 0.$$

Here $|x, z\rangle$ is the coherent state localized at $x \in \mathbb{T}^2$.

Theorem 7. *Let U_ϵ be as before. There exists $\delta_0 > 0$ with the following property. Let ψ_N be a sequence of eigenfunctions of U_ϵ that localizes at some point $a \in \mathbb{T}^2$. Then $r_\hbar \geq \hbar^{\frac{1}{2} - \delta_0}$.*

One can paraphrase this result as follows:

Eigenfunctions do it slowly (if they do it at all).

In other words, a sequence of eigenfunctions can concentrate on a point only if it does so slowly, meaning that the region in which it concentrates is "large", *i.e.* of linear size of order $\hbar^{\frac{1}{2} - \delta_0}$. The strategy of the proof is the following: we first prove an Egorov theorem for the system with control on the error term for time of order $\frac{2}{3\gamma_\epsilon} |\ln\hbar|$. We then use this result to get information on the propagation of coherent states, and finally use this to obtain the result. This result is of course far from optimal: this is largely due to the fact that we control the dynamics for insufficiently long times. Roughly speaking, and modulo some other technical problems, to show the desired result along those lines, one expects to need to control the dynamics up to times $(1 + \delta)\frac{1}{\gamma_\epsilon} |\ln\hbar|$. This is consistent with the observation that such time scales are used in the proof of Theorem 6 for the unperturbed case. An improvement on the above result has been announced in [17]: Nonnenmacher has announced he can take $\delta_0(\epsilon)$ in such a way that it converges to $1/2$ as ϵ goes to zero. To do this, he controls the relevant dynamics for times up to $(1 - \delta(\epsilon))\frac{1}{\gamma_\epsilon} |\ln\hbar|$.

A different approach to characterizing semi-classical measures has very recently been proposed in [1, 2]. In [1], the author investigated the semi-classical measures on manifolds of negative curvature and shows, modulo an as yet unproven hypothesis on the eigenfunctions, that their support can not have small topological entropy. That result implies in particular that they can not concentrate on a (finite union of) periodic orbit(s). Recently, Anantharaman has been able to get rid of the extra unproven assumption [2]. If it is confirmed that this last argument does indeed work, then I expect it to go through also for the perturbed cat maps, with the same result: no semi-classical measures concentrated on a finite set of periodic orbits. In addition, since Anantharaman obtains these results with a control on the quantum evolution up to times that are only of the form $\delta \frac{1}{\gamma_c} |\ln \hbar|$, this is very good news for quantum chaos in general: it would mean very strong results on the eigenfunction behaviour can be obtained with rather minimal control on the long time quantum evolution, contrary to the general intuition confirmed by the work described in this note.

Although I have tried to explain that Theorem 2 is particular to quantized CAT maps and that, in my opinion, strong scars on isolated hyperbolic periodic orbits are not likely to show up in other systems, for those wanting to prove unique quantum ergodicity in exponentially unstable systems, there is still a lesson to be learned from Theorem 2: any general strategy based on semi-classical analysis and ergodic properties of the underlying dynamical system alone will work for CAT maps as well. But Theorem 2 shows CAT maps are not uniquely ergodic.

Acknowledgments

It is a pleasure to thank S. Nonnenmacher and Z. Rudnick for helpful comments and remarks.

References

1. N. Anantharaman, *The eigenfunctions of the Laplacian do not concentrate on sets of small topological entropy*, preprint june 2004.
2. N. Anantharaman, private communication, march 2005.
3. M. Blank, G. Keller, C. Liverani, *Ruelle-Perron-Frobenius spectrum for Anosov maps*, Nonlinearity **15**, no. 6, 1905–1973 (2002).
4. J. M. Bouclet, S. De Bièvre, *Long time propagation and control on scarring for perturbed quantized hyperbolic toral automorphisms*, mp_arc 04-305, Annales H. Poincaré, to appear.
5. A. Bouzouina, S. De Bièvre, *Equipartition of the eigenfunctions of quantized ergodic maps on the torus*, Comm. Math. Phys. **178**, 83–105 (1996).
6. A. Bouzouina, S. De Bièvre, *Equidistribution des valeurs propres et ergodicité semi-classique de symplectomorphismes du tore quantifiés*, C. R. Acad. Sci. Paris, Série I **t.326**, 1021–1024 (1998).
7. P. Corvaja, Z. Rudnick, U. Zannier, *A lower bound for periods of matrices*, Comm. Math. Phys. **252**, 1-3, 535–541 (2004).
8. F. Bonecchi, S. De Bièvre, *Exponential mixing and $|\ln \hbar|$ time scales in quantized hyperbolic maps on the torus*, Comm. Math. Phys. **211**, 659–686 (2000).
9. F. Bonecchi, S. De Bièvre, *Controlling strong scarring for quantized ergodic toral automorphisms*, Duke Math. J, Vol. **117**, No. 3, 571–587 (2003).
10. F. Bonecchi, S. De Bièvre, *Controlling strong scarring for quantized ergodic toral automorphisms*, Section 5, mp_arc 02–81 (2002).
11. S. De Bièvre, *Quantum chaos: a brief first visit*, mp_ arc 01-207, Contemporary Mathematics **289**, 161–218 (2001).

12. S. De Bièvre, M. Degli Esposti, *Egorov Theorems and equipartition of eigen-functions for quantized sawtooth and Baker maps on the torus*, mp_arc 96-210, Ann. Inst. H. Poincaré **69**, 1, 1–30 (1998).

13. M. Degli Esposti, S. Graffi, S. Isola, *Stochastic properties of the quantum Arnold cat in the classical limit*, Commun. Math. Phys. **167**, 471–509 (1995).

14. M. Degli Esposti, S. Nonnenmacher, B. Winn, *Quantum variance and ergodicity for the baker's map*, preprint (2004), Commun. Math. Phys., to appear.

15. F. Faure, S. Nonnenmacher, S. De Bièvre, *Scarred eigenstates for quantum cats of minimal periods*, Commun. Math. Phys. **239**, 449–492 (2003).

16. F. Faure, S. Nonnenmacher, *On the maximal scarring for quantum cat map eigenstates*, Commun. Math. Phys. **245**, 201–214 (2004).

17. F. Faure, S. Nonnenmacher, contribution at the Workshop on Random Matrix theory and Arithmetic Aspects of Quantum Chaos, Newton Institute, Cambridge, june 2004.

18. J.H. Hannay, M.V. Berry, *Quantization of linear maps-Fresnel diffraction by a periodic grating*, Physica D **1**, 267–290 (1980).

19. E.J. Heller, *Bound-state eigenfunctions of classically chaotic hamiltonian systems: scars of periodic orbits*, Phys. Rev. Lett. **53**, 16, 1515–1518 (1984).

20. J.P. Keating, *Asymptotic properties of the periodic orbits of the cat maps*, Nonlinearity 4, 277–307 (1990).

21. J.P. Keating, F. Mezzadri, *Pseudo-symmetries of Anosov maps and Spectral statistics*, Nonlinearity **13**, 747–775 (2000).

22. P. Kurlberg, Z. Rudnick, *Hecke theory and equidistribution for the quantization of linear maps of the torus*, Duke Math. J. **103**, 1, 47–77 (2000).

23. P. Kurlberg, Z. Rudnick, *On quantum ergodicity for linear maps of the torus*, Comm. Math. Phys. **222**, 1, 201–227 (2001).

24. E. Lindenstrauss, *Invariant measures and arithmetic quantum unique ergodicity*, Annals of Math., to appear.

25. J. Marklof, S. O'Keefe, *Weyl's law and quantum ergodicity for maps with divided phase space*, mp_arc 04-122, Nonlinearity 18, 277–304 (2005).

26. S. Zelditch, *Quantum ergodicity and mixing of eigenfunctions*, math-ph/0503026, preprint (2005), Elsevier Encyclopedia of Mathematical Physics, to appear.

Level Repulsion and Spectral Type for One-Dimensional Adiabatic Quasi-Periodic Schrödinger Operators

Alexander Fedotov[1] and Frédéric Klopp[2]

[1] Department of Mathematical Physics, St Petersburg State University, 1, Ulianovskaja, 198904 St Petersburg-Petrodvorets, Russia
fedotov@svs.ru
[2] LAGA, Institut Galilée, U.M.R 7539 C.N.R.S, Université Paris-Nord, Avenue J.-B. Clément, F-93430 Villetaneuse, France
klopp@math.univ-paris13.fr

1 A Heuristic Description

This report is devoted to the study of the spectral properties of the family of one-dimensional quasi-periodic Schrödinger operators acting on $L^2(\mathbb{R})$ defined by

$$H_{z,\varepsilon}\psi = -\frac{d^2}{dx^2}\psi(x) + (V(x-z) + \alpha\cos(\varepsilon x))\psi(x)\,,\tag{1}$$

where

(H1) $V : \mathbb{R} \to \mathbb{R}$ is a non constant, locally square integrable, 1-periodic function;
(H2) ε is a small positive number chosen such that $2\pi/\varepsilon$ be irrational;
(H3) z is a real parameter indexing the operators;
(H4) α is a strictly positive parameter.

Quasi-periodic operators of the type (1) (and their discrete analogs) are standard models used to described the motion of an electron in a deterministic disordered system (see [3, 18, 20]).

It is well known that the spectral type of these operators depends on the energies at which one looks at them ([7]); Anderson transitions do occur in some regions ([10]) whereas other regions display "essentially" purely absolutely continuous ([9]) or purely singular spectrum ([8]). In the present note, we review an energy region where the spectrum of the operator displays energy level repulsion and show how this repulsion, more precisely its strength relates to the spectral nature of the operator. The material is taken from [5,6].

To describe the energy region where we work, consider the spectrum of the periodic Schrödinger operator (on $L^2(\mathbb{R})$)

$$H_0 = -\frac{d^2}{dx^2} + V(x)\tag{2}$$

A. Fedotov and F. Klopp: *Level Repulsion and Spectral Type for One-Dimensional Adiabatic Quasi-Periodic Schrödinger Operators*, Lect. Notes Phys. **690**, 383–402 (2006)
www.springerlink.com

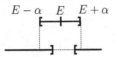

$$E - \alpha \quad E \quad E + \alpha$$

Fig. 1. Bands in interaction

We assume that two of its spectral bands are interacting through the perturbation $\alpha \cos$ i.e., that the relative position of the *spectral window* $\mathcal{F}(E) := [E - \alpha, E + \alpha]$ and the spectrum of the unperturbed operator H_0 is that shown in Fig. 1. In such an energy region, the spectrum is localized near two sequences of quantized energy values, say $(E_0^{(l)})_l$ and $(E_\pi^{(l')})_l'$ (see Theorem 1); each of these sequences is "generated" by one of the ends of the neighboring spectral bands of H_0. In [6], we study neighborhoods of such quantized energy values that are not resonant i.e. to neighborhoods of points $E_\mu^{(l)}$ that are not "too" close to the points $(E_\nu^{(l')})_l'$ for $\{\mu, \nu\} = \{0, \pi\}$. The distance between the two sequences influences the nature and location of the spectrum: a weak level repulsion arises due to "weakly resonant tunneling".

Similarly to what happens in the standard "double well" case (see [13, 19]), the resonant tunneling begins to play an important role when the two energies, each generated by one of the quantization conditions, are sufficiently close to each other.

In the resonant case, we find a strong relationship between the level repulsion and the nature of the spectrum. Recall that the latter is determined by the nature of the genralized eignefunctions i.e. the solutions to $(H_{z,\varepsilon} - E)\psi = 0$ (see [12]): localized or delocalized. Thus, it is very natural that the two characteristics are related: the slower the decay of the generalized eigenfunctions, the larger the overlap between generalized eigenfunction corresponding to close energy levels, hence, the larger the tunneling between these levels and thus the repulsion between them.

Let us now briefly describe the various situations we encounter. Let J be an interval of energies such that, for all $E \in J$, the spectral window $\mathcal{F}(E)$ covers the edges of two neighboring spectral bands of H_0 and the gap located between them (see Fig. 1 and assumption (TIBM)). Under this assumption, consider the *real* and *complex iso-energy curves* associated to (1). Denoted respectively by $\Gamma_\mathbb{R}$ and Γ, they are defined by

$$\Gamma_\mathbb{R} := \{(\kappa, \zeta) \in \mathbb{R}^2, \ \mathbf{E}(\kappa) + \alpha \cdot \cos(\zeta) = E\}, \tag{3}$$

$$\Gamma := \{(\kappa, \zeta) \in \mathbb{C}^2, \ \mathbf{E}(\kappa) + \alpha \cdot \cos(\zeta) = E\}. \tag{4}$$

where $\mathbf{E}(\kappa)$ be the dispersion relation associated to H_0. These curves are roughly depicted in Fig. 2. They are periodic both in the ζ and κ variables. Consider one of the periodicity cells of $\Gamma_\mathbb{R}$. It contains two tori. They are denoted by γ_0 and γ_π and shown in full lines. Integrating $1/2$ times the fundamental 1-form on Γ along γ_0 and γ_π, one defines the phases $\check{\Phi}_0$ and

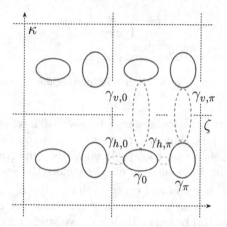

Fig. 2. The adiabatic phase space

$\check{\Phi}_\pi$ (see Theorem 1). Each of these phases defines a sequence of "quantized energies" in J, say, for $(l, m) \in \mathbb{N}^2$,

$$\frac{1}{\varepsilon}\check{\Phi}_0(E_0^{(l)}) = \frac{\pi}{2} + l\pi , \qquad (5)$$

$$\frac{1}{\varepsilon}\check{\Phi}_\pi(E_\pi^{(m)}) = \frac{\pi}{2} + m\pi . \qquad (6)$$

Each of the dashed lines in Fig. 2 represents a loop on Γ that connects certain connected components of $\Gamma_\mathbb{R}$; one can distinguish between the "horizontal" loops and the "vertical" loops. There are two special horizontal loops denoted by $\gamma_{h,0}$ and $\gamma_{h,\pi}$; the loop $\gamma_{h,0}$ (resp. $\gamma_{h,\pi}$) connects γ_0 to $\gamma_\pi - (2\pi, 0)$ (resp. γ_0 to γ_π). In the same way, there are two special vertical loops denoted by $\gamma_{v,0}$ and $\gamma_{v,\pi}$; the loop $\gamma_{v,0}$ (resp. $\gamma_{v,\pi}$) connects γ_0 to $\gamma_0 + (0, 2\pi)$ (resp. γ_π to $\gamma_\pi + (0, 2\pi)$). To each of these complex loops, one associates an action obtained by integrating $-i/2$ times the fundamental 1-form on Γ along the loop. For $a \in \{0, \pi\}$ and $b \in \{v, h\}$, we denote by $S_{a,b}$ the action associated to $\gamma_{a,b}$. For $E \in \mathbb{R}$, all these actions are real. One orients the loops so that they all be positive. Finally, we define tunneling coefficients as

$$t_{a,b} = e^{-S_{a,b}/\varepsilon}, \quad a \in \{0, \pi\}, \ b \in \{v, h\} .$$

In Theorem 1, we prove that, for $a \in \{0, \pi\}$, near each $E_a^{(l)}$, there is an exponentially small interval $I_a^{(l)}$ such that the spectrum of $H_{z,\varepsilon}$ in J is contained in the union of these intervals. The precise description of the spectrum in the interval $I_a^{(l)}$ depends on whether it intersects another such interval or not. Note that an interval of type $I_0^{(l)}$ can only intersect intervals of type $I_\pi^{(l)}$ and vice versa.

In the present report, we consider two intervals $I_0^{(l)}$ and $I_0^{(l')}$ that do intersect and describe the spectrum in the union of these two intervals. It is useful to keep in mind that this union is exponentially small (when ε goes to 0).

There are two main parameters controlling the spectral type:

$$\tau = 2\sqrt{\frac{t_{v,0}\, t_{v,\pi}}{t_{h,0}\, t_{h,\pi}}} \qquad \text{and} \qquad \rho = 2\sqrt{\frac{\max(t_{v,0}, t_{v,\pi})^2}{t_{h,0}\, t_{h,\pi}}}$$

As we are working inside an exponentially small interval, the point inside this interval at which we compute the tunneling coefficients does not really matter: over this interval, the variation of any of the actions is exponentially small.

We essentially distinguish three regimes:

- $\tau \gg 1$,
- $\rho \gg 1$ and $\tau \ll 1$,
- $\rho \ll 1$.

The symbol $\gg 1$ (resp. $\ll 1$) mean that the quantity is exponentially small (resp. large) in $\frac{1}{\varepsilon}$ as ε goes to 0, the exponential rate being arbitrary.

In each of the three cases, we consider two energies, say E_0 and E_π, satisfying respectively (5) and (6), and describe the evolution of the spectrum as E_0 and E_π become closer to each other. As the quantization conditions (5) and (6) show, this can be achieved by reducing ε somewhat. As noted above, when moving E_0 and E_π closer together, we stay in the same regime for τ and ρ.

In the case $\tau \gg 1$, when E_0 and E_π are still "far" away from each other, one sees two intervals containing spectrum, one located near each energy; they contain the same amount of spectrum i.e. the measure with respect to the density of states of each interval is $\varepsilon/2\pi$; and the Lyapunov exponent is positive on both intervals (see Fig. 5(a)). When E_0 and E_π approach each other, the picture does not change except for the fact that the intervals intersect, so only a single interval is seen (see Fig. 5(b)); its density of states measure is ε/π and the Lyapunov exponent is positive on this interval. There is no gap separating the intervals of spectrum generated by the two quantization conditions. This can be interpreted as a consequence of the positivity of the Lyapunov exponent : the states are well localized so the overlapping is weak and there no level repulsion. Nevertheless, the resonance has one effect: when the two intervals intersect, it gives rise to a sharp drop of the Lyapunov exponent (which stays positive) in the middle of the interval containing spectrum. Over a distance exponentially small in ε, the Lyapunov exponent drops by an amount of order one.

In the other extreme, in the case $\rho \ll 1$, the starting geometry of the spectrum is essentially the same as in the previous case, namely, two well separated intervals containing each an $\varepsilon/2\pi$ "part" of spectrum. The main difference is that each of these intervals now essentially only contains absolutely continuous spectrum. As the energies E_0 and E_π approach each other, so do the intervals until they roughly reach an interspacing of size $\sqrt{t_h}$; during this process, the size of the intervals which, at the start, was roughly of order $t_{v,0} + t_h$ and $t_{v,\pi} + t_h$ grew to reach the order of $\sqrt{t_h}$ (this number is much larger than any of the other two as $\rho \ll 1$). When E_0 and E_π move closer to each other, the intervals containing spectrum stay "frozen" at a distance of size $\sqrt{t_h}$ from each other, and their sizes do not vary noticeably either (see Fig. 6). They start moving and changing size again when E_0 and E_π again become separated by an interspacing of size at least $\sqrt{t_h}$. So, in this case, we see a very strong repulsion preventing the intervals of spectra from intersecting each other. The interspacing is quite similar to that observed in the case of the standard double well problem (see [13, 19]).

In the last case, when $\rho \gg 1$ and $\tau \ll 1$, we see an intermediate behavior. For the sake of simplicity, let us assume that $t_{v,\pi} \gg t_{v,0}$. Starting from the situation when E_0 and E_π are far apart, we see two intervals, one around each point E_0 and E_π and this as long as $|E_0 - E_\pi| \gg t_{v,\pi}$ (see Fig. 7(a)). As in the first case, the Lyapunov exponent varies by an amount of order one over each of the exponentially small intervals. The difference is that it need not stay positive: at the edges of the two intervals that are facing each other, it becomes small and even can vanish. These edges seem more prone to interaction. When now one moves E_0 towards E_π, the lacuna separating the two intervals stays open and starts moving with E_0; it becomes roughly centered at E_0, stays of fixed size (of order $t_h/t_{v,\pi}$) and moves along with E_0 as E_0 crosses E_π and up to a distance roughly $t_{v,\pi}$ on the other side of E_π (see Fig. 7(b)). Then, when E_0 moves still further away from E_π, it becomes again the center of some interval containing spectrum that starts moving away from the band centered at E_π. We see that, in this case as in the case of strong repulsion, there always are two intervals separated by a gap; both intervals contain a $\varepsilon/2\pi$ "part" of spectrum. But, now, the two intervals can become exponentially larger than the gap (in the case of strong interaction, the length of the gap was at least of the same order of magnitude as the lengths of the bands). Moreover, on both intervals, the Lyapunov exponent is positive near the outer edges i.e. the edges that are not facing each other; it can become small or even vanish on the inner edges. So, there may be some Anderson transitions within the intervals. We see here the effects of a weaker form of resonant tunneling and a weaker repulsion.

2 Mathematical Results

We now state our assumptions and results in a precise way.

2.1 The Periodic Operator

This section is devoted to the description of elements of the spectral theory of one-dimensional periodic Schrödinger operator H_0 that we need to present our results. For proofs and more details, we refer to [4, 11].

The Spectrum of H_0

The spectrum of the operator H_0 defined in (2) is a union of countably many intervals of the real axis, say $[E_{2n+1}, E_{2n+2}]$ for $n \in \mathbb{N}$, such that

$$E_1 < E_2 \leq E_3 < E_4 \ldots E_{2n} \leq E_{2n+1} < E_{2n+2} \leq \ldots ,$$
$$E_n \to +\infty, \quad n \to +\infty.$$

This spectrum is purely absolutely continuous. The points $(E_j)_j$ are the eigenvalues of the self-adjoint operator obtained by considering the differential polynomial (2) acting in $L^2([0, 2])$ with periodic boundary conditions (see [4]). For $n \in \mathbb{N}$, the intervals $[E_{2n+1}, E_{2n+2}]$ are the *spectral bands*, and the intervals (E_{2n}, E_{2n+1}), the *spectral gaps*. When $E_{2n} < E_{2n+1}$, one says that the n-th gap is *open*; when $[E_{2n-1}, E_{2n}]$ is separated from the rest of the spectrum by open gaps, the n-th band is said to be *isolated*. Generically all the gap are open.

From now on, to simplify the exposition, we suppose that

(O) all the gaps of the spectrum of H_0 are open.

The Bloch Quasi-Momentum

Let $x \mapsto \psi(x, E)$ be a non trivial solution to the periodic Schrödinger equation $H_0\psi = E\psi$ such that $\psi(x + 1, E) = \mu\,\psi(x, E)$, $\forall x \in \mathbb{R}$, for some $\mu \in \mathbb{C}^*$ independent of x. Such a solution is called a *Bloch solution* to the equation, and μ is the *Floquet multiplier* associated to ψ. One may write $\mu = \exp(ik)$ where, k is the *Bloch quasi-momentum* of the Bloch solution ψ.

It appears that the mapping $E \mapsto k(E)$ is analytic and multi-valued; its branch points are the points $\{E_n;\ n \in \mathbb{N}\}$. They are all of "square root" type.

The dispersion relation $k \mapsto \mathbf{E}(k)$ is the inverse of the Bloch quasi-momentum.

2.2 A "Geometric" Assumption on the Energy Region Under Study

Let us now describe the energy region where our study is valid.

Recall that the spectral window $\mathcal{F}(E)$ is the range of the mapping $\zeta \in \mathbb{R} \mapsto E - \alpha \cos(\zeta)$.

In the sequel, J always denotes a compact interval such that, for some $n \in \mathbb{N}^*$ and for all $E \in J$, one has

(TIBM) $[E_{2n}, E_{2n+1}] \subset \dot{\mathcal{F}}(E)$ and $\mathcal{F}(E) \subset]E_{2n-1}, E_{2n+2}[$.

where $\dot{\mathcal{F}}(E)$ is the interior of $\mathcal{F}(E)$ (see Fig. 1).

Remark 1. As all the spectral gaps of H_0 are assumed to be open, as their length tends to 0 at infinity, and, as the length of the spectral bands goes to infinity at infinity, it is clear that, for any non vanishing α, assumption (TIBM) is satisfied in any gap at a sufficiently high energy; it suffices that this gap be of length smaller than 2α.

2.3 The Definitions of the Phase Integrals and the Tunelling Coefficients

The Complex Momentum and its Branch Points

The phase integrals and the tunelling coefficients are expressed in terms of integrals of the *complex momentum*. Fix E in J. The complex momentum $\zeta \mapsto \kappa(\zeta)$ is defined by

$$\kappa(\zeta) = k(E - \alpha \cos(\zeta)) . \tag{7}$$

As k, κ is analytic and multi-valued. The set Γ defined in (4) is the graph of the function κ. As the branch points of k are the points $(E_i)_{i \in \mathbb{N}}$, the branch points of κ satisfy

$$E - \alpha \cos(\zeta) = E_j, \ j \in \mathbb{N}^* . \tag{8}$$

As E is real, the set of these points is symmetric with respect to the real axis, to the imaginary axis and it is 2π-periodic in ζ. All the branch points of κ lie on $\arccos(\mathbb{R})$ which consists of the real axis and all the translates of the imaginary axis by a multiple of π. As the branch points of the Bloch quasi-momentum, the branch points of κ are of "square root" type.

Due to the symmetries, it suffices to describe the branch points in the half-strip $\{\zeta; \ \text{Im}\zeta \geq 0, \ 0 \leq \text{Re}\zeta \leq \pi\}$. These branch points are described in detail in section 7.1.1 of [6]. In Fig. 3, we show some of them. The points ζ_j being defined by (8), one has $0 < \zeta_{2n} < \zeta_{2n+1} < \pi, \ 0 < \text{Im}\zeta_{2n+2} < \text{Im}\zeta_{2n+3} < \cdots$, $0 < \text{Im}\zeta_{2n-1} < \cdots < \text{Im}\zeta_1$.

The Contours

To define the phases and the tunneling coefficients, we introduce various integration contours in the complex ζ-plane.

These loops are shown in Fig. 3 and 4. The loops $\tilde{\gamma}_0$, $\tilde{\gamma}_\pi$, $\tilde{\gamma}_{h,0}$, $\tilde{\gamma}_{h,\pi}$, $\tilde{\gamma}_{v,0}$ and $\tilde{\gamma}_{v,\pi}$ are simple loops, respectively, going once around the intervals $[-\zeta_{2n}, \zeta_{2n}]$, $[\zeta_{2n+1}, 2\pi - \zeta_{2n+1}]$, $[-\zeta_{2n+1}, -\zeta_{2n}]$, $[\zeta_{2n}, \zeta_{2n+1}]$, $[\zeta_{2n-1}, \overline{\zeta_{2n-1}}]$ and $[\zeta_{2n+2}, \overline{\zeta_{2n+2}}]$.

On each of the above loops, one can fix a continuous branch of the complex momentum.

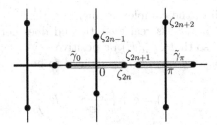

Fig. 3. The loops for the phase integrals

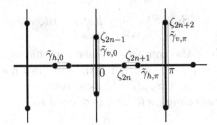

Fig. 4. The loops for the action integrals

Consider Γ, the complex isoenergy curve defined by (4). Define the projection $\Pi : (\zeta, \kappa) \in \Gamma \mapsto \zeta \in \mathbb{C}$. The fact that, on each of the loops $\tilde{\gamma}_0$, $\tilde{\gamma}_\pi$, $\tilde{\gamma}_{h,0}$, $\tilde{\gamma}_{h,\pi}$, $\tilde{\gamma}_{v,0}$ and $\tilde{\gamma}_{v,\pi}$, one can fix a continuous branch of the complex momentum implies that each of these loops is the projection on the complex plane of some loop in Γ (see [6]). The curves γ_0, γ_π, $\gamma_{h,0}$, $\gamma_{h,\pi}$, $\gamma_{v,0}$ and $\gamma_{v,\pi}$ represented in Fig. 2 are mapped onto the curves $\tilde{\gamma}_0$, $\tilde{\gamma}_\pi$, $\tilde{\gamma}_{h,0}$, $\tilde{\gamma}_{h,\pi}$, $\tilde{\gamma}_{v,0}$ and $\tilde{\gamma}_{v,\pi}$ respectively by the projector $\Pi : (\kappa, \zeta) \in \Gamma \to \zeta \in \Gamma$.

The Phase Integrals, the Action Integrals and the Tunneling Coefficients

Let $\nu \in \{0, \pi\}$. To the loop γ_ν, we associate the *phase integral* Φ_ν defined as

$$\Phi_\nu(E) = \frac{1}{2} \oint_{\tilde{\gamma}_\nu} \kappa \, d\zeta \,, \qquad (9)$$

where κ is a branch of the complex momentum that is continuous on $\tilde{\gamma}_\nu$. The function $E \mapsto \Phi_\nu(E)$ is real analytic and does not vanish on J. The loop $\tilde{\gamma}_\nu$ is oriented so that $\Phi_\nu(E)$ be positive. One shows that, for all $E \in J$,

$$\Phi_0'(E) < 0 \quad \text{and} \quad \Phi_\pi'(E) > 0 \,. \qquad (10)$$

To the loop $\gamma_{v,\nu}$, we associate the *vertical action integral* $S_{v,\nu}$ defined as

$$S_{v,\nu}(E) = -\frac{i}{2} \oint_{\tilde{\gamma}_{v,\nu}} \kappa d\zeta \,, \qquad (11)$$

where κ is a branch of the complex momentum that is continuous on $\tilde{\gamma}_{v,\nu}$. The function $E \mapsto S_{v,\nu}(E)$ is real analytic and does not vanish on J. The loop $\tilde{\gamma}_{v,\nu}$ is oriented so that $S_{v,\nu}(E)$ be positive.

The *vertical tunneling coefficient* is defined to be

$$t_{v,\nu}(E) = \exp\left(-\frac{1}{\varepsilon}S_{v,\nu}(E)\right) . \tag{12}$$

The index ν being chosen as above, we define *horizontal action integral* $S_{h,\nu}$ by

$$S_{h,\nu}(E) = -\frac{i}{2}\oint_{\tilde{\gamma}_{h,\nu}} \kappa(\zeta)\,d\zeta , \tag{13}$$

where κ is a branch of the complex momentum that is continuous on $\tilde{\gamma}_{h,\nu}$. The function $E \mapsto S_{h,\nu}(E)$ is real analytic and does not vanish on J. The loop $\tilde{\gamma}_{h,\nu}$ is oriented so that $S_{h,\nu}(E)$ be positive.

The *horizontal tunneling coefficient* is defined as

$$t_{h,\nu}(E) = \exp\left(-\frac{1}{\varepsilon}S_{h,\nu}(E)\right) . \tag{14}$$

As the cosine is even, one has

$$S_{h,0}(E) = S_{h,\pi}(E) \quad \text{and} \quad t_{h,0}(E) = t_{h,\pi}(E) . \tag{15}$$

Finally, one defines

$$S_h(E) = S_{h,0}(E) + S_{h,\pi}(E) \quad \text{and} \quad t_h(E) = t_{h,0}(E) \cdot t_{h,\pi}(E) . \tag{16}$$

In (9), (11), and (13), only the sign of the integral depends on the choice of the branch of κ; this sign was fixed by orienting the integration contour.

2.4 Ergodic Family

Before discussing the spectral properties of $H_{z,\varepsilon}$, we recall some general results from the spectral theory of ergodic operators.

As $2\pi/\varepsilon$ is supposed to be irrational, the operators defined by (1) form an ergodic family (see [17]).

The ergodicity has the following consequences:

1. the spectrum of $H_{z,\varepsilon}$ is almost surely independent of z ([18]); in fact, in the case of the quasi-periodic operators, it is independent of z ([1]);
2. the absolutely continuous spectrum and the singular spectrum are almost surely independent of z ([18]); in fact, in the case of the quasi-periodic operators, they are independent of z ([15]);
3. for almost all z, the discret spectrum is empty ([18]);

4. the Lyapunov exponent exists for almost all z and is independent of z ([18]); it is defined in the following way: let $x \mapsto \psi(x)$ be the solution to the Cauchy problem

$$H_{z,\varepsilon}\psi = E\psi, \quad \psi|_{x=0} = 0, \quad \psi'|_{x=0} = 1 \,,$$

the following limit (when it exists) defines the Lyapunov exponent:

$$\Theta(E) = \Theta(E,\varepsilon) := \lim_{x \to +\infty} \frac{\log\left(\sqrt{|\psi(x,E,z)|^2 + |\psi'(x,E,z)|^2}\right)}{|x|} \,.$$

5. the absolutely continuous spectrum is the essential closure of the set of E where $\Theta(E) = 0$ (the Ishii-Pastur-Kotani Theorem, see [18]);
6. the density of states exists for almost all z and is independent of z ([18]); it is defined in the following way: for $L > 0$, let $H_{z,\varepsilon;L}$ be the operator $H_{z,\varepsilon}$ restricted to the interval $[-L, L]$ with the Dirichlet boundary conditions; for $E \in \mathbb{R}$; the following limit (when it exists) defines the density of states:

$$N_\varepsilon(E) := \lim_{L \to +\infty} \frac{\#\{ \text{ eigenvalues of } H_{z,\varepsilon;L} \text{less then or equal to} E\}}{2L} \,;$$

7. the density of states is non decreasing; the spectrum of $H_{z,\varepsilon}$ is the set of points of increase of the density of states.

2.5 A Coarse Description of the Location of the Spectrum in J

Henceforth, we assume that the assumptions (H) and (O) are satisfied and that J is a compact interval satisfying (TIBM). We assume that

(T)
$$\max_{E \in J} \max(S_h(E), S_{v,0}(E), S_{v,\pi}(E)) < 2\pi \cdot \min_{E \in J} \min(\operatorname{Im}\zeta_{2n-2}(E), \operatorname{Im}\zeta_{2n+3}(E)) \,.$$

Note that (T) is verified if the spectrum of H_0 has two successive bands that are sufficiently close to each other and sufficiently far away from the remainder of the spectrum (this can be checked numerically on simple examples, see Sect. 2.9).
Define

$$\delta_0 := \frac{1}{2} \inf_{E \in J} \min(S_h(E), S_{v,0}(E), S_{v,\pi}(E)) > 0 \,. \tag{17}$$

To locate the spectrum roughly, we prove

Theorem 1 ([6]). *Fix $E_* \in J$. For ε sufficiently small, there exists $V_* \subset \mathbb{C}$, a neighborhood of E_*, and two real analytic functions $E \mapsto \check{\Phi}_0(E,\varepsilon)$ and $E \mapsto \check{\Phi}_\pi(E,\varepsilon)$, defined on V_* satisfying the uniform asymptotics*

$$\check{\Phi}_0(E,\varepsilon) = \Phi_0(E) + o(\varepsilon), \quad \check{\Phi}_\pi(E,\varepsilon) = \Phi_\pi(E) + o(\varepsilon) \quad when \ \varepsilon \to 0 \ , \quad (18)$$

such that, if one defines two finite sequences of points in $J \cap V_*$, say $(E_0^{(l)})_l :=$ $(E_0^{(l)}(\varepsilon))_l$ and $(E_\pi^{(l')})_{l'} := (E_\pi^{(l')}(\varepsilon))_{l'}$, by

$$\frac{1}{\varepsilon}\check{\Phi}_0(E_0^{(l)},\varepsilon) = \frac{\pi}{2} + \pi l \quad and \quad \frac{1}{\varepsilon}\check{\Phi}_\pi(E_\pi^{(l')},\varepsilon) = \frac{\pi}{2} + \pi l', \quad (l,l') \in \mathbb{N}^2 \ , \quad (19)$$

then, for all real z, the spectrum of $H_{z,\varepsilon}$ in $J \cap V_*$ is contained in the union of the intervals

$$I_0^{(l)} := E_0^{(l)} + [-e^{-\delta_0/\varepsilon}, e^{-\delta_0/\varepsilon}] \quad and \quad I_\pi^{(l')} := E_\pi^{(l')} + [-e^{-\delta_0/\varepsilon}, e^{-\delta_0/\varepsilon}]$$
$$(20)$$

that is

$$\sigma(H_{z,\varepsilon}) \cap J \cap V_* \subset \left(\bigcup_l I_0^{(l)}\right) \cup \left(\bigcup_{l'} I_\pi^{(l')}\right).$$

In the sequel, to alleviate the notations, we omit the reference to ε in the functions $\check{\Phi}_0$ and $\check{\Phi}_\pi$.

By (10) and (18), there exists $C > 0$ such that, for ε sufficiently small, the points defined in (19) satisfy

$$\frac{1}{C}\varepsilon \leq E_0^{(l)} - E_0^{(l-1)} \leq C\varepsilon \ , \quad (21)$$

$$\frac{1}{C}\varepsilon \leq E_\pi^{(l)} - E_\pi^{(l-1)} \leq C\varepsilon \ . \quad (22)$$

Moreover, for $\nu \in \{0,\pi\}$, in the interval $J \cap V_*$, the number of points $E_\nu^{(l)}$ is of order $1/\varepsilon$.

In the sequel, we refer to the points $E_0^{(l)}$ (resp. $E_\pi^{(l)}$), and, by extension, to the intervals $I_0^{(l)}$ (resp. $I_\pi^{(l)}$) attached to them, as of type 0 (resp. type π). By (21) and (22), the intervals of type 0 (resp. π) are two by two disjoints; any interval of type 0 (resp. π) intersects at most a single interval of type π (resp. 0).

2.6 A Precise Description of the Spectrum

As pointed out in the introduction, the present paper deals with the resonant case that is we consider two energies, say $E_0^{(l)}$ et $E_\pi^{(l')}$, that satisfy

$$|E_\pi^{(l')} - E_0^{(l)}| \leq 2e^{-\frac{\delta_0}{\varepsilon}} \ . \quad (23)$$

This means that the intervals $I_0^{(l)}$ and $I_\pi^{(l')}$ intersect each other. Moreover, by (21) and (22), these intervals stay at a distance at least $C^{-1}\varepsilon$ of all the other intervals of the sequences defined in Theorem 1. We now describe the spectrum of $H_{z,\varepsilon}$ in the union $I_0^{(l)} \cup I_\pi^{(l')}$.

To simplify the exposition, we set

$$E_0 := E_0^{(l)}, \qquad E_\pi := E_\pi^{(l')}, \qquad I_0 := I_0^{(l)}, \qquad \text{and} \qquad I_\pi := I_\pi^{(l')} . \qquad (24)$$

In the resonant case, the primary parameter controlling the location and the nature of the spectrum is

$$\tau = 2\sqrt{\frac{t_{v,0}(\bar{E})\, t_{v,\pi}(\bar{E})}{t_h(\bar{E})}} \quad \text{where} \quad \bar{E} = \frac{E_\pi + E_0}{2} . \qquad (25)$$

As tunneling coefficients are exponentially small, one typically has either $\tau \gg 1$ or $\tau \ll 1$. We will give a detailed analysis of these cases. More precisely, we fix $\delta > 0$ arbitrary and assume that either

$$\forall E \in V_* \cap J, \quad S_h(E) - S_{v,0}(E) - S_{v,\pi}(E) \geq \delta , \qquad (26)$$

or

$$\forall E \in V_* \cap J, \quad S_h(E) - S_{v,0}(E) - S_{v,\pi}(E) \leq -\delta . \qquad (27)$$

The case $\tau \asymp 1$ is more complicated and satisfied by less energies. We discuss it briefly later.

To describe our results, it is convenient to introduce the following "local variables"

$$\xi_\nu(E) = \frac{\check{\Phi}'_\nu(\bar{E})}{\varepsilon} \cdot \frac{E - E_\nu}{t_{v,\nu}(\bar{E})} \quad \text{where} \quad \nu \in \{0, \pi\} . \qquad (28)$$

When τ is Large

Let us now assume $\tau \gg 1$. The location of the spectrum in $I_0 \cup I_\pi$ is described by

Theorem 2 ([5]). *Assume we are in the case of Theorem 1. Assume (26) is satisfied. Then, there exist $\varepsilon_0 > 0$ and a non-negative function $\varepsilon \mapsto f(\varepsilon)$ tending to zero as $\varepsilon \to 0$ such that, for $\varepsilon \in (0, \varepsilon_0]$, the spectrum of $H_{z,\varepsilon}$ in $I_0 \cup I_\pi$ is located in two intervals $\check{I}_0$ and $\check{I}_\pi$ defined by*

$$\check{I}_0 = \{E \in I_0 : \ |\xi_0(E)| \leq 1 + f(\varepsilon)\} \text{ and } \check{I}_\pi = \{E \in I_\pi : \ |\xi_\pi(E)| \leq 1 + f(\varepsilon)\}.$$

Let $dN_\varepsilon(E)$ be the density of states measure of $H_{z,\varepsilon}$; then $\displaystyle\int_{\check{I}_0 \cup \check{I}_\pi} dN_\varepsilon(E) = \frac{\varepsilon}{\pi}$.

Moreover, if $\check{I}_0 \cap \check{I}_\pi = \emptyset$, then

$$\int_{\check{I}_0} dN_\varepsilon(E) = \int_{\check{I}_\pi} dN_\varepsilon(E) = \frac{\varepsilon}{2\pi} \qquad (29)$$

The Lyapunov exponent on $\check{I}_0 \cup \check{I}_\pi$ satisfies

$$\Theta(E, \varepsilon) = \frac{\varepsilon}{\pi} \log \left(\tau \sqrt{1 + |\xi_0(E)| + |\xi_\pi(E)|} \right) + o(1) , \qquad (30)$$

where $o(1) \to 0$ when $\varepsilon \to 0$ uniformly in E, E_0 and E_π.

By (29), if $\check{I}_0$ and $\check{I}_\pi$ are disjoint, they both contain spectrum of $H_{z,\varepsilon}$; if not, one only knows that their union contains spectrum.

Let us analyze the results of Theorem 2.

The location of the spectrum. By (28), the intervals $\check{I}_0$ and $\check{I}_\pi$ defined in Theorem 2 are respectively "centered" at the points E_0 and E_π. Their lengths are given by

$$|\check{I}_0| \underset{\varepsilon \to 0}{\sim} \frac{2\varepsilon}{|\check{\Phi}_0'(E_0)|} \cdot t_{v,0}(E_0) \quad \text{and} \quad |\check{I}_\pi| \underset{\varepsilon \to 0}{\sim} \frac{2\varepsilon}{|\check{\Phi}_\pi'(E_\pi)|} \cdot t_{v,\pi}(E_\pi) .$$

Depending on $|E_\pi - E_0|$, the picture of the spectrum in $I_0 \cup I_\pi$ is given by Fig. 5, case (a) and (b).

The repulsion effect observed in the non resonant case (see [6]) does not exist here: it is negligible with respect to the length of the intervals $\check{I}_0$ and $\check{I}_\pi$.

(a) $|E_\pi - E_0| \gg \max(t_{v,\pi}, t_{v,0})$ 　　　 (b) $|E_\pi - E_0| \ll \max(t_{v,\pi}, t_{v,0})$

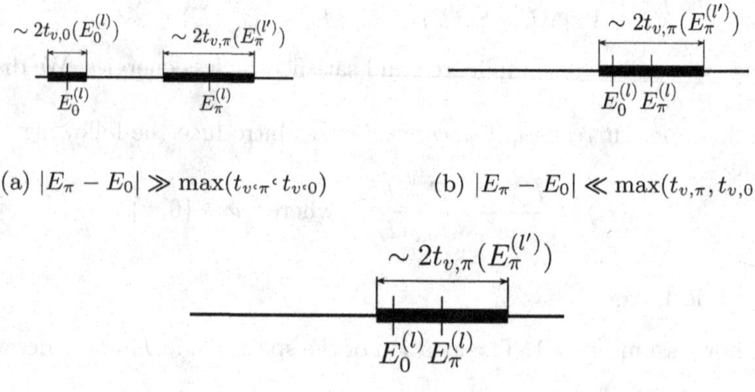

Fig. 5. The location of the spectrum for τ large

The nature of the spectrum. In the intervals $\check{I}_0$ and $\check{I}_\pi$, according to (26) and (30), the Lyapunov exponent is positive. Hence, by the Ishii-Pastur-Kotani Theorem ([17]), in both $\check{I}_0$ and $\check{I}_\pi$, the spectrum of $H_{z,\varepsilon}$ is singular.

The Lyapunov exponent $\Theta(E, \varepsilon)$ on the spectrum. The general formula (30) can be simplified in the following way:

$$\Theta(E, \varepsilon) = \frac{\varepsilon}{\pi} \log\left(\tau \sqrt{1 + |\xi_0(E)|}\right) + o(1) \quad \text{when} \quad E \in \check{I}_\pi ,$$

and

$$\Theta(E, \varepsilon) = \frac{\varepsilon}{\pi} \log\left(\tau \sqrt{1 + |\xi_\pi(E)|}\right) + o(1) \quad \text{when} \quad E \in \check{I}_0 .$$

If $|E_\pi - E_0| \gg \max(t_{v,\pi}, t_{v,0})$, then the Lyapunov exponent stays essentially constant on each of the intervals $\check{I}_0$ and $\check{I}_\pi$. On the other hand, if $|E_\pi - E_0| \ll$

$\max(t_{v,\pi}, t_{v,0})$, then, on these exponentially small intervals, the Lyapunov exponent may vary by a constant. To see this, let us take a simple example. Assume that $t_{v,0} \ll t_{v,\pi}$, more precisely, that there exists $\delta > 0$ such that

$$\forall E \in V_* \cap J, \quad S_{v,0}(E) > S_{v,\pi}(E) + \delta .$$

If E_0 and E_π coincide, then $\check{I}_0 \subset \check{I}_\pi$, and, near the center of $\check{I}_\pi$, the Lyapunov exponent assumes the value

$$\Theta(E, \varepsilon) = \frac{\varepsilon}{\pi} \log \tau + o(1) = \frac{1}{2\pi} \left(S_h(\bar{E}) - S_{v,\pi}(\bar{E}) - S_{v,0}(\bar{E}) \right) + o(1) ,$$

Near the edges of $\check{I}_\pi$, its value is given by

$$\Theta(E, \varepsilon) = \frac{\varepsilon}{\pi} \log \tau + \frac{\varepsilon}{2\pi} \log(t_{v,\pi}(\bar{E})/t_{v,0}(\bar{E})) + o(1)$$
$$= \frac{1}{2\pi} \left(S_h(\bar{E}) - 2S_{v,\pi}(\bar{E}) \right) + o(1) .$$

So the variation of the Lyapunov exponent is given by $\frac{1}{2\pi} \left(S_{v,0}(\bar{E}) - S_{v,\pi}(\bar{E}) \right)$ on an exponentially small interval. One sees a sharp drop of the Lyapunov exponent on the interval containing spectrum when going from the edges of $\check{I}_\pi$ towards E_π.

When τ is Small

We now assume that $\tau \ll 1$, i.e., that (27) is satisfied. Then, the spectral behavior depends on the value of the quantity $\Lambda_n(V)$ defined in [5]. We only recall that $\Lambda_n(V)$ depends solely on V and on the index of the gap separating the two interacting bands; moreover, it generically satisfies

$$\Lambda_n(V) > 1 . \tag{31}$$

Below, we only consider this generic case.

There are different possible "scenarii" for the spectral behavior when $\tau \ll 1$. Before describing them in detail, we start with a general description of the spectrum. We prove

Theorem 3 ([5]). *Assume we are in the case of Theorem 1. Assume that (27) and (31) are satisfied. Then, there exists $\varepsilon_0 > 0$ and a non negative function $\varepsilon \mapsto f(\varepsilon)$ tending to zero when $\varepsilon \to 0$ such that, for $\varepsilon \in]0, \varepsilon_0[$, the spectrum of $H_{z,\varepsilon}$ in $I_0^{(l)} \cup I_\pi^{(l')}$ is contained in $\Sigma(\varepsilon)$, the set of energies E satisfying*

$$\left| \tau^2 \xi_0(E) \xi_\pi(E) + 2\Lambda_n(V) \right| \le \left(2 + \tau^2 |\xi_0(E)| + \tau^2 |\xi_\pi(E)| \right) (1 + f(\varepsilon)) . \tag{32}$$

The set $\Sigma(\varepsilon)$ is the union of two disjoint compact intervals, say I_0 and I_π; both intervals are strictly contained inside the $(2e^{-\delta_0/\varepsilon})$-neighborhood of $\bar{E}$.

One checks the

Theorem 4 ([5]). *In the case of Theorem 1, if $dN_\varepsilon(E)$ denotes the density of states measure of $H_{z,\varepsilon}$, then*

$$\int_{I_0} dN_\varepsilon(E) = \int_{I_\pi} dN_\varepsilon(E) = \frac{\varepsilon}{2\pi} .$$

Hence, each of the intervals I_0 and I_π contains some spectrum of $H_{z,\varepsilon}$. This implies that, when $\tau \ll 1$ and $\Lambda_n(V) > 1$, there is a "level repulsion" or a "splitting" of resonant intervals.

As for the nature of the spectrum, one shows the following results. The behavior of the Lyapunov exponent is given by

Theorem 5 ([5]). *In the case of Theorem 1, on the set $\Sigma(\varepsilon)$, the Lyapunov exponent of $H_{z,\varepsilon}$ satisfies*

$$\Theta(E,\varepsilon) = \frac{\varepsilon}{2\pi} \log\left(\tau^2(|\xi_0(E)| + |\xi_\pi(E)|) + 1\right) + o(1) , \tag{33}$$

where $o(1) \to 0$ when $\varepsilon \to 0$ uniformly in E, E_0 and E_π.

For $c > 0$, one defines the set

$$I_c^+ = \left\{ E \in \Sigma(\varepsilon) : \ \varepsilon \log\left(\tau\sqrt{|\xi_0(E)| + |\xi_\pi(E)|}\right) > c \right\} . \tag{34}$$

Theorem 5 and the Ishii-Pastur-Kotani Theorem imply

Corollany 1. [5] *In the case of Theorem 1, for ε sufficiently small, the set $\Sigma(\varepsilon) \cap I_c^+$ only contains singular spectrum.*

Define

$$I_c^- = \left\{ E \in \Sigma(\varepsilon) : \ \varepsilon \log\left(\tau\sqrt{|\xi_0(E)| + |\xi_\pi(E)|}\right) < -c \right\} . \tag{35}$$

Theorem 3 implies that, for sufficiently small ε, the set I_c^- is contained in the set

$$\tilde{\Sigma}_{\mathrm{ac}}(\varepsilon) = \{ E \in \mathbb{R} : \ |\tau^2\xi_0(E)\xi_\pi(E) + 2\Lambda_n(V)| \le 2(1 + g(\varepsilon)) \} , \tag{36}$$

where $\varepsilon \mapsto g(\varepsilon)$ is independent of c and satisfies the estimate $g = o(1)$ as $\varepsilon \to 0$. The set $\tilde{\Sigma}_{\mathrm{ac}}(\varepsilon)$ consists of $\tilde{I}_0$ and $\tilde{I}_\pi$, two disjoint intervals, and the distance between these intervals is greater or equal to $C\,\varepsilon\sqrt{t_h(\bar{E})}$.

Let Σ_{ac} denote the absolutely continuous spectrum of $H_{z,\varepsilon}$. One shows

Theorem 6 ([5]). *Pick $\nu \in \{0, \pi\}$. In the case of Theorem 1, there exists $\eta > 0$ and $D \subset (0,1)$, a set of Diophantine numbers such that*

$$\frac{\mathrm{mes}\,(D \cap (0,\varepsilon))}{\varepsilon} = 1 + o\left(e^{-\eta/\varepsilon}\right) \ \text{when } \varepsilon \to 0 .$$

for $\varepsilon \in D$ sufficiently small, if $I_c^- \cap \tilde{I}_\nu \ne \emptyset$, then

$$\mathrm{mes}\,(\tilde{I}_\nu \cap \Sigma_{\mathrm{ac}}) = \mathrm{mes}\,(I_\nu)\,(1 + o(1)) ,$$

where $o(1) \to 0$ when $\varepsilon \to 0$ uniformly in E_0 and E_π.

Possible Scenarii When τ is Small

Assume that $\tau \ll 1$ and $\Lambda_n(V) > 1$. Essentially, there are two possible cases for the location and the nature of the spectrum of $H_{z,\varepsilon}$. Define

$$\rho := \left.\frac{\max(t_{v,\pi}, t_{v,0})}{\sqrt{t_h}}\right|_{E=\bar{E}} = \tau \left.\sqrt{\frac{\max(t_{v,\pi}, t_{v,0})}{\min(t_{v,\pi}, t_{v,0})}}\right|_{E=\bar{E}}. \tag{37}$$

Note that $\rho \geq \tau$. We only discuss the cases when ρ is exponentially small or exponentially large.

1. We now discuss the case $\rho \ll 1$. If $|E_\pi - E_0| \ll \varepsilon\sqrt{t_h(\bar{E})}$, $\Sigma(\varepsilon)$ is the union of two intervals of length roughly $\varepsilon\sqrt{t_h}$; they are separated by a gap of length roughly $\varepsilon\sqrt{t_h}$ (see Fig. 6); this gap is centered the point $\bar{E}$. The length of the intervals containing spectrum, as well as the length and center of the lacuna essentially do not change as the distance $E_\pi - E_0$ increases upto a size of order $\varepsilon\sqrt{t_h(\bar{E})}$; after that, the intervals containing spectrum begin to move away from each other.

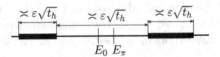

Fig. 6. When $\rho \ll 1$

As for the nature of the spectrum, when ρ is exponentially small and when $|E_\pi - E_0| \ll \varepsilon\sqrt{t_h(\bar{E})}$, the intervals containing spectrum are contained in the set I_c^-; so, most of the spectrum in these intervals is absolutely continuous (if ε satisfies the Diophantine condition of Theorem 6).

2. Consider the case $\rho \gg 1$. For sake of definiteness, assume that $t_{v,0} \ll t_{v,\pi}$. Then, there exists an interval, say I_π, that is asymptotically centered at E_π and that contains spectrum. The length of this interval is of order $\varepsilon t_{v,\pi}(\bar{E})$.

One can distiguish two cases:

1. if E_0 belongs to I_π and if the distance from E_0 to the edges of I_π is of the same order of magnitude as the length of I_π, then $\Sigma(\varepsilon)$ consists of the interval I_π without a "gap" of length roughly $\varepsilon t_h(\bar{E})/t_{v,\pi}(\bar{E})$ and containing E_0 (see Fig. 7(b)). Moreover, the distance from E_0 to any edge of the gap is also of order $\varepsilon t_h(\bar{E})/t_{v,\pi}(\bar{E})$.

2. if E_0 is outside I_π and at a distance from I_π at least of the same order of magnitude as the length of I_π, then $\Sigma(\varepsilon)$ consists in the union of I_π and an interval I_0 (see Fig. 7(a)). The interval I_0 is contained in neighborhood of E_0 of size roughly $\varepsilon^2 t_h(\bar{E})/|E_0 - E_\pi|$. The length of I_0 is of size $\varepsilon^2 t_h(\bar{E})/|E_0 - E_\pi| + \varepsilon t_{v,0}(\bar{E})$.

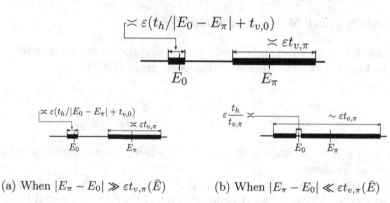

(a) When $|E_\pi - E_0| \gg \varepsilon t_{v,\pi}(\bar{E})$ (b) When $|E_\pi - E_0| \ll \varepsilon t_{v,\pi}(\bar{E})$

Fig. 7. The locus of the spectrum when $\tau \ll 1$ and $\rho \gg 1$

When ρ is exponentially large, the Lyapunov exponent may vary very quickly on the intervals containing spectrum. Consider the case $|E_\pi - E_0| \ll \varepsilon t_{v,\pi}(\bar{E})$. For E close to the gap surrounding E_0, $\tau^2 \xi_0(E)$ is of order 1 whereas $\tau^2 \xi_\pi(E)$ is exponentially small. Hence, Theorem 5 implies that $\Theta(E, \varepsilon) = o(1)$. On the other hand, at the external edges of the intervals containing spectrum, $\tau^2 \xi_0(E)$ is of size roughly ρ^2; this factor being exponentially large, at such energies, the Lyapunov exponent is positive and given by

$$\Theta(E, \varepsilon) = \frac{\varepsilon}{\pi} \log \rho + o(1) \,.$$

This phenomenon is similar to that observed for $\tau \gg 1$ except that, now, the Lyapunov exponent sharply drops to a value that is small and that may even vanish. On most of $\Sigma(\varepsilon)$, the Lyapunov exponent stays positive and, the spectrum is singular (by Corollary 1); near the lacuna surrounding E_0, neither Corollary 1, nor Theorem 6 apply. These zones are similar to the zones where asymptotic Anderson transitions were found in [10].

2.7 The Model Equation

To obtain the results described in Sect. 2.5 and 2.6, the study of the spectrum of $H_{z,\varepsilon}$ is reduced to the study of the finite difference equation:

$$\Psi_{k+1} = M(kh + z, E)\Psi_k, \quad \Psi_k \in \mathbb{C}^2, \quad k \in \mathbb{Z} \,, \tag{38}$$

where $h = \frac{2\pi}{\varepsilon} \bmod 1$, and $(z, E) \to M(z, E)$ is a function taking values in $SL(2, \mathbb{C})$ (the monodromy matrix, see [10]). The behavior of the solutions to (38) mimics that of those to $H_{z,\varepsilon}\psi = E\psi$ (see [10]). Equation (38) in which the matrix is replaced with its principal term is a model equation of our system. All the effects we have described can be seen when analyzing this model equation.

The asymptotic of M is decribed very precisely in [5,6]; here, we only write down its leading term. Assume additionally that $\tau^2(\bar{E}) \geq \min_{\nu \in \{0,\pi\}} t_{v,\nu}(\bar{E})$. Then, for $E \in I_0 \cup I_\pi$, M has a quite simple asymptotic

$$M(z,E) \sim M_0(z,E) := \begin{pmatrix} \tau^2 g_0(z+h/2)g_\pi(z) + \theta_n^{-1} & \tau g_0(z+h/2) \\ \theta_n \tau g_\pi(z) & \theta_n \end{pmatrix},$$
(39)

where

$$g_\nu = \xi_\nu + \sin(2\pi z), \quad \nu \in \{0,\pi\},$$

and θ_n is the solution of $2\Lambda_n = \theta_n + \theta_n^{-1}$ in $[1,+\infty[$.

Introducing the monodromy matrix for the Harper equation (see [2])

$$N_\nu(z) = \begin{pmatrix} \tau g_\nu(z) & -1 \\ 1 & 0 \end{pmatrix},$$

one notes that

$$M_0(z,E) = N_0(z+h/2,E)N_\pi(z,E).$$

So as suggested by the phase space picture, Fig. 2, at the energies under consideration, the system behaves like two interacting Harper equations.

2.8 When τ is of Order 1

When τ is of order 1, the principal term of the monodromy matrix asymptotics given by (39). If τ and $|\xi_0|$ and $|\xi_\pi|$ are of order of 1, the principal term does not contain any asymptotic parameter. This regime is similar to that of the asymptotic Anderson transitions found in [10]. If at least one of the "local variables" $\xi_\nu(E)$ becomes large, then, the spectrum can again be analyzed with the same precision as in the cases $\tau \gg 1$ and $\tau \ll 1$.

2.9 Numerical Computations

We now turn to some numerical results showing that the multiple phenomena described in Sect. 2.6 do occur.

All these phenomena only depend on the values of the actions S_h, $S_{v,0}$, $S_{v,\pi}$. We pick V to be a two-gap potential; for such potentials, the Bloch quasi-momentum k (see Sect. 2.1) is explicitly given by a hyper-elliptic integral ([14, 16]). The actions then become easily computable numerically. As the spectrum of $H_0 = -\Delta + V$ only has two gaps, we write $\sigma(H_0) = [E_1, E_2] \cup [E_3, E_4] \cup [E_5, +\infty)$. In the computations, we take the values

$$E_1 = 0, \ E_2 = 3.8571, \ E_3 = 6.8571, \ E_4 = 12.1004, \ \text{et } E_5 = 100.7092.$$

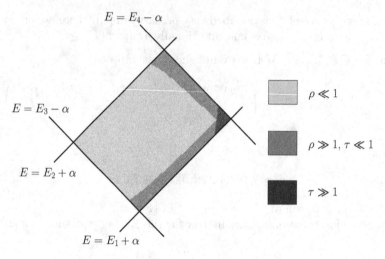

Fig. 8. Comparing τ and ρ to 1

On the Fig. 8, we represented the part of the (α, E)-plane where the condition (TIBM) is satisfied for $n = 1$, say Λ. Its boundary consists of the straight lines $E = E_1 + \alpha$, $E = E_2 + \alpha$, $E = E_3 - \alpha$ and $E = E_4 - \alpha$. The computations show that (T) is satisfied in the whole of Δ. As $n = 1$, one has $E_{2n-2} = -\infty$. It suffices to check (T) for $\zeta_{2n+3} = \zeta_5$. (T) can then be understood as a consequence of the fact that $E_5 - E_4$ is large.

On Fig. 8, we show the zones where τ and ρ are large and small. So, for carefully chosen α, all the phenomena described in Sect. 2.6, that is in Figs. 5, 6 and 7, do occur.

References

1. J. Avron and B. Simon. Almost periodic Schrödinger operators, II. the integrated density of states. *Duke Mathematical Journal*, 50:369–391, 1983.
2. V. Buslaev and A. Fedotov. Spectral properties of the monodromy matrix for Harper equation. In *Journées "Équations aux Dérivées Partielles" (Saint-Jean-de-Monts, 1996)*, pages Exp. No. IV, 11. École Polytech., Palaiseau, 1996.
3. H.L. Cycon, R.G. Froese, W. Kirsch, and B. Simon. *Schrödinger Operators*. Springer Verlag, Berlin, 1987.
4. M. Eastham. *The spectral theory of periodic differential operators*. Scottish Academic Press, Edinburgh, 1973.
5. A. Fedotov and F. Klopp. Strong resonant tunneling, level repulsion and spectral type for one-dimensional adiabatic quasi-periodic Schrödinger operator. In progress.
6. A. Fedotov and F. Klopp. Weakly resonant tunneling interactions for adiabatic quasi-periodic Schrödinger operators. Preprint math-ph/0408006, 2004. http://arxiv.org/abs/math-ph/0408006.

7. A. Fedotov and F. Klopp. The spectral theory of adiabatic quasi-periodic operators on the real line. *Markov Process. Related Fields*, 9(4):579–614, 2003.

8. A. Fedotov and F. Klopp. On the singular spectrum of one dimensional quasi-periodic Schrödinger operators in the adiabatic limit. *Ann. H. Poincaré*, 5:929–978, 2004.

9. A. Fedotov and F. Klopp. On the absolutely continuous spectrum of one dimensional quasi-periodic Schrödinger operators in the adiabatic limit. To appear in *Trans. A.M.S*, 2005.

10. Alexander Fedotov and Frédéric Klopp. Anderson transitions for a family of almost periodic Schrödinger equations in the adiabatic case. *Comm. Math. Phys.*, 227(1):1–92, 2002.

11. N.E. Firsova. On the global quasimomentum in solid state physics. In *Mathematical methods in physics (Londrina, 1999)*, pp. 98–141. World Sci. Publishing, River Edge, NJ, 2000.

12. D. Gilbert and D. Pearson. On subordinacy and analysis of the spectrum of one-dimensional Schrödinger operators. *Journal of Mathematical Analysis and its Applications*, 128:30–56, 1987.

13. B. Helffer and J. Sjöstrand. Multiple wells in the semi-classical limit I. *Communications in Partial Differential Equations*, 9:337–408, 1984.

14. A.R. It·s and V.B. Matveev. Hill operators with a finite number of lacunae. *Funkcional. Anal. i Priložen.*, 9(1):69–70, 1975.

15. Yoram Last and Barry Simon. Eigenfunctions, transfer matrices, and absolutely continuous spectrum of one-dimensional Schrödinger operators. *Invent. Math.*, 135(2):329–367, 1999.

16. H. McKean and P. van Moerbeke. The spectrum of Hill's equation. *Inventiones Mathematicae*, 30:217–274, 1975.

17. L. Pastur and A. Figotin. *Spectra of Random and Almost-Periodic Operators.* Springer Verlag, Berlin, 1992.

18. Leonid Pastur and Alexander Figotin. *Spectra of random and almost-periodic operators*, volume 297 of *Grundlehren der Mathematischen Wissenschaften [Fundamental Principles of Mathematical Sciences]*. Springer-Verlag, Berlin, 1992.

19. Barry Simon. Instantons, double wells and large deviations. *Bull. Amer. Math. Soc. (N.S.)*, 8(2):323–326, 1983.

20. M. Wilkinson. Critical properties of electron eigenstates in incommensurate systems. *Proc. Roy. Soc. London Ser. A*, 391(1801):305–350, 1984.

Low Lying Eigenvalues of Witten Laplacians and Metastability (After Helffer-Klein-Nier and Helffer-Nier)

Bernard Helffer[*]

Département de Mathématiques Bât. 425, Université Paris Sud, UMR CNRS 8628, F91405 Orsay Cedex (FRANCE)
Bernard.Helffer@math.u-psud.fr

Abstract. The aim of this lecture is to present the recent results obtained in collaboration with M. Klein and F. Nier on the low lying eigenvalues of the Laplacian attached to the Dirichlet form :

$$C_0^\infty(\Omega) \ni v \mapsto h^2 \int_\Omega |\nabla v(x)|^2 \ e^{-2f(x)/h} \ dx \ ,$$

where f is a C^∞ Morse function on $\overline{\Omega}$ and $h > 0$. We give in particular an optimal asymptotics as $h \to 0$ of the lowest strictly positive eigenvalue, which will hold under generic assumptions. We discuss also some aspects of the proof.

1 Main Goals and Assumptions

We are interested in the low lying, actually exponentially small, eigenvalues of the Dirichlet realization of the semiclassical Witten Laplacian on 0-forms

$$\Delta_{f,h}^{(0)} = -h^2 \Delta + |\nabla f(x)|^2 - h\Delta f(x) \ . \tag{1}$$

This question appears quite naturally when analyzing the asymptotic behavior as $t \to +\infty$ of $\exp -t\Delta_{f,h}^{(0)}$.

We would like to describe the recent results concerning the lowest strictly positive eigenvalue of this Laplacian. We will discuss briefly three cases :

- the case of $\mathbb{R}^n$, which was analyzed by Bovier-Eckhoff-Gayrard-Klein [1], Bovier-Gayrard-Klein [2], and by Helffer-Klein-Nier [6];
- the case of a compact riemannian manifold of dimension n ([6]);
- The case of a bounded set Ω in $\mathbb{R}^n$ with regular boundary treated by Helffer-Nier [8]) (in this case, we consider the Dirichlet realization of this operator).

In all these contributions, the goal is to get the optimal accuracy asymptotic formulas for the m_0 first eigenvalues of the Dirichlet realization of $\Delta_{f,h}^{(0)}$.

[*]Partially supported by the programme SPECT (ESF) and by the PROGRAMME HPRN-CT-2002-00277

B. Helffer: *Low Lying Eigenvalues of Witten Laplacians and Metastability (After Helffer-Klein-Nier and Helffer-Nier)*, Lect. Notes Phys. **690**, 403–415 (2006)
www.springerlink.com © Springer-Verlag Berlin Heidelberg 2006

This question was already addressed a long time ago, via a probabilistic approach, sometimes in relation with the problem of the simulated annealing. Here we would like to mention Freidlin-Wentcel [4], Holley-Kusuoka-Strook [11], Miclo [14], Kolokoltsov [13], Bovier-Eckhoff-Gayrard-Klein [1] and Bovier-Gayrard-Klein [2], but the proof, as mentioned for example in [13], of the optimal accuracy (except may be for the case of dimension 1) was left open.

The Witten Laplacian $\Delta^{(0)}_{f,h}$ is associated to the Dirichlet form

$$C^\infty_0(\Omega) \ni u \mapsto \int_\Omega |(h\nabla + \nabla f)u(x)|^2 \ dx \ .$$

Note that the probabilists look equivalently at:

$$C^\infty_0(\Omega) \ni v \mapsto h^2 \int_\Omega |\nabla v(x)|^2 \ e^{-2f(x)/h} \ dx \ .$$

Let us now describe the main assumptions. In the whole paper, we assume that:

Assumption 1 *The function f is a C^∞- function on $\overline{\Omega}$ and a Morse function on Ω.*

In the case when $\Omega = \mathbb{R}^n$, the operator defined on C^∞_0 is essentially selfadjoint in L^2 and if

Assumption 2

$$\liminf_{|x|\to+\infty} |\nabla f(x)|^2 > 0 \ ,$$

and

$$|D^\alpha_x f| \le C_\alpha(|\nabla f|^2 + 1) \ ,$$

for $|\alpha| = 2$,

is satisfied, Pearson's theorem permits to prove that the bottom of the essential spectrum is, for h small enough, strictly positive.

In the case with boundary, we will assume:

Assumption 3 *The function f has no critical points at the boundary and the function $f_{/\partial\Omega}$ is a Morse function on $\partial\Omega$.*

The last assumption, which appears to be generic, is more difficult to describe and will be presented in the next section.

2 Saddle Points and Labelling

The presentation of the results involves a right definition for the saddle points (or critical points of index 1). If for a point in Ω, we take the usual definition (the index at a critical point U being defined as the number of negative eigenvalues of the Hessian of f at U), we shall say that a point U at the boundary is a critical point of index 1 if U is a local minimum of $f_{/\partial\Omega}$ **and** if the external normal derivative of f is strictly positive.

Our statements also involve a labelling of the local minima, which is inspired by previous works by probabilists [1, 4, 14] and [2].

The existence of such a labelling is an assumption which is generically satisfied. This discussion can be done in the three cases. Let us focus on the case of a compact connected oriented Riemannian manifold $\overline{\Omega} = \Omega \cup \partial\Omega$ with boundary and that the function f satisfies Assumptions 1 and 3 and refer to the original papers [6, 8] (and references therein) for the other cases or for details. The set of critical points with index 1 is denoted by $\mathcal{U}^{(1)}$ and its cardinal by m_1. For the definition of the saddle set (or set of saddle points), we need some notations. For any $A, B \subset \overline{\Omega}$, $H(A, B)$ denotes the least height to be reached to go continuously from A to B. When A and B are closed nonempty subsets of $\overline{\Omega}$, one can show that $H(A, B)$ is a minimum. We next need two definitions.

Definition 1. *Let A and B be two closed subsets of $\overline{\Omega}$. We say that $Z \subset \overline{\Omega}$ is a saddle set for (A, B), if it is not empty and satisfies the following conditions:*

$$Z \subset \left(\mathcal{U}^{(1)} \cap f^{-1}(\{H(A, B)\}) \right) \ ,$$
$$\{C \in Conn \ (f^{-1}((-\infty, H(A, B)]) \setminus Z), \ C \cap A \neq \emptyset, C \cap B \neq \emptyset\} = \emptyset \ .$$

Definition 2. *Let A, B be closed nonempty disjoint subsets of $\overline{\Omega}$. The point $z \in \mathcal{U}^{(1)}$ is said to be a unique (one point)-saddle set[2] for the pair (A, B) if*

$$(\cap_{C \in \mathcal{C}(A,B)} C) \cap \left[\mathcal{U}^{(1)} \cap f^{-1}(H(A, B)) \right] = \{z\} \ ,$$

where $\mathcal{C}(A, B)$ denotes the set of closed connected sets $C \subset f^{-1}((-\infty, H (A, B)])$, such that $C \cap A \neq \emptyset$ and $C \cap B \neq \emptyset$.

We now give the main assumption which ensures that each exponentially small eigenvalue of $\Delta_{f,h}^{(0)}$ is simple, with a different asymptotic behavior.

Assumption 4 *There exists a labelling of the set of the local minima of f in Ω, $\mathcal{U}^{(0)} := \left\{ U_1^{(0)}, \ldots, U_{m_0}^{(0)} \right\}$, such that, by setting*

$$C_0 = \partial\Omega \quad and \quad C_k = \left\{ U_k^{(0)}, \ldots, U_1^{(0)} \right\} \cup C_0 \, , \ for \ k \geq 1 \, ,$$

we have:

---

[2]Or more shortly, a unique saddle point,

(i) For all $k \in \{1, \ldots, m_0\}$, $U_k^{(0)}$ is the unique minimizer of

$$H(U, C_k \setminus \{U\}) - f(U), \quad U \in C_k \setminus C_0 \, .$$

(ii) For all $k \in \{1, \ldots, m_0\}$, the pair $\left(\{U_k^{(0)}\}, C_{k-1}\right)$ admits a unique (one point)-saddle set $\{z_k^*\}$.

It is possible to check that this hypothesis is generically satisfied. More precisely, it is satisfied if all the critical values of f are distinct and all the quantities $f(U^{(1)}) - f(U^{(0)})$, with $U^{(1)} \in \mathcal{U}^{(1)}$ and $U^{(0)} \in \mathcal{U}^{(0)}$ are distinct.

By its definition, the point z_k^* is a critical point with index 1, $z_k^* \in \mathcal{U}^{(1)}$.

Definition 3. (The map j) If the critical points of index 1 are denoted by $U_j^{(1)}$, $j = 1, \ldots, m_1$, we define the application $k \to j(k)$ on $\{1, \ldots, m_0\}$ by:

$$U_{j(k)}^{(1)} = z_k^* \, . \tag{2}$$

With these definitions, one can prove :

Proposition 1. Under Assumption 4, the following properties are satisfied :
(a) The sequence $\left(f(U_{j(k)}^{(1)}) - f(U_k^{(0)})\right)_{k \in \{1, \ldots, m_0\}}$ is strictly decreasing.
(b) The application $j : \{1, \ldots, m_0\} \to \{1, \ldots, m_1\}$ is injective.

3 Rough Semi-Classical Analysis of Witten Laplacians and Applications to Morse Theory

3.1 Previous Results

It is known (see [10,17,18] and more recently [3,8]) that the Witten Laplacian on functions $\Delta_{f,h}^{(0)}$ admits in the interval $[0, h^{\frac{6}{5}}]$ and for $h > 0$ small enough exactly m_0 eigenvalues, where m_0 is the number of local minima of f in Ω.

This is easy to guess by considering, near each of the local minima $U_j^{(0)}$, of f the function

$$u_j^{wkb,(0)}(x) := \chi_j(x) \exp{-\frac{f(x) - f(U_j^{(0)})}{h}} \, , \tag{3}$$

where χ_j is a suitable cut-off function localizing near $U_j^{(0)}$ as suitable quasimode. This shows, via the minimax principle, that these eigenvalues are actually exponentially small as $h \to 0$.

Note that we consider the Dirichlet problem. So the first part of Assumption 3 implies that the eigenfunctions corresponding to low lying eigenvalues are localized far from the boundary.

3.2 Witten Laplacians on p-Forms

Although we are mainly interested in the operator $\Delta_{h,f}^{(0)}$, we will also meet in the proof other Laplacians. The spectral analysis can be extended (see Simon [17], Witten [18], Helffer-Sjöstrand [9], Chang-Liu [3]) to Laplacians on p-forms, $p \geq 1$. We recall that Witten is first introducing a dixtorted De Rham complex d :

$$d_{f,h} = \exp -\frac{f}{h}(hd) \exp \frac{f}{h} = hd + df \wedge , \tag{4}$$

The restriction of $d_{f,h}$ to p-forms is denoted by $d_{f,h}^{(p)}$. Witten then associates to this complex the Laplacian :

$$\Delta_{f,h} = (d_{f,h} + d_{f,h}^*)^2 , \tag{5}$$

where $d_{f,h}^*$ is its L^2-adjoint. By restriction to the p-forms, one gets the Witten Laplacian $\Delta_{f,h}^{(p)}$.

3.3 Morse Inequalities

In the compact case, the analysis of the low lying eigenvalues of the Witten Laplacians was the main point of the semi-classical proof suggested by Witten of the Morse inequalities [10, 17, 18].

Each of the Witten Laplacians is essentially selfadjoint and an analysis based on the harmonic approximation (consisting in replacing f by its quadratic approximation at a critical point of f) shows that the dimension of the eigenspace corresponding to $[0, h^{\frac{6}{5}}]$ is, for h small enough, equal to m_p the number of critical points of index p.

Note that the dimension of the kernel of $\Delta_{h,f}^{(p)}$ being equal to the Betti number b_p, this gives immediately the so called " Weak Morse Inequalities" :

$$b_p \leq m_p , \quad \text{for all } p \in \{0, \dots, n\} . \tag{6}$$

4 Main Result in the Case of $\mathbb{R}^n$

In the case of $\mathbb{R}^n$ and under Assumptions 1, 2 and 4, the main result is the following :

Theorem 5. *The first eigenvalues* $\lambda_k(h)$, $k \in \{2, \dots, m_0\}$, *of* $\Delta_{f,h}^{(0)}$ *have the following expansions :*

$$\lambda_k(h) = \frac{h}{\pi} |\widehat{\lambda}_1(U_{j(k)}^{(1)})| \sqrt{\frac{|\det(\mathrm{Hess} f(U_k^{(0)}))|}{|\det(\mathrm{Hess} f(U_{j(k)}^{(1)}))|}}$$
$$\times \exp -\frac{2}{h} \left(f(U_{j(k)}^{(1)}) - f(U_k^{(0)}) \right) \times (1 + r_1(h)) ,$$

with $r_1(h) = o(1)$.

Here the $U_k^{(0)}$ denote the local minima of f ordered in some specific way (see Sect. 2), the $U_{j(k)}^{(1)}$ are "saddle points" (critical points of index 1), attached to $U_k^{(0)}$ via the map j, and $\widehat{\lambda}_1(U_{j(k)}^{(1)})$ is the negative eigenvalue of $\mathrm{Hess} f(U_{j(k)}^{(1)})$.

Actually, the estimate

$$r_1(h) = \mathcal{O}(h^{\frac{1}{2}} |\log h|)\,,$$

is obtained in [2] (under weaker assumptions on f) and the complete asymptotics,

$$r_1(h) \sim \sum_{j \geq 1} r_{1j} h^j\,,$$

is proved in [6].

In the above statement, we have left out the case $k = 1$, which leads to a specific assumption (see Assumption 2) in the case of $\mathbb{R}^n$ for f at ∞. This implies that $\Delta_{f,h}^{(0)}$ is essentially selfadjoint and that the bottom of the essential spectrum is bounded below by some $\epsilon_0 > 0$ (independently of $h \in]0, h_0]$, h_0 small enough). If the function $\exp - \frac{f}{h}$ is in L^2, then

$$\lambda_1(h) = 0\;.$$

In this case, denoting by Π_0 the orthogonal projector on $\exp - \frac{f}{h}$, the main motivation for having a good control of $\lambda_2(h)$ is the estimate, for $t > 0$,

$$\exp -t\Delta_{h,f}^{(0)} - \Pi_0 ||_{\mathcal{L}(L^2(\Omega))} \leq \exp -t\lambda_2(h)\;.$$

In other words, the estimate of $\lambda_2(h)$ permits to measure the rate of the return to equilibrium.

Note finally that other examples like $f(x) = -(x^2 - 1)^2$ (with $n = 1$) are interesting and an asymptotic of $\lambda_1(h)$ can be given for this example.

5 About the Proof in the Case of $\mathbb{R}^n$

5.1 Preliminaries

The case of $\mathbb{R}^n$ requires some care (see [5, 12] or [7]) for controlling the problem at infinity. The approach given in [6] intensively uses, together with the techniques of [10], the two following facts :

- The Witten Laplacian is associated to a cohomology complex.
- The function $\exp - \frac{f(x)}{h}$ is, as a distribution, in the kernel of the Witten Laplacian on $0-$forms.

This permits to construct very easily and efficiently – and this is specific of the case of $\Delta_{f,h}^{(0)}-$ quasimodes. We note that we have between differential operators acting on C_0^∞ the relations

$$d_{f,h}^{(0)}\Delta_{f,h}^{(0)} = \Delta_{f,h}^{(1)}d_{f,h}^{(0)} , \tag{7}$$

and

$$\Delta_{f,h}^{(0)}d_{f,h}^{(0),*} = d_{f,h}^{(0),*}\Delta_{f,h}^{(1)} , \tag{8}$$

This shows in particular that if u is an eigenvector of $\Delta_{h,f}^{(0)}$ for an eigenvalue $\lambda \neq 0$, then $d_{h,f}u$ is an eigenvector of $\Delta_{f,h}^{(1)}$ for λ.

5.2 Witten Complex, Reduced Witten Complex

It is more convenient to consider the singular values of the restricted differential $d_{f,h} : F^{(0)} \to F^{(1)}$. The space $F^{(\ell)}$ is the m_ℓ-dimensional spectral subspace of $\Delta_{f,h}^{(\ell)}$, $\ell \in \{0,1\}$,

$$F^{(\ell)} = \operatorname{Ran} 1_{I(h)}(\Delta_{f,h}^{(\ell)}) ,$$

with $I(h) = [0, h^{\frac{6}{5}}]$ and the property

$$1_{I(h)}(\Delta_{f,h}^{(1)})d_{f,h} = d_{f,h}1_{I(h)}(\Delta_{f,h}^{(0)}) .$$

We will analyze :

$$\beta_{f,h}^{(\ell)} := (d_{f,h}^{(\ell)})_{/F^{(\ell)}} .$$

We will mainly concentrate on the case $\ell = 0$.

5.3 Singular Values

In order to exploit all the information which can be extracted from well chosen quasimodes, working with singular values s_j of $\beta_{f,h}^{(0)}$ happens to be more efficient than considering their squares, that is the eigenvalues $\lambda_j = s_j^2$ of $\Delta_{f,h}^{(0)}$. The main point[3] is probably that one can choose suitable approximate well localized "almost" orthogonal basis of $F^{(0)}$ and $F^{(1)}$ **separately** and that the errors appear "multiplicatively" when computing the singular values of $\beta_{f,h}^{(0)}$. By this we mean :

$$s_j = s_j^{app}(1 + \varepsilon_1(h)) ,$$

instead of additively

$$\lambda_j := s_j^2 = (s_j^{app})^2 + \varepsilon_2(h) ,$$

as for example in [10]. Here s_{app}^j will be explicitly obtained from the WKB analysis. In the first case, it is actually enough to prove that $\varepsilon_1(h) = \mathcal{O}(h^\infty)$.

---

[3]See also [15] for a pedagogical discussion,

In the second case, the analysis of [10] gives a control of $\varepsilon_2(h)$ in $\mathcal{O}(\exp -\frac{S}{h})$, with $S > 2\inf_{j,k}(f(U_j^{(1)}) - f(U_k^{(0)}))$, which is enough for estimating the highest low lying eigenvalue (see [7]) but could be unsatisfactory for the lowest strictly positive eigenvalue, as soon as the number of local minima is > 2. Although it is not completely hopeless to have a better control of $\varepsilon_2(h)$ by improving the analysis of [10] and introducing a refined notion of non resonant wells, the approach developed in [8] appears to be simpler.

6 The Main Result in the Case with Boundary

In the case with boundary, the function $\exp -\frac{f}{h}$, which is the only distribution in the kernel of $\Delta_{h,f}^{(0)}$, does not satisfy the Dirichlet condition, so the smallest eigenvalue can not be 0. The estimate of $\lambda_1(h)$ permits consequently to measure the decay of $\| \exp -t\Delta_{h,f}^{(0)} \| \leq \exp -t\lambda_1(h)$ as $t \to +\infty$.

For this case, a starting reference is the book by Freidlin-Wentzel [4], which says (in particular) that, if f has a unique critical point, corresponding to a non degenerate local minimum $U_{min}^{(0)}$, then the lowest eigenvalue $\lambda_1(h)$ of the Dirichlet realization $\Delta_{f,h}^{(0)}$ in Ω satisfies :

$$\lim_{h \to 0} -h \log \lambda_1(h) = 2 \inf_{x \in \partial\Omega} (f(x) - f(U_{min}^{(0)})) \ .$$

Other results are given in this book for the case of many local minima but they are again limited to the determination of logarithmic equivalents.

We have explained in Sect. 2 that, under Assumption 4, one can label the m_0 local minima and associate via the map j from the set of the local minima into the set of the m_1 (generalized) saddle points of the Morse functions in $\overline{\Omega}$ of index 1.

The main theorem of Helffer-Nier [8] is :

Theorem 6. *Under Assumptions 1, 3 and 4, there exists h_0 such that, for $h \in (0, h_0]$, the spectrum in $[0, h^{\frac{3}{2}})$ of the Dirichlet realization of $\Delta_{f,h}^{(0)}$ in Ω, consists of m_0 eigenvalues $\lambda_1(h) < \ldots < \lambda_{m_0}(h)$ of multiplicity 1, which are exponentially small and admit the following asymptotic expansions :*

$$\lambda_k(h) = \frac{h}{\pi} |\widehat{\lambda}_1(U_{j(k)}^{(1)})| \sqrt{\frac{|\det(\mathrm{Hess} f(U_k^{(0)}))|}{|\det(\mathrm{Hess} f(U_{j(k)}^{(1)}))|}} \left(1 + hc_k^1(h)\right)$$
$$\times \exp -\frac{2}{h} \left(f(U_{j(k)}^{(1)}) - f(U_k^{(0)})\right) , \quad \text{if } U_{j(k)}^{(1)} \in \Omega \ ,$$

and

$$\lambda_k(h) = \frac{2h^{1/2}|\nabla f(U_{j(k)}^{(1)})|}{\pi^{1/2}} \sqrt{\frac{|\det(\mathrm{Hess} f(U_k^{(0)}))|}{|\det(\mathrm{Hess} f|_{\partial\Omega}(U_{j(k)}^{(1)}))|}} \left(1 + hc_k^1(h)\right)$$
$$\times \exp -\frac{2}{h} \left(f(U_{j(k)}^{(1)}) - f(U_k^{(0)})\right) , \quad \text{if } U_{j(k)}^{(1)} \in \partial\Omega \ .$$

Here $c_k^1(h)$ admits a complete expansion :

$$c_k^1(h) \sim \sum_{m=0}^{\infty} h^m c_{k,m} \, .$$

This theorem extends to the case with boundary the previous results of [2] and [6] (see also the books [4] and [13] and references therein).

7 About the Proof in the Case with Boundary

As in [10], the proof is deeply connected with the analysis of the small eigenvalues of a suitable realization (which is **not** the Dirichlet realization) of the Laplacian on the 1-forms. In order to understand the strategy, three main points have to be explained.

7.1 Define the Witten Complex and the Associate Laplacian

The case of a compact manifold without boundary was treated in the foundational paper of Witten [18]. A finer (and rigorous) analysis is then given in [10] and further developments appear in [6]. The case with boundary creates specific new problems.

Our starting problem being the analysis of the Dirichlet realization of the Witten Laplacian on functions, we were let to find the right realization of the Witten Laplacian on 1-forms in the case with boundary in order to extend the commutation relation (7) in a suitable "strong" sense (at the level of the selfadjoint realizations).

The answer was actually present in the literature [3] in connection with the analysis of the relative cohomology and the proof of the Morse inequalities. Let us explain how we can guess the right condition by looking at the eigenvectors.

If u is eigenvector of the Dirichlet realization of $\Delta_{f,h}^{(0)}$, then by commutation relation, $d_{f,h}^{(0)} u$ (which can be identically 0) should be an eigenvector in the domain of the realization of $\Delta_{f,h}^{(1)}$. But $d_{f,h}^{(0)} u$ does not necessarily satisfy the Dirichlet condition in all its components, but only in its tangential components.

This is the natural condition that we keep in the definition of the variational domain to take for the quadratic form $\omega \mapsto \|d_{f,h}^{(1)}\omega\|^2 + \|d_{f,h}^{(0)\,*}\omega\|^2$.

The selfadjoint realization $\Delta_{f,h}^{(1)\,DT}$ obtained as the Friedrichs extension associated to the quadratic form gives the right answer.

Observing also that $d_{f,h}^{(0),*}(d_{f,h}^{(0)}u) = \lambda u$ (with $\lambda \neq 0$), we get the second natural (Neumann type)-boundary condition saying that a one form ω in the domain of the operator $\Delta_{f,h}^{(1)\,DT}$ should satisfy

$$d_{f,h}^{(0),*}\omega_{/\partial\Omega} = 0 \ . \tag{9}$$

So we have shown that the natural boundary conditions for the Witten Laplacians are (9) together with

$$\omega_{/\partial\Omega} = 0 \ . \tag{10}$$

The associated cohomology is called relative (see for example [16]).

7.2 Rough Localization of the Spectrum of this Laplacian on 1-Forms

The analysis of $\Delta_{f,h}^{(p)}$ was performed in [3], in the spirit of Witten's idea, extending the so called Harmonic approximation. But these authors, because they were interested in the Morse theory, used the possibility to add symplifying assumptions on f and the metric near the boundary. We emphasize that [8] treats the general case. May be one could understand what is going on at the boundary by analyzing the models corresponding to $f(x', x_n) = \frac{1}{2}|x'|^2 + \epsilon x_n$, with $\epsilon = \pm 1$ in $\mathbb{R}_+^n = \{x_n > 0\}$. The analysis in this case is easily reduced to the analysis of the one dimension case on $\mathbb{R}^+$ (together with the standard analysis of $\mathbb{R}^{n-1}$). The Dirichlet Laplacian to analyze is simply :

$$-h^2\frac{d^2}{dx^2} + 1 \ ,$$

on $\mathbb{R}^+$, which is strictly positive, but the Laplacian on 1-forms is

$$u(x)\,dx \mapsto (-h^2 u''(x) + u(x))\,dx \ ,$$

on $\mathbb{R}^+$, but with the boundary condition :

$$hu'(0) - \epsilon u(0) = 0 \ .$$

Depending on the sign of ϵ, the bottom of the spectrum is 0 if $\epsilon < 0$ or 1 if $\epsilon > 0$. This explains our definition of critical point of index 1 at the boundary.

7.3 Construction of WKB Solutions Attached to the Critical Points of Index 1

The construction of the approximate basis of $F^{(0)}$ and $F^{(1)}$ is obtained through WKB constructions. The constraints are quite different in the two cases. For $F^{(0)}$, we need rather accurate quasimodes but can take advantage of their simple structure given in (3). The difficulty is concentrated in the choice of χ_j. For $F^{(1)}$, it is enough to construct quasimodes lecalized in a small neighborhood of a critical point of index 1. This was done in [10] for the case without boundary, as an extension of previous constructions of [9].

The new point is the construction of WKB solutions near critical points of the restriction of the Morse function at the boundary, which is done in [8] for 1-forms. Let us explain the main lines of the construction.

The construction is done locally around a local minimum U_0 of $f|_{\partial\Omega}$ with $\partial_n f(U_0) > 0$. The function Φ is a local solution of the eikonal equation

$$|\nabla\Phi|^2 = |\nabla f|^2 ,$$

which also satisfies

$$\Phi = f \quad \text{on} \quad \partial\Omega$$

and

$$\partial_n \Phi = -\partial_n f \quad \text{on} \quad \partial\Omega$$

and we normalize f so that $f(U_0) = f(0) = 0$.

We first consider a local solution u_0^{wkb} near the point $x = 0$ of

$$e^{\frac{\Phi}{h}} \Delta_{f,h}^{(0)} u_0^{wkb} = \mathcal{O}(h^\infty) ,$$

with u_0^{wkb} in the form

$$u_0^{wkb} = a(x,h) e^{-\frac{\Phi}{h}} ,$$

$$a(x,h) \sim \sum_{j\geq 0} a_j(x) h^j ,$$

and the condition at the boundary

$$a(x,h) e^{-\frac{\Phi}{h}} = e^{-\frac{f}{h}} \quad \text{on} \quad \partial\Omega ,$$

which leads to the condition

$$a(x,h)\big|_{\partial\Omega} = 1 .$$

In order to verify locally the boundary condition for our future u_1^{wkb}, we substract $e^{-\frac{f}{h}}$ and still obtain

$$e^{\frac{\Phi}{h}} \Delta_f^{(0)} (u_0^{wkb} - e^{-\frac{f}{h}}) = \mathcal{O}(h^\infty) . \tag{11}$$

We now define the WKB solution u_1^{wkb} by considering :

$$u_1^{wkb} := d_{f,h} u_0^{wkb} = d_{f,h}(u_0^{wkb} - e^{-\frac{f}{h}}) .$$

The 1-form $u_1^{wkb} = d_{f,h} u_0^{wkb}$ satisfies locally the Dirichlet tangential condition on the boundary (10) and, using (11), (modulo $\mathcal{O}(h^\infty)$) the Neumann type condition (9) is satisfied. So u_1^{wkb} gives a good approximation for a ground state of a suitable realization of $\Delta_{f,h}^{(1)}$ in a neighborhood of this boundary critical point.

Acknowledgements

The author would like to thank F. Nier for fruitful collaboration on this subject and useful comments on the manuscript. He also thanks the organizers of the meeting J. Asch and A. Joye.

References

1. A. Bovier, M. Eckhoff, V. Gayrard, and M. Klein : Metastability in reversible diffusion processes I. Sharp asymptotics for capacities and exit times. J. of the Europ. Math. Soc. 6 (4), pp. 399–424 (2004).
2. A. Bovier, V. Gayrard, and M. Klein. Metastability in reversible diffusion processes II Precise asymptotics for small eigenvalues. J. Eur. Math. Soc. 7, pp. 69–99 (2005).
3. Kung Ching Chang and Jiaquan Liu. A cohomology complex for manifolds with boundary. Topological methods in non linear analysis. Vol. 5, pp. 325–340 (1995).
4. M.I. Freidlin and A.D. Wentzell. *Random perturbations of dynamical systems.* Transl. from the Russian by Joseph Szuecs. 2nd ed. Grundlehren der Mathematischen Wissenschaften. 260. New York (1998).
5. B. Helffer. *Semi-classical analysis, Witten Laplacians and statistical mechanics.* World Scientific (2002).
6. B. Helffer, M. Klein, and F. Nier. Quantitative analysis of metastability in reversible diffusion processes via a Witten complex approach. Proceedings of the Symposium on Scattering and Spectral Theory. Proceedings of the Symposium on Scattering and Spectral Theory. Matematica Contemporânea (Brazilian Mathematical Society), Vol. 26, pp. 41–86 (2004).
7. B. Helffer and F. Nier. *Hypoellipticity and spectral theory for Fokker-Planck operators and Witten Laplacians.* Lecture Notes in Mathematics 1862, Springer Verlag (2005).
8. B. Helffer and F. Nier. Quantitative analysis of metastability in reversible diffusion processes via a Witten complex approach : the case with boundary. To appear in Mémoires de la SMF (2006)
9. B. Helffer and J. Sjöstrand. Multiple wells in the semi-classical limit I, Comm. Partial Differential Equations 9 (4), pp. 337–408 (1984).
10. B. Helffer and J. Sjöstrand. Puits multiples en limite semi-classique IV -Etude du complexe de Witten -. Comm. Partial Differential Equations 10 (3), pp. 245–340 (1985).
11. R. Holley, S. Kusuoka, and D. Stroock. Asymptotics of the spectral gap with applications to the theory of simulated annealing. J. Funct. Anal. 83 (2), pp. 333–347 (1989).
12. J. Johnsen. On the spectral properties of Witten Laplacians, their range projections and Brascamp-Lieb's inequality. Integral Equations Operator Theory 36(3), pp. 288–324 (2000).
13. V.N. Kolokoltsov. *Semi-classical analysis for diffusions and stochastic processes.* Lecture Notes in Mathematics 1724. Springer Verlag, Berlin 2000.
14. L. Miclo. Comportement de spectres d'opérateurs à basse température. Bull. Sci. Math. 119, pp. 529–533 (1995).

15. F. Nier. Quantitative analysis of metastability in reversible diffusion processes via a Witten complex approach. Proceedings of the PDE conference in Forges-Les-Eaux (2004).

16. G. Schwarz. *Hodge decomposition. A method for solving boundary value problems.* Lect. Notes in Mathematics 1607, Springer (1995).

17. B. Simon. Semi-classical analysis of low lying eigenvalues, I.. Nondegenerate minima: Asymptotic expansions. Ann. Inst. H. Poincaré (Section Phys. Théor.) 38, pp. 296–307 (1983).

18. E. Witten. Supersymmetry and Morse inequalities. J. Diff. Geom. 17, pp. 661–692 (1982).

The Mathematical Formalism of a Particle in a Magnetic Field

Marius Măntoiu[1] and Radu Purice[2]

[1] "Simion Stoilow" Institute of Mathematics, Romanian Academy
   Marius.Mantoiu@imar.ro
[2] "Simion Stoilow" Institute of Mathematics, Romanian Academy
   Radu.Purice@imar.ro

1 Introduction

In this review article we develop a basic part of the mathematical theory involved in the description of a particle (classical and quantal) placed in the Euclidean space $\mathbb{R}^N$ under the influence of a magnetic field B, emphasising the structure of the family of observables.

The classical picture is known, see for example [21]; we present it here for the convenience of the reader, in a form well-fitted for the passage to the quantum counterpart. In doing this we shall emphasize a manifestly gauge invariant Hamiltonian description [27] that is less used, although it presents many technical advantages and a good starting point for quantization.

The main contribution concerns the quantum picture. Up to our knowledge, until recently the single right attitude towards defining quantum observables when a nonconstant magnetic field is present can be found in a remarkable old paper of Luttinger [14] (we thank Gh. Nenciu for pointing it out to us); still this was undeveloped and with a limited degree of generality.

In recent years, the solution to this problem appeared in two related forms: (1) a gauge covariant pseudodifferential calculus in [8,9,17,19] and (2) a C^*-algebraic formalism in [16] and [19]. We cite here also the results in [22], where a gauge independent perturbation theory is elaborated for the resolvent of a magnetic Schrödinger Hamiltonian, starting from an observation in [3].

For the classical picture, we define a perturbed symplectic form on phase space [27] and study the motions defined by classical Hamiltonians with respect to the associated perturbed Poisson algebra. The usual magnetic momenta appear then as momentum map for the associated 'symplectic translations'.

The quantum picture is treated in detail; two points of view are addopted: The first is to preserve (essentially) the same set of functions as observables, but with a different algebraic structure. The main input is a new, $(B, \hbar)$-dependent multiplication law associated to the perturbed symplectic form defined for the classical theory. This new product converges in a suitable sense to poinwise (classical) multiplication when $\hbar \to 0$. And it collapses for $B = 0$ to the symbol multiplication of Weyl and Moyal, familiar from pseudodifferential theory. It depends on no choice of a vector potential, so it

M. Măntoiu and R. Purice: *The Mathematical Formalism of a Particle in a Magnetic Field,*
Lect. Notes Phys. **690**, 417–434 (2006)
www.springerlink.com

is explicitely gauge invariant. Asside the pseudodifferential form, we present also a form comming from the theory of twisted crossed product C^*-algebras and justified by interpreting our physical system as a dynamical system given by an action twisted by the magnetic field.

The second point of view, more conventional, is in terms of self-adjoint operators in some Hilbert space. One achieves this by representing the previously mentioned intrinsic structures, and this is done by choosing vector potentials A generating the magnetic field B. For different but equivalent choices one gets unitarily equivalent representations, a form of what is commonly called "gauge covariance". The represented form is best-suited to the interpretation in terms of magnetic canonical commutation relations. A functional calculus is associated to this highly non-commutative family of operators. Actually, the twisted dynamical system mentioned above (a sort of twisted imprimitivity system) is equivalent to these commutation rules.

The limit $\hbar \to 0$ of the quantum system was studied in [18], in the framework of Rieffel's strict deformation quantization.

To show that the formalism is useful in applications, we dedicate a section to spectral theory for anisotropic magnetic operators, following [20]. This relies heavily on an affiliation result, saying that the resolvent family of a magnetic Schrödinger Hamiltonian belongs to a suitable C^*-algebra of magnetic pseudodifferential operators.

Recently, the usual pseudodifferential theory has been generalized to a groupoid setting, cf. [13, 23] and references therein; this is in agreement with modern trends in deformation quantization, cf. [10] for example. The right concept to include magnetic fields should be that of twisted groupoid, as appearing in [30], accompanied by the afferent C^*-algebras. Let us also mention here the possibility to use our general framework in dealing with nonabelian gauge theories.

2 The Classical Particle in a Magnetic Field

In this section we shall give a classical bakground for our quantum formalism. We use the setting and ideas in [21] but develop the gauge invariant Poisson algebra feature. We begin by very briefly recalling the usual Hamiltonian formalism for classical motion in a magnetic field and then change the point of view by perturbing the canonical symplectic structure.

2.1 Two Hamiltonian Formalisms

The basic fact provided by physical measurements is that the magnetic field in $\mathbb{R}^3$ may be described by a function $B : \mathbb{R}^3 \to \mathbb{R}^3$ with $divB = 0$, such that the motion $\mathbb{R} \ni t \mapsto q(t) \in \mathbb{R}^3$ of a classical particle (mass m and electric charge e) is given by the equation of motion defined by the *Lorentz* force:

$$m\ddot{q}(t) = e\dot{q}(t) \times B(q(t)) \tag{1}$$

where $\times$ is the antisymmetric vector product in $\mathbb{R}^3$ and the point denotes derivation with respect to time. An important fact about this equation of motion is that it can be derived from a Hamilton function, the price to pay being the necessity of a vector potential, i.e. a vector field $A : \mathbb{R}^3 \to \mathbb{R}^3$ such that $B = rotA$, that is unfortunately not uniquely determined.

Let us very briefly recall the essential facts concerning the Hamiltonian formalism. Given a smooth manifold X we associate to it its "phase space" defined as the cotangent bundle $\mathbb{T}^*X$ on which we have a canonical symplectic form, that we shall denote by σ. If we set $\Pi : \mathbb{T}[\mathbb{T}^*X] \to \mathbb{T}^*X$ and $\tilde{\pi} : \mathbb{T}^*X \to X$ the canonical projections and $\tilde{\pi}_* : \mathbb{T}[\mathbb{T}^*X] \to \mathbb{T}X$ the tangent map of $\tilde{\pi}$, then $\sigma := d\beta$ where $\beta(\boldsymbol{\xi}) := [\Pi(\boldsymbol{\xi})(\tilde{\pi}_*(\boldsymbol{\xi}))$, for $\boldsymbol{\xi}$ a smooth section in $\mathbb{T}[\mathbb{T}^*X]$. A Hamiltonian system is determined by a Hamilton function $\mathrm{h} : \mathbb{T}^*X \to \mathbb{R}$ (supposed to be smooth) such that the vector field associated to the law of motion of the system ($\mathbb{R} \ni t \mapsto \mathfrak{x}(t) \in \mathbb{T}^*X$) is given by the following first order differential equation $\boldsymbol{\xi} \lrcorner \sigma - d\mathrm{h} = 0$, where $\boldsymbol{\xi} \lrcorner \sigma$ is the one-form defined by $(\boldsymbol{\xi} \lrcorner \sigma)(\boldsymbol{\eta}) := \sigma(\boldsymbol{\xi}, \boldsymbol{\eta})$, for any $\boldsymbol{\eta}$ smooth section in $\mathbb{T}[\mathbb{T}^*X]$.

Let us take $X = \mathbb{R}^3$ such that all the above bundles are trivial and we have canonical isomorphisms $\mathbb{T}^*X \cong X \times X^*$ (that we shall also denote by Ξ) and $\mathbb{T}[\mathbb{T}^*X] \cong (X \times X^*) \times (X \times X^*)$, defined by the usual transitive action of translations on X; we can view any two sections $\boldsymbol{\xi}$ and $\boldsymbol{\eta}$ as functions $\boldsymbol{\xi}(q, p) = (x(q, p), k(q, p))$, $\boldsymbol{\eta}(q, p) = (y(q, p), l(q, p))$ and we can easily verify that $\sigma(\boldsymbol{\xi}, \boldsymbol{\eta}) = k \cdot y - l \cdot x$, with $\xi \cdot y$ the canonical pairing $X^* \times X \to \mathbb{R}$. Moreover, the equations of motion defined by a Hamilton function h become:

$$\begin{cases} \dot{q}_j = \partial \mathrm{h}/\partial p_j\,, \\ \dot{p}_j = -\partial \mathrm{h}/\partial q_j. \end{cases} \tag{2}$$

Then (1) may be written in the above form if one chooses a vector potential A such that $B = rotA$ and defines the Hamilton function

$$\mathrm{h}_A(q, p) := (2m)^{-1} \sum_{j=1}^{3} (p_j - eA_j(q))^2\,.$$

Although very useful, this Hamiltonian description has the drawback of involving the choice of a vector potential. Two different choices A and A' have to satisfy $rot(A - A') = 0$. Since $\mathbb{R}^3$ is simply connected, there exists a function $\varphi : \mathbb{R}^3 \to \mathbb{R}$ with $A' = A + \nabla\varphi$ and any such choice is admissible. We call these changes of descriptions "gauge transformations"; the "gauge group" is evidently $C^\infty(X)$ and the action of the gauge group is given by $\mathrm{h}_A \to \mathrm{h}_{A'}$.

An interesting fact is that we can actually obtain an explicitly gauge invariant description by using a perturbed symplectic form on $\mathbb{T}^*X$ [27]. For that it is important to notice that the magnetic field may in fact be described as a 2-form (a field of antisymmetric bilinear functions on $\mathbb{R}^3$), due to the obvious isomorphism between $\mathbb{R}^3$ and the space of antisymmetric matrices on $\mathbb{R}^3$ (just take $B_{jk} := \epsilon_{jkl}B_l$ with ϵ_{jkl} the completely antisymmetric tensor of rank 3 on $\mathbb{R}^3$). Thus from now on we shall consider the magnetic field B given by a smooth section of the vector bundle $\Lambda^2 X \to X$ (the fibre at x being $\mathbb{T}_x^* X \wedge \mathbb{T}_x^* X \cong [\mathbb{T}_x X \wedge \mathbb{T}_x X]^*$). Due to the canonical global trivialisation discussed above (defined by translations) we can view B as a smooth map $B : X \to X^* \wedge X^* \cong (X \wedge X)^*$. Then a vector potential is described by a 1-form $A : X \to X^*$ such that $B = dA$ where d is the exterior differential. This also allow us to consider the case $X = \mathbb{R}^N$ for any natural number N.

Any k-form on X may be considered as a k-form on $\mathbb{T}^* X$. Explicitely, using the projection $\tilde{\pi} : \mathbb{T}^* X \to X$, we may canonically define the pull-back $\tilde{\pi}^* B$ of B and the "perturbed symplectic form" on $\mathbb{T}^* X$ defined by the magnetic field B as $\sigma_B := \sigma + \tilde{\pi}^* B$.

Now let us briefly recall the construction of the Poisson algebra associated to a symplectic form. We start from the trivial fact that any nondegenerate bilinear form Σ on the vector space Ξ defines a canonical isomorphism $i_\Sigma : \Xi \to \Xi^*$ by the equality $[i_\Sigma(x)](y) := \Sigma(x, y)$. Then we define the following composition law on $C^\infty(X)$: $\{f, g\}_B := \sigma_B(i_{\sigma_B}^{-1}(df), i_{\sigma_B}^{-1}(dg))$, called the Poisson braket. The case $B = 0$ gives evidently the canonical Poisson braket $\{.,.\}$ on the cotangent bundle. A computation gives immediately

$$\{f, g\}_B = \sum_{j=1}^{N} \left(\partial_{p_j} f \, \partial_{q_j} g - \partial_{q_j} f \, \partial_{p_j} g \right) + e \sum_{j,k=1}^{N} B_{jk}(\cdot) \, \partial_{p_j} f \, \partial_{p_k} g . \qquad (3)$$

For the usual Hamilton function of the free classical particle $h(p) := (2m)^{-1} \sum_{j=1}^{N} p_j^2$, we can write down the Poisson form of the equation of motion:

$$\begin{cases} \dot{q}_j = \{h, q_j\}_B = \frac{1}{m} p_j , \\ \dot{p}_j = -\{h, p_j\}_B = \frac{e}{m} \sum_{k=1}^{N} B_{kj}(q) p_k, \end{cases} \qquad (4)$$

that combine to the equation of motion (1) defined by the Lorentz force.

We remark finally that in the present formulation the Hamilton function of the free particle $h(q, p) = (2m)^{-1} \sum p_j^2$ is no longer privileged; any Hamilton function is now a candidate for a Hamiltonian system in a magnetic field just by considering it on the phase space endowed with the magnetic symplectic form. The relativistic kinetic energy $h(p) := (p^2 + m^2)^{1/2}$ is a physically interesting example.

Remark. The real linear space $C^\infty(\Xi; \mathbb{R})$ endowed with the usual product of functions and the magnetic Poisson braket $\{.,.\}_B$ form a *Poisson algebra* (see [10, 18]), i.e. $(C^\infty(\Xi; \mathbb{R}), \cdot)$ is a real abelian algebra and $\{.,.\}_B : C^\infty(\Xi; \mathbb{R}) \times C^\infty(\Xi; \mathbb{R}) \to C^\infty(\Xi; \mathbb{R})$ is an antisymmetric bilinear composition law that satisfies the Jacobi identity and is a derivation with respect to the usual product.

2.2 Magnetic Translations

For the perturbed symplectic form on $\mathbb{T}^*X$, the usual translations are no longer symplectic. We intend to define "magnetic symplectic translations" and compute the associated momentum map. Using the canonical global trivialisation, we are thus looking for an action $X \ni x \mapsto \alpha_x \in \mathit{Diff}(X \times X^*)$ having the form $\alpha_x(q, p) = (q + x, p + \tau_x(q, p))$. A *group action* clearly imposes the 1-cocycle condition: $\tau_{x+y}(q, p) = \tau_x(q, p) + \tau_y(q + x, p + \tau_x(q, p))$. The symplectic condition reads: $(\alpha_{-x})^* \sigma_B = \sigma_B$. A simple computation gives us for any $(q, p) \in \Xi$:

$$[\alpha_{-x}]^* = \begin{pmatrix} 1 & 0 \\ [\tau_{-x}]_X^* & 1 + [\tau_{-x}]_{X^*}^* \end{pmatrix}^{\wedge 2} : \Lambda^2_{(q,p)}(\Xi) \to \Lambda^2_{(q+x,p+\tau_x(q,p))}(\Xi), \quad (5)$$

where we identified all the cotangent fibres

$$\mathbb{T}^*_{(q,p)}\Xi \cong \mathbb{T}^*_q X \oplus \mathbb{T}^*_p(\mathbb{T}^*_q X) \cong \mathbb{T}^*_q X \oplus \mathbb{T}^*_p X^* \qquad (6)$$

$$[\tau_{-x}]_X^* : \mathbb{T}^* X^* \to \mathbb{T}^* X, \qquad [\tau_{-x}]_{X^*}^* : \mathbb{T}^* X^* \to \mathbb{T}^* X^* . \qquad (7)$$

Finally we obtain:

$$\{[\alpha_{-x}]^* \sigma_B - \sigma_B\}\big|_{(q+x,p+\tau_x(q,p))} = \qquad (8)$$

$$\sum_{j,k=1}^{N} \{[\boldsymbol{T}_{-x}(q,p)]_{jk} dq_j \wedge dq_k + [\boldsymbol{S}_{-x}(q,p)]_{jk} dq_j \wedge dp_k\},$$

with $(\boldsymbol{T}_x(q,p))_{jk} =$

$$= (\partial/\partial q_j)(\tau_x(q,p))_k - (\partial/\partial q_k)(\tau_x(q,p))_j + eB(q)_{jk} - eB(q+x)_{jk}, \quad (9)$$

$$(\boldsymbol{S}_x(q,p))_{jk} = (\partial/\partial p_j)(\tau_x(q,p))_k . \qquad (10)$$

Asking for α_x to be symplectic implies that $\boldsymbol{S} = 0$, hence τ_x does not depend on p. If we fix a point $q_0 \in X$ we can define the function $a(x) := \tau_x(q_0) \in X^*$ and the condition imposed on $\tau_x(q)$ for having a group action leads to $\tau_x(q + q_0) = a(x + q) - a(q)$. Chosing $q_0 = 0$ and a vector potential A for B, the first equation in (9) implies $(\tau_x(q)) := eA(q + x) - eA(q)$.

Let us compute the associate differential action. We set $[(DA(q)) \cdot x]_j := \sum_k [\partial_k A_j(q)] x_k$ and for $x \in X$ we define the vector field in $\mathbb{T}(X \times X^*)$:

$$t_x(q,p) := (\partial/\partial t)\big|_{t=0} \alpha_{-tx}(q,p) = \qquad (11)$$

$$= (-x, (\partial/\partial t)\big|_{t=0} \tau_{-tx}(q)) = (-x, e(DA(q)) \cdot x) .$$

Let us find the associated momentum map. A computation using the definition above (see also [18]) gives: $[i_{\sigma_B}](x, l) = (l + ex \lrcorner B, -x)$, where $(x \lrcorner B)(y) := B(x, y)$. Then we obtain

$$[i_{\sigma_B}](t_x^B)_{(q,p)} = (e(DA(q)) \cdot x - ex \lrcorner B, x)_{(q,p)} = (-d(eA(q) \cdot x), x)_{(q,p)},$$

with $A(q) \cdot x = \sum_{j=1}^{N} A_j(q)x_j$. It follows then that $[i_{\sigma_B}](t_x^B) = d\gamma_x^A$, where $\gamma_x^A(q,p) := x \cdot p - eA(q) \cdot x$ and thus for any direction $\nu \in X$ ($|\nu| = 1$) we have defined the infinitesimal observable magnetic momentum along ν to be $\gamma_\nu^A(q,p) := \nu \cdot (p - eA(q))$. The momentum map ([21]) is thus given by

$$\mu^A : \mathbb{T}^* X \to X^*, \qquad [\mu^A(q,p)](x) := \gamma_x^A(q,p), \tag{12}$$

i.e. $\mu^A(q,p) = p - eA(q)$.

3 The Quantum Picture

A guide in guessing a quantum multiplication for observables is the Weyl-Moyal product of symbols, valid for $B = 0$ and underlying the Weyl form of pseudodifferential theory. A replacement of σ by σ_B, as suggested by Sect. 2.1, triggers a formalism which will be exposed in the following sections. Here we examine a way to extend the multiplication, put it into a form suited for dynamical systems and C^*-norms and study how unbounded observables may be expressed by means of bounded ones.

3.1 The Magnetic Moyal Product

The well-known formula of symbol composition in the usual Weyl quantization can be expressed in terms of the canonical symplectic form. Assume for simplicity that $f, g \in \mathcal{S}(\Xi)$; then Weyl and Moyal proposed the multiplication

$$(f \circ^\hbar g)(\xi) = (2/\hbar)^{2N} \int_\Xi d\eta \int_\Xi d\zeta \, \exp\{-(2i/\hbar)\sigma(\eta,\zeta)\} f(\xi - \eta)g(\xi - \zeta),$$

where $\xi = (q,p)$, $\eta = (y,k)$, $\zeta = (z,l)$. By a simple calculation, one gets

$$(f \circ^\hbar g)(\xi) = (2/\hbar)^{2N} \int_\Xi d\eta \int_\Xi d\zeta \, \exp\left\{-(i/\hbar) \int_{\mathcal{T}(\xi,\eta,\zeta)} \sigma\right\} f(\xi - \eta)g(\xi - \zeta),$$

in terms of the flux of σ through the triangle in phase space

$$\mathcal{T}(\xi,\eta,\zeta) :=< (q-y-z, p-k-l), (q+y-z, p+k-l), (q+z-y, p+l-k) > .$$

A magnetic field B is turned on, with components supposed of class $C_{\text{pol}}^\infty(X)$, i.e. indefinitely derivable and each derivative polynomially bounded. Taking into account the formalism of Sect. 2.1, it is natural to replace σ by σ_B:

$$(f \circ_B^\hbar g)(\xi) = (2/\hbar)^{2N} \int_\Xi d\eta \int_\Xi d\zeta \, \exp\left\{-(i/\hbar) \int_{T(\xi,\eta,\zeta)} \sigma_B\right\} f(\xi-\eta)g(\xi-\zeta) .$$

$$(13)$$

This leads readily to the formula.

$$(f \circ_B^\hbar g)(\xi) = \qquad\qquad (14)$$

$$= (2/\hbar)^{2N} \int_\Xi d\eta \int_\Xi d\zeta \, e^{-(2i/\hbar)\sigma(\eta,\zeta)} \exp\left\{-(i/\hbar) \int_{T(q,y,z)} B\right\} f(\xi-\eta)g(\xi-\zeta) ,$$

where the triangle $T(q,y,z) := < q - y - z, q + y - z, q + z - y >$ is the projection of $T(\xi,\eta,\zeta)$ on the configuration space. We call the composition law $\circ_B^\hbar : \mathcal{S}(\Xi) \times \mathcal{S}(\Xi) \to \mathcal{S}(\Xi)$ *the magnetic Moyal product*. It is well-defined, associative, non-commutative and satisfies $\overline{f \circ_B^\hbar g} = \overline{g} \circ_B^\hbar \overline{f}$. It offers a way to compose observables in a quantum theory of a particle placed in the magnetic field. It is expressed only in terms of B; no vector potential is needed.

3.2 The Magnetic Moyal Algebra

The $*$-algebra $\mathcal{S}(\Xi)$ is much too small for most of the applications. Extensions by absolutely convergent integrals still give rather poor results. One method to get much larger algebras (classes of Hörmander symbols) is by oscillatory integrals. This requires somewhat restricted conditions on the magnetic field, but leads to a powerful filtered symbolic calculus that we intend to develop in a forthcoming paper. Here we indicate an approach by duality.

So let us keep the mild assumption that the components of the magnetic field are $C_{\mathrm{pol}}^\infty(X)$-functions. The duality approach is based on the observation [17, Lem. 14] : For any f, g in the Schwartz space $\mathcal{S}(\Xi)$, we have

$$\int_\Xi d\xi \, (f \circ_B^\hbar g)(\xi) = \int_\Xi d\xi \, (g \circ_B^\hbar f)(\xi) = \int_\Xi d\xi \, f(\xi) g(\xi) = \langle \overline{f}, g \rangle \equiv (f, g) .$$

As a consequence, if f, g and h belong to $\mathcal{S}(\Xi)$, the equalities $(f \circ_B^\hbar g, h) = (f, g \circ_B^\hbar h) = (g, h \circ_B^\hbar f)$ hold. This suggests

Definition 1. *For any distribution $F \in \mathcal{S}'(\Xi)$ and any function $f \in \mathcal{S}(\Xi)$ we define*

$$(F \circ_B^\hbar f, h) := (F, f \circ_B^\hbar h), \quad (f \circ_B^\hbar F, h) := (F, h \circ_B^\hbar f) \quad \text{for all } h \in \mathcal{S}(\Xi) .$$

The expressions $F \circ_B^\hbar f$ and $f \circ_B^\hbar F$ are *a priori* tempered distributions. The Moyal algebra is precisely the set of elements of $\mathcal{S}'(\Xi)$ that preserves regularity by composition.

Definition 2. *The Moyal algebra* $\mathcal{M}(\Xi) \equiv \mathcal{M}_B^\hbar(\Xi)$ *is defined by*

$$\mathcal{M}(\Xi) := \left\{ F \in \mathcal{S}'(\Xi) \mid F \circ_B^\hbar f \in \mathcal{S}(\Xi) \text{ and } f \circ_B^\hbar F \in \mathcal{S}(\Xi) \text{ for all } f \in \mathcal{S}(\Xi) \right\}.$$

For two distributions F and G in $\mathcal{M}(\Xi)$, the Moyal product is extended by $(F \circ_B^\hbar G, h) := (F, G \circ_B^\hbar h)$ for all $h \in \mathcal{S}(\Xi)$.

The set $\mathcal{M}(\Xi)$ with this composition law and the complex conjugation $F \mapsto \overline{F}$ is a unital *-algebra. Actually, this extension by duality also gives compositions $\mathcal{M}(\Xi) \circ_B^\hbar \mathcal{S}'(\Xi) \subset \mathcal{S}'(\Xi)$ and $\mathcal{S}'(\Xi) \circ_B^\hbar \mathcal{M}(\Xi) \subset \mathcal{S}'(\Xi)$. An important result [17, Prop. 23] concerning the Moyal algebra is that it contains $C_{\mathrm{pol,u}}^\infty(\Xi)$, the space of infinitely derivable complex functions on Ξ having polynomial growth at infinity uniformly for all the derivatives.

This duality strategy is often substantiated in calculations by regularization techniques. Further properties of $\circ_B^\hbar$ and $\mathcal{M}(\Xi)$ can be found in [17].

3.3 The Twisted Crossed Product

One thing missing in the pseudodifferential setting is a "good norm" on suitable subclasses of $\mathcal{M}(\Xi)$. We can introduce some useful norms after a partial Fourier transformation $1 \otimes \mathcal{F} : \mathcal{S}(\Xi) \equiv \mathcal{S}(X \times X^*) \to \mathcal{S}(X \times X)$. Setting $(1 \otimes \mathcal{F})(f \circ_B^\hbar g) =: [(1 \otimes \mathcal{F})f] \diamond_B^\hbar [(1 \otimes \mathcal{F})f]$, one gets for $\varphi = (1 \otimes \mathcal{F})f$, $\psi = (1 \otimes \mathcal{F})g$ in $\mathcal{S}(X \times X)$ the multiplication law

$$\left(\varphi \diamond_B^\hbar \psi \right)(q; x) := \tag{15}$$

$$\int_X dy \; \varphi \left(q - \frac{\hbar}{2}(x - y); y \right) \psi \left(q + \frac{\hbar}{2}y; x - y \right) e^{-(i/\hbar)\Phi_B^\hbar(q,x,y)}$$

where $\Phi_B^\hbar(q, x, y)$ is the flux of B through the triangle defined by the points $q - \frac{\hbar}{2}x$, $q - \frac{\hbar}{2}x + \hbar y$ and $q + \frac{\hbar}{2}x$. The partial Fourier transformation also converts the complex conjugation $f \mapsto \overline{f}$ into the involution $\varphi \mapsto \varphi^\diamond$, with $\varphi^\diamond(q; x) := \overline{\varphi(q; -x)}$. Thus one gets a new *-algebra $\left(\mathcal{S}(X \times X), \diamond_B^\hbar, \diamond \right)$, isomorphic with the previous one. This also can be extended in various ways; in particular, there are Moyal type algebras $\mathcal{M}(X \times X) \equiv \mathcal{M}_B^\hbar(X \times X)$ in this setting too. But it is important to note that (15) is just a particular instance of a general mathematical object, *the twisted crossed product*. We give here the main ideas and refer to [25] and [26] for the full theory and to [16] and especially [19] for a comprehensive treatment of its relevance to quantum magnetic fields.

Let $\mathcal{A}$ be a unital C^*-algebra composed of bounded, uniformly continuous functions on X; this algebra is supposed to contain the "admissible" potentials. The idea behind this algebra is that for many problems it is more adequate to consider the whole algebra generated by a potential function and its translations. We shall always assume that $\mathcal{A}$ contains the constant functions as well as the ideal $C_0(X) := \{a : X \to \mathbb{C} \mid f \text{ is continuous and } a(x) \to 0 \text{ for } x \to \infty\}$ (in fact this hypothesis is not necessary everywhere) and is stable by translations, *i.e.* $\theta_x^\hbar(a) := a(\cdot + \hbar x) \in \mathcal{A}$ for all $a \in \mathcal{A}$ and $x \in X$.

Such a C^*-algebra will be called *admissible*. Thus, for any $\hbar \neq 0$, one can define the continuous action of X by automorphisms of $\mathcal{A}$:

$$\theta^\hbar : X \to \mathrm{Aut}(\mathcal{A}), \quad [\theta_x^\hbar(a)](y) := a(y + \hbar x) \,.$$

$\theta^\hbar$ is a group morphism and the maps $X \ni x \mapsto \theta_x^\hbar(a) \in \mathcal{A}$ are all continuous.
We suppose B to have components B_{jk} in $\mathcal{A}$ and we define the map:

$$(q, x, y) \mapsto \omega_B^\hbar(q; x, y) := e^{-(i/\hbar)\Gamma_B(<q, q+\hbar x, q+\hbar x+\hbar y>)} \,,$$

where $\Gamma_B(< q, q+\hbar x, q+\hbar x+\hbar y >)$ denotes the flux of the magnetic field B through the triangle defined by the vertices $q, q+\hbar x, q+\hbar x+\hbar y$ in X. It can be interpreted as a map $\omega_B^\hbar : X \times X \to C(X; \mathbb{T})$, $[\omega_B^\hbar(x, y)](q) := \omega_B^\hbar(q; x, y)$ with values in the set of continuous functions on X taking values in the 1-torus $\mathbb{T} := \{z \in \mathbb{C} \mid |z| = 1\}$. It is easy to see by Stokes Theorem and the equation $dB = 0$ that $\omega_B^\hbar$ satisfies *the 2-cocycle condition*

$$\omega_B^\hbar(x, y)\omega_B^\hbar(x + y, z) = \theta_x^\hbar[\omega_B^\hbar(y, z)]\omega_B^\hbar(x, y + z), \quad \forall x, y, z \in X \,.$$

It is also *normalized*, i.e. $\omega_B^\hbar(x, 0) = 1 = \omega_B^\hbar(0, x)$, $\forall x \in X$.

The quadruplet $(\mathcal{A}, \theta^\hbar, \omega_B^\hbar, X)$ is a *magnetic* example of *an abelian twisted C^*-dynamical system* $(\mathcal{A}, \theta, \omega, X)$. In the general case X is an abelian second countable locally compact group, $\mathcal{A}$ is an abelian C^*-algebra, θ is a continuous morphism from X to the group of automorphisms of $\mathcal{A}$ and ω is a continuous 2-cocycle with values in the group of all unitary elements of $\mathcal{A}$.

Given any abelian twisted C^*-dynamical system, a natural C^*-algebra can be defined. We recall its construction. Let $L^1(X; \mathcal{A})$ be the set of Bochner integrable functions on X with values in $\mathcal{A}$, with the L^1-norm $\|\varphi\|_1 := \int_X dx \, \|\varphi(x)\|_\mathcal{A}$. For any $\varphi, \psi \in L^1(X; \mathcal{A})$ and $x \in X$, we define the product

$$(\varphi \diamond \psi)(x) := \int_X dy \, \theta_{\frac{y-x}{2}}[\varphi(y)] \, \theta_{\frac{y}{2}}[\psi(x - y)] \, \theta_{-\frac{x}{2}}[\omega(y, x - y)]$$

and the involution $\phi^\diamond(x) := \theta_{-\frac{x}{2}}[\omega(x, -x)^{-1}]\phi(-x)^*$. In this way, one gets a Banach $*$-algebra.

Definition 3. *The enveloping C^*-algebra of $L^1(X, \mathcal{A})$ is called the twisted crossed product and is denoted by $\mathcal{A} \rtimes_\theta^\omega X$. It is the completion of $L^1(X; \mathcal{A})$ under the C^*-norm*

$$\| \varphi \| := \sup\{\| \pi(\varphi) \|_{B(\mathcal{H})} \mid \pi : L^1(X; \mathcal{A}) \to B(\mathcal{H}) \text{ representation}\} \,.$$

It is easy to see that, with $\theta = \theta^\hbar$, $\omega = \omega_B^\hbar$, one gets exactly the structure exposed above restricted to $\mathcal{S}(X \times X) \subset L^1(X; \mathcal{A})$. The C^*-algebra $\mathcal{A} \rtimes_{\theta^\hbar}^{\omega_B^\hbar} X$ will be denoted simply by $\mathfrak{C}_B^\hbar(\mathcal{A})$. In the magnetic case $\omega_B^\hbar(x, -x) = 1$.

After a partial Fourier transformation we get the C^*-algebra $\mathfrak{B}_B^\hbar(\mathcal{A}) := (1 \otimes \mathcal{F}^{-1})\mathfrak{C}_B^\hbar(\mathcal{A})$, which is another extension of the $*$-subalgebra $\mathcal{S}(\Xi)$ endowed with complex conjugation and the multiplication (14).

3.4 Abstract Affiliation

When working with a self-adjoint operator H in a Hilbert space $\mathcal{H}$, it might be useful to know that the functional calculus of H (its resolvent for example) belongs to some special C^*-algebra of $B(\mathcal{H})$. Our representation-free approach forces us to use an abstract version, borrowed from [1].

Definition 4. An observable affiliated to a C^*-algebra $\mathfrak{C}$ is a morphism Φ : $C_0(\mathbb{R}) \to \mathfrak{C}$.

Recall that a function $h \in C^\infty(X^*)$ is called an elliptic symbol of type $s \in \mathbb{R}$ if (with $\langle p \rangle := \sqrt{1 + p^2}$) $|(\partial^\alpha h)(p)| \leq c_\alpha \langle p \rangle^{s-|\alpha|}$ for all $p \in X^*, \alpha \in \mathbb{N}^N$ and there exist $R > 0$ and $c > 0$ such that $c \langle p \rangle^s \leq h(p)$ for all $p \in X^*$ and $|p| \geq R$. Such a function is naturally contained in $C^\infty_{\mathrm{pol,u}}(\Xi)$, thus in $\mathcal{M}(\Xi)$. For any $z \notin \mathbb{R}$, we also set $r_z : \mathbb{R} \to \mathbb{C}$ by $r_z(t) := (t - z)^{-1}$. $BC^\infty(X)$ is the space of all functions in $C^\infty(X)$ with bounded derivatives of any order.

Theorem 1. Assume that B is a magnetic field whose components belong to $\mathcal{A} \cap BC^\infty(X)$. Then each real elliptic symbol h of type $s > 0$ defines an observable $\Phi^\hbar_{B,h}$ affiliated to $\mathfrak{B}^\hbar_B(\mathcal{A})$, such that for any $z \notin \mathbb{R}$ one has

$$(h - z) \circ^\hbar_B \Phi^\hbar_{B,h}(r_z) = 1 = \Phi^\hbar_{B,h}(r_z) \circ^\hbar_B (h - z) . \qquad (16)$$

In fact one has $\Phi^\hbar_{B,h}(r_z) \in (1 \otimes \mathcal{F})(L^1(X; \mathcal{A})) \subset \mathcal{S}'(\Xi)$, so the compositions can be interpreted as $\mathcal{M}(\Xi) \times \mathcal{S}'(\Xi) \to \mathcal{S}'(\Xi)$ and $\mathcal{S}'(\Xi) \times \mathcal{M}(\Xi) \to \mathcal{S}'(\Xi)$.

The proof can be found in [20] and consists in starting with the usual inverse for function multiplication and control the corrections using the L^1-norm in the algebra $\mathfrak{C}^\hbar_B(\mathcal{A})$. This result is basic for our approach to spectral analysis for Hamiltonians with magnetic fields in Sections 6.1 and 6.2. A represented version will be found in Sect. 5.3.

4 The Limit $\hbar \to 0$

The quantum and classical descriptions we have given for a particle in a magnetic field, can be gathered into a common "continuous" structure indexed by the Plank' constant $\hbar \in [0, \hbar_0]$, by the procedure of strict deformation quantization. Our strategy follows [10] and the details may be found in our paper [18]. The main idea is to define for each value of $\hbar \in [0, \hbar_0]$ an algebra of bounded observables and using a common dense subalgebra, to prove that the family is in fact a continuous field of C^*-algebras (see [28, 29]).

So far we have defined for $\hbar > 0$ a C^*-algebra $\mathfrak{B}^\hbar_B(\mathcal{A})$ describing the observables of the quantum particle in a magnetic field B. Let us define now for $\hbar = 0$ the C^*-algebra $\mathfrak{B}^0_B(\mathcal{A}) := C(X^*; \mathcal{A})$ with the usual commutative product of functions $(f \circ^0_B g := fg)$ and the involution defined by complex conjugation. Setting $\mathcal{A}^\infty := \{a \in \mathcal{A} \cap C^\infty(X) \mid \partial^\alpha a \in \mathcal{A}, \forall \alpha \in \mathbb{N}^N\}$ one verifies that the linear space $\mathfrak{A} := \mathcal{S}(X^*; \mathcal{A})$ is closed for any Moyal product $\circ_B^\hbar$, also for $\hbar = 0$. For $\hbar \in [0, \hbar_0]$ we denote by $\|.\|_\hbar$ the C^*-norm in $\mathfrak{B}_B^\hbar(\mathcal{A})$.

Moreover let us remark that the real algebra $\mathfrak{A}_0 := \{f \in \mathfrak{A} \mid \overline{f} = f\}$ is a Poisson sub-algebra of $C^\infty(\Xi; \mathbb{R})$ endowed with the magnetic Poisson braket associated to the magnetic field B. It is easy to verify that one has the

- Completness condition: $\mathfrak{A} = \mathbb{C} \otimes \mathfrak{A}_0$ is dense in each C^*-algebra $\mathfrak{B}_B^\hbar(\mathcal{A})$.

The following convergences are proved by direct computation [18]:

- von Neumann condition: For f and g in $\mathfrak{A}_0$ one has

$$\lim_{\hbar \to 0} \| \frac{1}{2} \left(f \circ_B^\hbar g + g \circ_B^\hbar f \right) - fg \|_\hbar = 0 \,.$$

- Dirac condition: For f and g in $\mathfrak{A}_0$ one has

$$\lim_{\hbar \to 0} \| \frac{1}{i\hbar} \left(f \circ_B^\hbar g - g \circ_B^\hbar f \right) - \{f, g\}_B \|_\hbar = 0 \,.$$

An argument using a theorem in [24], concerning continuous fields of twisted crossed-products, allows to prove the following continuity result [18]:

- Rieffel condition: For $f \in \mathfrak{A}_0$ the map $[0, \hbar_0] \ni \hbar \mapsto \|f\|_\hbar \in \mathbb{R}$ is continuous.

Following [10,28,29] we say that we have a *strict deformation quantization* of the Poisson algebra $\mathfrak{A}_0$.

5 The Schrödinger Representation

A complete overview of the formalism is achieved only after representations in Hilbert spaces are also outlined. This will put forward magnetic potentials, but in a gauge covariant way. We obtain integrated forms of covariant representations as well as the magnetic version of pseudodifferential operators. Unbounded pseudodifferential operators have their resolvents in well-controlled C^*-algebras composed of bounded ones, as a consequence of Sect. 3.4; this is basic to the spectral results of Sect. 6.

5.1 Representations of the Twisted Crossed Product

Fortunately, non-degenerate representations of twisted crossed product C^*-algebras admit a complete classification. We recall that the representation $\rho : \mathfrak{C} \to B(\mathcal{H})$ of the C^*-algebra $\mathfrak{C}$ in the Hilbert space $\mathcal{H}$ is called *non-degenerate* if $\rho(\mathfrak{C})\mathcal{H}$ generates $\mathcal{H}$. Since $\mathcal{A} \rtimes_\theta^\omega X$ was obtained from the twisted C^*-dynamical system $(\mathcal{A}, \theta, \omega, X)$, one may expect that the representations of $\mathcal{A} \rtimes_\theta^\omega X$ can be deduced from a certain kind of Hilbert representations of the system $(\mathcal{A}, \theta, \omega, X)$.

Definition 5. *Given a twisted dynamical system* $(\mathcal{A}, \theta, \omega, X)$, *we call covariant representation a Hilbert space* $\mathcal{H}$ *together with two maps* $r : \mathcal{A} \to B(\mathcal{H})$ *and* $U : X \to \mathcal{U}(\mathcal{H})$ *satisfying:*

- r *is a non-degenerate representation,*
- U *is strongly continuous and* $U(x)U(y) = r[\omega(x,y)]U(x+y) \ \forall x, y \in X$,
- $U(x)r(a)U(x)^* = r[\theta_x(a)], \quad \forall x \in X, \ a \in \mathcal{A}$.

It can be shown that there is a one-to-one correspondence between covariant representations of $(\mathcal{A}, \theta, \omega, X)$ and non-degenerate representations of $\mathcal{A} \rtimes_\theta^\omega X$. The following evident statement will be needed.

Lemma 1. *For* $(\mathcal{H}, r, U)$ *covariant representation of* $(\mathcal{A}, \theta, \omega, X)$, *the map* $r \rtimes U$ *defined on* $L^1(X; \mathcal{A})$ *by the formula*

$$(r \rtimes U)\varphi := \int_X dx \, r \left[\theta_{x/2}\big(\varphi(x)\big) \right] U(x)$$

extends to a representation of $\mathcal{A} \rtimes_\theta^\omega X$, *called the integrated form of* (r, U).

For our magnetic C^*-dynamical systems one constructs covariant representations by choosing vector potentials. We shall call them and their integrated forms *Schrödinger representations*, inspired by the case $B = 0$. For A such that $B = dA$ and for points $x, y \in X$, we define $\Gamma_A([x,y]) := \int_{[x,y]} A$ the circulation of A through the segment $[x,y] := \{sx + (1-s)y \mid s \in [0,1]\}$. By Stokes Theorem we have

$$\Gamma_B(< q, q + \hbar x, q + \hbar x + \hbar y >) =$$

$$= \Gamma_A([q, q + \hbar x])\Gamma_A([q + \hbar x, q + \hbar x + \hbar y])\Gamma_A([q + \hbar x + \hbar y, q]),$$

leading to

$$\omega_B^\hbar(q; x, y) = \lambda_A^\hbar(q; x)\lambda_A^\hbar(q + \hbar x; y) \left[\lambda_A^\hbar(q; x + y) \right]^{-1}, \qquad (17)$$

where we set $\lambda_A^\hbar(q; x) := \exp\{-(i/\hbar)\Gamma_A([q, q + \hbar x])\}$. We define $\mathcal{H} := L^2(X)$, $r : \mathcal{A} \to B[L^2(X)]$, $r(a) :=$ the operator of multiplication by $a \in \mathcal{A}$ and

$$[U_A^\hbar(x)u](q) := \lambda_A^\hbar(q; x)u(q + \hbar x), \quad \forall q, x \in X, \ \forall u \in L^2(X).$$

It follows easily that $(\mathcal{H}, r, U_A^\hbar)$ is a covariant representation of $(\mathcal{A}, \theta^\hbar, \omega_B^\hbar, X)$. The integrated form associated to $(\mathcal{H}, r, U_A^\hbar)$ is $\mathfrak{Rep}_A^\hbar \equiv r \rtimes U_A^\hbar : \mathfrak{C}_B^\hbar(\mathcal{A}) \to B\left[L^2(X)\right]$, given explicitly on $L^1(X; \mathcal{A})$ by

$$\left[\mathfrak{Rep}_A^\hbar(\varphi)u \right](x) = \hbar^{-N} \int_X dy \, e^{(i/\hbar)\Gamma_A([x,y])} \varphi\left(\frac{x+y}{2}, \frac{y-x}{\hbar} \right) u(y). \qquad (18)$$

5.2 Pseudodifferential Operators

Let us compose $\mathfrak{Rep}_A^\hbar$ with the partial Fourier transformation in order to get a representation $\mathfrak{Op}_A^\hbar := \mathfrak{Rep}_A^\hbar \circ (1 \otimes \mathcal{F}) : \mathfrak{B}_B^\hbar(\mathcal{A}) \to B(\mathcal{H})$. A calculation on suitable subsets of $\mathfrak{B}_B^\hbar(\mathcal{A})$ (on $\mathcal{S}(\Xi)$ for example) gives the explicit action

$$\left[\mathfrak{Op}_A^\hbar(f)u\right](x) = \tag{19}$$

$$= \hbar^{-N} \int_X \int_{X^*} dy \, dk \, e^{(i/\hbar)(x-y)\cdot k} e^{-(i/\hbar)\Gamma_A([x,y])} f\left(\frac{x+y}{2}, k\right) u(y) \, .$$

We call $\mathfrak{Op}_A^\hbar(f)$ *the magnetic pseudodifferential operator associated to the symbol f*. A posteriori, one may say that *la raison d'être* of the composition (13) is to ensure the equality: $\mathfrak{Op}_A^\hbar(f)\mathfrak{Op}_A^\hbar(g) = \mathfrak{Op}_A^\hbar(f \circ_B^\hbar g)$. One also has $\mathfrak{Op}_A^\hbar(f)^* = \mathfrak{Op}_A^\hbar(\overline{f})$. Some properties of $\mathfrak{Op}_A^\hbar$ can be found in [17] and [19].

Now it is easy to see what gauge covariance is at the level of the two representations $\mathfrak{Rep}_A^\hbar$ and $\mathfrak{Op}_A^\hbar$. If two 1-forms A and A' are equivalent ($A' = A + d\rho$) then one will get unitarily equivalent representations:

$$\mathfrak{Op}_{A'}^\hbar(f) = e^{(i/\hbar)\rho}\mathfrak{Op}_A^\hbar(f)e^{(-i/\hbar)\rho} \quad \text{and} \quad \mathfrak{Rep}_{A'}^\hbar(\varphi) = e^{(i/\hbar)\rho}\mathfrak{Rep}_A^\hbar(\varphi)e^{(-i/\hbar)\rho} \, .$$

We refer to [17] for a comparaison with a quantization procedure $f \mapsto \mathfrak{Op}^{\hbar,A}(f)$, combining (in an inappropriate order) the usual, non-magnetic calculus with the minimal coupling rule $(x,p) \mapsto (x, p - A(x))$. It is *not* gauge-covariant, so that it is not suitable as a real quantization procedure.

Finally let us quote a result linking $\mathcal{M}(\Xi)$ with $\mathfrak{Op}_A^\hbar$ [17, Prop. 21] : For any vector potential A in $C_{\text{pol}}^\infty(X)$, $\mathfrak{Op}^A$ is an isomorphism of *-algebras between $\mathcal{M}(\Xi)$ and $\mathcal{L}[\mathcal{S}(X)] \cap \mathcal{L}[\mathcal{S}'(X)]$, where $\mathcal{L}[\mathcal{S}(X)]$ and $\mathcal{L}[\mathcal{S}'(X)]$ are, respectively, the spaces of linear continuous operators on $\mathcal{S}(X)$ and $\mathcal{S}'(X)$.

5.3 A New Justification: Functional Calculus

We give here a new justification of our formalism. It is obvious that if one gives some convincing reason for working with (19), then the remaining part can be deduced as a necessary consequence, by reversing the arguments.

Let us accept that our quantum particle placed in a magnetic field is described by the family of elementary operators $Q_1, \ldots, Q_N; (\Pi_A^\hbar)_1, \ldots, (\Pi_A^\hbar)_N$, where Q_j is the operator of multiplication by x_j and $(\Pi_A^\hbar)_j := P_j^\hbar - A_j = -i\hbar\partial_j - A_j$ is the j'th component of the magnetic momentum defined by a vector potential A with $dA = B$ (these may be considered as quantum observables associated to the position and the momentum map for the translation group). Then $\mathfrak{Op}_A^\hbar$ should be a functional calculus $f \mapsto \mathfrak{Op}_A^\hbar(f) \equiv f(Q, \Pi_A^\hbar)$ for this family of non-commuting self-adjoint operators. The scheme is: (i) consider the commutation relations satisfied by $Q, \Pi_A^\hbar$, (ii) condense them in a global, exponential form, (iii) define $\mathfrak{Op}_A^\hbar(f)$ by decomposing f as a continuous linear combination of exponentials. We mention that exactly this argument leads to the usual Weyl calculus ($B = 0$).

So let us take into account the following commutation relations, easy to check: $i[Q_j, Q_k] = 0$, $i[\Pi_{A,j}^\hbar, Q_k] = \hbar\delta_{j,k}$, $i[\Pi_{A,j}^\hbar, \Pi_{A,k}^\hbar] = \hbar B_{kj}(Q)$, $\forall j, k = 1, \ldots, N$. A convenient global form may be given in terms of *the magnetic Weyl system*. Recall the unitary group $\left(e^{iQ \cdot p}\right)_{p \in X^*}$ of the position as well as *the magnetic translations* $\left(U_A^\hbar(q) := e^{iq \cdot \Pi_A^\hbar}\right)_{q \in X}$, given explicitly in the Hilbert space $\mathcal{H} := L^2(X)$ by

$$U_A^\hbar(x) = e^{-(i/\hbar)\Gamma_A([Q,Q+\hbar x])} e^{ix \cdot P^\hbar} , \tag{20}$$

which is just another way to write (17). The family $\left(U_A^\hbar(x)\right)_{x \in X}$ satisfies

$$U_A^\hbar(x) U_A^\hbar(x') = \omega_B^\hbar(Q; x, x') U_A^\hbar(x + x'), \quad x, x' \in X ,$$

where we set $\omega_B^\hbar(q; x, x') := e^{-(i/\hbar)\Gamma_B(<q,q+\hbar x, q+\hbar x+\hbar x'>)}$.

Now the magnetic Weyl system is the family $\left(W_A^\hbar(q,p)\right)_{(q,p) \in \Xi}$ of unitary operators in $\mathcal{H}$ given by

$$W_A^\hbar(q,p) := e^{-i\sigma\left((q,p),(Q,\Pi_A^\hbar)\right)} = e^{-i(\hbar/2)q \cdot p} e^{-iQ \cdot p} U_A^\hbar(x)$$

and it satisfies for all $(q,p), (q',p') \in \Xi$

$$W_A^\hbar(q,p) W_A^\hbar(q',p') = e^{(i/2)\sigma\left((q,p),(q',p')\right)} \omega_B^\hbar(Q; q, q') W_A^\hbar(q + q', p + p') .$$

To construct $\mathfrak{Op}_A^\hbar(f) \equiv f(Q, \Pi_A^\hbar))$ one does not dispose of a spectral theorem. Having the functional calculus with a C_0-group in mind, one proposes

$$\mathfrak{Op}_A^\hbar(f) := \int_\Xi d\xi \, (\mathfrak{F}_\Xi f)(\xi) W_A^\hbar(\xi) ,$$

where $(\mathfrak{F}_\Xi f)(\xi) := \int_\Xi d\eta \, e^{-i\sigma(\xi,\eta)} f(\eta)$ is the symplectic Fourier transform (with a suitable Haar measure). Some simple replacements lead to (19). Details concerning this construction may be found in [19] together with an analysis of the role of the algebra $\mathcal{A}$.

5.4 Concrete Affiliation

If $\mathcal{H}$ is a Hilbert space and $\mathfrak{C}$ is a C^*-subalgebra of $B(\mathcal{H})$, then a self-adjoint operator H in $\mathcal{H}$ defines an observable Φ_H affiliated to $\mathfrak{C}$ if and only if $\Phi_H(\eta) := \eta(H)$ belongs to $\mathfrak{C}$ for all $\eta \in C_0(\mathbb{R})$. A sufficient condition is that $(H - z)^{-1} \in \mathfrak{C}$ for some $z \in \mathbb{C}$ with $\Im m z \neq 0$. Thus an observable affiliated to a C^*-algebra is the abstract version of the functional calculus of a self-adjoint operator. By combining Theorem 1 with the representations introduced above one gets

Corollary 1. *We are in the framework of Theorem 1. Let A be a continuous vector potential that generates B. Then* $\mathfrak{Op}_A^{\hbar}(\mathrm{h})$ *defines a self-adjoint operator* $\mathrm{h}(\Pi_A^{\hbar})$ *in* $\mathcal{H}$ *with domain given by the image of the operator* $\mathfrak{Op}_A^{\hbar}\left[(\mathrm{h}-z)^{-1}\right]$ *(which do not depend on* $z \notin \mathbb{R}$*). This operator is affiliated to* $\mathfrak{Op}_A^{\hbar}\left[\mathfrak{B}_B^{\hbar}(\mathcal{A})\right] = \mathfrak{Rep}_A^{\hbar}\left[\mathfrak{B}_B^{\hbar}(\mathcal{A})\right]$.

6 Applications to Spectral Analysis

It seems to be common knowledge the fact that "the essential spectrum of partial differential operators depend only on the behaviour at infinity of the coefficients". But precise and general results emerged quite recently; some references are [1, 4–7, 11, 12], [15]. We review here a Theorem of [20] under simplifying assumptions (a scalar potential V can be easily added). Compared with the nice results of [7], it is much better if B (and V) is bounded, but we cannot say anything when B is unbounded towards infinity, case generously treated in [7]. The theory is in terms of C^*-algebras, quasi-orbits of some dynamical systems and asymptotic Hamiltonians associated to these quasi-orbits. The same asymptotic Hamiltonians play a role in localisation results (leading to non-propagation properties for the evolution group), extracted in an abridged form from [20] and [2].

6.1 The Essential Spectrum

We give a description of the essential spectrum of observables affiliated to the C^*-algebra $\mathfrak{B}_B^{\hbar}(\mathcal{A})$. For the generalised magnetic Schrödinger operators of Theorem 1, this is expressed in terms of the spectra of so-called *asymptotic operators*. The affiliation criterion and the algebraic formalism introduced above play an essential role in the proof of this result. We start by recalling some definitions in relation with topological dynamical systems.

By Gelfand theory, the abelian C^*-algebra $\mathcal{A}$ is isomorphic to the C^*-algebra $C(S_{\mathcal{A}})$, where $S_{\mathcal{A}}$ is the spectrum of $\mathcal{A}$. Since $\mathcal{A}$ was assumed unital and contains $C_0(X)$, $S_{\mathcal{A}}$ is a compactification of X. We shall therefore identify X with a dense open subset of $S_{\mathcal{A}}$. By stability under translations, the group law $\theta : X \times X \to X$ extends then to a continuous map $\tilde{\theta} : X \times S_{\mathcal{A}} \to S_{\mathcal{A}}$. Thus the complement $F_{\mathcal{A}}$ of X in $S_{\mathcal{A}}$ is closed and invariant; it is the space of a compact topological dynamical system. For any $\mathfrak{x} \in F_{\mathcal{A}}$, let us call the set $\{\tilde{\theta}(x, \mathfrak{x}) \mid x \in X\}$ *the orbit generated by* $\mathfrak{x}$, and its closure a *quasi-orbit*. Usually there exist many elements of $F_{\mathcal{A}}$ that generate the same quasi-orbit. In the sequel, we shall often encounter the restriction $a|_F$ of an element $a \in \mathcal{A} \equiv C(S_{\mathcal{A}})$ to a quasi-orbit F. Naturally $a|_F$ is an element of $C(F)$, but this algebra can be realized as a subalgebra of $BC_u(X)$. By a slight abuse of notation, we shall identify $a|_F$ with a function defined on X, thus inducing a multiplication operator in $\mathcal{H}$.

The calculation of the essential spectrum may be performed at an abstract level, *i.e.* without using any representation, (see [20] where a potential V is also included). We present, for convenience, a represented version.

Theorem 2. *Let B be a magnetic field whose components belong to $\mathcal{A} \cap BC^\infty(X)$. Assume that $\{F_\nu\}_\nu$ is a covering of $F_{\mathcal{A}}$ by quasi-orbits. Then for each real elliptic symbol* h *of type $s > 0$, if A, A_ν are continuous vector potentials respectively for B, $B_\nu \equiv B_{F_\nu}$, one has*

$$\sigma_{\text{ess}}[\text{h}(\Pi_A^\hbar)] = \overline{\bigcup_\nu \sigma[\text{h}(\Pi_{A_\nu}^\hbar)]}. \tag{21}$$

The operators $\text{h}(\Pi_{A_\nu}^\hbar)$ are the asymptotic operators mentioned earlier. All the spectra in (21) are only depending on the respective magnetic fields. Examples may be found in [20], see also [15]. Some related results may be found in the recent paper [11].

6.2 A Non-Propagation Result

We finally describe, following [20], how the localization results proved in [2] in the case of Schrödinger operators without magnetic field can be extended to the situation where a magnetic field is present. Once again, the algebraic formalism and the affiliation criterion introduced above play an essential role in the proofs. For any quasi-orbit F, let $\mathfrak{N}_F$ be the family of sets of the form $W = \mathcal{W} \cap X$, where $\mathcal{W}$ is any element of a base of neighbourhoods of F in $S_{\mathcal{A}}$. We write χ_W for the characteristic function of W.

Theorem 3. *Let B be a magnetic field whose components belong to $\mathcal{A} \cap BC^\infty(X)$ and let* h *be a real elliptic symbol of type $s > 0$. Assume that $F \subset F_{\mathcal{A}}$ is a quasi-orbit. Let A, A_F be continuous vector potentials for B and B_F, respectively. If $\eta \in C_0(\mathbb{R})$ with $\text{supp}(\eta) \cap \sigma\left[\text{h}(\Pi_{A_F}^\hbar)\right] = \emptyset$, then for any $\epsilon > 0$ there exists $W \in \mathfrak{N}_F$ such that $\left\| \chi_W(Q) \, \eta\left[\text{h}(\Pi_A^\hbar)\right] \right\| \leq \epsilon$. In particular, the following inequality holds uniformly in $t \in \mathbb{R}$ and $u \in \mathcal{H}$:*

$$\left\| \chi_W(Q) \, e^{-it\text{h}(\Pi_A^\hbar)} \, \eta\left[\text{h}(\Pi_A^\hbar)\right] u \right\| \leq \epsilon \|u\|.$$

The last statement of this theorem gives a precise meaning to the notion of non-propagation. We refer to [2] for physical explanations and interpretations of this result as well as for some examples.

Acknowledgements

This work was deeply influenced by Vladimir Georgescu to whom we are grateful. Part of the results described in this paper were obtained while the authors were visiting the University of Geneva; many thanks are due to Werner Amrein for his kind hospitality. We are greateful to Serge Richard for enjoyable collaboration. M.M. thanks Joseph Avron for a stimulating discussion and encouragements in developing this work. We also acknowledge partial support from the CERES Program of the Romanian Ministery of Education and Research.

References

1. W.O. Amrein, A. Boutet de Monvel, V. Georgescu: C_0-Groups, Commutator Methods and Spectral Theory of N-Body Hamiltonians, (Birkhäuser Verlag, 1996).
2. W.O. Amrein, M. Măntoiu, R. Purice: Ann. Henri Poincaré **3**, 1215 (2002).
3. H. D. Cornean, G. Nenciu: Ann. Henri Poincaré **1**, 203 (2000).
4. V. Georgescu, A. Iftimovici: preprint mp-arc01-99.
5. V. Georgescu, A. Iftimovici: Commun. Math. Phys. **228**, 519 (2002).
6. V. Georgescu, A. Iftimovici: C^*-Algebras of Quantum Hamiltonians. In Operator Algebras and Mathematical Physics, Constanţa, 2001, ed by J.-M. Combes, J. Cuntz, G. A. Elliott et al, (Theta Foundation, 2003).
7. B. Helffer, A. Mohamed: Ann. Inst. Fourier **38**, 95 (1988).
8. M. V. Karasev, T. A. Osborn: J. Math. Phys. **43**, 756 (2002).
9. M. V. Karasev, T. A. Osborn: J. Phys. A **37**, 2345 (2004).
10. N. P. Landsman: Mathematical Topics Between Classical and Quantum Mechanics, (Springer-Verlag, New-York, 1998), 529 pp.
11. Y. Last, B. Simon: preprint mp-arc 05-112.
12. R. Lauter, V. Nistor: Analysis of Geometric Operators on Open Manifolds: a Groupoid Approach. In Quantization of Singular Symplectic Quotients, ed by N. P. Landsman, M. Pflaum and M. Schlichenmaier, (Birkhäuser, Basel, 2001).
13. R. Lauter, B. Monthubert and V. Nistor: Documenta Math. **5**, 625 (2000).
14. J. M. Luttinger: Phys. Rev. **84**, 814 (1951).
15. M. Măntoiu: J. reine angew. Math. **550**, 211 (2002).
16. M. Măntoiu, R. Purice: The Algebra of Observables in a Magnetic Field. In: Mathematical Results in Quantum Mechanics (Taxco, 2001), ed by R. Weder, P. Exner and B. Grébert (Amer. Math. Soc., Providence, RI, 2002) pp 239–245.
17. M. Măntoiu, R. Purice: J. Math.Phys. **45**, 1394 (2004).
18. M. Măntoiu, R. Purice: Strict deformation quantization for a particle in a magnetic field, to appear in J. Math. Phys. 2005.
19. M. Măntoiu, R. Purice, S. Richard: preprint mp-arc 04-76.
20. M. Măntoiu, R. Purice, S. Richard: preprint mp-arc 05-84.
21. J. E. Marsden, T. S. Raţiu: Introduction to Mechanics and Symmetry, 2nd edn, (Springer-Verlag, Berlin, New York, 1999), 582 pp.
22. G. Nenciu: preprint mp-arc 00-96.
23. V. Nistor, A. Weinstein, P. Xu: Pacific J. Math., **189**, 117 (1999).
24. M. Nilsen: Indiana Univ. Math. J. **45**, 436 (1996).
25. J.A. Packer, I. Raeburn: Math. Proc. Camb. Phil. Soc. **106**, 293 (1989).
26. J.A. Packer, I. Raeburn: Math. Ann. **287**, 595 (1990).
27. G. D. Raikov, M. Dimassi: Cubo Mat. Educ. **3**, 317 (2001).

28. M. Rieffel: Math. Ann. **283**, 631 (1989).
29. M. Rieffel: Memoirs of the AMS, **106**, 93 pp (1993).
30. J. Renault: *A Groupoid Approach to C^*-Algebras*, (Springer, Berlin, 1980) 160 pp.

Fractal Weyl Law for Open Chaotic Maps

Stéphane Nonnenmacher

Service de Physique Théorique, CEA/DSM/PhT, Unité de recherche associée au CNRS, CEA/Saclay , 91191 Gif-sur-Yvette, France
snonnenmacher@cea.fr

1 Introduction

We summarize our work in collaboration with Maciej Zworski [16], on the semiclassical density of resonances for a quantum open system, in the case when the associated classical dynamics is uniformly hyperbolic, and the set of *trapped* trajectories is a *fractal repeller*. The system we consider is not a Hamiltonian flow, but rather a "symplectic map with a hole" on a compact phase space (the 2-torus). Such a map can be considered as a model for the Poincaré section associated with a scattering Hamiltonian on $\mathbb{R}^2$, at some positive energy; the "hole" represents the points which never return to the Poincaré section, that is, which are scattered to infinity. We then quantize this open map, obtaining a sequence of subunitary operators, the eigenvalues of which are interpreted as resonances.

We are especially interested in the asymptotic density of "long-living resonances", representing metastable states which decay in a time bounded away from zero (as opposed to "short resonances", associated with states decaying instantaneously). Our results (both numerical and analytical) support the conjectured *fractal Weyl law*, according to which the number of long-living resonances scales as $\hbar^{-d}$, where d is the (partial) fractal dimension of the trapped set.

1.1 Generalities on Resonances

A Hamiltonian dynamical system (say, $H(q,p) = p^2 + V(q)$ on $\mathbb{R}^{2n}$) is said to be "closed" at the energy E when the energy surface Σ_E is a compact subset of the phase space. The associated quantum operator $H_\hbar = -\hbar^2 \Delta + V(q)$ then admits discrete spectrum near the energy E (for small enough $\hbar$). Furthermore, if E is nondegenerate (meaning that the flow of H has no fixed point on Σ_E), then the semiclassical density of eigenvalues is given by the celebrated Weyl's law [11]:

$$\#\left\{\operatorname{Spec}(H) \cap [E - \delta, E + \delta]\right\} = \frac{1}{(2\pi\hbar)^n} \iint_{|H(q,p)-E|<\delta} \mathrm{d}q\,\mathrm{d}p + \mathcal{O}(\hbar^{1-n}).$$

$$(1)$$

S. Nonnenmacher: *Fractal Weyl Law for Open Chaotic Maps*, Lect. Notes Phys. **690**, 435–450 (2006)
www.springerlink.com

This formula connects the density of quantum eigenvalues with the geometry of the classical energy surface Σ_E. It shows that the number of resonances in an interval of type $[E + C\hbar, E - C\hbar]$ is of order $\mathcal{O}(\hbar^{1-n})$. Intuitively, this Weyl law means that one quantum state is associated with each phase space cell of volume $(2\pi\hbar)^n$.

When Σ_E is non-compact, or even of infinite volume, the spectral properties of $H_\hbar$ are different. Consider the case of a scattering situation, when the potential $V(q)$ is of compact support: for any $E > 0$, Σ_E is unbounded, and $H_\hbar$ admits absolutely continuous spectrum on $[0, \infty)$. However, one can meromorphically continue the resolvent $(z - H_\hbar)^{-1}$ across the real axis from the upper half-plane into the lower half-plane. In general, this continuation will have discrete poles $\{z_j = E_j - \Im\gamma_j\}$ with "widths" $\gamma_j > 0$, which are the *resonances* of $H_\hbar$.

Physically, each resonance is associated with a *metastable state*: a (not square-integrable) solution of the Schrödinger equation at the energy z_j, which decays like $e^{-t\gamma_j/\hbar}$ when $t \to +\infty$. In spectroscopy experiments, one measures the energy dependence of some scattering cross-section $\sigma(E)$. Each resonance z_j imposes a Lorentzian component $\frac{\gamma_j}{(E-E_j)^2+\gamma_j^2}$ on $\sigma(E)$; a resonance z_j will be detectable on the signal $\sigma(E)$ only if its Lorentzian is well-separated from the ones associated with nearby resonances of comparable widths, therefore iff $|E_j' - E_j| \gg \gamma_j$. This condition of "well-separability" is NOT the one we will be interested in here. We will rather consider the order of magnitude of each resonance lifetime $\hbar/\gamma_j$, independently of the nearby ones, in the semiclassical régime: a resonant state will be "visible", or "long-living", if $\gamma_j = \mathcal{O}(\hbar)$. Our objective will be to count the number of resonances z_j in boxes of the type $\{|E_j - E| \leq C\hbar, \gamma_j \leq C\hbar\}$, or equivalently $\{|z_j - E| \leq C\hbar\}$.

1.2 Trapped Sets

Since resonant states are "invariant up to rescaling", it is natural to relate them, in the semiclassical spirit, to invariant structures of the classical dynamics. For a scattering system, the set of points (of energy E) which don't escape to infinity (either in past or future) is called the *trapped set* at energy E, and denoted by $K(E)$. The textbook example of a radially-symmetric potential shows that this set may be empty (if $V(r)$ decreases monotonically from $r = 0$ to $r \to \infty$), or have the same dimension as Σ_E (if $V(r)$ has a maximum $V(r_0) > 0$ before decreasing as $r \to \infty$).

For $n = 2$ degrees of freedom, the geometry of the trapped set can be more complex. Let us consider the well-known example of 2-dimensional scattering by a set of non-overlapping disks [5,9] (a similar model was studied in [2,26]).

When the scatterer is a single disk, the trapped set is obviously empty.

The scattering by two disks admits a single trapped periodic orbit, bouncing back and forth between the disks. Since the evolution between two bounces is "trivial", it is convenient to represent the scattering system through the *bounce map* on the reduced phase space (position along the boundaries × velocity angle). This map is actually defined only on a fraction of this phase space, namely on those points which will bounce again at least once. For the 2-disk system, this map has a unique periodic point (of period 2), which is of hyperbolic nature due to the curvature of the disks. The trapped set K of the map ("reduced" trapped set) reduces to this pair of points; it lies at the intersection of the forward trapped set Γ_- (points trapped as $t \to +\infty$) and the backward trapped set Γ_+ (points trapped as $t \to -\infty$).

The addition of a third disk generates a complex bouncing dynamics, for which the trapped set is a *fractal repeller* [9]. We will explain in the next section how such a structure arises in the case of the open baker's map. As in the 2-disk case, the bounce map is uniformly hyperbolic; each forward trapped point $x \in \Gamma_-$ admits a stable manifold $W_-(x)$ (and vice-versa for $x \in \Gamma_+$). One can show that Γ_- is fractal along the unstable direction W_+: $\Gamma_- \cap W_+$ has a Hausdorff dimension $0 < d < 1$ which depends on the positions and sizes of the disks. Due to time-reversal symmetry, $\Gamma_+ \cap W_-$ has the same Hausdorff dimension. Finally, the reduced trapped set $K = \Gamma_+ \cap \Gamma_-$ is a fractal of dimension $2d$, which contains infinitely many periodic orbits. The unreduced trapped set $K(E) \subset \Sigma_E$ has one more dimension corresponding to the direction of the flow, so it is of dimension $D = 2d + 1$.

1.3 Fractal Weyl Law

We now relate the geometry of the trapped set $K(E)$, to the density of resonances of the quantized Hamiltonian H_h in boxes $\{|z - E| \leq C\hbar\}$. The following conjecture (which dates back at least to the work of Sjöstrand [21]) relates this density with the "thickness" of the trapped set.

Conjecture 1. Assume that the trapped set $K(E)$ at energy E has dimension $2d_E + 1$. Then, the density of resonances near E grows as follows in the semiclassical limit:

$$\forall r > 0, \quad \frac{\#\{\mathrm{Res}(H_\hbar) \cap \{z \,:\, |z - E| < r\,\hbar\}\}}{\hbar^{-d_E}} \xrightarrow{h \to 0} c_E(r)\,, \qquad (2)$$

for a certain "shape function" $0 \leq c_E(r) < \infty$.

We were voluntarily rather vague on the concept of "dimension" (a fractal set can be characterized by many different dimensions). In the case of a closed system, $K(E)$ has dimension $2n - 1$, so we recover the Weyl law (1). If $K(E)$ consists in one unstable periodic orbit, the resonances form a (slightly deformed) rectangular lattice of sides $\propto \hbar$, so each $\hbar$-box contains at most finitely many resonances [22].

For intermediate situations ($0 < d_E < n - 1$), one has only been able to prove one half of the above estimate, namely the *upper bound* for this resonance counting [10, 21, 24, 27]. The dimension appearing in these upper bounds is the *Minkowski dimension* defined by measuring ϵ-neighborhoods of $K(E)$. In the case we will study, this dimension is equal to the Hausdorff one. Some lower bounds for the resonance density have been obtained as well [23], but are far below the conjectured estimate.

Several numerical studies have attempted to confirm the above estimate for a variety of scattering Hamiltonians [10, 12–14], but with rather inconclusive results. Indeed, it is numerically demanding to compute resonances. One method is to "complex rotate" the original Hamiltonian into a non-Hermitian operator, the eigenvalues of which are the resonances. Another method uses the (approximate) relationship between, on one side, the resonance spectrum of $H_\hbar$, one the other side, the set of zeros of some semiclassical zeta function, which is computed from the knowledge of classical periodic orbits [5, 14]. In the case of the geodesic flow on a convex co-compact quotient of the Poincaré disk (which has a fractal trapped set), the resonances of the Laplace operator are *exactly* given by the zeros of Selberg's zeta function. Even in that case, it has been difficult to check the asymptotic Weyl law (2), due to the necessity to reach sufficiently high values of the energy [10].

1.4 Open Maps

Confronted with these difficulties to deal with open Hamiltonian systems, we decided to study semiclassical resonance distributions for toy models which have already proven efficient to modelize closed systems. In the above example of obstacle scattering, the *bounce map* emerged as a way to simplify the description of the classical dynamics. It acts on a reduced phase space, and gets rid of the "trivial" evolution between bounces. The exact quantum problem also reduces to analyzing an operator acting on wavefunctions on the disk boundaries, but this operator is infinite-dimensional, and extracting its resonances is not a simple task [2, 9].

Canonical maps on the 2-torus were often used to mimic closed Hamiltonian systems; they can be quantized into unitary matrices, the eigenphases of which are to be compared with the eigenvalues $e^{-iE_j/\hbar}$ of the propagator $e^{-iH_\hbar/\hbar}$ (see e.g. [7] and references therein for a mathematical introduction on quantum maps).

We therefore decided to construct a "toy bounce map" on $\mathbb{T}^2$, with dynamics similar to the original bounce map, and which can be easily quantized into an $N \times N$ *subunitary matrix* (where $N = (2\pi\hbar)^{-1}$). This matrix is then easily diagonalized, and its subunitary eigenvalues $\{\lambda_j\}$ should be compared with the set $\{e^{-iz_j/\hbar}\}$, where the z_j are the resonances of $H_\hbar$ near some positive energy E. We cannot prove any direct correspondence between, on one side the eigenvalues of our quantized map, on the other side resonances of a *bona fide* scattering Hamiltonian. However, we expect a semiclassical property like the fractal Weyl law to be *robust*, in the sense that it should be shared by all types of "quantum models". To support this claim, we notice that the usual, "closed" Weyl law is already (trivially) satisfied by quantized maps: the number of eigenphases θ_j on the unit circle (corresponding to an energy range $\Delta E = 2\pi\hbar$) is exactly $N = (2\pi\hbar)^{-1}$, which agrees with the Weyl law (1) for $n = 2$ degrees of freedom. Testing Conjecture 1 in the framework of quantum maps should therefore give a reliable hint on its validity for more realistic Hamiltonian systems.

Schomerus and Tworzydło recently studied the quantum spectrum of an open chaotic map on the torus, namely the open kicked rotator [20]; they obtain a good agreement with a fractal Weyl law for the resonances (despite the fact that the geometry of the trapped set is not completely understood for that map). The authors also provide a heuristic argument to explain this Weyl law. We believe that this argument, upon some technical improvement, could yield a rigorous proof of the upper bound for the fractal Weyl law in case of maps.

We preferred to investigate that problem using one of the best understood chaotic maps on $\mathbb{T}^2$, namely the baker's map.

2 The Open Baker's Map and Its Quantization

2.1 Classical Closed Baker

The (closed) baker's map is one of the simplest examples of uniformly hyperbolic, strongly chaotic systems (it is a perfect model of Smale's horseshoe). The "3-baker's map" B on $\mathbb{T}^2 \equiv [0,1) \times [0,1)$ is defined as follows:

$$\mathbb{T}^2 \ni (q,p) \mapsto B(q,p) = \begin{cases} (3q, \frac{p}{3}) & \text{if } 0 \le q < 1/3\,, \\ (3q-1, \frac{p+1}{3}) & \text{if } 1/3 \le q < 2/3\,, \\ (3q-2, \frac{p+2}{3}) & \text{if } 2/3 \le q < 1. \end{cases} \tag{3}$$

This map preserves the symplectic form $dq \wedge dp$ on $\mathbb{T}^2$, and is invertible. Compared with a generic Anosov map, it has the particularity to be linear by parts, and its linearized dynamics (well-defined away from its lines of discontinuities) is independent of the point $x \in \mathbb{T}^2$. As a consequence, the stretching exponent is constant on $\mathbb{T}^2$, as well as the unstable/stable directions (horizontal/vertical).

This map admits a very simple Markov partition, made of the three vertical rectangles $R_j = \{q \in [j/3, (j+1)/3), p \in [0,1)\}$, $j = 0, 1, 2$ (see Fig. 1). Any bi-infinite sequence of symbols $\ldots \epsilon_{-2}\epsilon_{-1} \cdot \epsilon_0\epsilon_1\epsilon_2 \ldots$ (where each $\epsilon_i \in \{0,1,2\}$) will be associated with the *unique* point x s.t. $B^t(x) \in R_{\epsilon_t}$ for all $t \in \mathbb{Z}$. This is the point of coordinates (q,p), where q and p admit the ternary decompositions

$$q = 0 \cdot \epsilon_0\epsilon_1 \ldots \overset{\text{def}}{=} \sum_{i \ge 1} \frac{\epsilon_{i-1}}{3^i}\,, \qquad p = 0 \cdot \epsilon_{-1}\epsilon_{-2} \ldots\,.$$

The baker's map B simply acts as a *shift* on this symbolic sequence:

$$B(x = \ldots \epsilon_{-2}\epsilon_{-1} \cdot \epsilon_0\epsilon_1\epsilon_2 \ldots) = \ldots \epsilon_{-2}\epsilon_{-1}\epsilon_0 \cdot \epsilon_1\epsilon_2 \ldots . \tag{4}$$

2.2 Opening the Classical Map

We explained above that the bounce maps associated with the 2- or 3-disk systems were defined only on parts of the reduced phase space, namely on those points which bounce at least one more time. The remaining points, which escape to infinity right after the bounce, have no image through the map.

Hence, to open our baker's map B, we just decide to restrict it on a subset $S \subset \mathbb{T}^2$, or equivalently we send points in $\mathbb{T}^2 \setminus S$ to infinity. We obtain an Anosov map "with a hole", a class of dynamical systems recently studied in the literature [4]. The study is simpler when the hole corresponds to a Markov rectangle [3], so this is the choice we will make (we expect the fractal Weyl law to hold for an arbitrary hole as well). Let us choose for the hole the second Markov rectangle R_1, so that $S = R_0 \cup R_2$. Our open map $C = B_{\upharpoonright S}$ reads (see Fig. 1):

$$C(q,p) = \begin{cases} (3q, \frac{p}{3}) & \text{if } q \in R_0 , \\ (3q - 2, \frac{p+2}{3}) & \text{if } q \in R_2. \end{cases} \tag{5}$$

This map is canonical on S, and its inverse C^{-1} is defined on the set $C(S)$.

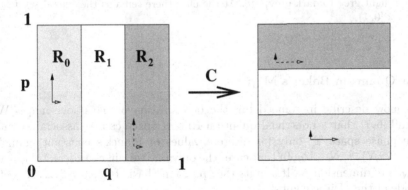

Fig. 1. Open baker's map C. The points in the middle rectangle are sent to infinity.

Our choice for S coincides with the points $x = (q,p)$ satisfying $\epsilon_0(x) \in \{0,2\}$ (equivalently, points s.t. $\epsilon_0(x) = 1$ are sent to infinity through C). This allows us to characterize the trapped sets very easily:

– the forward trapped set Γ_- (see fig. 2) is made of the points x which will never fall in the strip R_1 for times $t \geq 0$: these are the points s.t.

$\epsilon_i \in \{0, 2\}$ for all $i \geq 0$, with no constraint on the ϵ_i for $i < 0$. This set is of the form $\Gamma_- = Can \times [0, 1)$, where Can is the standard 1/3-Cantor set on the unit interval. As a result, the intersection $\Gamma_- \cap W_+ \equiv Can$ has the Hausdorff (or Minkowski) dimension $d = \frac{\log 2}{\log 3}$.

- the backward trapped set Γ_+ is made of the points satisfying $\epsilon_i \in \{0, 2\}$ for all $i < 0$, and is given by $[0, 1) \times Can$.
- the full trapped set $K = Can \times Can$.

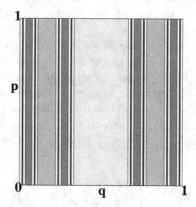

Fig. 2. Iterative construction of the forward trapped set Γ_- for the open baker's map C: we remove from $\mathbb{T}^2$ the points leaving to infinity at times $t = 1, 2, 3, 4$ (from light grey to dark grey) etc. At the end, there remains the fractal set $\Gamma_- = Can \times [0, 1)$.

2.3 Quantum Baker's Map

We now describe in some detail the quantization of the above maps. We recall [6, 7] that a nontrivial quantum Hilbert space can be associated with the phase space $\mathbb{T}^2$ only for discrete values of Planck's constant, namely $\hbar = (2\pi N)^{-1}$, $N \in \mathbb{N}_0$. In that case (the only one we will consider), this space $\mathcal{H}_N$ is of dimension N. It admits the "position" basis $\{Q_j, j = 0, \ldots, N-1\}$ made of the "Dirac combs"

$$Q_j(q) = \frac{1}{\sqrt{N}} \sum_{\nu \in \mathbb{Z}} \delta\left(q - \frac{j}{N} - \nu\right).$$

This basis is connected to the "momentum" basis $\{P_k, k = 0, \ldots, N-1\}$ through the discrete Fourier transform:

$$\langle P_k | Q_j \rangle = (\mathcal{F}_N)_{kj} = \frac{e^{-2i\pi Nkj}}{\sqrt{N}}, \quad j, k \in \{0, \ldots, N-1\}, \tag{6}$$

where the Fourier matrix F_N is unitary. Balazs and Voros [1] proposed to quantize the closed baker's map B as follows, when N is a multiple of 3 (a condition we will always assume): in the position basis, it takes the block form

$$B_N = \mathcal{F}_N^{-1} \begin{pmatrix} \mathcal{F}_{N/3} & & \\ & \mathcal{F}_{N/3} & \\ & & \mathcal{F}_{N/3} \end{pmatrix} . \tag{7}$$

This matrix is obviously unitary, and exactly satisfies the Van Vleck formula (the semiclassical expression for a quantum propagator, in terms of the classical generating function). In the semiclassical limit $N \to \infty$, it was shown [8] that these matrices classically propagate Gaussian coherent states supported far enough from the lines of discontinuities. As usual, discontinuities of the classical dynamics induce diffraction effects at the quantum level, which have been partially analyzed for the baker's map [25] (in particular, diffractive orbits have to be taken into account in the Gutzwiller formula for $\mathrm{tr}(B_N^t)$). We believe that these diffractive effects should only induce lower-order corrections to the Weyl law (9).

We are now ready to quantize our open baker's map C of (5): since the classical map sends points in R_1 to infinity and acts through B on $S = R_0 \cup R_2$, the quantum propagator should kill states microsupported on R_1, and act as B_N on states microsupported on S. Therefore, in the position basis we get the subunitary matrix

$$C_N = \mathcal{F}_N^{-1} \begin{pmatrix} \mathcal{F}_{N/3} & & \\ & 0 & \\ & & \mathcal{F}_{N/3} \end{pmatrix} . \tag{8}$$

A very similar open quantum baker was constructed in [18], as a quantization of Smale's horseshoe. In Fig. 3 (left) we represent the moduli of the matrix elements $(C_N)_{nm}$. The largest elements are situated along the "tilted diagonals" $n = 3m$, $n = 3(m - 2N/3)$, which correspond to the projection on the q-axis of the graph of C. Away from these "diagonals", the amplitudes of the elements decrease relatively slowly (namely, like $1/|n - 3m|$). This slow decrease is due to the diffraction effects associated with the discontinuities of the map.

2.4 Resonances of the Open Baker's Map

We numerically diagonalized the matrices C_N, for larger and larger Planck's constants N. First of all, we notice that the subspace $\mathrm{Span}\{Q_j,\ j = N/3, \ldots, 2N/3 - 1\}$, made of position states in the "hole", is in the kernel of C_N. Therefore, it is sufficient to diagonalize the matrix obtained by removing the corresponding lines and columns. Upon a slight modification of the quantization procedure [17], one obtains for C_N a matrix covariant w.r.to parity, allowing for a separation of the even and odd eigenstates, and therefore reducing the dimension of each part by 2. This is the quantization we used for

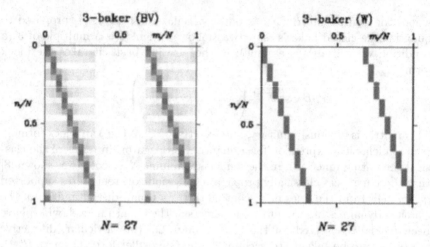

Fig. 3. Graphical representation of the matrices C_N (8) and $\widetilde{C}_N$ (10). Each grey square represents the modulus of a matrix element (white=0, black=$1/\sqrt{3}$)

our numerics: we only plot the even-parity resonances (the distribution of the odd-parity ones is very similar). In Fig. 4 we show the even-parity spectra of the matrix C_N for $N = 3^5$ and $N = 3^8$. Although we could not detect exact null states for the reduced matrix, many among the $N/3$ eigenvalues had very small moduli: for large values of N, the spectrum of C_N accumulates near the origin. This accumulation is an obvious consequence of the fractal Weyl law we want to test:

Conjecture 2. For any radius $1 > r > 0$ and $N \in \mathbb{N}_0$, $3|N$, let us denote

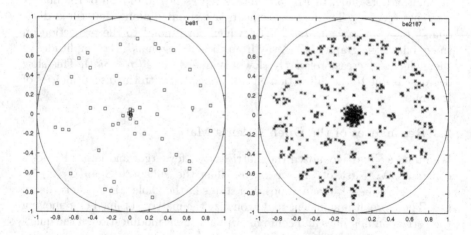

Fig. 4. Even-parity spectrum of the matrices C_N for $N/3 = 81, 2187$

$$n(N,r) \overset{\text{def}}{=} \#\{\lambda \in \text{Spec}(C_N) \cap \{|\lambda| \ge r\}\}.$$

In the semiclassical limit, this counting function behaves as

$$\frac{n(N,r)}{N^{\frac{\log 2}{\log 3}}} \overset{N \to \infty}{\longrightarrow} c(r), \tag{9}$$

with a "shape function" $0 \le c(r) < \infty$.

To test this conjecture, we proceed in two ways:
- In a first step, we select some discrete values for r, and plot $n(N,r)$ for an arbitrary sequence of N, in a log-log plot (see Fig. 5). We observe that the

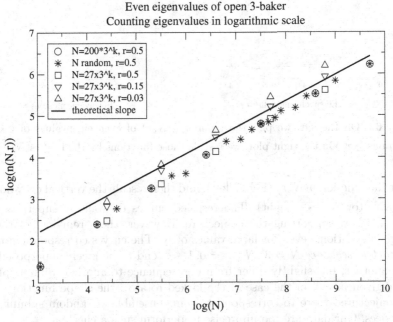

Fig. 5. Checking the N-dependence of $n(N,r)$ for various values of r, along geometric and arbitrary sequences for N. The thick curve has the slope $\log 2 / \log 3$

slope of the data nicely converges towards the theoretical one $\frac{\log 2}{\log 3}$ (thick line), all the more so along geometric subsequences $N = 3^k N_o$, and for relatively large values of the radius ($r = 0.5$). For the smaller value $r = 0.03$, the annulus $\{|z| \ge r\}$ still contains "too many resonances" and the asymptotic régime is not yet reached.
- In a second step, confident that $n(N,r)$ scales like $N^{\frac{\log 2}{\log 3}}$, we try to extract the shape function $c(r)$. For an arbitrary sequence of values of N, we

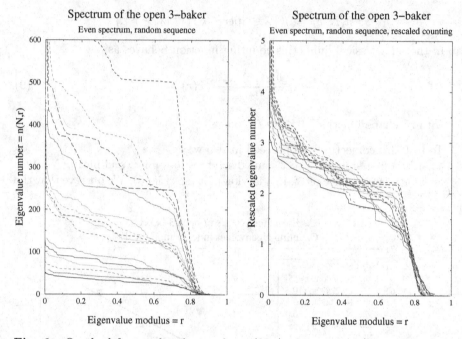

Fig. 6. On the left, we plot the number $n(N, r)$ of even eigenvalues of C_N of modulus $\geq r$. On the right plot, we rescale those functions by the factors $N^{-\frac{\log 2}{\log 3}}$

plot the function $n(N, r)$ (Fig. 6, left), and then rescale the vertical coordinate by a factor $N^{-\frac{\log 2}{\log 3}}$ (right). The rescaled curves do roughly superpose on one another, supporting the conjecture. However, there remains relatively large fluctuations, even for large values of N. The curves corresponding to a *geometric sequence* $N = 3^k N_o$, $k = 0, 1, \ldots$ tend to be nicely superposed to one another, but slightly differ from one sequence to another. Similar plots were given in [20] in the case of the kicked rotator; the shape function $c(r)$ is conjectured there to correspond to some ensemble of random subunitary matrices. Our data are too unprecise to perform such a check.

The fact that the spectra of the matrices C_N "behave nicely" along geometric sequences, while they fluctuate more strongly between successive values of N, is not totally unexpected (similar phenomena had been noticed for the quantizations B_N of the closed baker [1]). In view of Fig. 6, our conjecture (9) may be too strict if we apply it to a general sequence of N. At least, it seems to be satisfied along geometric sequences $\left\{ 3^k N_o, \, k \in \mathbb{N} \right\}$, with shape functions $c_{N_o}(r)$ slightly depending on the sequence.

3 A Solvable Toy Model for the Quantum Baker

3.1 Description of the Toy Model

In an attempt to get some analytical grip on the resonances, we tried to simplify the quantum matrix C_N, keeping only its "backbone" along the tilted diagonals and removing the off-diagonal components. We obtained the "toy-of-the-toy model" given by the following matrices (the moduli of the components are shown on right plot of Fig. 3):

$$\widetilde{C}_{N=9} = \frac{1}{\sqrt{3}} \begin{pmatrix} 1 & 0 & 0 & 0 & 0 & 1 & 0 & 0 \\ 1 & 0 & 0 & 0 & 0 & \omega^2 & 0 & 0 \\ 1 & 0 & 0 & 0 & 0 & \omega & 0 & 0 \\ 0 & 1 & 0 & 0 & 0 & 0 & 1 & 0 \\ 0 & 1 & 0 & 0 & 0 & 0 & \omega^2 & 0 \\ 0 & 1 & 0 & 0 & 0 & 0 & \omega & 0 \\ 0 & 0 & 1 & 0 & 0 & 0 & 0 & 1 \\ 0 & 0 & 1 & 0 & 0 & 0 & 0 & \omega^2 \\ 0 & 0 & 1 & 0 & 0 & 0 & 0 & \omega \end{pmatrix}, \quad \omega = e^{2\pi i/3} . \quad (10)$$

From this example, it is pretty clear how one constructs $\widetilde{C}_N$ for N an arbitrary multiple of 3. A similar quantization of the closed 2-baker was introduced in [19].

Before describing the spectra of these matrices, we describe their propagation properties. Removing the "off-diagonal" elements, we have eliminated the effects of diffraction due to the discontinuities of C. However, this elimination is so abrupt that it modifies the semiclassical transport. Indeed, a coherent state situated at a point x away from the discontinuities will not be transformed by $\widetilde{C}_N$ into a single coherent state (as does C_N), but rather into a *linear combination* of 3 coherent states, shifted vertically by $1/3$ from one another. Therefore, the matrices $\widetilde{C}_N$ do not quantize the open baker C of (5), but rather the following *multivalued* ("ray-splitting") map:

$$\widetilde{C}(q,p) = \begin{cases} (3q, \frac{p}{3}) \cup (3q, \frac{p+1}{3}) \cup (3q, \frac{p+2}{3}) & \text{if } q \in R_0 , \\ (3q-2, \frac{p}{3}) \cup (3q-2, \frac{p+1}{3}) \cup (3q-2, \frac{p+2}{3}) & \text{if } q \in R_2. \end{cases} \quad (11)$$

This modification of the classical dynamics is rather annoying. Still, the dynamics $\widetilde{C}$ shares some common features with that of C: the forward trapped set for $\widetilde{C}$ is the same as for C, that is the set Γ_- described in Fig. 2. On the other hand, the backward trapped set is now the full torus $\mathbb{T}^2$.

3.2 Interpretation of $\widetilde{C}_N$ as a Walsh-Quantized Baker

A possible way to avoid this modified classical dynamics is to interpret $\widetilde{C}_N$ as a "Walsh-quantized map" (this interpretation makes sense when $N = 3^k$,

$k \in \mathbb{N}$). To introduce this Walsh formalism, let us first write the Hilbert space as a tensor product $\mathcal{H}_N = (\mathbb{C}^3)^{\otimes k}$, where we take the ternary decomposition of discrete positions $\frac{j}{N} = 0 \cdot \epsilon_0 \epsilon_1 \cdots \epsilon_{k-1}$ into account. If we call $\{e_0,\ e_1,\ e_2\}$ the canonical basis of $\mathbb{C}^3$, each position state $Q_j \in \mathcal{H}_N$ can be represented as the tensor product state

$$Q_j = e_{\epsilon_0} \otimes e_{\epsilon_1} \otimes \cdots \otimes e_{\epsilon_{k-1}} .$$

In the language of quantum computing, each tensor factor $\mathbb{C}^3$ is the Hilbert space of a "qutrit" associated with a certain scale [19].

The Walsh Fourier transform is a modification of the discrete Fourier transform (6), which first appeared in signal theory, and has been recently used as a toy model for harmonic analysis [15]. Its major advantage is the possibility to construct states compactly supported in both position and "Walsh momentum". In our finite-dimensional framework, this Walsh transform is the matrix

$$(W_N)_{jj'} = 3^{-k/2} \exp\left(-\frac{2i\pi}{3} \sum_{\ell+\ell'=k-1} \epsilon_\ell(Q_j)\, \epsilon_{\ell'}(Q'_j)\right), \quad j,j' = 0,\ldots,N-1 ,$$

and acts as follows on tensor product states:

$$W_N\,(v_0 \otimes v_1 \otimes \cdots v_{k-1}) = \mathcal{F}_3 v_{k-1} \otimes \cdots \mathcal{F}_3 v_1 \otimes \mathcal{F}_3 v_0 , \quad v_\ell \in \mathbb{C}^3,\ \ell = 0,\ldots,k-1 .$$

Now, in the case $N = 3^k$, our toy model $\widetilde{C}_N$ can be expressed as

$$\widetilde{C}_N = W_N^{-1} \begin{pmatrix} W_{N/3} & & \\ & 0 & \\ & & W_{N/3} \end{pmatrix} .$$

One can show that "Walsh coherent states" are propagated through $\widetilde{C}_N$ according to the map C. Hence, as opposed to what happens in "standard" quantum mechanics, $\widetilde{C}_N$ Walsh-quantizes the open baker C.

3.3 Resonances of $\widetilde{C}_{N=3^k}$

We now use the very peculiar properties of the matrices $\widetilde{C}_{3^k}$ to analytically compute their spectra. From the expressions in last section, one can see that the toy model $\widetilde{C}_N$ acts very simply on tensor product states:

$$\widetilde{C}_N\, v_0 \otimes v_1 \otimes \cdots v_{k-1} = v_1 \otimes \cdots v_{k-1} \otimes \mathcal{F}_3^{-1} \pi_{02} v_0 , \tag{12}$$

where π_{02} projects $\mathbb{C}^3$ orthogonally onto $\mathrm{Span}\,\{\,e_0,\ e_2\,\}$. Like its classical counterpart, $\widetilde{C}_N$ realizes a symbolic shift between the different scales. It also sends the first symbol ϵ_0 to the "end of the queue", after a projection and a Fourier transform. The projection π_{02} kills the states Q_j localized in the

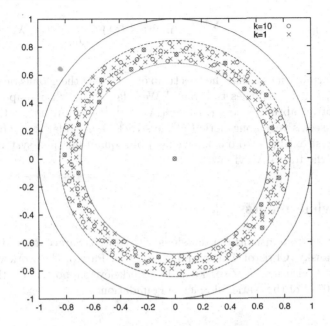

Fig. 7. Spectra of the matrices $\widetilde{C}_N$ for $N = 3^{10}$ (circles), $N = 3^{15}$ (crosses). The large circles have radii $|\lambda| = 1$, $|\lambda_+|$, $|\lambda_+\lambda_-|^{\frac{1}{2}}$, $|\lambda_-|$

rectangle R_1. The vector $\mathcal{F}_3^{-1}e_{\epsilon_0}$ in the last qutrit induces a localization in the momentum direction, near the momentum $p = 0 \cdot \epsilon_0$.

By iterating this expression k times, we see that the operator $(\widetilde{C}_N)^k$ acts independently on each tensor factor $\mathbb{C}^3$, through the matrix $\mathcal{F}_3^{-1}\pi_{02}$. The latter has three eigenvalues:

- it kills the state e_1, implying that $(\widetilde{C}_N)^k$ kills any state Q_j for which at least one of the symbols $\epsilon_\ell(Q_j)$ is equal to 1. These $3^k - 2^k$ position states are localized "outside" of the trapped set Γ_-, which explains why they are killed by the dynamics.
- its two remaining eigenvalues $\lambda_\pm$ have moduli $|\lambda_+| \approx 0.8443$, $|\lambda_-| \approx 0.6838$. They build up the ($2^k$-dimensional) nontrivial spectrum of $\widetilde{C}_N$, which has the form of a "lattice" (see Fig. 7):

Proposition 1. For $N = 3^k$, the nonzero spectrum of $\widetilde{C}_N$ is the set

$$\{\lambda_+\} \cup \{\lambda_-\} \bigcup \left\{ e^{2i\pi\frac{j}{k}} \lambda_+^{1-p/k} \lambda_-^{p/k} \; : \; 1 \le p \le k-1, \; 0 \le j \le k-1 \right\}.$$

Most of these eigenvalues are highly degenerate (they span a subspace of dimension 2^k). When $k \to \infty$, the highest degeneracies occur when $p/k \approx 2$, which results in the following asymptotic distribution:

$$\forall f \in C(\mathbb{R}^2), \quad \lim_{k \to \infty} \frac{1}{2^k} \sum_{\lambda \in \mathrm{Spec}(\widetilde{C}_{3^k}) \backslash 0} \mathrm{mult}(\lambda)\, f(\lambda) = \int_0^{2\pi} f(|\lambda_-\lambda_+|^{1/2}, \theta)\, \frac{d\theta}{2\pi} .$$

The last formula shows that the spectrum of $\widetilde{C}_N$ along the geometric sequence $\{N = 3^k,\ k \in \mathbb{N}\}$ satisfies the fractal Weyl law (9), with a shape function in form of an abrupt step: $c(r) = \Theta(|\lambda_+\lambda_-|^{1/2} - r)$. Although the above spectrum seems very nongeneric (lattice structure, singular shape function), it is the first example (to our knowledge) of a quantum open system proven to satisfy the fractal Weyl law.

Acknowledgments

We benefited from insightful discussions with Marcos Saraceno, André Voros, Uzy Smilansky, Christof Thiele and Terry Tao. Part of the work was done while I was visiting M. Zworski in UC Berkeley, supported by the grant DMS-0200732 of the National Science Foundation.

References

1. N.L. Balazs and A. Voros *The quantized baker's transformation*, Ann. Phys. **190** (1989), 1–31.
2. R. Blümel and U. Smilansky, *A simple model for chaotic scattering*, Physica **D 36** (1989), 111–136.
3. N. Chernov and R. Markarian, *Ergodic properties of Anosov maps with rectangular holes*, Boletim Sociedade Brasileira Matematica **28** (1997), 271–314.
4. N. Chernov, R. Markarian and S. Troubetzkoy, *Conditionally invariant measures for Anosov maps with small holes*, Ergod. Th. Dyn. Sys. **18** (1998), 1049–1073.
5. P. Cvitanović and B. Eckhardt, *Periodic-orbit quantization of chaotic systems*, Phys. Rev. Lett. **63** (1989) 823–826
6. S. De Bièvre, *Recent results on quantum map eigenstates*, these Proceedings.
7. M. Degli Esposti and S. Graffi, editors *The mathematical aspects of quantum maps*, volume 618 of Lecture Notes in Physics, Springer, 2003.
8. M. Degli Esposti, S. Nonnenmacher and B. Winn, *Quantum variance and ergodicity for the baker's map*, to be published in Commun. Math. Phys. (2005), arXiv:math-ph/0412058.
9. P. Gaspard and S.A. Rice, *Scattering from a classically chaotic repellor*, J. Chem. Phys. **90** (1989), 2225–2241; ibid, *Semiclassical quantization of the scattering from a classically chaotic repellor*, J. Chem. Phys. **90** (1989), 2242–54; ibid, *Exact quantization of the scattering from a classically chaotic repellor*, J. Chem. Phys. **90** (1989), 2255–2262; Errata, J. Chem. Phys. **91** (1989), 3279–3280.
10. L. Guillopé, K. Lin, and M. Zworski, *The Selberg zeta function for convex cocompact Schottky groups*, Comm. Math. Phys., **245**(2004), 149–176.

11. V. Ivrii, *Microlocal Analysis and Precise Spectral Asymptotics*, Springer Verlag, 1998.
12. K. Lin, *Numerical study of quantum resonances in chaotic scattering*, J. Comp. Phys. **176**(2002), 295–329.
13. K. Lin and M. Zworski, *Quantum resonances in chaotic scattering*, Chem. Phys. Lett. **355**(2002), 201–205.
14. W. Lu, S. Sridhar, and M. Zworski, *Fractal Weyl laws for chaotic open systems*, Phys. Rev. Lett. **91**(2003), 154101.
15. C. Muscalu, C. Thiele, and T. Tao, *A Carleson-type theorem for a Cantor group model of the Scattering Transform*, Nonlinearity **16** (2003), 219–246.
16. S. Nonnenmacher and M. Zworski, *Distribution of resonances for open quantum maps*, preprint, 2005.
17. M. Saraceno, *Classical structures in the quantized baker transformation*, Ann. Phys. (NY) **199** (1990), 37–60.
18. M. Saraceno and R.O. Vallejos, *The quantized D-transformation*, Chaos **6** (1996), 193–199.
19. R. Schack and C.M. Caves, *Shifts on a finite qubit string: a class of quantum baker's maps*, Appl. Algebra Engrg. Comm. Comput. **10** (2000) 305–310
20. H. Schomerus and J. Tworzydło, *Quantum-to-classical crossover of quasi-bound states in open quantum systems*, Phys. Rev. Lett. **93** (2004), 154102.
21. J. Sjöstrand, *Geometric bounds on the density of resonances for semiclassical problems*, Duke Math. J., **60** (1990), 1–57.
22. J. Sjöstrand, *Semiclassical resonances generated by a non-degenerate critical point*, Lecture Notes in Math. **1256**, 402–429, Springer (Berlin), 1987.
23. J. Sjöstrand and M. Zworski, *Lower bounds on the number of scattering poles*, Commun. PDE **18** (1993), 847–857.
24. J. Sjöstrand and M. Zworski, *Geometric bounds on the density of semiclassical resonances in small domains*, in preparation (2005).
25. F. Toscano, R.O. Vallejos and M. Saraceno, *Boundary conditions to the semiclassical traces of the baker's map*, Nonlinearity **10** (1997), 965–978.
26. G. Troll and U. Smilansky, *A simple model for chaotic scattering*, Physica **D** **35** (1989), 34–64.
27. M. Zworski, *Dimension of the limit set and the density of resonances for convex co-compact Riemann surfaces*, Inv. Math. **136** (1999), 353–409.

Spectral Shift Function
for Magnetic Schrödinger Operators

Georgi Raikov

Facultad de Ciencias, Universidad de Chile, Las Palmeras 3425, Santiago, Chile
graykov@uchile.cl

1 Introduction

In this survey article based on the papers [7, 10], and [8], we consider the
3D Schrödinger operator with constant magnetic field of intensity $b > 0$,
perturbed by an electric potential V which decays fast enough at infinity,
and discuss various asymptotic properties of the corresponding spectral shift
function. More precisely, let $H_0 = H_0(b) := (i\nabla + \mathbf{A})^2 - b$ be the unperturbed
operator, essentially self-adjoint on $C_0^\infty(\mathbb{R}^3)$. Here $\mathbf{A} = \left(-\frac{bx_2}{2}, \frac{bx_1}{2}, 0\right)$ is the
magnetic potential which generates the constant magnetic field $\mathbf{B} = \operatorname{curl} \mathbf{A} =
(0, 0, b)$, $b > 0$. It is well-known that $\sigma(H_0) = \sigma_{\mathrm{ac}}(H_0) = [0, \infty)$ (see [1]),
where $\sigma(H_0)$ stands for the spectrum of H_0, and $\sigma_{\mathrm{ac}}(H_0)$ for its absolutely
continuous spectrum. Moreover, the so-called Landau levels $2bq$, $q \in \mathbb{Z}_+ :=
\{0, 1, \ldots\}$, play the role of thresholds in $\sigma(H_0)$.

For $\mathbf{x} = (x_1, x_2, x_3) \in \mathbb{R}^3$ we denote by $X_\perp = (x_1, x_2)$ the variables on the
plane perpendicular to the magnetic field. Throughout the paper we assume
that the electric potential V satisfies

$$V \not\equiv 0, \quad V \in C(\mathbb{R}^3), \quad |V(\mathbf{x})| \le C_0 \langle X_\perp \rangle^{-m_\perp} \langle x_3 \rangle^{-m_3}, \quad \mathbf{x} = (X_\perp, x_3) \in \mathbb{R}^3, \tag{1}$$

with $C_0 > 0$, $m_\perp > 2$, $m_3 > 1$, and $\langle x \rangle := (1 + |x|^2)^{1/2}$, $x \in \mathbb{R}^d$, $d \ge 1$. Some
of our results hold under a more restrictive assumption than (1), namely

$$V \not\equiv 0, \quad V \in C(\mathbb{R}^3), \quad |V(\mathbf{x})| \le C_0 \langle \mathbf{x} \rangle^{-m_0}, \quad m_0 > 3, \quad \mathbf{x} \in \mathbb{R}^3. \tag{2}$$

Note that (2) implies (1) with any $m_3 \in (0, m_0)$ and $m_\perp = m_0 - m_3$. In
particular, we can choose $m_3 \in (1, m_0 - 2)$ so that $m_\perp > 2$.
On the domain of H_0 define the operator $H = H(b) := H_0 + V$. Obviously,
$\inf \sigma(H) \le \inf \sigma(H_0) = 0$. Moreover, if (1) holds, then for $E < \inf \sigma(H)$
we have $(H - E)^{-1} - (H_0 - E)^{-1} \in S_1$ where S_1 denotes the trace class.
Hence, there exists a unique function $\xi = \xi(\cdot; H, H_0) \in L^1(\mathbb{R}; (1 + E^2)^{-1}dE)$
vanishing on $(-\infty, \inf \sigma(H))$, such that *the Lifshits-Krein trace formula*

$$\operatorname{Tr}\left(f(H) - f(H_0)\right) = \int_{\mathbb{R}} \xi(E; H, H_0) f'(E) dE$$

holds for each $f \in C_0^\infty(\mathbb{R})$ (see the original works [20, 22], the survey article
[5], or Chapter 8 of the monograph [44]). The function $\xi(\cdot; H, H_0)$ is called

G. Raikov: *Spectral Shift Function for Magnetic Schrödinger Operators*, Lect. Notes Phys.
690, 451–465 (2006)
www.springerlink.com

the spectral shift function (SSF) for the operator pair (H, H_0). If $E < 0 = \inf \sigma(H_0)$, then the spectrum of H below E could be at most discrete, and for almost every $E < 0$ we have

$$\xi(E; H, H_0) = -N(E; H) \tag{3}$$

where $N(E; H)$ denotes the number of eigenvalues of H lying in the interval $(-\infty, E)$, and counted with their multiplicities. On the other hand, for almost every $E \in [0, \infty)$, the SSF $\xi(E; H, H_0)$ is related to the scattering determinant $\det S(E; H, H_0)$ for the pair (H, H_0) by *the Birman-Krein formula*

$$\det S(E; H, H_0) = e^{-2\pi i \xi(E; H, H_0)}$$

(see [4] or [44, Sect. 8.4]). A survey of various asymptotic results concerning the SSF for numerous quantum Hamiltonians is contained in [39].

A priori, the SSF $\xi(E; H, H_0)$ is defined for almost every $E \in \mathbb{R}$. In this article we will identify this SSF with a representative of its equivalence class which is well-defined on $\mathbb{R} \setminus 2b\mathbb{Z}_+$, bounded on every compact subset of $\mathbb{R} \setminus 2b\mathbb{Z}_+$, and continuous on $\mathbb{R} \setminus (2b\mathbb{Z}_+ \cup \sigma_{\mathrm{pp}}(H))$ where $\sigma_{\mathrm{pp}}(H)$ denotes the set of the eigenvalues of H. In the case of V of definite sign this representative is described explicitly in Subsect. 2.2 below (see, in particular, (5)). In the case of non-sign-definite V its description can be found in [7, Sect. 3]. In the present article we investigate the behaviour of the SSF in several asymptotic regimes:

- First, we analyse the singularities of the SSF at the Landau levels. In other words, we fix $b > 0$ and $q \in \mathbb{Z}_+$, and investigate the behaviour of $\xi(2bq + \lambda; H, H_0)$ as $\lambda \to 0$.
- Further, we study the strong-magnetic-field asymptotics of the SSF, i.e. the behaviour of the SSF as $b \to \infty$. Here we distinguish between the asymptotics far from the Landau levels, and the asymptotics near a given Landau level.
- Finally, we obtain a Weyl type formula describing the high energy asymptotics of the SSF.

The paper is organised as follows. Section 2 contains some auxiliary results: in Subsect. 2.1 we introduce some basic concepts and notations used throughout the paper, while In Subsect. 2.2 we describe the representation of the SSF in the case of perturbations of fixed sign, due to A. Pushnitski (see [29]). In Sect. 3 we formulate our main results, and discuss briefly on them. More precisely, Subsect. 3.1 contains the results on the singularities of the SSF at the Landau levels, Subsect. 3.2 is devoted to the strong magnetic field asymptotics of the SSF, and Subsect. 3.3 – to its high energy behaviour. We refer the reader to the original articles [7, 10], and [8], for the proofs of the results presented in this survey.

2 Auxiliary Results

2.1 Notations and Preliminaries

In this subsection we give definitions and introduce notations used throughout the paper. We denote by S_∞ the class of linear compact operators acting in a given Hilbert space. Let $T = T^* \in S_\infty$. Denote by $\mathbb{P}_I(T)$ the spectral projection of T associated with the interval $I \subset \mathbb{R}$. For $s > 0$ set

$$n_\pm(s; T) := \operatorname{rank} \mathbb{P}_{(s,\infty)}(\pm T) .$$

For an arbitrary (not necessarily self-adjoint) operator $T \in S_\infty$ put

$$n_*(s; T) := n_+(s^2; T^*T), \quad s > 0 .$$

Further, we denote by S_p, $p \in (0, \infty)$, the Schatten-von Neumann class of compact operators for which the functional $\|T\|_p := \left(p \int_0^\infty s^{p-1} n_*(s; T)\, ds \right)^{\frac{1}{p}}$ is finite. It is well-known that if $p \in [1, \infty)$, then S_p are Banach spaces with norm $\| \cdot \|_p$. Finally, if T is a bounded linear operator acting in a given Hilbert space, we set $\operatorname{Re} T := \frac{1}{2}(T + T^*)$, $\operatorname{Im} T := \frac{1}{2i}(T - T^*)$.

2.2 A. Pushnitski's Representation of the SSF

In this subsection we introduce a representation of the SSF $\xi(\cdot; H, H_0)$ in the case of sign-definite potentials V

Let $I \in \mathbb{R}$ be a Lebesgue measurable set. Set $\mu(I) := \frac{1}{\pi} \int_I \frac{dt}{1+t^2}$.

Lemma 1. [29, Lemma 2.1] Let $T_1 = T_1^* \in S_\infty$ and $T_2 = T_2^* \in S_1$. Then

$$\int_{\mathbb{R}} n_\pm(s_1 + s_2; T_1 + t T_2)\, d\mu(t) \le n_\pm(s_1; T_1) + \frac{1}{\pi s_2}\|T_2\|_1, \quad s_1, s_2 > 0 . \quad (4)$$

Suppose that V satisfies (1), and $\pm V \ge 0$; in this case we will write $H = H_\pm = H_0 \pm |V|$. For $z \in \mathbb{C}$, $\operatorname{Im} z > 0$, set $T(z) := |V|^{1/2}(H_0 - z)^{-1}|V|^{1/2} \in S_\infty$.

Proposition 1. [7, Lemma 4.2] Assume that (1) holds, and $E \in \mathbb{R} \setminus 2b\mathbb{Z}_+$. Then the operator limit $T(E + i0) := \operatorname{n} - \lim_{\delta \downarrow 0} T(E + i\delta)$ exists, and we have

$$\|T(E + i0)\| \le C_1 \left(\operatorname{dist}(E, 2b\mathbb{Z}_+)\right)^{-1/2}$$

with C_1 independent of E and b. Moreover, $0 \le \operatorname{Im} T(E + i0) \in S_1$, and if $E < 0$ then $\operatorname{Im} T(E + i0) = 0$, while for $E \in (0, \infty) \setminus 2b\mathbb{Z}_+$ we have

$$\|\operatorname{Im} T(E + i0)\|_1 = \operatorname{Tr} \operatorname{Im} T(E + i0) = \frac{b}{4\pi} \sum_{l=0}^{[\frac{E}{2b}]} (E - 2bl)^{-1/2} \int_{\mathbb{R}^3} |V(\mathbf{x})| d\mathbf{x}$$

where $[x]$ denotes the integer part of $x \in \mathbb{R}$.

By Lemma 1 and Proposition 1, the quantity

$$\tilde{\xi}(E; H_\pm, H_0) = \pm \int_{\mathbb{R}} n_\mp(1; \operatorname{Re} T(E + i0) + t \operatorname{Im} T(E + i0)) \, d\mu(t) \quad (5)$$

is well-defined for every $E \in \mathbb{R} \setminus 2b\mathbb{Z}_+$, and bounded on every compact subset of $\mathbb{R} \setminus 2b\mathbb{Z}_+$. Moreover, by [7, Proposition 2.5], $\tilde{\xi}(\cdot; H_\pm, H_0)$ is continuous on $\mathbb{R} \setminus \{2b\mathbb{Z}_+ \cup \sigma_{\mathrm{pp}}(H_\pm)\}$. On the other hand, the general result of [29, Theorem 1.2] implies $\tilde{\xi}(E; H_\pm, H_0) = \xi(E; H_\pm, H_0)$ for almost every $E \in \mathbb{R}$. As explained in the introduction, in the case of sign-definite perturbations we will identify the SSF $\xi(E; H_\pm, H_0)$ with $\tilde{\xi}(E; H_\pm, H_0)$, while in the case of non-sign-definite perturbations, we will identify it with the generalisation of $\tilde{\xi}(E; H_\pm, H_0)$ described in [7, Sect. 3] on the basis of the general results of [14] and [30], using the concept of the index of orthogonal projections (see [2]).

Here it should be underlined that in contrast to the case $b = 0$, we cannot rule out the possibility that the operator H has infinite discrete spectrum, or eigenvalues embedded in the continuous spectrum by imposing short-range conditions on V. First, it is well-known that if V satisfies

$$V(\mathbf{x}) \le -C\chi(\mathbf{x}), \quad \mathbf{x} \in \mathbb{R}^3 , \quad (6)$$

where $C > 0$, and χ is the characteristic function of a non-empty open subset of $\mathbb{R}^3$, then the discrete spectrum of H is infinite (see [1, Theorem 5.1], [38, Theorem 2.4]). Further, assume that V is axisymmetric, i.e. $V = V(|X_\perp|, x_3)$. It is well-known (see e.g. [1]) that in this case the operators H_0 and H are unitarily equivalent to the orthogonal sums $\sum_{m \in \mathbb{Z}} \oplus H^{(m)}$ and $\sum_{m \in \mathbb{Z}} \oplus H_0^{(m)}$ respectively, where the operators

$$H_0^{(m)} := -\frac{1}{\varrho} \frac{\partial}{\partial \varrho} \varrho \frac{\partial}{\partial \varrho} - \frac{\partial^2}{\partial x_3^2} + \left(\frac{b\varrho}{2} + \frac{m}{\varrho} \right)^2 - b, \quad H^{(m)} := H_0^{(m)} + V(\varrho, x_3) ,$$

$$(7)$$

are self-adjoint in $L^2(\mathbb{R}_+ \times \mathbb{R}; \varrho d\varrho dx_3)$. Assume moreover that V satisfies (6). Then the operator $H^{(m)}$ with $m \ge 0$ has at least one eigenvalue in the interval $(2bm - \|V\|_{L^\infty(\mathbb{R}^3)}, 2bm)$, and hence the operator H has infinitely many eigenvalues embedded in its continuous spectrum (see [1, Theorem 5.1]). Suppose now that V is axisymmetric and satisfies the estimate

$$V(X_\perp, x_3) \le -C\chi_\perp(X_\perp)\langle x_3 \rangle^{-m_3}, \quad (X_\perp, x_3) \in \mathbb{R}^3 , \quad (8)$$

where $C > 0$, $\chi_\perp$ is the characteristic function of a non-empty open subset of $\mathbb{R}^2$, and $m_3 \in (0, 2)$ which is compatible with (1) if $m_3 \in (1, 2)$. Then, using the argument of the proof of [1, Theorem 5.1] and the variational principle, we can easily check that for each $m \ge 0$ the operator $H^{(m)}$ has infinitely many discrete eigenvalues which accumulate to the infimum $2bm$ of its essential spectrum. Hence, if V is axisymmetric and satisfies (8), then below each Landau level $2bq$, $q \in \mathbb{Z}_+$, there exists an infinite sequence of finite-multiplicity eigenvalues of H, which converges to $2bq$. Note however that the claims in [10, p. 385] and [8, p. 3457] that [1, Theorem 5.1] implies the same phenomenon for axisymmetric non-positive potentials compactly supported in $\mathbb{R}^3$, are not justified. The challenging and interesting problem about the accumulation at a given Landau level of embedded eigenvalues and/or resonances of H will be considered in a future work. Finally, we note that generically the only possible accumulation points of the eigenvalues of H are the Landau levels (see [1, Theorem 4.7], [13, Theorem 3.5.3 (iii)]). Further information on the location of the eigenvalues of H can be found in [7, Proposition 2.6].

3 Main Results

3.1 Singularities of the SSF at the Landau Levels

Introduce the Landau Hamiltonian

$$h(b) := \left(i\frac{\partial}{\partial x_1} - \frac{bx_2}{2}\right)^2 + \left(i\frac{\partial}{\partial x_2} + \frac{bx_1}{2}\right)^2 - b\,,$$

essentially self-adjoint on $C_0^\infty(\mathbb{R}^2)$. It is well-known that $\sigma(h(b)) = \cup_{q=0}^\infty \{2bq\}$ and each eigenvalue $2bq$, $q \in \mathbb{Z}_+$, has infinite multiplicity (see e.g. [1]). Denote by $p_q = p_q(b)$ the orthogonal projection onto the eigenspace $\text{Ker}\,(h(b) - 2bq)$, $q \in \mathbb{Z}_+$. The estimates of the SSF for energies near the Landau level $2bq$, $q \in \mathbb{Z}_+$, will be given in the terms of traces of functions of Toeplitz operators $p_q U p_q$ where $U : \mathbb{R}^2 \to \mathbb{R}$ decays in a certain sense at infinity.

Lemma 2. [31, Lemma 5.1], [10, Lemma 3.1] *Let* $U \in L^r(\mathbb{R}^2)$, $r \geq 1$, *and* $q \in \mathbb{Z}_+$. *Then* $p_q U p_q \in S_r$.

Assume that (1) holds. For $X_\perp \in \mathbb{R}^2$ set $W(X_\perp) := \int_\mathbb{R} |V(X_\perp, x_3)| dx_3$. Since V satisfies (1), we have $W \in L^1(\mathbb{R}^2)$, and Lemma 2 with $U = W$ implies $p_q W p_q \in S_1$, $q \in \mathbb{Z}_+$. Evidently, $p_q W p_q \geq 0$, and it follows from $V \not\equiv 0$ and $V \in C(\mathbb{R}^2)$, that $\text{rank}\, p_q W p_q = \infty$ for all $q \in \mathbb{Z}_+$ (see below Lemma 5). If, moreover, V satisfies (2), then $0 \leq W(X_\perp) \leq C_0'\langle X_\perp\rangle^{-m_0+1}$, $X_\perp \in \mathbb{R}^2$, with $C_0' = C_0 \int_\mathbb{R} \langle x\rangle^{-m_0} dx$.

In the following two theorems we assume that V has a definite sign. As in Subsect. 2.2, if $\pm V \geq 0$, we will write $H = H_\pm = H_0 \pm |V|$.

Theorem 1. [10, Theorem 3.1] *Assume that (2) is valid, and* $\pm V \geq 0$. *Let* $q \in \mathbb{Z}_+$, $b > 0$. *Then the asymptotic estimates*

$$\xi(2bq - \lambda; H_+, H_0) = O(1)\,,$$

$$-n_+((1-\varepsilon)2\sqrt{\lambda}; p_q W p_q) + O(1) \leq \xi(2bq - \lambda; H_-, H_0) \leq$$

$$- n_+((1+\varepsilon)2\sqrt{\lambda}; p_q W p_q) + O(1)\,, \tag{9}$$

hold as $\lambda \downarrow 0$ *for each* $\varepsilon \in (0,1)$.

Suppose that V satisfies (1). For $\lambda \geq 0$ define the matrix-valued function

$$\mathcal{W}_\lambda = \mathcal{W}_\lambda(X_\perp) := \begin{pmatrix} w_{11} & w_{12} \\ w_{21} & w_{22} \end{pmatrix}, \quad X_\perp \in \mathbb{R}^2 ,$$

where

$$w_{11} := \int_\mathbb{R} |V(X_\perp, x_3)| \cos^2\left(\sqrt{\lambda}x_3\right)dx_3 ,$$

$$w_{12} = w_{21} := \int_\mathbb{R} |V(X_\perp, x_3)| \cos\left(\sqrt{\lambda}x_3\right)\sin\left(\sqrt{\lambda}x_3\right)dx_3 ,$$

$$w_{22} := \int_\mathbb{R} |V(X_\perp, x_3)| \sin^2\left(\sqrt{\lambda}x_3\right)dx_3 .$$

It is easy to check that for $\lambda \geq 0$ and $q \in \mathbb{Z}_+$ the operator $p_q\mathcal{W}_\lambda p_q :$ $L^2(\mathbb{R}^2)^2 \to L^2(\mathbb{R}^2)^2$ satisfies $0 \leq p_q\mathcal{W}_\lambda p_q \in S_1$, and rank $p_q\mathcal{W}_\lambda p_q = \infty$.

Theorem 2. [10, Theorem 3.2] *Assume that (2) is valid, and $\pm V \geq 0$. Let $q \in \mathbb{Z}_+$, $b > 0$. Then the asymptotic estimates*

$$\pm\frac{1}{\pi}\text{Tr arctan}\left(((1 \pm \varepsilon)2\sqrt{\lambda})^{-1}p_q\mathcal{W}_\lambda p_q\right) + O(1) \leq \xi(2bq + \lambda; H_\pm, H_0) \leq$$

$$\pm\frac{1}{\pi}\text{Tr arctan}\left(((1 \mp \varepsilon)2\sqrt{\lambda})^{-1}p_q\mathcal{W}_\lambda p_q\right) + O(1) \tag{10}$$

hold as $\lambda \downarrow 0$ for each $\varepsilon \in (0,1)$.

Relations (9) and (10) allow us to reduce the analysis of the behaviour as $\lambda \to 0$ of $\xi(2bq + \lambda; H_\pm, H_0)$, to the study of the asymptotic distribution of the eigenvalues of Toeplitz-type operators $p_q U p_q$. The following three lemmas concern the spectral asymptotics of such operators.

Lemma 3. [31, Theorem 2.6] *Let the function $0 \leq U \in C^1(\mathbb{R}^2)$ satisfy the estimates*

$$U(X_\perp) = u_0(X_\perp/|X_\perp|)|X_\perp|^{-\alpha}(1 + o(1)), \quad |X_\perp| \to \infty ,$$

$$|\nabla U(X_\perp)| \leq C_1\langle X_\perp\rangle^{-\alpha-1}, \quad X_\perp \in \mathbb{R}^2 ,$$

where $\alpha > 0$, and u_0 is a continuous function on $\mathbb{S}^1$ which is non-negative and does not vanish identically. Then for each $q \in \mathbb{Z}_+$ we have

$$n_+(s; p_q U p_q) = \frac{b}{2\pi}\left|\left\{X_\perp \in \mathbb{R}^2 | U(X_\perp) > s\right\}\right| (1 + o(1))$$

$$= \psi_\alpha(s)\,(1 + o(1)), \quad s \downarrow 0 ,$$

where $|.|$ denotes the Lebesgue measure, and

$$\psi_\alpha(s) := s^{-2/\alpha}\frac{b}{4\pi}\int_{\mathbb{S}^1} u_0(t)^{2/\alpha}dt, \quad s > 0 . \tag{11}$$

Lemma 4. [38, Theorem 2.1, Proposition 4.1] *Let* $0 \leq U \in L^\infty(\mathbb{R}^2)$. *Assume that*

$$\ln U(X_\perp) = -\mu |X_\perp|^{2\beta}(1 + o(1)), \quad |X_\perp| \to \infty,$$

for some $\beta \in (0, \infty)$, $\mu \in (0, \infty)$. *Then for each* $q \in \mathbb{Z}_+$ *we have*

$$n_+(s; p_q U p_q) = \varphi_\beta(s)(1 + o(1)), \quad s \downarrow 0,$$

where

$$\varphi_\beta(s) := \begin{cases} \frac{b}{2\mu^{1/\beta}} |\ln s|^{1/\beta} & \text{if } 0 < \beta < 1, \\ \frac{1}{\ln(1 + 2\mu/b)} |\ln s| & \text{if } \beta = 1, \\ \frac{\beta}{\beta - 1} (\ln|\ln s|)^{-1} |\ln s| & \text{if } 1 < \beta < \infty, \end{cases} \quad s \in (0, e^{-1}). \quad (12)$$

Lemma 5. [38, Theorem 2.2, Proposition 4.1] *Let* $0 \leq U \in L^\infty(\mathbb{R}^2)$. *Assume that the support of* U *is compact, and there exists a constant* $C > 0$ *such that* $U \geq C$ *on an open non-empty subset of* $\mathbb{R}^2$. *Then for each* $q \in \mathbb{Z}_+$ *we have*

$$n_+(s; p_q U p_q) = \varphi_\infty(s)(1 + o(1)), \quad s \downarrow 0,$$

where

$$\varphi_\infty(s) := (\ln|\ln s|)^{-1} |\ln s|, \quad s \in (0, e^{-1}). \quad (13)$$

Employing Lemmas 3, 4, 5, we easily find that asymptotic estimates (9) and (10) entail the following

Corollary 1. [10, Corollaries 3.1 – 3.2] *Let (2) hold with* $m_0 > 3$.
i) Assume that the hypotheses of Lemma 3 hold with $U = W$ *and* $\alpha > 2$. *Then*

$$\xi(2bq - \lambda; H_-, H_0) = -\frac{b}{2\pi} \left| \left\{ X_\perp \in \mathbb{R}^2 | W(X_\perp) > 2\sqrt{\lambda} \right\} \right| (1 + o(1))$$
$$= -\psi_\alpha(2\sqrt{\lambda})(1 + o(1)), \quad \lambda \downarrow 0, \quad (14)$$

$$\xi(2bq + \lambda; H_\pm, H_0) = \pm \frac{b}{2\pi^2} \int_{\mathbb{R}^2} \arctan\left((2\sqrt{\lambda})^{-1} W(X_\perp)\right) dX_\perp (1 + o(1))$$
$$= \pm \frac{1}{2\cos(\pi/\alpha)} \psi_\alpha(2\sqrt{\lambda})(1 + o(1)), \quad \lambda \downarrow 0.$$

the function ψ_α *being defined in (11).*
ii) Assume that the hypotheses of Lemma 4 hold with $U = W$. *Then*

$$\xi(2bq - \lambda; H_-, H_0) = -\varphi_\beta(2\sqrt{\lambda})(1 + o(1)), \quad \lambda \downarrow 0, \quad \beta \in (0, \infty),$$

the functions φ_β *being defined in (12). If, in addition,* V *satisfies (1) for some* $m_\perp > 2$ *and* $m_3 > 2$, *we have*

$$\xi(2bq + \lambda; H_\pm, H_0) = \pm \frac{1}{2} \varphi_\beta(2\sqrt{\lambda})(1 + o(1)), \quad \lambda \downarrow 0, \quad \beta \in (0, \infty).$$

iii) Assume that the hypotheses of Lemma 5 hold with $U = W$. Then

$$\xi(2bq - \lambda; H_-, H_0) = -\varphi_\infty(2\sqrt{\lambda})\,(1 + o(1)), \quad \lambda \downarrow 0\,,$$

the function φ_∞ being defined in (13). If, in addition, V satisfies (1) for some $m_\perp > 2$ and $m_3 > 2$, we have

$$\xi(2bq + \lambda; H_\pm, H_0) = \pm\frac{1}{2}\,\varphi_\infty(2\sqrt{\lambda})\,(1 + o(1)), \quad \lambda \downarrow 0\,,$$

the function φ_∞ being defined in (13).

In particular, we find that

$$\lim_{\lambda \downarrow 0} \frac{\xi(2bq - \lambda; H_-, H_0)}{\xi(2bq + \lambda; H_-, H_0)} = \frac{1}{2\cos\frac{\pi}{\alpha}} \tag{15}$$

if W has a power-like decay at infinity (more precisely, if the assumptions of Corollary 1 i) hold), or

$$\lim_{\lambda \downarrow 0} \frac{\xi(2bq - \lambda; H_-, H_0)}{\xi(2bq + \lambda; H_-, H_0)} = \frac{1}{2} \tag{16}$$

if W decays exponentially or has a compact support (more precisely, if the assumptions of Corollary 1 ii) - iii) are fulfilled). Relations (15) and (16) could be interpreted as analogues of the classical Levinson formulae (see e.g. [39]).

Remarks: i) Since the ranks of $p_q W p_q$ and $p_q \mathcal{W}_\lambda p_q$ are infinite, the quantities $n_+(s2\sqrt{\lambda}; p_q W p_q)$ and $\mathrm{Tr}\arctan((s2\sqrt{\lambda})^{-1}p_q\mathcal{W}_\lambda p_q)$ tend to infinity as $\lambda \downarrow 0$ for every $s > 0$. Therefore, Theorems 1 and 2 imply that the SSF $\xi(\cdot; H_\pm, H_0)$ has a singularity at each Landau level. The existence of singularities of the SSF at strictly positive energies is in sharp contrast with the non-magnetic case $b = 0$ where the SSF $\xi(E; -\Delta + V, -\Delta)$ is continuous for $E > 0$ (see e.g. [39]). The main reason for this phenomenon is the fact that the Landau levels play the role of thresholds in $\sigma(H_0)$ while the free Laplacian $-\Delta$ has no strictly positive thresholds in its spectrum.

It is conjectured that the singularity of the SSF $\xi(\cdot; H_\pm(b), H_0(b))$, $b > 0$, at a given Landau level $2bq$, $q \in \mathbb{Z}_+$, could be related to a possible accumulation of resonances and/or eigenvalues of H at $2bq$. Here it should be recalled that in the case $b = 0$ the high energy asymptotics (see [27]) and the semi-classical asymptotics (see [28]) of the derivative of the SSF for appropriate compactly supported perturbations of the Laplacian, are related by the Breit-Wigner formula to the asymptotic distribution near the real axis of the resonances defined as poles of the meromorphic continuation of the resolvent of the perturbed operator.

ii) In the case $q = 0$, when by (3) we have $\xi(-\lambda; H_-, H_0) = -N(-\lambda; H_-)$ for $\lambda > 0$, asymptotic relations of the type of (14) have been known since long ago (see [42], [41], [17,31,43]). An important characteristic feature of the methods used in [31], and later in [38], is the systematic use, explicit or implicit, of the connection between the spectral theory of the Schrödinger operator with constant magnetic field, and the theory of Toeplitz operators acting in holomorphic spaces of Fock-Segal-Bargmann type and the related pseudodifferential operators with generalised anti-Wick symbols (see [3,12,15,40]). Various important aspects of the interaction between these two theories have been discussed in [37] and [7, Sect. 9]. The Toeplitz-operator approach turned to be especially fruitful in [38] where electric potentials decaying rapidly at infinity (i.e. decaying exponentially, or having compact support) were considered (see Lemmas 4 - 5). It is shown in [11] that the precise spectral asymptotics for the Landau Hamiltonian perturbed by a compactly supported electric potential U of fixed sign recovers the logarithmic capacity of the support of U.

iii) Let us mention several other existing extensions of Lemmas 3 – 5. Lemmas 3 and 5 have been generalised to the multidimensional case where p_q is the orthogonal projection onto a given eigenspace of the Schrödinger operator with constant magnetic field of full rank, acting in $L^2(\mathbb{R}^{2d})$, $d > 1$ (see [31] and [25] respectively). Moreover, Lemma 5 has been generalised in [25] to a relativistic setting where p_q is an eigenprojection of the Dirac operator. Finally, in [36] Lemmas 3 – 5 have been extended to the case of the 2D Pauli operator with variable magnetic field from a certain class including the almost periodic fields with non-zero mean value (in this case the role of the Landau levels is played by the origin), and electric potentials U satisfying the assumptions of Lemmas 3 – 5. In the case of compactly supported U of definite sign, [11] contains a more precise version of the corresponding result of [36], involving again the logarithmic capacity of the support of U.

iv) To the author's best knowledge, the singularities at the Landau levels of the SSF for the 3D Schrödinger operator in constant magnetic field have been investigated for the first time in [10]. However, it is appropriate to mention here the article [19] where axisymmetric potentials V have been considered. For a *fixed* magnetic quantum number $m \in \mathbb{Z}$ the authors of [19] studied the behaviour of the SSF $\xi(E; H^{(m)}, H_0^{(m)})$ for the operator pair $\left(H^{(m)}, H_0^{(m)} \right)$ defined in (7) at energies E near the Landau level $2m$ if $m > 0$, and near the origin if $m \leq 0$, and deduced analogues of the classical Levinson formulae, concerning the pair $\left(H^{(m)}, H_0^{(m)} \right)$. Later, the methods in [19] were developed in [23] and [24]. However, it is not possible to recover the results of Theorem 1, Theorem 2 and Corollary 1 from the results of [19], [23], and [24] even in the case of axisymmetric V.

v) Finally, [16] contains general bounds on the SSF for appropriate pairs of magnetic Schrödinger operators. These bounds are applied in order to deduce Wegner estimates of the integrated density of states for some random alloy-type models.

3.2 Strong Magnetic Field Asymptotics of the SSF

Our first theorem in this subsection treats the asymptotics as $b \to \infty$ of $\xi(\cdot; H(b), H_0(b))$ far from the Landau levels.

Theorem 3. [7, Theorem 2.1] *Let (1) hold. Assume that $\mathcal{E} \in (0, \infty) \setminus 2\mathbb{Z}_+$, and $\lambda \in \mathbb{R}$. Then*

$$\xi(\mathcal{E}b + \lambda; H(b), H_0(b)) = \frac{b^{1/2}}{4\pi^2} \sum_{l=0}^{[\mathcal{E}/2]} (\mathcal{E} - 2l)^{-1/2} \int_{\mathbb{R}^3} V(\mathbf{x}) d\mathbf{x} + O(1), \quad b \to \infty .$$

(17)

The following two theorems concern the asymptotics of the SSF near a given Landau level. In order to formulate our next theorem, we introduce the following self-adjoint operators

$$\chi_0 := -d^2/dx_3^2, \quad \chi = \chi(X_\perp) := \chi_0 + V(X_\perp, .), \quad X_\perp \in \mathbb{R}^2 ,$$

which are defined on the Sobolev space $\mathrm{H}^2(\mathbb{R})$, and depend on the parameter $X_\perp \in \mathbb{R}^2$. If (1) holds, then $(\chi(X_\perp) - \lambda_0)^{-1} - (\chi_0 - \lambda_0)^{-1} \in S_1$ for each $X_\perp \in \mathbb{R}^2$ and $\lambda_0 < \inf \sigma(\chi(X_\perp))$. Hence, the SSF $\xi(.; \chi(X_\perp), \chi_0)$ is well-defined. Set $\Lambda := \min_{X_\perp \in \mathbb{R}^2} \inf \sigma(\chi(X_\perp))$. Evidently, $\Lambda \in [-C_0, 0]$. Moreover, $\Lambda = \lim_{b \to \infty} \inf \sigma(H(b))$ (see [1, Theorem 5.8]).

Proposition 2. (cf. [7, Proposition 2.2]) *Assume that (1) holds.*
i) For each $\lambda \in \mathbb{R} \setminus \{0\}$ we have $\xi(\lambda; \chi(.), \chi_0) \in L^1(\mathbb{R}^2)$.
ii) The function $(0, \infty) \ni \lambda \mapsto \int_{\mathbb{R}^2} \xi(\lambda; \chi(X_\perp), \chi_0) dX_\perp$ is continuous, while the non-increasing function

$$(-\infty, 0) \ni \lambda \mapsto \int_{\mathbb{R}^2} \xi(\lambda; \chi(X_\perp), \chi_0) dX_\perp = -\int_{\mathbb{R}^2} N(\lambda; \chi(X_\perp)) dX_\perp$$

(see (3)), is continuous at the point $\lambda < 0$ if and only if

$$|\{X_\perp \in \mathbb{R}^2 | \lambda \in \sigma(\chi(X_\perp))\}| = 0 .$$

(18)

iii) Assume $\pm V \geq 0$. If $\lambda > \Lambda$, $\lambda \neq 0$, then $\pm \int_{\mathbb{R}^2} \xi(\lambda; \chi(X_\perp), \chi_0) dX_\perp > 0$.

Remark: The third part of Proposition 2 is not included in [7, Proposition 2.2], but it follows easily from A. Pushnitski's representation of the SSF (see [29]), and the hypotheses $V \not\equiv 0$ and $V \in C(\mathbb{R}^3)$.

Theorem 4. [7, Theorem 2.3] *Assume that (1) holds. Let $q \in \mathbb{Z}_+$, $\lambda \in \mathbb{R} \setminus \{0\}$. If $\lambda < 0$, suppose also that (18) holds. Then we have*

$$\lim_{b \to \infty} b^{-1} \xi(2bq + \lambda; H(b), H_0(b)) = \frac{1}{2\pi} \int_{\mathbb{R}^2} \xi(\lambda; \chi(X_\perp), \chi_0) \, dX_\perp .$$

(19)

By Proposition 2 iii), if $\pm V \geq 0$, then the r.h.s. of (19) is different from zero if $\lambda > \Lambda$, $\lambda \neq 0$. Unfortunately, we cannot prove the same for general non-sign-definite potentials V. On the other hand, it is obvious that for arbitrary V we have $\int_{\mathbb{R}^2} \xi(\lambda; \chi(X_\perp), \chi_0) dX_\perp = 0$ if $\lambda < \Lambda$. The last theorem of this subsection contains a more precise version of (19) for the case $\lambda < \Lambda$.

Theorem 5. [7, Theorem 2.4] *Let (1) hold.*
i) Let $\lambda < \Lambda$. Then for sufficiently large $b > 0$ we have $\xi(\lambda; H(b), H_0(b)) = 0$.
ii) Let $q \in \mathbb{Z}_+$, $q \geq 1$, $\lambda < \Lambda$. Assume that the partial derivatives of $\langle x_3 \rangle^{m_3} V$ with respect to the variables $X_\perp \in \mathbb{R}^2$ exist, and are uniformly bounded on $\mathbb{R}^3$. Then we have

$$\lim_{b \to \infty} b^{-1/2} \xi(2bq + \lambda; H(b), H_0(b)) = \frac{1}{4\pi^2} \sum_{l=0}^{q-1} (2(q-l))^{-1/2} \int_{\mathbb{R}^3} V(\mathbf{x}) d\mathbf{x} . \quad (20)$$

Remarks: i) Relations (17), (19), and (20) can be unified into a single asymptotic formula. In order to see this, notice that a general result on the high-energy asymptotics of the SSF for 1D Schrödinger operators (see e.g. [39]) implies, in particular, that

$$\lim_{E \to \infty} E^{1/2} \xi(E; \chi(X_\perp), \chi_0) = \frac{1}{2\pi} \int_{\mathbb{R}} V(X_\perp, x_3) \, dx_3, \quad X_\perp \in \mathbb{R}^2 .$$

Then relation (17) with $0 < \mathcal{E} \notin 2\mathbb{Z}_+$, or relations (19) and (20) with $\mathcal{E} = 2q$, $q \in \mathbb{Z}_+$, entail

$\xi(\mathcal{E}b + \lambda; H(b), H_0(b))$

$$= \frac{b}{2\pi} \sum_{l=0}^{[\mathcal{E}/2]} \int_{\mathbb{R}^2} \xi(b(\mathcal{E} - 2l) + \lambda; \chi(X_\perp), \chi_0) dX_\perp (1 + o(1)), \quad b \to \infty . \quad (21)$$

ii) By (3) for $\lambda < 0$ we have $\xi(\lambda; H(b), H_0) = -N(\lambda; H(b))$. The asymptotics as $b \to \infty$ of the counting function $N(\lambda; H_0(b))$ with $\lambda < 0$ fixed, has been investigated in [32] under considerably less restrictive assumptions on V than in Theorems 3–5. The asymptotic properties as $\lambda \uparrow 0$, and as $\lambda \downarrow \Lambda$ if $\Lambda < 0$, of the asymptotic coefficient $-\frac{1}{2\pi} \int_{\mathbb{R}^2} N(\lambda; \chi(X_\perp)) dX_\perp$ which appears at the r.h.s. of (19) in the case of a negative perturbation, have been studied in [33]. The asymptotic distribution of the discrete spectrum for the 3D magnetic Pauli and Dirac operators in strong magnetic fields has been considered in [34] and [35] respectively. The main purpose in [32, 34], and [35] was to obtain the main asymptotic term (without any remainder estimates) of the corresponding counting function under assumptions close to the minimal ones which guarantee that the Hamiltonians are self-adjoint, and the asymptotic coefficient is well-defined. Other results which again describe the asymptotic distribution of the discrete spectrum of the Schrödinger and Dirac operator in strong magnetic fields, but contain also sharp remainder estimates, have been obtained [9,17], and [18] under assumptions on V which, naturally, are more restrictive than those in [32,34], and [35].

iii) Generalizations of asymptotic relation (17) in several directions can be found in [26]. In particular, [26, Theorem 4] implies that if $V \in \mathcal{S}(\mathbb{R}^3)$, then the SSF $\xi(\mathcal{E}b+\lambda; H(b), H_0(b))$, $\mathcal{E} \in (0, \infty) \setminus 2\mathbb{Z}_+$, $\lambda \in \mathbb{R}$, admits an asymptotic expansion of the form

$$\xi(\mathcal{E}b + \lambda; H(b), H_0(b)) \sim \sum_{j=0}^{\infty} c_j b^{\frac{1-2j}{2}}, \quad b \to \infty .$$

iv) Together with the pointwise asymptotics as $b \to \infty$ of the SSF for the pair $(H_0(b), H(b))$ (see (17), (19), or (20)), it also is possible to consider its *weak asymptotics*, i.e. the asymptotics of the convolution of the SSF with an arbitrary $\varphi \in C_0^\infty(\mathbb{R})$. Results of this type are contained in [6].

3.3 High Energy Asymptotics of the SSF

Theorem 6. [8, Theorem 2.1] *Assume that V satisfies (1). Then we have*

$$\lim_{E \to \infty, E \in \mathcal{O}_r} E^{-1/2}\xi(E; H, H_0) = \frac{1}{4\pi^2} \int_{\mathbb{R}^3} V(\mathbf{x})d\mathbf{x}, \quad r \in (0, b) , \quad (22)$$

where $\mathcal{O}_r := \{E \in (0, \infty) | \text{dist}(E, 2b\mathbb{Z}_+) \geq r\}$.

Remarks: i) It is essential to avoid the Landau levels in (22), i.e. to suppose that $E \in \mathcal{O}_r$, $r \in (0, b)$, as $E \to \infty$, since by Theorems 1 - 2, the SSF has singularities at the Landau levels, at least in the case $\pm V \geq 0$.

ii) For $E \in \mathbb{R}$ set

$$\xi_{\text{cl}}(E) := \int_{T^*\mathbb{R}^3} \left(\Theta(E - |\mathbf{p} + \mathbf{A}(\mathbf{x})|^2) - \Theta(E - |\mathbf{p} + \mathbf{A}(\mathbf{x})|^2 - V(\mathbf{x})) \right) d\mathbf{x}d\mathbf{p} =$$

$$\frac{4\pi}{3} \int_{\mathbb{R}^3} \left(E_+^{3/2} - (E - V(\mathbf{x}))_+^{3/2} \right) d\mathbf{x}$$

where $\Theta(s) := \begin{cases} 0 & \text{if } s \leq 0 , \\ 1 & \text{if } s > 0, \end{cases}$ is the Heaviside function. Note that $\xi_{\text{cl}}(E)$ is independent of the magnetic field $b \geq 0$. Evidently, under the assumptions of Theorem 6 we have $\lim_{E \to \infty} E^{-1/2}\xi_{\text{cl}}(E) = 2\pi \int_{\mathbb{R}^3} V(\mathbf{x})d\mathbf{x}$. Hence, if $\int_{\mathbb{R}^3} V(\mathbf{x})d\mathbf{x} \neq 0$, then (22) is equivalent to

$$\xi(E; H, H_0) = (2\pi)^{-3}\xi_{\text{cl}}(E)(1 + o(1)), \quad E \to \infty, \quad E \in \mathcal{O}_r, \quad r \in (0, b) .$$

iii) As far as the author is informed, the high-energy asymptotics of the SSF for 3D Schrödinger operators in constant magnetic fields was investigated for the first time in [8]. Nonetheless, in [19] the asymptotic behaviour as $E \to \infty$, $E \in \mathcal{O}_r$, of the SSF $\xi(E; H^{(m)}, H_0^{(m)})$ for the operator pair $(H^{(m)}, H_0^{(m)})$ (see (7)) with fixed $m \in \mathbb{Z}$ has been been investigated. It does not seem possible to deduce (22) from the results of [19] even in the case of axisymmetric V.

Acknowledgements

I would like to thank my co-authors Vincent Bruneau, Claudio Fernández, and Alexander Pushnitski for having let me present our joint results in this survey article. The major part of this work has been done during my visit to the Institute of Mathematics of the Czech Academy of Sciences in December 2004. I am sincerely grateful to Miroslav Engliš for his warm hospitality. The partial support by the Chilean Science Foundation *Fondecyt* under Grant 1050716 is acknowledged.

References

1. J. AVRON, I. HERBST, B. SIMON, *Schrödinger operators with magnetic fields. I. General interactions*, Duke Math. J. **45** (1978), 847–883.
2. J. AVRON, R. SEILER, B. SIMON, *The index of a pair of projections*, J. Funct. Anal. **120** (1994), 220–237.
3. F.A. BEREZIN, M.A. SHUBIN, *The Schrödinger Equation*. Kluwer Academic Publishers, Dordrecht, 1991.
4. M.Š. BIRMAN, M.G. KREĬN, *On the theory of wave operators and scattering operators*, Dokl. Akad. Nauk SSSR **144** (1962), 475–478 (Russian); English translation in Soviet Math. Doklady **3** (1962).
5. M.Š. BIRMAN, D.R. YAFAEV, *The spectral shift function. The papers of M. G. Kreĭn and their further development*, Algebra i Analiz **4** (1992), 1–44 (Russian); English translation in St. Petersburg Math. J. **4** (1993), 833–870.
6. V. BRUNEAU, M. DIMASSI, *Weak asymptotics of the spectral shift function in strong constant magnetic field*, Preprint 2004 (to appear in *Math. Nachr.*)
7. V. BRUNEAU, A. PUSHNITSKI, G.D. RAIKOV, *Spectral shift function in strong magnetic fields*, Algebra i Analiz **16** (2004), 207 - 238; see also St. Petersburg Math. Journal **16** (2005), 181–209.
8. V. BRUNEAU, G. RAIKOV, *High energy asymptotics of the magnetic spectral shift function*, J. Math. Phys. **45** (2004), 3453–3461.
9. M. DIMASSI, *Développements asymptotiques de l'opérateur de Schrödinger avec champ magnétique fort*, Commun. Part. Diff. Eqns **26** (2001), 595–627.
10. C. FERNÁNDEZ, G.D. RAIKOV, *On the singularities of the magnetic spectral shift function at the Landau levels*, Ann. Henri Poincaré **5** (2004), 381–403.
11. N. FILONOV, A. PUSHNITSKI, *Spectral asymptotics for Pauli operators and orthogonal polynomials in complex domains*, Preprint 2005; available at http://arxiv.org/pdf/math.SP/0504044. (to appear in Comm. Math. Phys.)
12. V.FOCK, *Bemerkung zur Quantelung des harmonischen Oszillators im Magnetfeld*, Z. Physik **47** (1928), 446–448.
13. C. GÉRARD, I. LABA, *Multiparticle Quantum Scattering in Constant Magnetic Fields*, Mathematical Surveys and Monographs, **90**, AMS, Providence, RI, 2002.
14. F. GESZTESY, K. MAKAROV, *The Ξ operator and its relation to Krein's spectral shift function*, J. Anal. Math. **81** (2000), 139–183.
15. B.C. HALL, *Holomorphic methods in analysis and mathematical physics*, In: First Summer School in Analysis and Mathematical Physics, Cuernavaca Morelos, 1998, 1–59, Contemp.Math. **260**, AMS, Providence, RI, 2000.

16. D. HUNDERTMARK, R. KILLIP, S. NAKAMURA, P. STOLLMANN, I. VESELIĆ, *Bounds on the spectral shift function and the density of states*, Comm. Math. Phys. **262** (2006), 489–503.

17. V. IVRII, *Microlocal Analysis and Precise Spectral Asymptotics*, Springer monographs in Math. Springer, Berlin, 1998.

18. V. IVRII, *Sharp spectral asymptotics for magnetic Schrödinger operator with irregular potential.*, Russian Journal of Mathematical Physics, **11** (2004), 415–428.

19. V. KOSTRYKIN, A. KVITSINSKY, S. MERKURIEV, *Potential scattering in constant magnetic field: spectral asymptotics and Levinson formula*, J. Phys. A **28** (1995), 3493–3509.

20. M.G. KREIN, *On the trace formula in perturbation theory*, Mat. Sb. **33** (1953), 597–626 (Russian).

21. L. LANDAU, *Diamagnetismus der Metalle*, Z. Physik **64** (1930), 629–637.

22. I.M. LIFSHITS, *On a problem in perturbation theory*, Uspekhi Mat. Nauk **7** (1952), 171–180 (Russian).

23. M. MELGAARD, *New approach to quantum scattering near the lowest Landau threshold for a Schrödinger operator with a constant magnetic field*, Few-Body Systems **32** (2002), 1–22.

24. M. MELGAARD, *Quantum scattering near the lowest Landau threshold for a Schrödinger operator with a constant magnetic field*, Cent. Eur. J. Math. **1**, (2003) 477–509.

25. M. MELGAARD, G. ROZENBLUM, *Eigenvalue asymptotics for weakly perturbed Dirac and Schrödinger operators with constant magnetic fields of full rank*, Commun. Part. Diff. Eqns **28** (2003), 697–736.

26. L. MICHEL, *Scattering amplitude and scattering phase for the Schrödinger equation with strong magnetic field*, J. Math. Phys. **46** (2005) 043514, 18 pp.

27. V. PETKOV, M. ZWORSKI, *Breit-Wigner approximation and the distribution of resonances* Comm. Math. Phys. **204** (1999) 329–351; Erratum: Comm. Math. Phys. **214** (2000) 733–735.

28. V. PETKOV, M. ZWORSKI, *Semi-classical estimates on the scattering determinant*, Ann. Henri Poincaré **2** (2001) 675–711.

29. A. PUSHNITSKIĬ, *A representation for the spectral shift function in the case of perturbations of fixed sign*, Algebra i Analiz **9** (1997), 197–213 (Russian); English translation in St. Petersburg Math. J. **9** (1998), 1181–1194.

30. A. PUSHNITSKI, *The spectral shift function and the invariance principle*, J. Funct. Anal. **183** (2001), 269–320.

31. G. D. RAIKOV, *Eigenvalue asymptotics for the Schrödinger operator with homogeneous magnetic potential and decreasing electric potential. I. Behaviour near the essential spectrum tips*, Commun. P.D.E. **15** (1990), 407–434; Errata: Commun. P.D.E. **18** (1993), 1977–1979.

32. G.D. RAIKOV, *Eigenvalue asymptotics for the Schrödinger operator in strong constant magnetic fields*, Comm. Partial Differential Equations 23 (1998), no. 9–10, 1583–1619.

33. G.D. RAIKOV, *Asymptotic properties of the magnetic integrated density of states*, Electron. J. Diff. Eq. **1999**, No. 13, 27 pp. (1999).

34. G.D. RAIKOV, *Eigenvalue asymptotics for the Pauli operator in strong nonconstant magnetic fields*, Ann.Inst.Fourier, **49** (1999), 1603–1636.

35. G.D. RAIKOV, *Eigenvalue asymptotics for the Dirac operator in strong constant magnetic fields*, Math.Phys.Electr.J., **5** (1999), No.2, 22 pp.

36. G.D. RAIKOV, *Spectral asymptotics for the perturbed 2D Pauli operator with oscillating magnetic fields. I. Non-zero mean value of the magnetic field*, Markov Process. Related Fields **9** (2003) 775–794.

37. G.D. RAIKOV, M. DIMASSI, *Spectral asymptotics for quantum Hamiltonians in strong magnetic fields*, Cubo Mat. Educ. **3** (2001), 317 - 391.

38. G.D. RAIKOV, S. WARZEL, *Quasi-classical versus non-classical spectral asymptotics for magnetic Schrödinger operators with decreasing electric potentials*, Rev. Math. Phys. **14** (2002), 1051–1072.

39. D. ROBERT, *Semiclassical asymptotics for the spectral shift function*, In: Differential Operators and Spectral theory, AMS Translations Ser. 2 **189**, 187–203, AMS, Providence, RI, 1999.

40. M.A. SHUBIN, *Pseudodifferential Operators and Spectral Theory*, Berlin etc.: Springer-Verlag. (1987).

41. A.V. SOBOLEV, *Asymptotic behavior of energy levels of a quantum particle in a homogeneous magnetic field, perturbed by an attenuating electric field. I*, Probl. Mat. Anal. **9** (1984), 67–84 (Russian); English translation in: J. Sov. Math. **35** (1986), 2201–2212.

42. S.N. SOLNYSHKIN, *Asymptotic behaviour of the energy of bound states of the Schrödinger operator in the presence of electric and homogeneous magnetic fields*, Probl. Mat. Fiziki, Leningrad University **10** (1982), 266–278 (Russian); Engl. transl. in Selecta Math. Soviet. **5** (1986), 297–306.

43. H. TAMURA, *Asymptotic distribution of eigenvalues for Schrödinger operators with homogeneous magnetic fields*, Osaka J. Math. **25** (1988), 633–647.

44. D.R. YAFAEV, *Mathematical scattering theory. General theory.* Translations of Mathematical Monographs, **105** AMS, Providence, RI, 1992.

Counting String/M Vacua

Steve Zelditch

Department of Mathematics, Johns Hopkins University, Baltimore, MD 21218, USA*
zelditch@math.jhu.edu

Abstract. We report on some recent work with M. R. Douglas and B. Shiffman on vacuum statistics for flux compactifications in string/M theory.

1 Introduction

According to string/M theory, the vacuum state of our universe is a 10 dimensional spacetime of the form $M^{3,1} \times X$, where $M^{3,1}$ is Minkowski space and X is a small 3-complex dimensional Calabi-Yau manifold X known as the "small" or "extra" dimensions [6, 28]. The *vacuum selection problem* is that there are many candidate vacua for the Calabi-Yau 3-fold X. Here, we report on recent joint work with B. Shiffman and M. R. Douglas devoted to counting the number of supersymmetric vacua of type IIb flux compactifications [12–14]. We also describe closely related the physics articles of Ashok-Douglas and Denef-Douglas [2, 8, 11] on the same problem.

At the time of writing of this article, vacuum statistics is being intensively investigated by many string theorists (see for instance [1, 7, 10, 19, 30] in addition to the articles cited above). One often hears that the number of possible vacua is of order 10^{500} (see e.g. [4]). This large figure is sometimes decried (at this time) as a blow to predictivity of string/M theory or extolled as giving string theory a rich enough "landscape" to contain vacua that match the physical parameters (e.g. the cosmological constant) of our universe. However, it is very difficult to obtain sufficiently accurate results on vacuum counting to justify the claims of 10^{500} total vacua, or even the existence of one vacuum which is consistent with known physical parameters. The purpose of our work is to develop methods and results relevant to accurate vacuum counting.

From a mathematical viewpoint, supersymmetric vacua are critical points

$$\nabla_G W(Z) = 0 \tag{1}$$

of certain holomorphic sections W_G called *flux superpotentials* of a line bundle $\mathcal{L} \to \mathcal{C}$ over the moduli space $\mathcal{C}$ of complex structures on $X \times T^2$ where $T^2 = \mathbb{R}^2/\mathbb{Z}^2$. Flux superpotentials depend on a choice of flux $G \in$

---

*Research partially supported by NSF grant DMS-0302518

$H^3(X, \mathbb{Z} \oplus \sqrt{-1}\mathbb{Z})$. There is a constraint on G called the "tadpole constraint', so that G is a lattice point lying in a certain hyperbolic shell $0 \leq Q[G] \leq L$ in $H^3(X, \mathbb{C})$ (16). Our goal is to count all critical points of all flux superpotentials W_G in a given compact set of $\mathcal{C}$ as G ranges over such lattice points. Thus, counting vacua in $K \subset \mathcal{C}$ is a combination of an equidistribution problem for projections of lattice points and an equidistribution problem for critical points of random holomorphic sections.

The work we report on gives a rigorous foundation for the program initiated by M. R. Douglas [11] to count vacua by making an approximation to the Gaussian ensembles the other two authors were using to study statistics of zeros of random holomorphic sections (cf. [3, 27]). The results we describe here are the first rigorous results on counting vacua in a reasonably general class of models (type IIb flux compaticifications). They are admittedly still in a rudimentary stage, in particular because they are asymptotic rather than effective. We will discuss the difficulties in making them effective below.

This report is a written version of our talk at the QMath9 conference in Giens in October, 2004. A more detailed expository article with background on statistical algebraic geometry as well as string theory is given in [32], which was based on the author's AMS address in Atlanta, January 2005.

2 Type IIb Flux Compactifications of String/M Theory

The string/M theories we consider are type IIb string theories compactified on a complex 3-dimensional Calabi-Yau manifold X with flux [2, 18–20, 23]. We recall that a Calabi-Yau 3-fold is a compact complex manifold X of dimension 3 with trivial canonical bundle K_X, i.e. $c_1(X) = 0$ [21, 22]. Such X possesses a unique Ricci flat Kähler metric in each Kähler class. In what follows, we fix the Kähler class, and then the CY metrics correspond to the complex structures on X. We denote the moduli space of complex structures on X by $\mathcal{M}_{\mathbb{C}}$. In addition to the complex structure moduli on X there is an extra parameter τ called the dilaton axion, which ranges over complex structure moduli on $T^2 = \mathbb{R}^2/\mathbb{Z}^2$. Hence, the full configuration space $\mathcal{C}$ of the model is the product

$$\mathcal{C} = \mathcal{M}_{\mathbb{C}} \times \mathcal{E}, \quad (Z = (z, \tau); \ z \in \mathcal{M}_{\mathbb{C}}, \tau \in \mathcal{E}) \tag{2}$$

where $\mathcal{E} = \mathcal{H}/SL(2, \mathbb{Z})$ is the moduli space of complex 1-tori (elliptic curves). One can think of $\mathcal{C}$ as a moduli space of complex structures on the CY 4-fold $X \times T^2$.

By "flux" is meant a complex integral 3-form

$$G = F + iH \in H^3(X, \mathbb{Z} \oplus \sqrt{-1}\mathbb{Z}) . \tag{3}$$

The *flux superpotential* $W_G(Z)$ corresponding to G is defined as follows: On a Calabi-Yau 3-fold, the space $H_z^{3,0}(X)$ of holomorphic $(3,0)$-forms for each complex structure z on X has dimension 1, and we denote a holomorphically varying family by $\Omega_z \in H_z^{3,0}(X)$. Given G as in (3) and $\tau \in \mathcal{H}$, physicists define the superpotential corresponding to G, τ by:

$$W_G(z, \tau) = \int_X (F - \tau H) \wedge \Omega_z \,. \tag{4}$$

This is not well-defined as a function on $\mathcal{C}$, since Ω_z is not unique and τ corresponds to the holomorphically varying form $\omega_\tau = dx + \tau dy \in H_\tau^{1,0}(T^2)$ which is not unique either. To be more precise, we define W_G to be a holomorphic section of a line bundle $\mathcal{L} \to \mathcal{C}$, namely the dual line bundle to the Hodge line bundle $H_{z,\tau}^{4,0} = H_z^{3,0}(X) \otimes H^{1,0}(T^2) \to \mathcal{C}$. We form the 4-form on $X \times T^2$

$$\tilde{G} = F \wedge dy + H \wedge dx$$

and define a linear functional on $H_z^{3,0}(X) \otimes H_\tau^{1,0}(T^2)$ by

$$\langle W_G(z, \tau), \Omega_z \wedge \omega_\tau \rangle = \int_{X \times T^2} \tilde{G} \wedge \Omega_z \wedge \omega_\tau \,. \tag{5}$$

When $\omega_\tau = dx + \tau dy$ we obtain the original formula. As $Z = (z, \tau) \in \mathcal{C}$ varies, (5) defines a holomorphic section of the line bundle $\mathcal{L}$ dual to $H_z^{3,0} \otimes H_\tau^{1,0} \to \mathcal{C}$.

The Hodge bundle carries a natural Hermitian metric

$$h_{WP}(\Omega_z \wedge \omega_\tau, \Omega_z \wedge \omega_\tau) = \int_{X \times T^2} \Omega_z \wedge \omega_\tau \wedge \overline{\Omega_z \wedge \omega_\tau}$$

known as the Weil-Petersson metric, and an associated metric (Chern) connection by ∇_{WP}. The Kähler potential of the Weil-Petersson metric on $\mathcal{M}_{\mathbb{C}}$ is defined by

$$\mathcal{K} = \ln\left(-i \ln \int_X \Omega \wedge \overline{\Omega}\right) \,. \tag{6}$$

There is a similar definition on $\mathcal{E}$ and we take the direct sum to obtain a Kähler metric on $\mathcal{C}$. We endow $\mathcal{L}$ with the dual Weil-Petersson metric and connection. The hermitian line bundle $(H^{4,0}, h_{WP}) \to \mathcal{M}_{\mathbb{C}}$ is a positive line bundle, and it follows that $\mathcal{L}$ is a negative line bundle.

The vacua we wish to count are the classical vacua of the effective supergravity Lagrangian of the string/M model, which is derived by "integrating out" the massive modes (cf. [28]). The only term relevant of the Lagrangian to our counting problem is the scalar potential [31]

$$V_G(Z) = |\nabla W_G(Z)|^2 - 3|W(Z)|^2 \,, \tag{7}$$

where the connection and hermitian metric are the Weil-Petersson ones. We only consider the supersymmetric vacua here, which are the special critical points Z of V_G satisfying (1).

3 Critical Points and Hessians of Holomorphic Sections

We see that type IIb flux compactifications involve holomorphic sections of hermitian holomorphic line bundles over complex manifolds. Thus, counting flux vacua is a problem in complex geometry. In this section, we provide a short review from [12, 13].

Let $L \to M$ denote a holomorphic line bundle over a complex manifold, and endow L with a hermitian metric h. In a local frame e_L over an open set $U \subset M$, one defines the Kähler potential K of h by

$$|e_L(Z)|_h^2 = e^{-K(Z)} . \tag{8}$$

We write a section $s \in H^0(M, L)$ locally as $s = fe_L$ with $f \in \mathcal{O}(U)$. We further choose local coordinates z. In this frame and local coordinates, the covariant derivative of a section s takes the local form

$$\nabla s = \sum_{j=1}^m \left(\frac{\partial f}{\partial Z_j} - f \frac{\partial K}{\partial Z_j} \right) dZ_j \otimes e_{\mathcal{L}} = \sum_{j=1}^m e^K \frac{\partial}{\partial Z_j} \left(e^{-K} f \right) dZ_j \otimes e_{\mathcal{L}} . \tag{9}$$

The critical point equation $\nabla s(Z) = 0$ thus reads,

$$\frac{\partial f}{\partial Z_j} - f \frac{\partial K}{\partial Z_j} = 0 .$$

It is important to observe that although s is holomorphic, ∇s is not, and the critical point equation is only C^∞ and not holomorphic. This is due to the factor $\frac{\partial K}{\partial Z_j}$, which is only smooth. Connection critical points of s are the same as ordinary critical points of $\log |s(Z)|_h$. Thus, the critical point equation is a system of real equations and the number of critical points varies with the holomorphic section. It is not a topological invariant, as would be the number of zeros of m sections in dimension m, even on a compact complex manifold. This is one reason why counting critical points, hence vacua, is so complicated.

We now consider the Hessian of a section at a critical point. The Hessian of a holomorphic section s of a general Hermitian holomorphic line bundle $(L, h) \to M$ at a critical point Z is the tensor

$$D\nabla W(Z) \in T^* \otimes T^* \otimes L$$

where D is a connection on $T^* \otimes L$. At a critical point Z, $D\nabla s(Z)$ is independent of the choice of connection on T^*. The Hessian $D\nabla W(Z)$ at a critical point determines the complex symmetric matrix H^c (which we call the "complex Hessian'). In an adapted local frame (i.e. holomorphic derivatives vanish at Z_0) and in Kähler normal coordinates, it takes the form

$$H^c := \begin{pmatrix} H' & H'' \\ \overline{H''} & \overline{H'} \end{pmatrix} = \begin{pmatrix} H' & -f(Z_0)\Theta \\ -\overline{f(z_0)\Theta} & \overline{H'} \end{pmatrix} , \tag{10}$$

whose components are given by

$$H'_{jq} = \left(\frac{\partial}{\partial Z_j} - \frac{\partial K}{\partial Z_j}\right)\left(\frac{\partial}{\partial Z_q} - \frac{\partial K}{\partial Z_q}\right) f(Z_0) , \qquad (11)$$

$$H''_{jq} = -\left. f\frac{\partial^2 K}{\partial Z_j \partial \bar{Z}_q}\right|_{Z_0} = -f(Z_0)\Theta_{jq} . \qquad (12)$$

Here, $\Theta_h(z_0) = \sum_{j,q}\Theta_{jq}dZ_j \wedge d\bar{Z}_q$ is the curvature.

4 The Critical Point Problem

We can now define the critical point equation (1) precisely. We define a supersymmetric vacuum of the flux superpotential W_G corresponding to the flux G of (3) to be a critical point $\nabla_{WP}W_G(Z) = 0$ of W_G relative to the Weil-Petersson connection on $\mathcal{L}$.

We obtain a local formula by writing $W_G(Z) = f_G(Z)e_Z$ where e_Z is local frame for $\mathcal{L} \to \mathcal{C}$. We choose the local frame e_Z to be dual to $\Omega_z \otimes \omega_\tau$, and then $f_G(z,\tau)$ is given by the formula (4). The $\mathcal{E}$ component of ∇_{WP} is $\frac{\partial}{\partial\tau} - \frac{1}{\tau-\bar{\tau}}$. The critical point equation is the system:

$$\begin{cases} \int_X (F - \tau H) \wedge \{\frac{\partial\Omega_z}{\partial z_j} + \frac{\partial K}{\partial z_j}\Omega_z\} = 0 , \\ \\ \int_X (F - \bar{\tau}H) \wedge \Omega_z = 0 , \end{cases} \qquad (13)$$

where K is from (6).

Using the *special geometry* of $\mathcal{C}$ ([5,29]), one finds that the critical point equation is equivalent to the following restriction on the Hodge decomposition of $H^3(X,\mathbb{C})$ at z:

$$\nabla_{WP}W_G(z,\tau) = 0 \iff F - \tau H \in H_z^{2,1} \oplus H_z^{0,3} . \qquad (14)$$

Here, we recall that each complex structure $z \in \mathcal{M}_\mathbb{C}$ gives rise to a Hodge decomposition

$$H^3(X,\mathbb{C}) = H_z^{3,0}(X) \oplus H_z^{2,1}(X) \oplus H_z^{1,2}(X) \oplus H_z^{0,3}(X) \qquad (15)$$

into forms of type (p,q). In the case of a CY 3-fold, $h^{3,0} = h^{0,3} = 1$, $h^{1,2} = h^{2,1}$ and $b_3 = 2 + 2h^{2,1}$.

Next, we specify the tadpole constraint. We define the real symmetric bilinear form on $H^3(X,\mathbb{C})$ by

$$Q(\psi,\varphi) = i^3 \int_X \psi \wedge \bar{\varphi} . \qquad (16)$$

The Hodge-Riemann bilinear relations for a 3-fold say that the form Q is definite in each $H_z^{p,q}(X)$ for $p + q = 3$ with sign alternating $+ - + -$ as one moves left to right in (15). The tadpole constraint is that

$$Q[G] = i^3 \int_X G \wedge \bar{G} \leq L \, . \tag{17}$$

Here, L is determined by X in a complicated way (it equals $\chi(Z)/24$ where Z is CY 4-fold which is an elliptic fibration over X/g, where $\chi(Z)$ is the Euler characteristic and where g is an involution of X). Although Q is an indefinite symmetric bilinear form, we see that $Q \gg 0$ on $H_z^{2,1}(X) \oplus H_z^{0,3}$ for any complex structure z.

We now explain the sense in which we are dealing with a lattice point problem. The definition of W_G makes sense for any $G \in H^3(X, \mathbb{C})$, so we obtain a real (but not complex) linear embedding $H^3(X, \mathbb{C}) \subset H^0(\mathcal{C}, \mathcal{L})$. Let us denote the image by $\mathcal{F}$ and call it the space of complex-valued flux superpotentials with dilaton-axion. The set of W_G with $G \in H^3(X, \mathbb{Z} \oplus \sqrt{-1}\mathbb{Z})$ is then a lattice $\mathcal{F}_{\mathbb{Z}} \subset \mathcal{F}$, which we will call the lattice of quantized (or integral) flux superpotentials.

Each integral flux superpotential W_G thus gives rise to a discrete set of critical points $Crit(W_G) \subset \mathcal{C}$, any of which could be the vacuum state of the universe. Moreover, the flux G can be any element of $H^3(X, \mathbb{Z} \oplus \sqrt{-1}\mathbb{Z})$ satisfying the tadpole constraint (17). Thus, the set of possible vacua is the union

$$\text{Vacua}_L = \bigcup_{G \in H^3(X, \mathbb{Z} \oplus \sqrt{-1}\mathbb{Z}), \ 0 \leq Q[G] \leq L} Crit(W_G) \, . \tag{18}$$

Our purpose is to count the number of vacua $\#\text{Vacua}_L \cap K$ in any given compact subset $K \subset \mathcal{C}$.

More generally, we wish to consider the sums

$$N_\psi(L) = \sum_{N \in H^3(X, \mathbb{Z} \oplus \sqrt{-1}\mathbb{Z}) : Q[N] \leq L} \langle C_N, \psi \rangle \, , \tag{19}$$

where

$$\langle C_N, \psi \rangle = \sum_{(z, \tau) : \nabla N(z, \tau) = 0} \psi(N, z, \tau) \, , \tag{20}$$

and where ψ is a reasonable function on the incidence relation

$$\mathcal{I} = \{(W; z, \tau) \in \mathcal{F} \times \mathcal{C} : \nabla W(z, \tau) = 0\} \, . \tag{21}$$

We often write $Z = (z, \tau) \in \mathcal{C}$. Points (W, Z) such that Z is a degenerate critical point of W cause problems. They belong to the *discriminant variety* $\widetilde{\mathcal{D}} \subset \mathcal{I}$ of singular points of the projection $\pi : \mathcal{I} \to \mathcal{F}$. We note that $\pi^{-1}(W) = \{(W, Z) : Z \in Crit(W)\}$. This number is constant on each component of $\mathcal{F} \backslash \mathcal{D}$ where $\mathcal{D} = \pi(\widetilde{\mathcal{D}})$ but jumps as we cross over $\mathcal{D}$.

To count critical points in a compact subset $\mathcal{K} \subset \mathcal{C}$ of moduli space, we would put $\psi = \chi_K(z, \tau)$. We often want to exclude degenerate critical points and then use test functions $\psi(W, Z)$ which are homogeneous of degree 0 in W and vanish on $\widetilde{\mathcal{D}}$ Another important example is the cosmological constant $\psi(W, z, \tau) = V_W(z, \tau)$, i.e. the value of the potential at the vacuum, which is homogeneous of degree 2 in W.

5 Statement of Results

We first state an initial estimate which is regarded as "trivial' in lattice counting problems. In pure lattice point problems it is sharp, but we doubt that it is sharp in the vacuum counting problem because of the "tilting" of the projection $\mathcal{I} \rightarrow \mathcal{C}$. We denote by χ_Q the characteristic function of the hyperbolic shell $0 < Q_Z[W] < 1 \subset \mathcal{F}$ and by χ_{Q_Z} the characteristic function of the elliptic shell $0 < Q_Z[W] < 1 \subset \mathcal{F}_Z$.

Proposition 1. *Suppose that $\psi(W, Z) = \chi_K$ where $K \subset \mathcal{I}$ is an open set with smooth boundary. Then:*

$$\mathcal{N}_\psi(L) = L^{b_3} \left[\int_{\mathcal{C}} \int_{\mathcal{F}_Z} \psi(W, Z) |\det H^c W(Z)| \chi_{Q_Z} dW \, dV_{WP} + R_K(L) \right] ,$$

where

1. *If $\overline{K}$ is disjoint from the $\widetilde{\mathcal{D}}$, then $R_K(L) = O\left(L^{-1/2}\right)$.*
2. *If $\overline{K}$ is a general compact set (possibly intersecting the discriminant locus), then $R_K(L) = O\left(L^{-1/2}\right)$*

Here, $b_3 = \dim H_3(X, \mathbb{R})$, $Q_{z,\tau} = Q|_{\mathcal{F}_{z,\tau}}$ and $\chi_{Q_{z,\tau}}(W)$ is the characteristic function of $\{Q_{z,\tau} \leq 1\} \subset \mathcal{F}_{z,\tau}$, $H^c W(Z)$ is the complex Hessian of W at the critical point Z in the sense of (10). We note that the integral converges since $\{Q_Z \leq 1\}$ is an ellipsoid of finite volume. This is an asymptotic formula which is a good estimate on the number of vacua when L is large (recall that L is a topological invariant determined by X).

The reason for assumption (1) is that number of critical points and the summand $\langle C_W, \psi \rangle$ jump across $\mathcal{D}$, so in $N_\psi(L)$ we are summing a discontinuous function. This discontinuity could cause a relatively large error term in the asymptotic counting. However, superpotentials of physical interest have non-degenerate supersymmetric critical points. Their Hessians at the critical points are 'fermionic mass matrices', which in physics have only non-zero eigenvalues (masses), so it is reasonable assume that $supp\psi$ is disjoint from $\mathcal{D}$.

Now we state the main result.

Theorem 1 *Suppose $\psi(W, z, \tau) \in C_b^\infty(\mathcal{F} \times \mathcal{C})$ is homogeneous of degree 0 in W, with $\psi(W, z, \tau) = 0$ for $W \in \mathcal{D}$. Then*

$$\mathcal{N}_\psi(L) = L^{b_3} \left[\int_{\mathcal{C}} \int_{\mathcal{F}_{z,\tau}} \psi(W, z, \tau) |\det H^c W(z, \tau)| \chi_{Q_{z,\tau}}(W) dW dV_{WP}(z, \tau) \right.$$

$$\left. + O\left(L^{-\frac{2b_3}{2b_3+1}}\right) \right].$$

Here, C_b^∞ denotes bounded smooth functions.

There is a simple generalization to homogeneous functions of any degree such as the cosmological constant. The formula is only the starting point of a number of further versions which will be presented in Sect. 8 in which we "push-forward" the dW integral under the Hessian map, and then perform an Itzykson-Zuber-Harish-Chandra transformation on the integral. The latter version gets rid of the absolute value and seems to most useful for numerical studies. Further, one can use the special geometry of moduli space to simplify the resulting integral. Before discussing them, we pause to compare our results to the expectations in the string theory literature.

6 Comparison to the Physics Literature

The reader following the developments in string theory may have encountered discussions of the "string theory landscape" (see e.g. [4, 30]). The multitude of superpotentials and vacua is a problem for the predictivity of string theory. It is possible that a unique vacuum will distinguish itself in the future, but until then all critical points are candidates for the small dimensions of the universe, and several groups of physicists are counting or enumerating them in various models (see e.g. [7, 8, 10]).

The graph of the scalar potential energy may be visualized as a landscape [30] whose local minima are the possible vacua. It is common to hear that there are roughly 10^{500} possible vacua. This heuristic figure appears to originate in the following reasoning: assuming $b_3 \sim 250$, the potential energy $V_G(Z)$ is a function roughly 500 variables (including fluxes G). The critical point equation for a function of m variables is a system of m equations. Naively, the number of solutions should grow like d^m where d is the number of solutions of the jth equation with the other variables held fixed. This would follow from Bézout's formula if the function was a polynomial and if we were counting complex zeros. Thus, if the "degree" of V_G were a modest figure of 10 we would obtain the heuristic figure.

Such an exponential growth rate of critical points in the number of variables also arises in estimates of the number of metastable states (local minima of the Hamiltonian) in the theory of spin glasses. In fact, an integral similar to that in Theorem 1 arises in the formula for the expected number of local minima of a random spin glass Hamiltonian. Both heuristic and rigorous calculations lead to an exponential growth rate of the number of local minima as the number of variables tends to infinity (see e.g. [17] for a mathematical discussion and references to the literature). The mathematical similarity of the problems at least raises the question whether the number of string/M vacua should grow exponentially in the number $2b_3$ of variables (G, Z), i.e. in the "topological complexity" of the Calabi-Yau manifold X.

Our results do not settle this problem, and indeed it seems to be a difficult question. Here are some of the difficulties: First, in regard to the Bézout estimate, the naive argument ignores the fact that the critical point equation is a real C^∞ equation, not a holomorphic one and so the Bézout estimate could be quite inaccurate. Moreover, a flux superpotential is not a polynomial and it is not clear what "degree" it has, as measured by its number critical points. In simple examples (see e.g. [2, 8, 10], the superpotentials do not have many critical points and it is rather the large number of fluxes satisfying the tadpole constraint which produces the leading term L^{b_3}. This is why the flux G has to be regarded as one of the variables if one wants to rescue the naive counting argument. In addition, the tadpole constraint has a complicated dimensional dependence. It induces a constraint on the inner integral in Theorem 1 to an ellipse in b_3 dimensions, and the volume of such a domain shrinks at the rate $1/(b_3)!$. Further, the volume of the Calabi-Yau moduli space is not known, and could be very small. Thus, there are a variety of competing influences on the growth rate of the number of vacua in b_3 which all have a factorial dependence on the dimension.

To gain a better perspective on these issues, it is important to estimate the integral giving the leading coefficient and the remainder in Theorem 1. The inner integral is essentially an integral of a homogeneous function of degree b_3 over an ellipsoid in b_3 dimensions, and is therefore very sensitive to the size of b_3. The full integral over moduli space carries the additional problem of estimating its volume. Further, one needs to estimate how large L is for a given X. Without such effective bounds on L, it is not even possible to say whether any vacua exist which are consistent with known physical quantities such as the cosmological constant.

7 Sketch of Proofs

The proof of Theorem 1 is in part an application of a lattice point result to the lattice of flux superpotentials. In addition, it uses the formalism on the density of critical points of Gaussian random holomorphic sections in [12]. The lattice point problem is to study the distribution of radial projections of lattice points in the shell $0 \leq Q[G] \leq L$ on the surface $Q[G] = 1$. Radial projections arise because the critical point equation $\nabla W_G = 0$ is homogeneous in G.

Thus, we consider the model problem: Let $\mathbf{Q} \subset \mathbb{R}^n$ $(n \geq 2)$ be a smooth, star-shaped set with $0 \in \mathbf{Q}^\circ$ and whose boundary has a non-degenerate second fundamental form. Let $|X|_\mathbf{Q}$ denote the norm of $X \in \mathbb{R}^n$ defined by $\mathbf{Q} = \{X \in \mathbb{R}^n : |X|_\mathbf{Q} < 1\}$. In the following, we denote the large parameter by $\sqrt{L}$ to maintain consistency with Theorem 1.

Theorem 2 *[14] If f is homogeneous of degree 0 and $f|_{\partial Q} \in C_0^\infty(\partial Q)$, then*

$$S_f(L) := \sum_{k \in \mathbb{Z}^n \cap \sqrt{L}\mathbf{Q} \setminus \{0\}} f(k) = L^{\frac{n}{2}} \int_\mathbf{Q} f\, dX + O(L^{\frac{n}{2} - \frac{n}{n+1}}), \quad L \to \infty .$$

Although we have only stated it for smooth f, the method can be generalized to $f|_{\partial Q} = \chi_K$ where K is a smooth domain in ∂Q [33]. However, the remainder then depends on K and reflects the extent to which projections of lattice points concentrate on $\partial K \subset \partial Q$. The asymptotics are reminiscent of the the result of van der Corput, Hlawka, Herz and Randol on the number of lattice points in dilates of a convex set, but as of this time of writing we have not located any prior studies of the radial projection problem. Number theorists have however studied the distribution of lattice points lying exactly on spheres (Linnik, Pommerenke). We also refer the interested reader to [15] for a recent article counting lattice points in certain rational cones using methods of automorphic forms, in particular L-functions. We thank B. Randol for some discussions of this problem; he has informed the author that the result can also be extended to more general kinds of surfaces with degenerate second fundamental forms.

Applying Theorem 2 to the string/M problem gives that

$$\mathcal{N}_\psi(L) = L^{b_3}\left[\int_{\{Q[W]\leq 1\}} \langle C_W, \psi\rangle\, dW + O\left(L^{-\frac{2b_3}{2b_3+1}}\right)\right]. \qquad (22)$$

We then write (22) as an integral over the incidence relation (21) and change the order of integration to obtain the leading coefficient

$$\int_{\{Q[W]\leq 1\}} \langle C_W, \psi\rangle\, dW$$

$$= \int_{\mathcal{C}}\int_{\mathcal{F}_{z,\tau}} \psi(W, z, \tau)|\det H^c W(z, \tau)|\chi_{Q_{z,\tau}}\, dW\, dV_{WP}(z, \tau) \qquad (23)$$

in Theorem 1. Heuristically, the integral on the left side is given by

$$\int_{\mathcal{F}}\int_{\mathcal{C}} \psi(W, Z)|\det H^c W(Z)|\delta(\nabla W(z))\chi_Q(Z)\, dW\, dV_{WP}(Z). \qquad (24)$$

The factor $|\det H^c W(Z)|$ arises in the pullback of δ under $\nabla W(Z)$ for fixed W, since it weights each term of (20) by $\frac{1}{|\det H^c W(Z)|}$. We obtain the stated form of the integral in (23) by integrating first in W and using the formula for the pull-back of a δ function under a linear submersion. That formula also contains another factor $\frac{1}{\det A(Z)}$ where $A(Z) = \nabla_{Z'_j}\nabla_{Z''_k}\Pi_Z(Z', Z'')|_{Z'=Z''=Z}$, where Π_Z is the Szegö kernel of $\mathcal{F}_Z$, i.e. the orthogonal projection onto that subspace. Using special geometry, the matrix turns out to be just I and hence the determinant is one.

8 Other Formulae for the Critical Point Density

In view of the difficulty of estimating the leading term in Theorem 1, it is useful to have alternative expressions. We now state two of them.

The first method is to change variables to the Hessian $H^c W(Z)$ under the Hessian map

$$H_Z : \mathcal{S}_Z \to \mathrm{Sym}(m, \mathbb{C}) \oplus \mathbb{C}, \quad H_Z(W) = H^c W(Z), \tag{25}$$

where $m = \dim \mathcal{C} = h^{2,1} + 1$. It turns out that Hessian map is an isomorphism to a real b_3-dimensional space $\mathcal{H}_Z \oplus \mathbb{C}$, where

$$\mathcal{H}_Z = \mathrm{span}_{\mathbb{R}} \left\{ \begin{pmatrix} 0 & e_j \\ e_j^t & \mathcal{F}^j(z) \end{pmatrix}, \begin{pmatrix} 0 & ie_j \\ ie_j^t & -i\mathcal{F}^j(z) \end{pmatrix} \right\}_{j=1,\dots,h^{2,1}} . \tag{26}$$

Here, e_j is the j-th standard basis element of $\mathbb{C}^{h^{2,1}}$ and $\mathcal{F}^j(z) \in \mathrm{Sym}(h^{2,1}, \mathbb{C})$ is the matrix $\left(\mathcal{F}_{ik}^j(z) \right)$ whose entries define the so-called "Yukawa couplings" (see [5, 29] for the definition). We define the positive definite operator $C_Z : \mathcal{H}_Z \to \mathcal{H}_Z$ by:

$$(C_Z^{-1} H_Z W, H_Z W) = Q_Z(W, \overline{W}) . \tag{27}$$

The entries in C_Z are quadratic expressions in the $\mathcal{F}_{ik}^j$ (see [14]).

Theorem 3 *We have:*

$$\mathcal{K}^{\mathrm{crit}}(Z) = \frac{1}{b_3! \det C_Z'} \int_{\mathcal{H}_Z \oplus \mathbb{C}} \left| \det H^* H - |x|^2 I \right| \, e^{-(C_Z^{-1} H, H) + |x|^2)} \, dH \, dx \,,$$

$$= \frac{1}{\det C_Z'} \int_{\mathcal{H}_Z \oplus \mathbb{C}} \left| \det H^* H - |x|^2 I \right| \chi_{C_Z}(H, x) dH dx,$$

where χ_{C_Z} *is the characteristic function of the ellipsoid* $\{ (C_Z H, H) + |x|^2) \leq 1 \} \subset \mathcal{H}_Z$.

Finally, we give formula of Itzykson-Zuber type as in [13, Lemma 3.1], which is useful in that it has a fixed domain of integration.

Theorem 4 *Let* $\Lambda_Z = C_Z \oplus I$ *on* $\mathcal{H}_Z \oplus \mathbb{C}$ *and let* P_Z *denote the orthogonal projection from* $\mathrm{Sym}(m, \mathbb{C})$ *onto* $\mathcal{H}_Z$. *Then:*

$$\mathcal{K}^{\mathrm{crit}}(Z) = c_m \lim_{\varepsilon' \to 0^+} \int_{\mathbb{R}^m} \lim_{\varepsilon \to 0^+} \int_{\mathbb{R}^m} \int_{\mathrm{U}(m)}$$

$$\times \frac{\Delta(\xi) \, \Delta(\lambda) \, | \prod_j \lambda_j | \, e^{i\langle \xi, \lambda \rangle} e^{-\varepsilon |\xi|^2 - \varepsilon' |\lambda|^2}}{\sqrt{\det \left[i\Lambda_Z P_Z \rho(g)^* \widehat{D}(\xi) \rho(g) + I \right]}} \, dg \, d\xi \, d\lambda \,,$$

where:

- $m = h^{2,1} + 1$, $c_m = \frac{(-i)^{m(m-1)/2}}{2^m \pi^{2m} \prod_{j=1}^m j!}$;
- $\Delta(\lambda) = \Pi_{i<j}(\lambda_i - \lambda_j)$,
- dg *is unit mass Haar measure on* $\mathrm{U}(m)$,

- $\widehat{D}(\xi)$ *is the Hermitian operator on* $\mathrm{Sym}(m, \mathbb{C}) \oplus \mathbb{C}$ *given by*

$$\widehat{D}(\xi)((H_{jk}), x) = \left(\left(\frac{\xi_j + \xi_k}{2} H_{jk} \right), -\left(\sum_{q=1}^{m} \xi_q \right) x \right),$$

- ρ *is the representation of* $\mathrm{U}(m)$ *on* $\mathrm{Sym}(m, \mathbb{C}) \oplus \mathbb{C}$ *given by*

$$\rho(g)(H, x) = (gHg^t, x).$$

- $\mathcal{H}_Z$ *is a real (but not complex) subspace of* $\mathrm{Sym}(m, \mathbb{C})$.

The proof is similar to the one in [13], but we sketch the proof here to provide a published reference. Some care must be taken since the Gaussian integrals are over real but not complex spaces of complex symmetric matrices.

Proof. We first rewrite the integral in Theorem 3 as a Gaussian integral over $\mathcal{H}_Z \oplus \mathbb{C}$ (viewed as a real vector space):

$$\mathcal{K}^{\mathrm{crit}}(Z) = \int_{\mathcal{H}_Z \oplus \mathbb{C}} |\det H^* H - |x|^2 I| \, \chi_{\{\langle \Lambda_Z^{-1} H, H \rangle \le 1\}} dH dx$$

$$= \frac{\pi^m \sqrt{\det \Lambda_Z}}{b_3!} \mathcal{I}(Z),$$

where

$$\mathcal{I}(Z) = \frac{1}{\pi^m \sqrt{\det \Lambda_Z}} \int_{\mathcal{H}_Z \times \mathbb{C}} |\det(HH^* - |x|^2 I)|$$

$$\times \exp\left(-\langle \Lambda_Z^{-1}(H, x), (H, x) \rangle \right) dH dx . \qquad (28)$$

Here, H is a complex $m \times m$ symmetric matrix, so $H^* = \overline{H}$. The inner product in the exponent is the real part of the Hilbert-Schmidt inner product, $\langle A, B \rangle = \mathrm{Re}\,\mathrm{Tr}\,AB^*$.

As in [13], we rewrite the integral as

$$\mathcal{I}(Z) = \lim_{\varepsilon' \to 0} \lim_{\varepsilon \to 0} \mathcal{I}_{\varepsilon, \varepsilon'}(Z),$$

where $\mathcal{I}_{\varepsilon, \varepsilon'}(Z)$ is the absolutely convergent integral,

$$\mathcal{I}_{\varepsilon, \varepsilon'}(Z) = \frac{1}{(2\pi)^{m^2} \pi^m \sqrt{\det \Lambda}}$$

$$\int_{\mathcal{H}_m} \int_{\mathcal{H}_m} \int_{\mathcal{H}_Z \times \mathbb{C}} |\det P| e^{-\varepsilon \mathrm{Tr} \Xi^* \Xi - \varepsilon' \mathrm{Tr} P^* P} e^{i \langle \Xi, P - HH^* + |x|^2 I \rangle_{HS}}$$

$$\times \exp\left(-\langle \Lambda_Z^{-1}(H, x), (H, x) \rangle \right) dH \, dx \, dP \, d\Xi . \qquad (29)$$

Here, $\mathcal{H}_m$ denotes the space of all Hermitian matrices of rank m, and $\langle, \rangle_{HS}$ is the Hilbert-Schmidt inner product $\mathrm{Tr}\,AB^*$. Formula 29 is valid, since as $\varepsilon \to 0$, the $d\Xi$ integral converges to the delta function $\delta_{HH^*-|x|^2 I}(P)$. Then, as $\varepsilon' \to 0$, the dP integral evaluates the integrand at $P = HH^* - |x|^2 I$ and we retrieve the original integral $\mathcal{I}(Z)$.

By the same manipulations as in [13], we obtain:

$$\mathcal{I}_{\varepsilon,\varepsilon'}(Z) = \frac{(-i)^{m(m-1)/2}}{(2\pi)^m (\prod_{j=1}^m j!) \pi^m \sqrt{\det \Lambda_Z}}$$

$$\int_{U(m)} \int_{\mathcal{H}_Z \times \mathbb{C}} \int_{\mathbb{R}^m} \int_{\mathbb{R}^m} \Delta(\lambda) \Delta(\xi) \; |\det(D(\lambda))| \; e^{i\langle \lambda, \xi \rangle} e^{-\varepsilon(|\xi|^2 + |\lambda|^2)}$$

$$\times \; e^{i\langle D(\xi), |x|^2 I - gHH^*g^* \rangle_{HS} - \langle \Lambda^{-1}(H,x),(H,x) \rangle} \; d\xi \, d\lambda \, dH \, dx \, dg \; . \quad (30)$$

Further we observe that the $dH\,dx$ integral is a Gaussian integral. Simplifying the phase as in [13] using

$$\langle D(\xi), gHH^*g^* - |x|^2 I \rangle_{HS} = Tr(D(\xi)gHg^t \bar{g} H^* g^*) - Tr D(\xi) |x|^2$$

$$= \left\langle \widehat{D}(\xi) \rho(g)(H,x), \rho(g)(H,x) \right\rangle_{HS}$$

where $\widehat{D}(\xi)$ and $\rho(g)$ are as in the statement of the theorem, the $\mathcal{H}_Z \times \mathbb{C}$ integral becomes

$$\mathcal{I}_{\xi,g}(Z) := \int_{\mathcal{H}_Z \times \mathbb{C}} \exp \Big[-i \left\langle \widehat{D}(\xi) \rho(g)(H,x), \rho(g)(H,x) \right\rangle_{HS}$$

$$- \langle \Lambda_Z^{-1}(H,x), (H,x) \rangle \Big] \, dH \, dx \; . \quad (31)$$

The only new points in the calculation are that this Gaussian integral is over the Hessian space $\mathcal{H}_Z$ rather than over the full space of complex symmetric matrices of this rank, and that it is a real subspace a complex vector space. Hence the Gaussian integral is a real one albeit with a complex quadratic form. We denote by $\mathcal{P}_Z$ the orthogonal projection

$$\mathcal{P}_Z : \text{Sym}(m, \mathbb{C}) \to \mathcal{H}_Z$$

and then we have:

$$\frac{1}{\pi^m \sqrt{\det \Lambda_Z}} \mathcal{I}_{\xi,g}(Z) = \frac{1}{\sqrt{\det \Lambda_Z}} \frac{1}{\sqrt{\det[i\mathcal{P}_Z \rho(g)^* \widehat{D}(\xi) \rho(g) + \Lambda_Z^{-1}]}}$$

$$= \frac{1}{\sqrt{\det \left[i\Lambda_Z \mathcal{P}_Z \rho(g)^* \widehat{D}(\xi) \rho(g) + I_m \right]}} \; . \quad (32)$$

Substituting (32) into (30), we obtain the desired formula. We now recall that $\Lambda = C' \oplus 1$. It follows that

$$\Lambda_Z \mathcal{P}_Z \rho(g)^* \widehat{D}(\xi) \rho(g) + I_m = \left(C'_Z \mathcal{P}_Z \rho(g)^* D(\xi) \rho(g) + I_{h^{21}} \right) \oplus \left(1 - \sum_{q=1}^m \xi_q \right),$$

where

$$D(\xi)(H_{jk})) = \left(\frac{\xi_j + \xi_k}{2} H_{jk} \right).$$

Hence, its determinant equals

$$\left(1 - \sum_{q=1}^{h^{21}} \xi_q \right) \det(C_Z' \mathcal{P}_Z \rho(g)^* D(\xi) \rho(g) + I_{h^{21}}).$$

9 Black Hole Attractors

We close this survey with a discussion of a simpler problem analogous to counting flux vacua that arises in the quantum gravity of black holes [16,28], namely counting solutions of the black-hole attractor equation. For a mathematical introduction to this equation, we refer the reader to [26]. The attractor equation is the same as the critical point equation for flux superpotentials except that $\mathcal{C} = \mathcal{M}$ and $G \in H^3(X, \mathbb{R})$. Physically, $\mathcal{N}_\psi(S)$ counts the so-called duality-inequivalent, regular, spherically symmetric BPS black holes with entropy $S \leq S_*$. The charge of a black hole is an element $Q = N^\alpha \Sigma_\alpha \in H^3(X, \mathbb{Z})$. The central charge $\mathcal{Z} = \langle Q, \Omega \rangle$ plays the role of the superpotential.

There are two main differences to the vacuum counting problems for flux superpotentials. First, the reality of the flux G in the black-hole attractor equation $\nabla W_G(z) = 0$ forces $G \in H_z^{3,0} \oplus H_z^{0,3}$ rather than $G \in H_z^{2,1} \oplus H_z^{0,3}$ as in the flux vacua equation. The space $H^{3,0} \oplus H^{0,3}$ is only 2-dimensional and that drastically simplifies the problem. Second, by a well-known computation due to Strominger, the Hessian $D\nabla G(z)$ of $|\mathcal{Z}|^2$ at a critical point is always a scalar multiple $x\Theta$ of the curvature form of the line bundle, which is the Weil-Petersson $(1,1)$ form.

We now state the analogue of Theorem 3 in the black hole attractor case (see also [8]). The new feature is that the image of Hessian map from the space $\mathcal{S}_z$ of W_G with a critical point at z is the one-dimensional space of Hessians of the form

$$\begin{pmatrix} 0 & -x\Theta \\ -\overline{x\Theta} & 0 \end{pmatrix}, \tag{33}$$

and hence the pushforward under the Hessian map truly simplifies the integral in Theorem 3. The formula for the black-hole density becomes

$$\mathcal{K}_{\gamma,\nabla}^{crit}(z) = \int_{\mathbb{C}} |x|^{2b_3} \chi_{Q_z}(x) dx.$$

We note that the difficult absolute value in Theorem 3 simplifies to a perfect square in the black hole density formula and can therefore be evaluated as a

Gaussian integral. Additionally, the one-dimensionality of the space of Hessians has removed the complexity of the b_3-dimensional integral in the flux vacuum setting.

We can further simplify the integral by removing Q_z, which is a scalar multiple of the Euclidean $|x|^2$. The scalar multiple involves the orthogonal projection $\Pi_{S_z}(z, w)$ onto the space of S_z for the inner product Q_z. If we change variables $x \rightarrow \sqrt{\Pi_{S_z}(z, z)}$, we get

$$\mathcal{K}^{\text{crit}}_{\gamma, \nabla}(z) = |\Pi_{S_z}(z, z)| \int_{\mathbb{C}} |x|^{2b_3} e^{-\langle x, x \rangle} dx.$$

In Kähler normal coordinates, use of special geometry shows that $\Pi_z(z, z) = 1$. A simple calculation shows:

Proposition 2. *The density of extremal black holes is given by:*

$$\mathcal{K}^{\text{crit}}_{\gamma, \nabla}(z) = \frac{1}{b_3} dV_{WP} \implies \mathcal{N}_\psi(L) \sim L^{b_3} Vol_{WP}(\mathcal{M}).$$

The analogy between the black hole density and flux vacuum critical point density should be taken with some caution since the simplifying features are likely to have over-simplified the problem. We therefore mention another modified flux vacuum problem in which the off-diagonal entries $x\Theta$ of the Hessian matrix vanish, so that the Hessian matrix is purely holomorphic and $|\det H^*H - |x|^2\Theta| = |\det H^*H|$ again becomes a perfect square which can be evaluated by the Wick method. Namely, if one uses a flat meromorphic connection ∇ rather than the Weil-Petersson connection, the curvature vanishes away from the polar divisor. The Weil-Petersson connection arises naturally in string/M theory [6], but one may view a meromorphic connection as an approximation in which the "Planck mass" is infinitely large. In any case, it would be interesting to evaluate the density of critical points relative to meromorphic connections since they are more calculable and should have the same complexity as those for Weil-Petersson connections.

References

1. N. Arkani-Hamed, S. Dimopoulos and Shamit Kachru, Predictive Landscapes and New Physics at a TeV (hep-th/0501082).
2. S. Ashok and M. Douglas, Counting Flux Vacua, J. High Energy Phys. 0401 (2004) 060 hep-th/0307049.
3. P. Bleher, B. Shiffman, and S. Zelditch, Universality and scaling of correlations between zeros on complex manifolds. Invent. Math. 142 (2000), no. 2, 351–395.
4. R. Bousso and J. Polchinski, The String Theory Landscape, Scientific American, September 2004 issue.
5. P. Candelas, X. C. de la Ossa, Moduli space of Calabi-Yau manifolds. Nuclear Phys. B 355 (1991), no. 2, 455–481.

6. P. Candelas, G. Horowitz, A. Strominger, and E. Witten, Vacuum configurations for superstrings. Nuclear Phys. B 258 (1985), no. 1, 46–74.

7. J. P. Conlon and F. Quevedo, On the explict construction and statistics of Calabi-Yau vacua, Jour. High Energy Physics 0410:039 (2004) (e-Print Archive: hep-th/0409215).

8. F. Denef and M. R. Douglas, Distributions of flux vacua, J. High Energy Phys. JHEP05(2004)072 (arxiv preprint hep-th/0404116.)

9. F. Denef and M. R. Douglas, Enumerating flux vacua with enhanced symmetries. J. High Energy Phys. 2005, no. 2, 037, e-Print Archive: hep-th/0411183.

10. O. DeWolfe, A. Giryavets, S. Kachru and W. Taylor, Enumerating flux vacua with enhanced symmetries. J. High Energy Phys. 2005, no. 2, 037, (hep-th/0411061).

11. M. R. Douglas, The statistics of string/M theory vacua, J. High Energy Phys. 2003, no. 5, 046, 61 pp. (hep-th/0303194).

12. M. R. Douglas, B. Shiffman and S. Zelditch, Critical points and supersymmetric vacua I, Communications in Mathematical Physics 252, Numbers 1-3 (2004), pp. 325 - 358 (arxiv preprint arxiv.org/math.CV/0402326).

13. M. R. Douglas, B. Shiffman and S. Zelditch, Critical Points and supersymmetric vacua II: asymptotics and extremal metrics, math.CV/0406089, to appear in Jour. Diff. Geom.

14. M. R. Douglas, B. Shiffman and S. Zelditch, Critical Points and supersymmetric vacua III: string/M models, to appear in Comm. Math. Phys.

15. W. Duke and O. Imamoglu, Lattice points in cones and Dirichlet series, IMRN 53 (2004)

16. S. Ferrara, G. W. Gibbons, and R. Kallosh, Black holes and critical points in moduli space. Nuclear Phys. B 500 (1997), no. 1-3, 75–93.

17. Y. V. Fyodorov, Complexity of random energy landscapes, glass transition, and absolute value of the spectral determinant of random matrices. Phys. Rev. Lett. 92 (2004), no. 24, 240601.

18. S.B. Giddings, S. Kachru, and J. Polchinski, Hierarchies from fluxes in string compactifications. Phys. Rev. D (3) 66 (2002), no. 10, 106006.

19. A. Giryavets, S. Kachru and P. K. Tripathy, On the taxonomy of flux vacua, JHEP 0408:002,2004 e-Print Archive: hep-th/0404243.

20. A. Giryavets, S. Kachru, P. K. Tripathy, and S. P. Trivedi, Flux Compactifications on Calabi-Yau threefolds, JHEP 0404:003,2004 e-Print Archive: hep-th/0312104.

21. M. Gross, Calabi-Yau manifolds and mirror symmetry. Calabi-Yau manifolds and related geometries, in [22], 69–159.

22. M. Gross, D. Huybrechts, and D. Joyce, *Calabi-Yau manifolds and related geometries.* Lectures from the Summer School held in Nordfjordeid, June 2001. Universitext. Springer-Verlag, Berlin, 2003.

23. S. Gukov, C. Vafa, and E. Witten, CFT's from Calabi-Yau four-folds. Nuclear Phys. B 584 (2000), no. 1–2, 69–108.

24. S. Kachru, R. Kallosh, A. Linde, and S. P. Trivedi, de Sitter Vacua in String Theory, Phys.Rev. D68 (2003) 046005 (hep-th/0301240).

25. S. Kachru, M. Schulz, and S. Trivedi, Moduli Stabilization from Fluxes in a Simple IIB Orientifold, JHEP 0310 (2003) 007 (hep-th/0201028).

26. S.D. Miller and G. Moore, Landau-Siegel zeros and black hole entropy, e-Print Archive: hep-th/9903267.

27. B. Shiffman and S. Zelditch, Distribution of zeros of random and quantum chaotic sections of positive line bundles. Comm. Math. Phys. 200 (1999), no. 3, 661–683.

28. A. Strominger, Kaluza-Klein compactifications, supersymmetry and Calabi-Yau manifolds, p. 1091-1115 of *Quantum fields and strings: a course for mathematicians. Vol. 1, 2.* Material from the Special Year on Quantum Field Theory held at the Institute for Advanced Study, Princeton, NJ, 1996–1997. Edited by Pierre Deligne, Pavel Etingof, Daniel S. Freed, Lisa C. Jeffrey, David Kazhdan, John W. Morgan, David R. Morrison and Edward Witten. American Mathematical Society, Providence, RI; Institute for Advanced Study (IAS), Princeton, NJ, 1999.

29. A. Strominger, Special geometry. Comm. Math. Phys. 133 (1990), no. 1, 163–180.

30. L. Susskind, The anthropic landscape of string theory, e-Print Archive: hep-th/0302219.

31. J. Wess and J. Bagger, *Supersymmetry and Supergravity*, Second Edition, Princeton Series in Physics, Princeton U. Press (1992).

32. S. Zelditch, Random Complex Geometry and Vacua, or: How to count universes in string/M theory (preprint, 2005).

33. S. Zelditch, Angular distribution of lattice points (in preparation).

Lecture Notes in Physics

For information about earlier volumes
please contact your bookseller or Springer
LNP Online archive: springerlink.com